Organic
Reaction Mechanisms

Organic
Reaction Mechanisms

A Step by Step Approach

Second Edition

MICHAEL EDENBOROUGH

UK   Taylor & Francis Ltd, 1 Gunpowder Square, London EC4A 3DE
USA   Taylor & Francis Inc., 325 Chestnut Street, 8th Floor, Philadelphia, PA 19106

British Library Cataloguing in Publication Data
A catalogue record for this book is available from the British Library.

ISBN 0-7484-06417 (paper)

Library of Congress Cataloguing in Publication data are available

Cover design by Amanda Barragry Design
Typeset in Times 10/12pt by Santype International Ltd, Salisbury, Wiltshire
Printed in Great Britain by T. J. International Ltd., Padstow, Cornwall

Contents

Contents

Preface to the First Edition

This book is designed to give the student a solid foundation in the basic principles that are involved in writing organic mechanisms so that further study may be undertaken with confidence.

In Part I, the basic principles that are common to all mechanisms are explained in detail. Some of the exercises may seem very easy, but they are all important, because they instil a precision and discipline that is essential in dealing with more complicated mechanisms. Fluency in writing mechanisms comes from practice, and in particular from recognising when a proposed intermediary is impossible or highly unlikely.

In Part II, examples of the main mechanistic pathways are given. No attempt has been made to cover exhaustively all the synthetic routes or reactions that could be encountered by a first year undergraduate. However, by reaching an understanding of why a particular mechanism is followed in any given situation, an appreciation is necessarily developed of the major synthetic methods that may be deployed in producing organic chemicals. There is an emphasis upon understanding the stereochemistry of the interactions between reagents and substrate and the consequences on the geometry of the product, because such an appreciation is of fundamental importance in comprehending an organic mechanism fully.

Part III consists of several appendices, which cover all the essential information and simple odds and ends that are often left untaught because it has been erroneously assumed that the student already knows the point in question.

There may only be one name on the front cover, but every book is the result of the hard work of many people. First, I gratefully acknowledge the large amount of time and effort that resulted in the invaluable comments made by my friends Drs Ruth M. Dixon and Deborah J. G. Mackay, and thank my parents for the valuable criticisms which they made. Without their efforts this book would have been the poorer. I would also like to thank my publishers who have been of great assistance at every stage in the production of this book. In particular I would like to mention Ms Ann Berne, Dr Jeremy Lucas and Ms Wendy Mould.

Even after every effort has been made to eliminate all possible errors, ambiguities and omissions, it is still likely that some such faults remain within the text. I would welcome any comments from readers, in particular those that point out mistakes so that they may be rectified, but also those comments that suggest additions or other alterations that would make the book of greater value to the reader.

Michael Edenborough
Saint Valentine's Day, 1994
Covent Garden, London

Preface to the Second Edition

The second edition of this book has been made possible by the enthusiasm with which the first edition was received. Many people have taken the trouble to write to me, care of Taylor & Francis, and all their comments have been most useful in revising the text. Apart from revisiting the whole text, I have taken the opportunity to include in this edition a full index and comprehensive running heads in order to help the reader locate the exact position of the information that is sought. Furthermore, I have also included a summary at the end of each chapter that is designed to crystallise the points made in each chapter. I hope that each of these changes will increase the value of the work, and the ease with which it can be used.

I would like to thank Taylor & Francis, who have been of great assistance at every stage in the production of this book. In particular, I would like to mention Tony Moore, Mariangela Palazzi-Williams, Andrew Carrick, Luke Hacker, Rachel Brazear and Ann Berne. I would also like to thank Wendy Mould for the sterling work that she has done in copy-editing a preliminary draft of this second edition. Last, but by no means least, I would like to thank Dr Richard Bowen, who has read the entire text, and has been foremost in assisting me in revising the present work: without all of his efforts, the present edition would be the poorer. Lastly, the title of the book has been changed from *Writing Organic Reaction Mechanisms: A Practical Guide* to *Organic Reaction Mechanisms: A Step by Step Approach*, as this better describes the contents of the revised edition.

I would welcome any comments that readers might have on this edition.

Michael Edenborough
All Saints' Day, 1998
Islington, London

Basic Principles

1

Introduction

1.1 The aims of this book

The principal aim of this book is to teach you the basics of organic reaction mechanisms. Accordingly, the essentials will be taught in a simple, direct manner that avoids all unnecessary complications. In brief, this is a guide for all students who wish to achieve, with the minimum amount of effort, a thorough grounding in the subject. Thus, once you have worked your way through this book, you will be able to write the mechanism for any organic reaction that you will be likely to meet on an undergraduate chemistry course. From this starting point, further study can be undertaken with confidence.

To be able to achieve this thorough grounding only a few fundamental ideas need to be learnt and practised thoroughly. However, only to learn the ideas is not enough: practice engenders fluency and this is what is needed in order to be able to propose a mechanism under, for example, examination conditions. Many of the errors that students make when writing organic mechanisms are very simple and stem more from a lack of practice of the basic skills of the art, than from a lack of understanding. Once the basic skills have been mastered, they may then be applied to a wide range of reactions.

Of these skills, there are two that are pre-eminent: first, being able to count correctly the number of electrons around any given atom in a molecule; and, secondly, being aware of the position of one group relative to another during a reaction sequence. In fact, it is barely an exaggeration to say that so long as you can count to eight and tell the difference between left and right, then you ought to be able to propose correctly a mechanism for almost any organic reaction.

Yet, in order to write a mechanism for an organic reaction, not all topics need to be understood to the same level. For example, thermodynamics and kinetics

3

will be introduced without any detailed explanation, because a full under-standing is not needed in order to use effectively the principles expounded in those topics. An analogy may be drawn with the photographer who uses a special lens so as to obtain a certain effect. He may know how to use the lens and how to arrange matters so to obtain the effect that he desires. However, he might have no detailed knowledge of the physics of refraction upon which the lens depends in order to work. This knowledge of physics, although interesting, is not essential to produce a successful picture. Similarly, where appropriate, some topics will be dealt with more lightly, extracting the useful points without embarking upon a full explanation, which would be unnecessary in order to utilise the idea profitably at this level.

A further aim of this book is to give you *confidence* in writing organic mechanisms, so that you can write down something sensible when confronted with a novel problem. Once you have worked your way through this book, you will be able to do this, because you will have acquired an understanding of the electronic characteristics of many systems and gained experience of a large number of reactions, i.e. you will have fully assimilated the fundamentals of the subject. However, this assimilation will only come from the practice that results from writing down what you think is the answer to each of the set problems that are contained in the text, and then comparing your answer to the solution that is given later. After using this book, you should know the majority of what has been covered, because you will have been repeatedly using the information and ideas at every stage, and so repeatedly reinforcing your knowledge at every step. Thus, there should be little need for revision. However, summaries have been included at the end of each substantive chapter to act as an *aide-mémoire* to the full text. In addition, the glossaries and other appendices are intended to form a useful reference point for the basic definitions and ideas that have been covered in this book.

1.2 What is organic chemistry?

Organic chemistry originally referred to the study of chemicals that were derived from nature, i.e. an organic source. It was believed that these chemicals were different from those that were mineral in origin, i.e. the inorganic chemicals. It was proposed that this difference was so fundamental that it would be impossible to make an organic chemical from components that were derived only from mineral sources, because the latter did not contain the essential vital force that was required to form organic compounds.

The fallacy of this argument was demonstrated in 1828 when Wöhler heated ammonium cyanate, an inorganic salt, and produced urea, hitherto considered purely an organic compound. This showed that there was no essential difference between organic and inorganic compounds. Thus, there was no vital force that was characteristic of organic compounds.

However, the two major branches of chemistry are still studied separately. This is because there is a useful distinction to be drawn between the two. Yet, this distinction is no longer based on the presence or absence of a vital force, but rather is partly derived from the practical consequences of the stability of the homoatomic carbon/carbon bond. One such consequence is that the chemistry of carbon-containing compounds often involves the study of molecules that contain many such bonds. Thus, because of the multitude and variety of these compounds, it is convenient to study them separately. In contrast, inorganic chemistry is often concerned with the structure and utility of molecules that contain heteroatomic bonds, either ionic or covalent. This means that in an inorganic molecule, there are often two or more elements that could be said to be of equal prominence. In addition, there still remains the point that carbon compounds form the vast majority of the constituent components of living beings, while non-carbon compounds comprise the environment in which those beings exist. So the distinction is the same, but the justification is different.

A present-day definition of organic chemistry would be one that includes the study of carbon compounds, and in particular those compounds that possess covalent carbon/carbon and carbon/hydrogen bonds. This definition is very general, and as such there are many exceptions to it. For example, carbonates and carbides are normally considered under the chemistry of the cation with which they are associated. This is because, even though they contain covalent bonds, they are often ionic solids with high melting points, and these are characteristic properties of inorganic compounds.

To highlight the pragmatic approach of the above definition, the compounds that contain the cyanide unit are studied under whichever heading is most appropriate. So ionic cyanides are studied under inorganic chemistry, while compounds in which the cyanide group is covalently bonded to the rest of the molecule are usually studied under organic chemistry.

1.3 Organic synthesis

Industrial organic chemistry is primarily concerned with the manufacture of useful compounds, such as pharmaceuticals, or commercial chemicals, such as glues or plastics. All these products are made from simpler compounds, i.e. they have been synthesised. Synthesis may be defined as the rearrangement of the constituent atoms within the starting materials to give the target molecule. This involves the making and breaking of covalent bonds. Not only is the elemental composition (i.e. the molecular formula) of the resultant product important, but so is the spatial arrangement of the atoms (i.e. the structural formula).

In addition, many organic compounds, while having the same structural formula, differ in the precise spatial arrangement of the various parts of the molecule relative to each other; that is, they can exist in one of several forms, which are called diastereomers. Diastereomers have identical chemical formulae, but differ in the absolute spatial relationship of one part of the

5

molecule to another. In biological systems, such molecules abound. Many of them are only active when they exist in the correct form: thus, even though a compound may have exactly the correct elemental composition, its biological activity may be reduced from that which was expected, because it is not stereochemically pure. Sometimes the wrong isomer may have no activity, but on other occasions an isomer with the incorrect stereochemistry may have harmful effects. For example, the drug thalidomide was produced as a racemic mixture, i.e. as a mixture that contained equal quantities of both the left- and right-handed enantiomers. It was the presence of one enantiomer that led to the adverse effects upon the developing foetuses within the pregnant women who had taken the drug. Accordingly, it is clear that those reactions that control the precise spatial arrangement of the constituent atoms within the target molecule are of great commercial value, not only because they avoid waste, i.e. more of the product of the correct diastereomer is produced, but also because they ensure that there are no contaminants that might have adverse side effects.

1.4 Organic mechanisms

Those covalent bonds that are made and broken in the synthesis of any organic product are, like most other covalent bonds, formed by pairs of electrons. These electron pairs occupy molecular orbitals within the molecules. Organic mechanisms chart the rearrangement of the constituent atoms of the starting materials into the target molecule by indicating the movement of the electrons from one molecular orbital to another. This allows the mechanistic chemist not only to predict the molecular formula of the new product, but also to predict the final spatial arrangement of the atoms relative to one another, i.e. the stereochemistry of the product.

By studying these mechanisms, an understanding of the principal synthetic routes may be gained, which in turn may lead to the optimisation of existing syntheses and the development of novel synthetic methods.

1.5 Two examples

This section gives a foretaste of the types of reaction that you should be able to tackle once you have worked your way through this book. The two examples will probably appear somewhat daunting at this stage, but they are given here in order to introduce some of the concepts that will be explained in more detail later, and to show how an apparently complicated reaction may be broken down into a sequence of simple steps.

At an elementary level, organic chemistry is often explained by allowing the adjacent groups, as shown on the two-dimensional representations of the reagents, to interact with each other. A common example of this is the formation of an ester from the reaction of an alcohol with a carboxylic acid. The hydrogen of the alcohol hydroxyl group and the whole of the hydroxyl group of the carboxylic acid are encircled and so bind together to form water, leaving the truncated alcohol and acid to combine together to form an ester. This is often referred to as lasso chemistry.

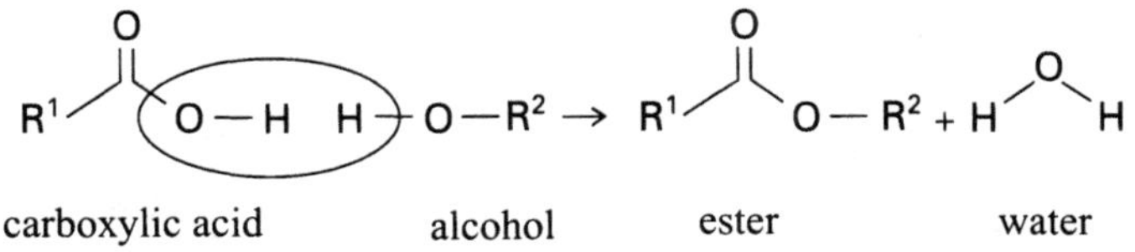

carboxylic acid   alcohol   ester   water

The simplicity of the idea that the most proximate groups react with each other is appealing. However, it does not lead to a very sound understanding of how the bonds within the molecules rearrange themselves in order to form new compounds. Also, it is of little help in predicting what will happen when there is a choice of groups that are all near enough to react with each other. For example, in the following reaction sequence, it is not immediately obvious how the epoxide ring has been formed.

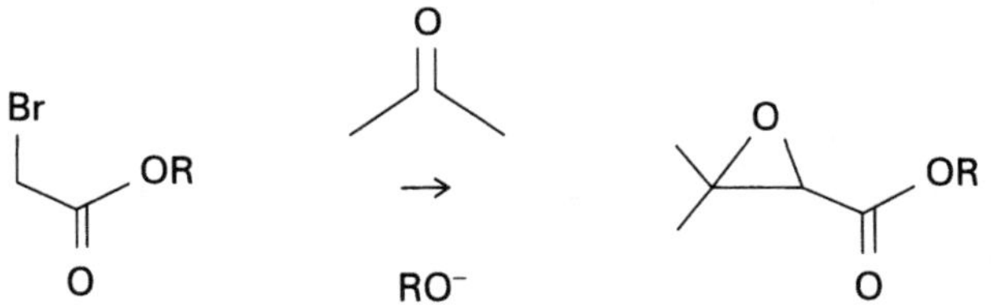

Here, an α-bromocarboxylate, which has been treated with a base, reacts with a carbonyl containing compound to form an epoxide species.

As an aside, it should be noted that it is traditional in organic chemistry to name a reaction after the person who first published it or after the reagents or products that first characterised the reaction, and then to use that name as a convenient way of referring to that type of reaction on all future occasions. This is a bit daunting to the uninitiated, but it does serve the useful purpose that a given reaction type may be readily identified. This approach will be used in this book to identify the reactions under consideration. The reaction we have just looked at is called the Darzens condensation.

If we now look at each step of the reaction sequentially, showing the individual movement of the electron pairs within the reacting molecules, some understanding of what is actually happening might be gleaned.

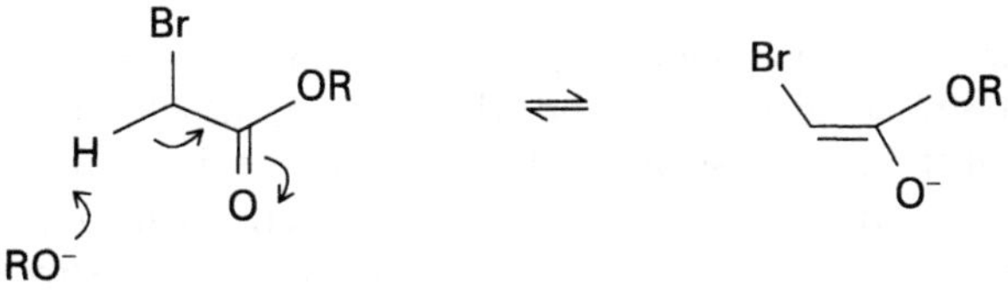

Here, the base has removed the hydrogen atom that was attached to the carbon bearing the bromine atom and the ester group. The resultant anion, which is stabilised by the ester group, then attacks the slightly positive, $\delta+$, carbon of the carbonyl group. The movement of the electron pairs is indicated by the use of a curved arrow, which is why this type of representation is often referred to as curly arrow chemistry.

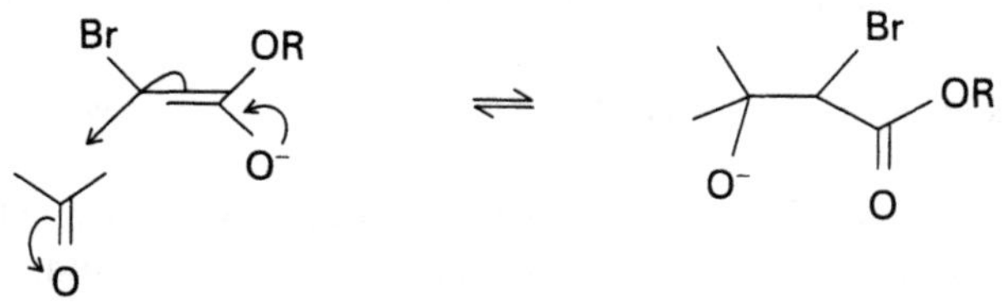

The last part of this reaction comprises the attack by the oxygen anion on the $\delta+$ carbon that bears the bromine atom, which in turn leaves as the bromide anion so that the carbon never has more than four covalent bonds around its nucleus at any one time. This internal attack results in the formation of the three-membered epoxide ring.

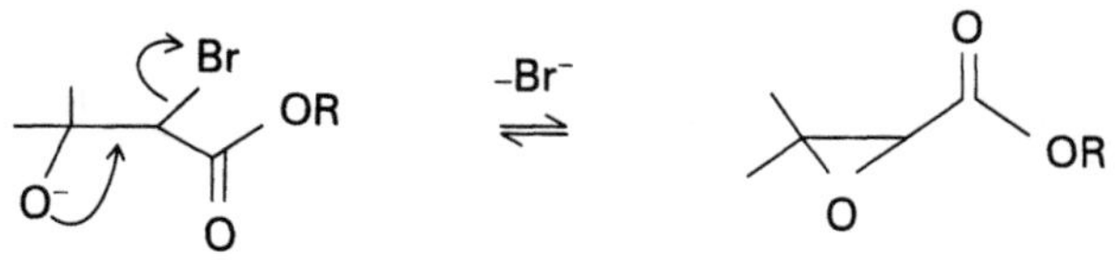

In summary, there was an initial acid/base reaction, which resulted in the formation of an anion with the negative charge on a carbon atom, which was stabilised by a neighbouring group. This anion then attacked another molecule to form a further anion, which in turn reacted, but this time internally, to form a cyclic compound.

When this epoxide product is treated with aqueous acid, a rearrangement of the carbon skeleton occurs. This is the glycidic acid rearrangement, which is named after the product formed in the archetypal reaction.

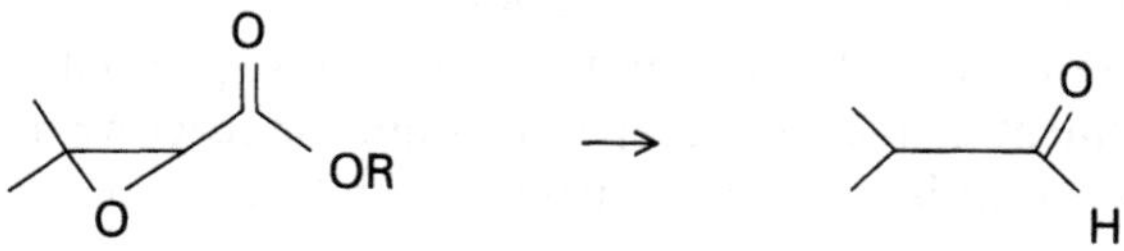

The second part of this synthetic sequence is slightly more complicated. The ester is hydrolysed to give the carboxylate anion, which then undergoes a fragmentation of the carbon skeleton that involves the departure of a molecule of carbon dioxide, and finally there is a tautomeric rearrangement to give the product. However, each step follows the principles introduced above, i.e. acid/base reactions, the stabilisation of charged species, and the internal movements of electrons.

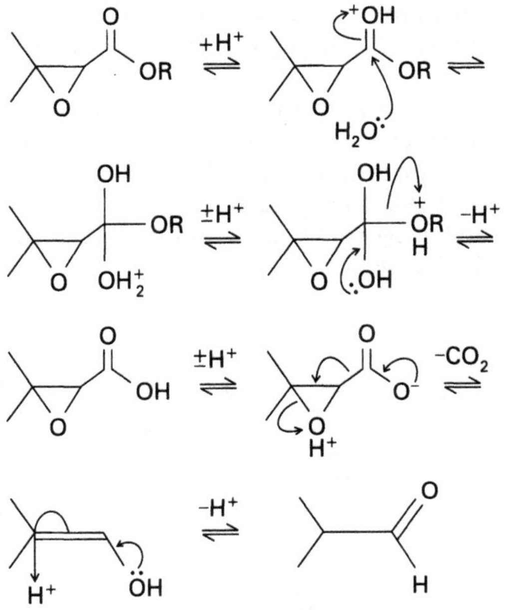

The second example highlights the attention that must be paid to the exact spatial arrangement of the atoms within the starting materials, and in particular to any intermediate products that might be formed.

2R,3S   2R,3S

Lasso chemistry seems to be able to solve this problem well.

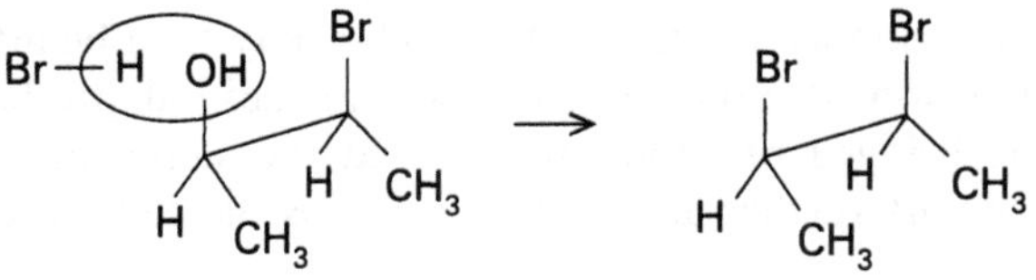

However, this analysis would predict that when (2S,3S)-3-bromobutan-2-ol reacts with HBr, the stereochemistry of the resultant product would be

9

(2*S*,3*S*)-2,3-dibromobutane, i.e. an optically active product. In fact, the product is a mixture, which comprises an equal amount of (2*S*,3*S*)-2,3-dibromobutane and (2*R*,3*R*)-2,3-dibromobutane, and which displays no optical activity. The suggested product and the product that is actually formed have exactly the same elemental composition, but the detailed stereochemistry is different.

By taking into account the experimental observation of the stereochemistry of the product, and assuming that the two reactions proceed by the same mechanistic route, it is apparent that the mechanism cannot proceed along the pathway indicated by the lasso approach.

In contrast, the following representation of the mechanism accounts for the stereochemistry in both cases.

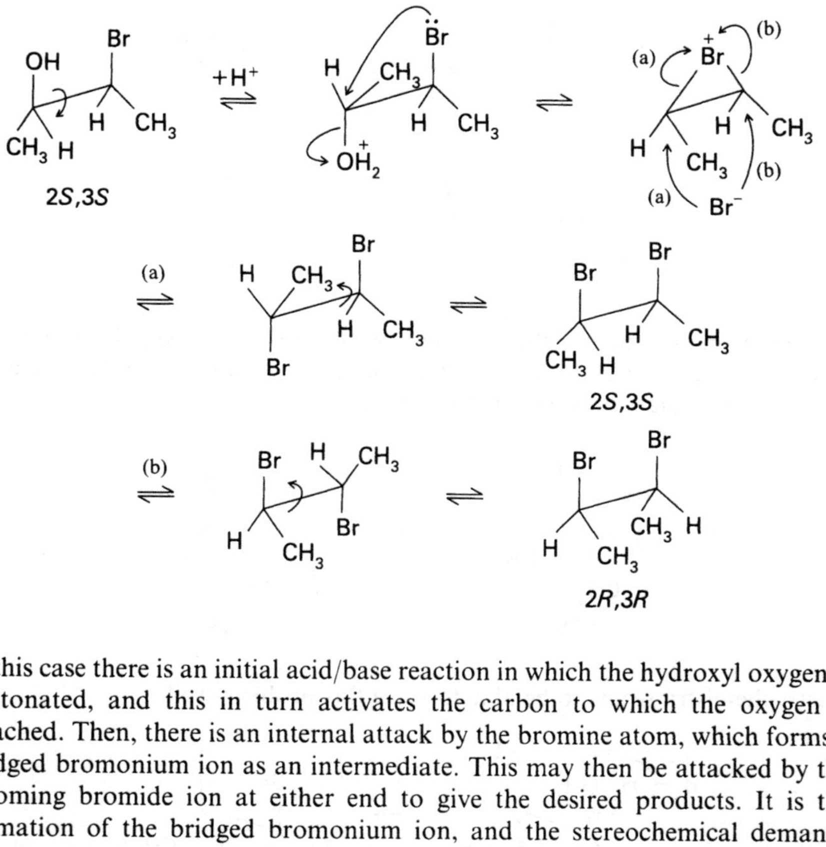

In this case there is an initial acid/base reaction in which the hydroxyl oxygen is protonated, and this in turn activates the carbon to which the oxygen is attached. Then, there is an internal attack by the bromine atom, which forms a bridged bromonium ion as an intermediate. This may then be attacked by the incoming bromide ion at either end to give the desired products. It is the formation of the bridged bromonium ion, and the stereochemical demands thereby placed on any further attack, that explains the stereochemistry of the products.

These reactions may look complicated, but as we have seen, they are formed from simple parts. By the end of this book, you will be able to write such mechanisms for yourself with confidence and ease.

1.6 The structure of this book

This book is divided into three parts: Basic Principles; Mechanisms; and Appendices. Within the first part, all the basic ideas that are needed in order to write organic mechanisms are discussed and explained. The areas covered are: electron counting; covalent bonding and polarisation; shapes of molecules; stabilisation of charged species; thermodynamic and kinetic considerations; and acid/base characteristics. In each case, the underlying principles will be highlighted and many of the common errors and misunderstandings will be explained so that not only do you know what to do, but you also know what not to do and why.

The second part of the book uses the basic principles that have been introduced in the first part and applies them to the major mechanistic types, so that you may see how these ideas are used in practice. The mechanisms are considered in increasing order of complexity, namely: substitution reactions; addition reactions; elimination reactions; rearrangement reactions; and redox reactions.

Each section commences with a short introduction in order to set the background for the work to come. However, the majority of the text in the first two parts is in the form of an interactive question and answer format. Generally, a point is introduced in an instructional paragraph and then a question will be posed around that new piece of information. Maximum benefit will be gained if each question is attempted, and your proposed answer is written down, before moving onto the next paragraph where the solution will be given. Often, there will be some further elaboration that will include pointers on how to avoid the errors or misunderstandings that are commonly encountered when dealing with that particular sort of problem.

Sufficient information will always be given to ensure that the correct solution may be deduced without having to guess the answer: that is, so long as you have understood what has been taught before and you have remembered it you will be able to answer the question in hand. Information, once given, is continually used in future paragraphs, as the later paragraphs build upon, and refer to, the earlier ones. Accordingly, by the time you have worked your way to the end of a particular chapter, you should have assimilated everything dealt with therein.

This book is not intended to cover comprehensively either every synthetic method that is available or every mechanism that is possible. Rather, the object is to teach the basic principles of the tools that are required to write correctly the fundamental organic mechanisms and to do so with confidence. This will allow you to understand and appreciate the chemistry that is contained in the standard textbooks on organic chemistry. Once this has been achieved, then more advanced work can be commenced if so desired.

In the third part of this book, there are various appendices, within which you will find a working definition and explanation of every common term, notation

11

or abbreviation that you will have encountered in this book and most of the terms that you will meet in an undergraduate chemistry course. There is also a brief overview of stereochemical terminology and a short explanation as to the use of oxidation numbers in covalent molecules. These sections may be used as a quick source of information while using the main text, or later, as a ready reference.

Electron Counting

2.1 Introduction

The ability to count correctly the number of electrons around any given atom is central to being able to write a plausible reaction mechanism in organic chemistry.

The level of ability required is not great. In fact, essentially, one only needs to be able to count to two when considering hydrogen atoms, and eight when considering carbon, nitrogen or oxygen atoms.

Yet, most of the errors that are found in the scripts of first year university students and second year A-level students, result either from placing too many electrons around an atom, e.g. CH_5, or by placing the wrong charge on an atom, e.g. R_3NH. Both of these errors have their origins in a failure to count correctly the number of electrons on the central atom, and then to realise the consequences that must inevitably flow from such an addition. So, in the first case, a carbon atom may never (or at least very, very rarely) have more than eight electrons around it; while in the second case, a nitrogen atom that is sharing eight electrons between four covalent bonds must bear a single positive charge.

The first thing to learn is the skill of counting the electrons around just a single atom or ion. This is very simple, but absolute fluency in doing this flawlessly is essential before progressing to molecules in which the electrons are shared between different atoms.

2.2 Atoms

Atoms are formed from protons, neutrons and electrons. The protons and neutrons together constitute the nucleus; while the electrons occupy orbitals, which are centred on the nucleus. The number of protons within the nucleus determines the element. This number is called the atomic number. The sum of the number of neutrons and protons is called the atomic mass number. The number of neutrons within the nucleus of a particular element is not fixed, although usually there are approximately the same number of neutrons as there are protons. For a given element, atoms that contain different numbers of neutrons are called isotopes, and they have different physical properties, e.g. a different atomic mass number, and a different stability of the nucleus. However, most importantly for the chemist, the chemical properties of different isotopes are the same. The smallest unit of an element is the atom.

Neutrons are electrically neutral, while protons carry a single positive charge and electrons carry a single negative charge. In an atom, the number of electrons always equals the number of protons. Thus, every element in the atomic state is electrically neutral.

The element that has only one proton is hydrogen, and its atomic number is one. The commonest isotope of hydrogen does not possess a neutron in its nucleus, but only contains a solitary proton. More strictly, this isotope should be called protium in order to distinguish it from the other isotopes of hydrogen that contain either one or two neutrons within the nucleus, and which are called deuterium and tritium respectively.

As was indicated in Chapter 1, the bulk of this text is designed to be interactive. Thus, in order to ensure that the information that has just been given is assimilated, the text will require you to do certain things, like drawing out the electronic structure of an atom or ion, or writing down a mechanism. The maximum benefit can only be gained if you actually follow these directions, and do what is requested. It really is for your own good. Thus, to start with, draw a diagram that indicates the composition in terms of protons, neutrons and electrons of the three isotopes of hydrogen mentioned above.

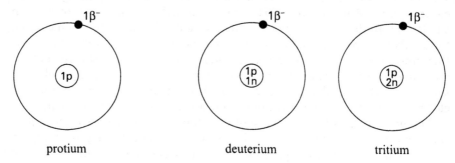

protium   deuterium   tritium

For each isotope, the diagram you have drawn should show a single electron in an orbital centred on the nucleus; and within the nucleus there should be one proton with zero, one or two neutrons depending on whether the diagram is meant to represent the protium, deuterium or tritium isotope.

Normally, the various isotopes of an element are not distinguished by separate names, but instead they are distinguished by having the different atomic mass numbers as a raised prefix to the atomic symbol. Hydrogen is unique in that each isotope has a separate name, which may also be represented by a separate symbol, namely H, D or T, although the conventional ^1H, ^2H or ^3H may also be used. When referring to an element, it is usual to imply that the most stable isotope, or the naturally occurring mixture of isotopes, is meant unless otherwise stated.

Helium is the element that has the atomic number of two. There are two common isotopes that contain one and two neutrons respectively. Draw two diagrams to represent these two isotopes, and also write down the conventional atomic symbols that represent these isotopes.

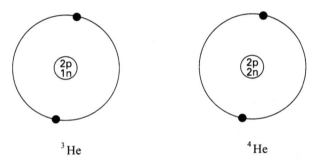

^3He ^4He

You should have drawn a nucleus containing two protons, with either one or two neutrons depending on the isotope. In each case you should also have indicated two electrons occupying an orbital centred on the nucleus so as to ensure that the atom is electrically neutral. The conventional symbols are ^3He and ^4He respectively.

The chemical properties of an element are determined by its electrical properties, and so chemists are particularly interested in the placement of positive and negative charges. If an element loses or gains one or more electrons then that element will no longer be electrically neutral, because the number of protons will not equal the number of electrons that surround the nucleus. The resultant charged species is called an ion. If the ion is negative, it is an anion; if it is positive, it is a cation.

The hydrogen atom may gain or lose one electron to form either an anion or a cation respectively. Draw two diagrams, one to represent the anion and the other the cation of hydrogen.

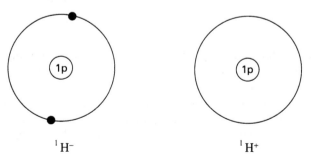

$^1H^-$ $^1H^+$

15

The anion of hydrogen has one proton and two electrons, and is called a hydride anion (or more strictly, a protiide anion); while the cation of hydrogen is a bare proton. Ions of elements are represented by the atomic symbol of the element with the appropriate overall charge as a raised postscript. So these ions would be H^- and H^+ respectively.

The commonest isotope of helium is ^4He, and this may lose one or two electrons to form two different cations. Draw a diagram to represent these two cations.

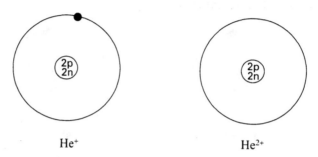

He^+ He^{2+}

The cation of helium that results from the loss of one electron carries a single positive charge as there is only one electron around a nucleus which contains two neutrons and two protons. The cation that results from the loss of two electrons carries a double positive charge, and is a bare nucleus consisting of two neutrons and two protons. This species is called an alpha particle.

If there is an unpaired electron, or electrons, in an atom or ion, then it is called a radical. Draw the electronic configurations of all the possible radical species of hydrogen and helium that contain only one electron.

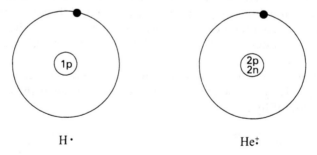

$H\cdot$ He^+

The hydrogen atom has only one electron, and so is a radical; while the monopositive cation of helium is the only species of that element that has an odd number of electrons. The hydrogen cation has no electrons, while the hydrogen anion and the helium atom each have a pair of electrons, and so all of these species are not radicals. In order to highlight the point that a radical is being considered, the unpaired electron is sometimes indicated by a single dot after the atomic symbol, e.g. $H^\bullet$.

The pairing of electrons is important, because it confers a degree of stability on the species concerned. Electrons in an atom are contained within atomic orbitals, each of which may only hold a maximum of two electrons. The electron in a hydrogen atom is contained in the first principal quantum shell, which is indicated by 1. Within this principal quantum shell, there is only one type of subshell, and this is represented by the letter "s". The electronic configuration of hydrogen may be written as $1s^1$. The raised postscript indicates that there is only one electron in that particular orbital.

The two electrons in helium are both contained within the first principal quantum shell. Write down the electronic configuration of helium.

The electronic configuration of helium is $1s^2$.

Each subshell represents a different orbital, and so may contain a maximum of two electrons. Write down the electronic configurations of the ions of hydrogen and helium that have been discussed above, i.e. H^+, H^-, He^+ and He^{2+}.

The electronic configurations are as follows: the cation of hydrogen is $1s^0$, and the anion of hydrogen is $1s^2$; the cations of helium are $1s^1$ and $1s^0$ for the monopositive and dipositive cations respectively.

As each orbital may contain a maximum of two electrons, the helium atom has a full complement at $1s^2$, as does the hydrogen anion. There is only one subshell in the first principal quantum level, and so there can only be two elements that have electrons only in this level. That is why there are only two first row elements of the Periodic Table.

In the second principal quantum shell, there are four subshells, one s and three p subshells, each of which may contain two electrons, which gives a total of eight electrons in all at the second level. This is why there are eight second row elements. It is possible to add an electron to, or remove electrons from, an orbital that is not within the highest principal quantum level that is normally occupied in the neutral atom. However, this requires a lot of energy and so only occurs very rarely. Write down the electronic configuration of the mononegative anion of helium, and so suggest why this species is not readily formed.

He^- has the electronic configuration of $1s^2, 2s^1$.

The electron that was added to the helium atom to form the anion could not fit into the s subshell of the first principal quantum shell, because it already had two electrons and so was full. Instead, this extra electron occupied the next available orbital, which was the s subshell of the second principal quantum level. This requires a lot of energy, and so it is difficult. Hence, the He^- anion, which it should also be noted is a radical species, is not normally found.

The second row elements, lithium to neon, sequentially fill the available four subshells in the second principal quantum level, starting with the s subshell and then progressing to the three p subshells. The latter three each have the same energy, i.e. they are degenerate, and are distinguished by the subscripts x, y and z. If electrons are placed in degenerate orbitals, they fill them so as to minimise the doubling-up of electrons, because this reduces the electron/electron repulsion. This is called Hund's rule. Given this information, now write down the electronic configurations of the second row elements.

Li	$1s^2, 2s^1$
Be	$1s^2, 2s^2$
B	$1s^2, 2s^2, 2p_x^1$
C	$1s^2, 2s^2, 2p_x^1, 2p_y^1$
N	$1s^2, 2s^2, 2p_x^1, 2p_y^1, 2p_z^1$
O	$1s^2, 2s^2, 2p_x^2, 2p_y^1, 2p_z^1$
F	$1s^2, 2s^2, 2p_x^2, 2p_y^2, 2p_z^1$
Ne	$1s^2, 2s^2, 2p_x^2, 2p_y^2, 2p_z^2$

As the loss or gain of electrons normally only occurs from the highest principal quantum level that is occupied in the atom, it is this level that is concerned with the chemistry of the element. Accordingly, the highest principal quantum level that is normally occupied in the neutral atom is called the valence shell, the name of which reflects its involvement in the chemistry of the atom. Usually only the electrons within the valence shell are indicated, and the electrons within the core shells are assumed to be in order.

Suggest the electronic configurations for the anions of the second row elements formed by the gain of one electron.

Li^-	$1s^2, 2s^2$
Be^-	$1s^2, 2s^2, 2p_x^1$
B^-	$1s^2, 2s^2, 2p_x^1, 2p_y^1$
C^-	$1s^2, 2s^2, 2p_x^1, 2p_y^1, 2p_z^1$
N^-	$1s^2, 2s^2, 2p_x^2, 2p_y^1, 2p_z^1$
O^-	$1s^2, 2s^2, 2p_x^2, 2p_y^2, 2p_z^1$
F^-	$1s^2, 2s^2, 2p_x^2, 2p_y^2, 2p_z^2$

In each case, the ion is mononegative, because one electron has been added to the neutral atom and so the number of electrons now exceeds the number of protons by one. The electronic configuration of the mononegative ion corresponds to that of the element one to the right in the Periodic Table. Note that as neon already has eight electrons, no more may be added to it without a great deal of energy being expended, because the extra electron would have to be accommodated within the third principal quantum level, i.e. $1s^2, 2s^2, 2p_x^2, 2p_y^2, 2p_z^2, 3s^1$. Thus, the Ne^- ion is very difficult to form.

Now write down the electronic configuration of the anion that results from the addition of two electrons to the oxygen atom, and the electronic configuration of the two ions that result from the removal of one and two electrons from the beryllium atom.

O^{2-} $1s^2, 2s^2, 2p_x^2, 2p_y^2, 2p_z^2$
Be^+ $1s^2, 2s^1$
Be^{2+} $1s^2, 2s^0$

The electron configuration of the dinegative anion corresponds to that of the neutral element two positions to the right of the original element, i.e. in this case the neon atom. In contrast, the electronic configuration of the monocation corresponds to that of the element one to the left of the original element; while the dication mimics that two to the left.

It is difficult to form certain ions of the second row elements, e.g. Ne^-, F^{2-} and Li^{2+}, because in each case it would be necessary either to remove (in the case of lithium), or to add (in the case of neon and fluorine) an electron from, or to, a different principal quantum level than the one in which the outermost electrons of the neutral atom were accommodated.

As each element in the second row has four subshells in the second principal quantum level, and as each subshell may only accommodate a maximum of two electrons, the result is that no element in the second row may have more than eight electrons in the second principal quantum shell. This is called the Octet rule, and is one of the most fundamental rules concerning the electron distribution around second row elements. Under normal conditions, it is never violated.

The third row elements have more subshells in their valance shell, i.e. the third principal quantum shell, than the second row elements, namely one s subshell, three p subshells and five d subshells. This means that they may accommodate a maximum of 18 electrons. As far as the organic chemist is concerned, the important consequence of this is that the elements phosphorus and sulphur may expand their octet and have five or six bonds respectively, instead of being limited to just four bonds as is the case with carbon and other second row elements.

Now that we can count the unshared electrons around an atom, be it neutral or charged, we will look at molecules in which the electrons are shared between more than one atom. Initially, we will look at neutral molecules, and then ones that carry a full unit of charge.

2.3 Uncharged molecules

In an atom, any charge that is present may easily be determined by the difference between the number of electrons that are accommodated in the orbitals around it and the number of protons within the nucleus. The matter is more complicated for molecules, because there are two or more nuclei to consider. Yet in principle, the same counting process is involved.

The simplest molecule is formed from two atoms of hydrogen, to give the hydrogen molecule, H_2. This consists of two hydrogen nuclei joined by one covalent bond formed from two electrons, one donated by each hydrogen atom. Each hydrogen atom now shares two electrons accommodated within a single molecular orbital, which encompasses the whole molecule. Each molecular orbital, like each atomic orbital, may only hold a maximum of two electrons. For this reason, the maximum number of normal covalent bonds that a hydrogen atom can form is one. What is the charge on each hydrogen atom in H_2?

Each hydrogen atom is neutral, because each hydrogen nucleus sees the two electrons on average for only half of the time. That is, in a single covalent bond between two identical atoms, each bonded atom shares the electrons equally with the other nucleus. Thus, two electrons shared by two nuclei give rise to an effective single negative unit for each nucleus. So, with respect to each hydrogen atom, subtracting the one positive charge of the proton leaves an overall zero charge.

The overall charge on an atom is, of course, the difference between the number of protons in its nucleus and the total number of electrons in the core and valence shell. When discussing atoms in molecules, however, it is easier to consider only the electrons in the valence shell, for it is assumed that the electrons in the core shell are fully occupied. The charge on any given atom can then be calculated according to the following formula. First, find the number of electrons that would surround the uncharged, isolated atom in its valence shell. Then, subtract the total number of electrons that are actually in the valence shell of that atom within the molecule. Finally, add half the number of electrons that that atom shares with any other atom in the molecule. The result is the charge on that atom within the molecule.

Now, let us see how this formula works in practice. First, for the elements lithium to neon, write down the number of electrons that surround the neutral atom in the valence shell.

For lithium to neon, the number increases from one to eight sequentially.

Now, if one considers the diatomic molecule of fluorine, F_2, there is one covalent bond joining the two fluorine atoms. Draw the dot and cross structure of this molecule, using a dot to represent the electrons that originate from the valence shell of one fluorine atom, and a cross to represent the electrons that originate from the valence shell of the other fluorine atom. Then, calculate the charge on each fluorine atom within this molecule.

The charge on each fluorine atom in the diatomic fluorine molecule is zero. Normally, every fluorine atom has seven electrons in the valence shell, which are combined into three lone pairs and one unpaired electron. It is this unpaired electron that combines with the unpaired electron of another fluorine atom to form the single covalent bond. Each fluorine atom now has eight electrons around its nucleus. Hence, each fluorine nucleus is involved in one single covalent bond, which contains two electrons, which are accommodated in a single molecular orbital that envelops the two nuclei. These two electrons are shared equally between the two fluorine nuclei. Using the above formula: there are seven electrons normally present in the valence shell of the elemental atom, minus the eight that are actually present, plus half of the two that are shared in the single covalent bond, which results in an overall zero charge, i.e. $7 - 8 + \frac{1}{2}(2) = 0$.

Neon has eight electrons in its valence shell and chemically it is a very stable species. This stability is due to the atom having a full valence shell for a second row element, i.e. a complete octet. When atoms combine to form molecules, they strive to obtain a full complement of electrons in the valence shell, which would result in them acquiring this stability.

Carbon has four electrons in its valence shell and needs four more electrons to reach the stable octet of electrons. It may do this by combining with four hydrogen atoms to form a methane molecule. Write down a dot and cross representation of this molecule and calculate the charge that is present on each of the five atoms.

$$
\begin{array}{c}
H \\
\text{x} \; \bullet \\
H \; \overset{\bullet}{\underset{\text{x}}{\bullet}} \; C \; \overset{\text{x}}{\underset{\bullet}{\text{x}}} \; H \\
\bullet \; \text{x} \\
H
\end{array}
$$

All five atoms have zero charge. The four hydrogens are in identical environments and so not surprisingly they have the same charge. Hydrogen has one electron in its valence shell in its uncharged, atomic state. Now for each hydrogen atom, there are two electrons that form a single covalent bond between it and the carbon atom. This bond may be represented by a molecular orbital that encompasses the central carbon nucleus and the hydrogen nucleus in question. Moreover, both electrons are being shared between the central carbon nucleus and the hydrogen nucleus. Thus, each hydrogen atom bears a zero charge, i.e. $1 - 2 + \frac{1}{2}(2) + 0$. The carbon nucleus has a total of eight electrons accommodated in four molecular orbitals. All of the electrons are shared with the hydrogen atoms. It normally has four electrons in the atomic state in its valence shell. Hence, it also bears a zero overall charge, i.e. $4 - 8 + \frac{1}{2}(8) = 0$.

Draw the dot and cross representation of the dihydride of oxygen, i.e. water, and so calculate the charge that is resident on each of the three atoms.

$$
\begin{array}{c}
\text{x} \; \text{x} \\
\overset{\text{x}}{\underset{\text{x}}{}} \; O \; \overset{\bullet}{\underset{\text{x}}{}} \; H \\
\text{x} \; \bullet \\
H
\end{array}
$$

Each atom is neutral. The same counting procedure applies to the hydrogens as it did in methane. Oxygen has six electrons in its atomic state in its valence shell. In the water molecule, the oxygen atom has a total of eight electrons accommodated around its nucleus in the valence shell, and so it has a stable octet. Four of these electrons are involved in the two single covalent bonds with the hydrogen atoms. Hence, the charge on the oxygen atom is zero, i.e. $6 - 8 + \frac{1}{2}(4) + 0$. Note that of the eight electrons that surround the oxygen nucleus, four are in two molecular orbitals forming the two covalent bonds, while the other four are in two lone pairs. The latter four electrons are not shared with any other atom.

Draw the dot and cross representation of the trihydride of nitrogen, i.e. ammonia, and so calculate the charge that is resident on each of the four atoms.

$$H \overset{\times}{\underset{\times}{\bullet}} N \overset{\times\,\times}{\underset{\bullet\,\times}{\overset{\times}{\bullet}}} H$$

$$H$$

Again, each atom is neutral. The nitrogen atom is involved in three molecular orbitals, which each contain two electrons, and which form the three covalent bonds to the hydrogen atoms. Furthermore, there is also one lone pair of electrons centred on the nitrogen atom, which makes a total of eight electrons centred on that atom. In the atomic state of nitrogen, there are five electrons in the valence shell. The nitrogen atom is involved in three covalent bonds, hence giving a resultant charge of zero, $5 - 8 + \frac{1}{2}(6) = 0$.

Now that you are fully familiar with the number of electrons that can be accommodated in the valence shell of the common elements when they are combined in neutral species, we may proceed to the next step, which is the counting of electrons in molecular species that carry a whole unit of charge and the determination of which atom has gained or lost an electron.

2.4 Molecules with whole charges

This is the area in which most mistakes are made. Most mechanisms that will be examined in the second part of this book deal with intermediates that carry one or more whole charges. This arises because most mechanisms start by splitting a covalent bond unequally; thus, one part has an extra electron, while the other part has a corresponding deficiency. This means that each part carries a charge.

When drawing atoms or molecules that carry a whole electric charge, the charge itself is indicated by a plus or minus sign, which in some texts is enclosed within a circle. Here, however, it will be represented by a plain + or − sign, which is placed as a superscript, usually after the atomic symbol that bears the charge.

You are already familiar with the notation that an electron is represented by a dot, ·. Sometimes a short solid bar, —, is used to represent a pair of electrons. These bars are placed around the atomic symbol to form an open box, i.e. one that is not joined up at the corners. Thus, neon, which has four electron pairs, can be represented as shown below.

$$
{}^{x}_{x} \overset{x\,x}{\underset{x\,x}{\text{Ne}}} {}^{x}_{x} \quad \equiv \quad \overline{} \\ | \text{ Ne } | \\ \overline{}
$$

This notation often leads to confusingly unclear diagrams, and so is best avoided. Thus, it will not be mentioned further in this text.

When a molecular species is indicated with a net charge, the charge is often shown at the end of the group symbol, and so does not necessarily indicate the atom on which the charge actually resides.

So long as the electrons are counted carefully on every atom, mistakes may be easily avoided. With practice, it will become second nature as to what should be the correct charge on any given species.

Let us start with the dihydride of oxygen, namely water. This may be ionised to give a hydroxide anion and a hydrogen cation. Suggest the equation for this reaction, and calculate the charge on each species.

$$
H_2O \rightleftharpoons OH^- + H^+
$$

The hydroxide anion bears a single negative charge, while the hydrogen cation bears a single positive charge. We have already dealt with how to calculate the charge on the hydrogen cation. Within the hydroxide ion, the hydrogen is involved in one single covalent bond, which contains two electrons, and hence it is neutral, i.e. $1 - 2 + \frac{1}{2}(2) = 0$. The uncharged atomic oxygen has six electrons in its valence shell. However, in the hydroxide anion, eight electrons are accommodated around it, of which only two are involved in a single covalent bond with the hydrogen atom. Hence, there is a single unit of negative charge resident upon the oxygen atom, i.e. $6 - 8 + \frac{1}{2}(2) = -1$. Draw a dot and cross representation of the hydroxide anion.

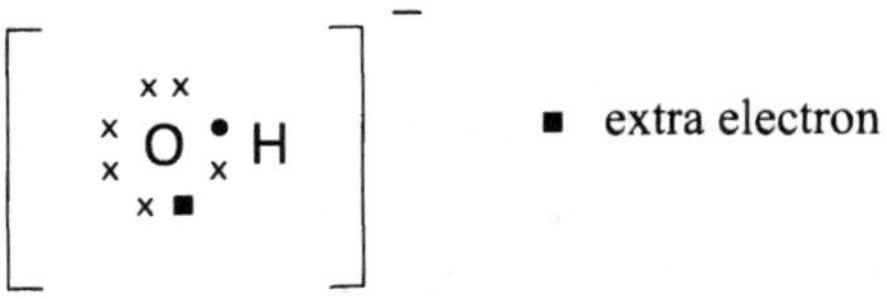

$$
\left[\; {}^{x}_{x} \overset{x\,x}{\underset{x\,\blacksquare}{O}} {}^{\bullet}\text{H} \; \right]^{-} \qquad \blacksquare \text{ extra electron}
$$

There are three lone pairs on the oxygen and one single covalent bond, and so the oxygen atom bears a single negative charge. When an oxygen atom is involved in two single covalent bonds, it has two lone pairs of electrons. When it has only one single bond and three lone pairs, then there is, in effect, an extra electron, which accounts for the single negative charge.

Propose the equation for the addition of a hydrogen cation to a water molecule to give a hydroxonium ion.

$$H_2O + H^+ \rightleftharpoons H_3O^+$$

Now draw a dot and cross representation of the hydroxonium ion.

$$\left[\begin{array}{c} \overset{x\,x}{} \\ H\,\overset{x}{\underset{x}{\,}}O\,\overset{\bullet}{\underset{x}{\,}}H \\ \overset{x\,\bullet}{} \\ H \end{array} \right]^{+}$$

Calculate the charge that is resident on each atom in the hydroxonium ion.

For each of the three hydrogen atoms the charge is zero, i.e. $1 - 2 + \frac{1}{2}(2) = 0$; while on the oxygen it is plus one. There are eight electrons around the oxygen, compared with the six electrons found in the valence shell of the uncharged atom. The oxygen atom is involved in three single covalent bonds, each of which accommodate two electrons. This means that there is a deficit of one electron on the oxygen atom, and so it carries a charge of plus one, i.e. $6 - 8 + \frac{1}{2}(6) = +1$. Note that there is now only one lone pair of electrons. The second lone pair, which normally exists in the neutral molecule, is involved in the dative bond to the hydrogen cation.

Write the equation that represents the further addition of another hydrogen cation to the hydroxonium ion. Draw a dot and cross structure for the resultant ion.

$$H_3O^+ + H^+ \rightleftharpoons H_4O^{2+}$$

$$\left[\begin{array}{c} H \\ \overset{x\,x}{} \\ H\,\overset{x}{\underset{x}{\,}}O\,\overset{\bullet}{\underset{x}{\,}}H \\ \overset{x\,\bullet}{} \\ H \end{array} \right]^{2+}$$

Calculate the charge that is resident on each atom in this ion.

For each of the four hydrogens the charge is again zero; while on the oxygen, it is plus two. There are eight electrons accommodated in the valence shell of the oxygen atom. However, these eight electrons are now involved in four single covalent bonds with the four hydrogen atoms. Thus, there is a deficit of two electrons, and so the oxygen carries a charge of plus two, i.e. $6 - 8 + \frac{1}{2}(8) = +2$. Note that there are no lone pairs of electrons left, because the original two are now both involved in dative bonds to the two extra hydrogen cations.

For each of the three equations that have just been given, add up the charges on the left hand side of the equation and those on the right.

$$H_2O \rightleftharpoons OH^- + H^+ \qquad\qquad 0 \qquad\qquad 0$$
$$H_2O + H^+ \rightleftharpoons H_3O^+ \qquad\qquad +1 \qquad\qquad +1$$
$$H_3O^+ + H^+ \rightleftharpoons H_4O^{2+} \qquad\qquad +2 \qquad\qquad +2$$

On every occasion the total charge is the same before the reaction as it is after the reaction. This observation is embodied in the principle called the Conservation of Charge, which is never broken. This principle states that the sum of the charges of the reactants always equals the sum of the charges of the products. It is a common error to violate this principle by losing or gaining charges.

A common violation of this principle occurs when the wrong charge is placed on a molecular species. We will now look at a few typical mistakes. Consider what is wrong with the charged species, H_4O^+.

If you are having problems, try to write a balanced equation for the formation of this species from known species, and then try to draw the dot and cross structure.

Using the principle of the Conservation of Charge:

$$H_4O^{2+} + \beta^- = H_4O^+ \qquad\qquad +1 \qquad\qquad +1$$

The symbol, β^-, is commonly used to represent an electron. Thus, if an electron is added to the H_4O^{2+} species in order to reduce the overall charge to plus one, on drawing out the required electronic structure it becomes apparent that there are now nine electrons attempting to be accommodated around the oxygen. This would violate the Octet rule. Thus, this species cannot exist under normal circumstances. When calculating the charge on a molecular species, both the Octet rule and the principle of the Conservation of Charge must be followed.

If one of the hydrogen atoms in water is replaced by a general alkyl group, R, then a general alcohol is formed. If the remaining hydrogen is subsequently replaced, then an ether is formed. Propose the general formulae for these two types of compounds.

alcohol	ROH
ether	R_2O or ROR

How many lone pairs are there on the oxygen in the alcohol and in the ether?

There are two lone pairs in each case on the oxygen atom.

When a hydrogen ion is added to a species, the process is called protonation. Write a balanced equation for the protonation of a general alcohol and a general ether.

$$ROH + H^+ \rightleftharpoons ROH_2{}^+$$
$$R_2O + H^+ \rightleftharpoons R_2OH^+$$

Notice that in each case the oxygen now has three groups joined to it by single covalent bonds, and so there is only one lone pair of electrons remaining. Suggest the general formulae for the ionic and neutral compounds of oxygen that contain eight electrons around the oxygen atom.

$$O^{2-}, XO^-, X_2O, X_3O^+ \text{ and } X_4O^{2+}$$

These general formulae describe the ionic and neutral states of all the common species of oxygen compounds. The middle three species are the ones that are common in organic chemistry. The dianion commonly occurs in inorganic compounds as the ionic oxide. The dication is rare under all circumstances, since the high charge density upon the oxygen reduces its stability.

In summary, if oxygen has one covalent bond then it carries a charge of minus one; with two covalent bonds it is neutral; and with three covalent bonds it has a charge of plus one.

The element to the left of oxygen in the Periodic Table is nitrogen. This has five electrons in its valence shell in the elemental state, and so in order to make up the octet it becomes involved in three covalent bonds. In the trihydride of nitrogen, namely ammonia, each of the constituent atoms bears a zero charge.

Write an equation for the protonation of ammonia, and then calculate the charge on each atom. Draw the resultant ion.

$$NH_3 + H^+ \rightleftharpoons NH_4{}^+$$

$$
\left[
\begin{array}{c}
H \\
\overset{\times\times}{H \overset{\times}{\underset{\bullet}{\,}} N \overset{\bullet}{\underset{\times}{\,}} H} \\
\overset{\times\,\bullet}{H}
\end{array}
\right]^+
$$

The charge on the nitrogen is now plus one, i.e. $5 - 8 + \frac{1}{2}(8) = +1$; while it is still zero on each of the hydrogen atoms. The lone pair of the ammonia is now utilised in the dative bond to the hydrogen ion. This cation is called the ammonium ion. Note that the principle of Conservation of Charge has been obeyed in the equation, and also that the nitrogen has only eight electrons around it, and so it still adheres to the Octet rule.

Write an equation for the deprotonation of the ammonia molecule. Calculate the charge on each atom and draw a dot and cross representation for the resultant species.

$$NH_3 \rightleftharpoons NH_2{}^- + H^+$$

$$\left[\begin{array}{c} \times\times \\ \times\ N\ \bullet\ H \\ \blacksquare\ \times \\ \times\ \bullet \\ H \end{array} \right]^-$$

■ extra electron

The charge on each of the hydrogen atoms is still zero, while on the nitrogen it is now minus one, i.e. $5 - 8 + \frac{1}{2}(4) = -1$. This ion is called the amide ion and has two lone pairs of electrons. Again, note that the overall charge has been conserved and that the nitrogen still has eight electrons accommodated in orbitals around it. However, as there are now only two single covalent bonds, there is an overall excess of one electron and so the molecule bears a single negative charge.

Give reasons why the $NH_5{}^{2+}$ ion does not exist under normal circumstances.

By using the principle of the Conservation of Charge, the following equation may be suggested for the formation of the $NH_5{}^{2+}$ ion:

$$NH_4{}^+ + H^+ = NH_5{}^{2+}$$

However, adding a hydrogen cation does not add any electrons, and so there are still eight electrons left around the nitrogen. Yet these eight electrons are required to bond five hydrogen atoms to the central nitrogen atom. This would make available less than two electrons for each bond. Even though this is possible, it does not occur under normal circumstances, and so for the present purposes we may assume that this molecule does not exist, because there are too few electrons to hold all the hydrogens in place.

Suggest why the species $NH_5{}^+$ does not exist under normal circumstances.

By using the principle of the Conservation of Charge, the following equation may be suggested for the formation of the $NH_5{}^+$ ion:

$$NH_4{}^+ + H = NH_5{}^+$$

If a hydrogen atom is added to the ammonium ion, then an extra electron would be added to the eight that already are accommodated around the nitrogen and this would bring the total to nine, which would violate the Octet rule. These examples show why nitrogen usually has a maximum of four covalent bonds radiating from it.

Now suggest why the neutral species NH_4 does not exist.

By using the principle of the Conservation of Charge again, the following equation may be suggested for the formation of the NH_4 species:

$$NH_4{}^+ + \beta^- = NH_4$$

Adding one electron to the eight that already are accommodated around the nitrogen would violate the Octet rule; so for this reason this species does not exist.

One or more of the hydrogen atoms may be replaced by alkyl groups. Suggest the general formulae for the neutral nitrogen compounds with one, two and three alkyl groups.

$$RNH_2, \ R_2NH \text{ and } R_3N$$

Notice that there are only three groups in total, and each species has one, and only one, lone pair. The univalent $-NH_2$ group is called the amino group. Unlike the alkyl derivatives of oxygen where the mono- and di-substituted oxygen compounds were given different names, i.e. alcohols and ethers respectively, all three types of substituted nitrogen species are classed as amines. However, they are subdivided into primary, secondary and tertiary amines, depending on whether the nitrogen bears one, two or three alkyl groups respectively.

Write equations to indicate the products that result from the mono-protonation of each type of amine.

$$RNH_2 + H^+ \rightleftharpoons RNH_3^+$$
$$R_2NH + H^+ \rightleftharpoons R_2NH_2^+$$
$$R_3N + H^+ \rightleftharpoons R_3NH^+$$

Now, write equations to indicate the products that result from the mono-deprotonation of each type of amine.

$$RNH_2 \rightleftharpoons RNH^- + H^+$$
$$R_2NH \rightleftharpoons R_2N^- + H^+$$

The tertiary amine cannot lose a proton from the nitrogen, because it has none to lose: it can only be protonated.

Propose the generalised formulae for the ionic and neutral compounds of nitrogen that contain eight electrons around the nitrogen atom.

$$N^{3-}, \ XN^{2-}, \ X_2N^-, \ X_3N \text{ and } X_4N^+$$

The trianion is found in inorganic compounds, while the last three species are common in organic chemistry.

In summary, if nitrogen is involved in two covalent bonds then it carries a charge of minus one; if it is involved in three covalent bonds then it is neutral, and if it is involved in four covalent bonds then it bears a charge of plus one.

The element to the left of nitrogen is carbon. This has four electrons in its valence shell in the uncharged atomic state. Thus, in order to achieve the octet it must combine with four other atoms, and so form four single covalent bonds. Draw the dot and cross structure of the tetrahydride of carbon, i.e. methane.

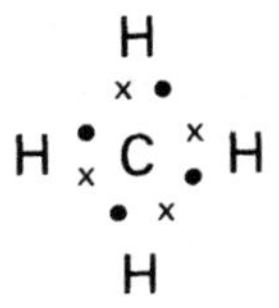

$$\begin{array}{c} H \\ \overset{x}{\underset{}{\bullet}} \\ H \overset{\bullet}{\underset{x}{\;}} C \overset{x}{\underset{x}{\;}} H \\ \underset{\bullet \;\; x}{\;} \\ H \end{array}$$

Now write an equation for the deprotonation of methane, and draw the structure of the resultant species.

$$CH_4 \rightleftharpoons CH_3^- + H^+$$

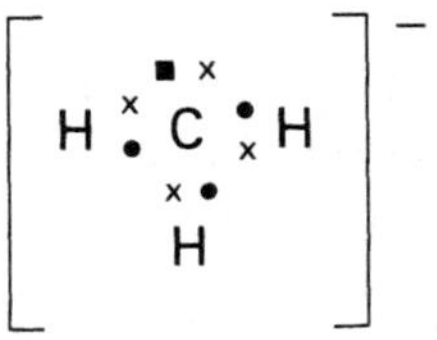

$$\left[\begin{array}{c} \blacksquare \; x \\ H \overset{x}{\underset{\bullet}{\;}} C \overset{\bullet}{\underset{x}{\;}} H \\ \underset{x \; \bullet}{\;} \\ H \end{array} \right]^- \qquad \blacksquare \; \text{extra electron}$$

Notice that the carbon atom has one lone pair of electrons and is involved in three single covalent bonds. This carbon species is called a methyl carbanion, and it bears a single negative charge, i.e. $4 - 8 + \frac{1}{2}(6) = -1$.

If, instead of a proton being removed from the methane molecule, a hydride anion is removed, then a positive carbon species will be formed. Write an equation for the formation of this species and draw its structure.

$$CH_4 \rightleftharpoons CH_3^+ + H^-$$

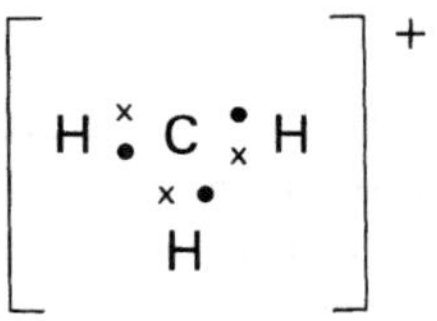

$$\left[\begin{array}{c} H \overset{x}{\underset{\bullet}{\;}} C \overset{\bullet}{\underset{x}{\;}} H \\ \underset{x \; \bullet}{\;} \\ H \end{array} \right]^+$$

This carbon species is called a methyl carbonium ion or a carbocation. It is different from the charged species that we have looked at before, because it only has six electrons accommodated around the central atom. These electrons occupy three single covalent bonds. The central atom is thus electron deficient, and bears a single positive charge, i.e. $4 - 6 + \frac{1}{2}(6) = +1$. The notional fourth orbital is empty, and as such is called an empty orbital.

Suggest reasons why the species CH_5^+ and CH_5 do not exist under normal circumstances. (In passing, it ought to be borne in mind, that 'normal circumstances' has a wide range of meanings, which depends upon to whom one is talking. For example, the cation CH_5^+ is very common in chemical ionisation mass spectrometry. However, for present purposes, i.e. school and undergraduate texts, it is rare.)

By using the principle of the Conservation of Charge, the following equations can be suggested for the formation of the CH_5^+ ion and CH_5 species:

$$CH_4 + H^+ = CH_5^+$$
$$CH_4 + H = CH_5$$

In the first case, the carbon would now have eight electrons accommodated around it. Yet these eight electrons are required to bond five hydrogen atoms to the central carbon atom. This would make available less than two electrons for each bond. Even though this is possible, it does not occur under normal circumstances, and so for the present purposes we may assume that this molecule does not exist, because there are too few electrons to hold all the hydrogens in place. In the second case, the proposed molecule would have nine electrons, and so would violate the Octet rule. So, under normal circumstances pentavalent carbon atoms are not found, regardless of whether they are charged or neutral.

So far we have looked at the CH_3^- and CH_3^+ ions: the first has eight electrons, while the second has six electrons accommodated around the carbon. Suggest a structure for the trihydride of carbon that has seven electrons around the carbon, and calculate the charge that is resident on the carbon. Suggest an equation for its formation from methane.

$$CH_4 \rightleftharpoons CH_3^{\cdot} + H^{\cdot}$$

$$H \overset{x}{\underset{\bullet}{\times}} \overset{x}{\underset{x}{C}} \overset{\bullet}{\underset{x}{\bullet}} H$$
$$H$$

The $CH_3^{\cdot}$ species is neutral. It has seven electrons accommodated around the carbon, six of which are in three single covalent bonds to the three hydrogens. This leaves a single unpaired electron on the carbon. Molecules that have unpaired electrons are called radicals, and this particular radical is called the methyl radical.

Propose the general formulae of the charged and neutral trisubstituted carbon species that have been discussed so far, and indicate the number of electrons around each carbon.

$$CX_3^-, \; CX_3^{\cdot} \; and \; CX_3^+$$

The central carbon atom bears eight, seven and six electrons respectively. These three species are illustrative of the principal electronic configurations that a carbon atom adopts in organic chemistry.

There is one more carbon species that is encountered less frequently. Propose an equation for the production of the species that results from the loss of the X^- anion from the CX_3^- species. Draw its dot and cross structure.

$$CX_3^- \rightleftharpoons :CX_2 + X^-$$

$$X \overset{x}{\underset{\bullet}{\cdot}} C \overset{x}{\underset{}{\times}}$$

$$\underset{X}{\overset{x \bullet}{}}$$

In this species, the carbon atom has only six electrons accommodated around it, four of which are involved in two single covalent bonds, while the other two are not involved in bonding with other atoms. Thus, the carbon atom is electron deficient but uncharged, i.e. $4 - 6 + \frac{1}{2}(4) = 0$. This type of carbon species is called a carbene and is highly reactive.

In summary, we have looked at species that have gained or lost electrons, and so the charge that they bear must be either zero or an integral positive or negative number. There have been no fractional charges. Further, we have used the principle of the Conservation of Charge to ensure that all the electrons have been accounted for when suggesting equations for the formation of the different species. We have also found that elements in the second row of the Periodic Table cannot accommodate more than eight electrons in the valence shell, although in some circumstances they can have fewer than eight valence electrons, i.e. the Octet rule is never broken.

In the next chapter we will look at partial charges and their distribution.

2.5 Summary of electron counting

Most errors result either from placing too many electrons around an atom, e.g. CH_5, or by placing the wrong charge on an atom, e.g. R_3NH.

The smallest unit of an element is the atom. Atoms are formed from protons, neutrons and electrons. The protons and neutrons together constitute the nucleus; the electrons occupy orbitals, which are centred on the nucleus. The number of protons within the nucleus determines the element. This number is called the atomic number. The sum of the number of neutrons and protons is called the atomic mass number.

Usually, there are approximately the same number of neutrons as there are protons. For a given element, atoms that contain different numbers of neutrons are called isotopes, and they have different physical properties. However, the chemical properties of different isotopes are the same. When referring to an element, it is usual to imply that the most stable isotope, or the naturally occurring mixture of isotopes, is meant unless otherwise stated. Usually, different isotopes are distinguished by having the different atomic mass numbers as a raised prefix to the atomic symbol, e.g. ^{17}O and ^{18}O.

The chemical properties of an element are determined by the element's electrical properties. In an atom, the number of electrons always equals the number of protons, and so every element in the atomic state is electrically neutral. If an element loses or gains one or more electrons then that element will no longer be electrically neutral, because the number of protons will not equal the number of electrons that surround the nucleus. The resultant charged species is called an ion. If the ion is negative, it is an anion; if it is positive, it is a cation. If there is an unpaired electron in an atom or ion, it is called a radical.

Electrons in an atom are contained within atomic orbitals, each of which may only hold a maximum of two electrons. The principal quantum shell is indicated by a number, e.g. 1, 2, 3, etc. Within each shell, there may exist a number of subshells: s, p, d and f, depending upon which shell is being considered. There is only one s subshell, while there are three p subshells, which are degenerate: p_x, p_y and p_z. Hund's rule states that if electrons are placed in degenerate orbitals, they fill them so as to minimise the doubling-up of electrons. The number of electrons within each subshell is indicated by a raised postscript, e.g. for nitrogen, the electronic configuration is represented as $1s^2$, $2s^2$, $2p_x^1$, $2p_y^1$, $2p_z^1$.

The outermost principal quantum level that is normally occupied in the neutral atom is called the valence shell. There are only two first row elements, because there is only one subshell within the first principal quantum level, and so only space for two electrons. However, there are eight second row elements, because there are a total of four subshells within the second principal quantum level, which permit eight electrons to be accommodated. Usually only the electrons within the valence shell are indicated and the electrons within the core shell, i.e. non-valence shell, are assumed to be in order.

Hydrogen is the first element, and so has an atomic number of one. Its electronic configuration is $1s^1$. There are three isotopes: protium, deuterium and tritium, which may be represented by the symbols ^1H, ^2H or ^3H, or alternatively, H, D or T. The cation of hydrogen is a bare proton, H^+ ($1s^0$) while the anion of hydrogen is called a hydride anion, H^- ($1s^2$) (or more strictly, a protiide anion).

Helium, $1s^2$, is the element that has an atomic number of two. There are two common isotopes that contain one and two neutrons, namely ^3He and ^4He respectively. The dication, He^{2+} ($1s^0$), is called an alpha particle.

The electronic configurations of carbon, nitrogen and oxygen are $1s^2$, $2s^2$, $2p_x^1$, $2p_y^1$; $1s^2$, $2s^2$, $2p_x^1$, $2p_y^1$, $2p_z^1$; and $1s^2$, $2s^2$, $2p_x^2$, $2p_y^1$, $2p_z^1$, respectively.

In an atom, the difference between the number of orbiting electrons and protons within the nucleus determines the charge that resides on that atom or ion, as the case may be.

The Octet rule states that there may never be more than eight electrons in the valence shell of a second row element. This is equivalent to four bonds, four lone pairs or a mixture of the two. Third row elements are not constrained by the Octet rule, and so may have more than four bonds around them, e.g. phosphorous and sulphur can have five and six bonds, respectively.

In a molecule, the same principle of counting the electrons and protons determines the location and the magnitude of any charge that might be present.

A molecular orbital, like an atomic orbital, can accommodate only two electrons.

The charge on any given atom, whether isolated or in a molecule, can be calculated by the following formula: subtract the number of electrons that are actually present in the valence shell from the number normally present in the neutral uncharged atomic species. Then, add half the number of electrons which that atom shares with any other atom in the molecule. The result is the charge on that atom. Thus, the charge on an atom equals (the number of normal valence electrons) minus (the number of electrons actually present in the valence shell) plus half (the number of electrons shared by that atom).

The simplest molecule is hydrogen, H_2. This consists of two hydrogen nuclei joined by one covalent bond formed from two electrons contained within a single molecular orbital, into which each hydrogen atom has donated one electron. Each hydrogen atom is neutral, because each atom shares equally the two electrons within the molecular orbital.

The Conservation of Charge states that the sum of the charges of the reactants always equals the sum of the charges of the products.

Water, H_2O, can lose a proton, i.e. be deprotonated, and so split into a hydroxide anion, OH^-, and a hydrogen cation, H^+. The oxygen atom in the hydroxide anion bears a single negative charge, i.e. $6 - 8 + \frac{1}{2}(2) = -1$. Alternatively, water can accept a hydrogen cation, i.e. be protonated, to form the hydroxonium cation, H_3O^+. In this case the oxygen atom bears a single positive charge, i.e. $6 - 8 + \frac{1}{2}(6) = +1$.

In the series of oxygen compounds O^{2-}, XO^-, X_2O, X_3O^+ and X_4O^{2+}, the oxygen has four, three, two, one and no lone pairs respectively, and bears a charge of $-2, -1, 0, +1$ and $+2$ respectively. Only the middle three compounds are common in organic chemistry. In the monocation and dication, there are one or two dative bonds respectively. The dication is very rare, while the dianion only commonly exists in inorganic compounds. Where the substituent in the neutral compound is an alkyl group, then the resultant compounds, namely ROH and R_2O, are called alcohols and ethers respectively. H_4O^+ is a non-existent species, because there are nine electrons on the oxygen, which would violate the Octet rule.

In the series of nitrogen compounds N^{3-}, XN^{2-}, X_2N^-, X_3N and X_4N^+, the nitrogen has four, three, two, one and no lone pairs respectively, and bears a charge of $-3, -2, -1, 0$, and $+1$ respectively. Only the last three compounds are common in organic chemistry. Where the substituent is an alkyl group, the monoanion is called an amine anion, while the monocation is called an alkyl ammonium cation. The $-NR_2$ group is called an amino group. Non-existent species include: NH_5^{2+} (because there are eight electrons trying to hold five hydrogens to the nitrogen atom, i.e. less than the normal two electrons for each covalent bond), NH_5^+ and NH_4 (because, in each case, there are nine electrons on the nitrogen, which would violate the Octet rule).

In the series of carbon compounds CX_3^-, $CX_3^{\cdot}$, CX_3^+ and $:CX_2$, the carbon has eight, seven, six and six electrons respectively, and bears a charge of

-1, 0, $+1$ and 0 respectively. The second species is a radical, while the last is a carbene. The monocation is called a carbonium ion or carbocation, while the monoanion is called a carbanion. The carbanion has one lone pair, while the radical, carbocation and the carbene are all electron deficient. The carbocation has one empty orbital. Unpaired electrons exist in carbenes and carbon radicals, which partially accounts for their reactivity. Non-existent species include: CH_5^+ (because there are eight electrons trying to hold five hydrogens to the carbon atom, i.e. less than the normal two electrons for each covalent bond); and CH_5 (because there are nine electrons on the carbon, which would violate the Octet rule).

Bond Polarisation and Fission

3.1 Introduction

In this chapter we will look at covalent bonds and the manner in which they may be broken. Prior to discussing the complete breaking of a bond, it is natural to investigate bonds that exhibit a degree of polarisation, i.e. ones that are on the way towards cleaving.

3.2 Partially charged species

In a homoatomic bond, i.e. one between two atoms that are the same, the electrons in the covalent bond are shared equally between the two nuclei. This means that there is no permanent charge separation, i.e. one end is not permanently positive, nor the other permanently negative. Suggest two homoatomic molecules formed from elements selected from the first two rows of the Periodic Table.

e.g. H–H and F–F

While there is no permanent separation of charge in such homoatomic molecules, there may be a temporary separation of the charge present in the covalent bond. The electron density within the molecular orbital is subject to random fluctuations, which result in a temporary separation of charge called the London effect. These fluctuations average to zero over time.

There is another way in which such temporary charge separation may occur. Suggest what would happen if a homoatomic molecule were brought close to a fixed positive charge.

The positive charge would attract the electrons within the bond, and so the end closer to the fixed positive charge would become negative. The other end of the bond would correspondingly become positive, because it is now electron deficient. Hence, a charge separation would have been induced by the approach of the molecule to a fixed charge.

The resultant charged species is called a dipole, because there are two poles, one positive and the other negative. If the original molecule is neutral, then the negative and positive charges that have been separated must sum to zero. Only small amounts of charge are separated, and not full units. The sign δ is used to indicate this small amount of charge, and it is used as a raised postscript with the appropriate sign, i.e. $\delta+$ or $\delta-$.

Draw the dihydrogen molecule that is next to a fixed positive charge, and indicate the charge separation that has been induced.

$$H \overset{\delta+}{\rule{3cm}{0.4pt}} H \overset{\delta-}{} \quad [+]$$

The hydrogen closest to the fixed positive charge will bear a small negative charge, while the other hydrogen will bear an equally small positive charge. The sum of these charges is obviously zero. However, the sum of the moduli of these charges, i.e. the magnitude of the charge without consideration of whether it is positive or negative, is less than one.

Where there are two different elements in a diatomic molecule, a heteroatomic bond results. Each element has a different affinity for electrons, and this is called its electronegativity. The greater an element's attraction for electrons, the greater its electronegativity. In Chapter 2, the differences in the electronegativities of the various atoms were ignored when we looked at the hydrides of carbon, nitrogen and oxygen. Now, we will study these, and other, molecules more closely in order to see what is the effect of these differences. In the diatomic molecule of hydrogen fluoride, suggest which atom carries the small negative charge, bearing in mind that fluorine is the most electronegative element in the Periodic Table.

$$H \overset{\delta+}{\longrightarrow} F \overset{\delta-}{}$$

The fluorine atom is more electronegative than the hydrogen atom, and so attracts the electrons within the single covalent bond more strongly than the hydrogen atom. This dipole is permanent, and is indicated on a diagram of the molecule by an arrowhead on the line that represents the single covalent bond. The arrowhead points towards the negative end of the bond. This attraction of electrons towards one end of a single covalent bond is called induction. A group that attracts electrons is said to exert a negative inductive effect and vice versa. The symbol for this property is I.

Show the inductive effect on the following bonds: H–N, H–O and H–C.

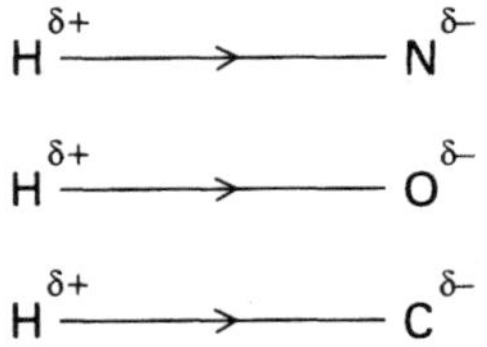

$$H \overset{\delta+}{\longrightarrow} N^{\delta-}$$

$$H \overset{\delta+}{\longrightarrow} O^{\delta-}$$

$$H \overset{\delta+}{\longrightarrow} C^{\delta-}$$

In all cases the hydrogen forms the positive end of the covalent bond. The numbers that represent the relative magnitudes of the electronegativities of oxygen, nitrogen, carbon and hydrogen are 3.5, 3.0, 2.5 and 2.1 respectively. Hence, suggest which of the above bonds is the most polarized and which is the least.

The O–H bond is the most polarised, while the C–H bond is the least. In each case the charge separation is still only a small fraction of an electrical unit, and the sum of these small charges equals zero. The difference between the electronegativities of carbon and hydrogen is fairly small, and for most purposes the polarisation of the C–H bonds may be ignored.

The electronegativities of sulphur and chlorine are 2.5 and 3.5 respectively. Suggest what will be the polarisation in the H–S bond and the H–Cl molecule.

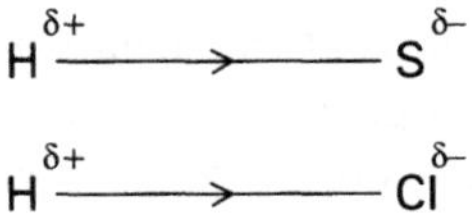

$$H \overset{\delta+}{\longrightarrow} S^{\delta-}$$

$$H \overset{\delta+}{\longrightarrow} Cl^{\delta-}$$

The degree of charge separation found in the H–S bond is similar to that present in the H–C bond, while the H–Cl bond is quite strongly polarised.

The strong polarisation of the single covalent bond found in the H–F, H–Cl and H–O bonds means that the hydrogen is quite positive. In each of these cases, there is a lone pair of electrons that exists on the other atom, i.e. the heteroatom. (A heteroatom is any other atom apart from hydrogen or carbon.) A lone pair of electrons is a region of space that is rich in electrons. Suggest how in the case of HCl, there may be an interaction between two different molecules of HCl.

$$H \overset{\delta+}{\longrightarrow} Cl^{\delta-} \cdots\cdots H \overset{\delta+}{\longrightarrow} Cl^{\delta-}$$

The dotted line indicates a weak intermolecular Coulombic interaction, i.e. an electrostatic attraction, which is called a hydrogen bond, between the $\delta+$ hydrogen and an electron-rich lone pair on the chlorine atom of the other HCl molecule.

Indicate the polarisation that is present for each of the single covalent bonds that exist between carbon and nitrogen, oxygen and chlorine respectively.

$$C \overset{\delta+}{\underset{}{\longrightarrow}} N^{\delta-}$$

$$C \overset{\delta+}{\underset{}{\longrightarrow}} O^{\delta-}$$

$$C \overset{\delta+}{\underset{}{\longrightarrow}} Cl^{\delta-}$$

In all these cases, the carbon is at the positive end. This means that it is liable to be attacked by a negative species, or that it is on the way to becoming a carbonium ion, which, it will be remembered, is the name for a carbon species that bears a positive charge.

Carbon sometimes forms covalent bonds with metals such as lithium and magnesium, the electronegativities of which are 1.0 and 1.2 respectively. Indicate the polarisation that is present in these bonds.

$$C^{\delta-} \overset{}{\underset{}{\longleftarrow}} Li^{\delta+}$$

$$C^{\delta-} \overset{}{\underset{}{\longleftarrow}} Mg^{\delta+}$$

In these cases, the carbon atom is at the negative end of the bond. This means that it is liable to be attacked by positive species, or that the carbon is on the way to becoming a carbanion. This is the case for organometallic bonds in general.

So far we have only considered each bond in isolation, but, in all except the simplest of molecules, there will be a chain of atoms that will interact. Suggest what will happen in the $CH_3CH_2CH_2Cl$ molecule.

$$CH_3 \text{---} CH_2 \overset{\delta\delta+}{\longrightarrow} CH_2 \overset{\delta+}{\longrightarrow} Cl^{\delta-}$$

The C–Cl bond is polarised as expected, but now the first carbon in the chain bears a permanent partial positive charge. This attracts electrons from the next C–C bond in the chain. However, this attraction is much smaller than the initial one, and to indicate this the symbol $\delta\delta$ is used, i.e. very small. The total of all the small and very small charges must still sum to zero. The transmission of this effect is very inefficient, and so after the third carbon in the chain the effect is negligible. Generally, only the effect on the carbon that is bonded directly to the polarising element needs to be considered.

The polarisation along the bond is characterised by both its magnitude and its direction, i.e. it is a vector. This may be illustrated by considering a molecule of boron trichloride, which has a symmetrical planar trigonal shape. First, consider an isolated B–Cl bond and determine the direction of polarisation, given that the electronegativity of boron is 2.0. Then suggest what is the overall polarisation for the BCl$_3$ molecule.

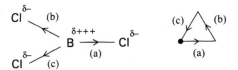

Each B–Cl bond is polarised towards the chlorine atom. However, by vector addition of the three dipoles, it is apparent that they cancel each other out, so that there is no overall polarisation of the molecule.

Now, if we consider the water molecule, which by experiment has been shown to exist with a bond angle, HOH, of about 105°, suggest what will be the overall polarisation of this molecule.

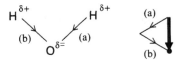

In this case the dipoles created by the polarisation of the two oxygen/hydrogen bonds do not sum to zero, but instead there is a resultant vector that bisects the HOH bond angle and is directed towards the oxygen.

The single bond between the elements that has been considered so far is called a σ bond. Notice that for atomic orbitals, Roman letters are used, while for molecular orbitals, Greek letters are used to characterise the various types. Furthermore, a σ molecular orbital is often formed from the combination of two s atomic orbitals. So far we have only looked at the polarisation of this σ bond, which in all cases remained intact. This was the case even if there was a permanent charge separation. Now we will look at what happens when the charge separation is taken a stage further and the bond breaks.

3.3 Bond fission

In a homoatomic bond, such as between two fluorine atoms, there is no permanent charge separation. If the bond in the fluorine molecule should break, what do you think would be the likely products? Write an equation for your suggested answer.

$$F_2 \rightleftharpoons 2F^\bullet$$

It is not unreasonable to suppose that the symmetrical fluorine molecule would break into two equal parts, each of which is a fluorine atom. What is the charge on each fluorine atom after the bond has broken, and how many electrons are there around each atom?

The charge is zero, and there are seven electrons around each atom. Using the principle of the Conservation of Charge the products must have a net zero charge, because that is the charge on the fluorine molecule. The fluorine atom has seven electrons in its valence shell, six of which exist in three lone pairs, while the last is an unpaired electron; hence, the fluorine atom is a radical.

This type of bond breakage is called homolytic cleavage, because the bond breaks equally between the atoms that formed it. This type of cleavage is represented by two curly arrows, each with only half of the arrowhead present, sometimes referred to as fishhook arrows. The tail of the arrow indicates the source of the electrons that are moving, and is usually an electron-rich area, i.e. a bond or a lone pair. The head of the arrow is the target to which the electrons are moving, and is usually deficient in electrons, i.e. an empty orbital, or a positive charge, or an electronegative atom. Half an arrowhead is used to represent one electron, while two electrons moving together are represented by a complete arrowhead.

Draw the appropriate arrows to represent the homolytic cleavage of the fluorine molecule.

$$F \overset{\frown}{\underset{\smile}{\text{——}}} F \; \rightleftharpoons \; F\bullet \; + \; F\bullet$$

In this case, there are two arrows, each of which has its tail in the middle of the bond that is going to break. One arrow with half an arrowhead is going to one fluorine atom, while the other is going to the other. The half arrowhead indicates that only one electron is involved in each movement. This shows that the two electrons, which originally formed the single bond between the atoms, will be divided equally between the two fluorine atoms, with one electron going to each atom.

For the hydrogen chloride molecule, draw the bond polarisation that would be expected.

$$\overset{\delta+}{H} \longrightarrow \overset{\delta-}{Cl}$$

If the polarisation continued to the limit, which element would you expect to gain the electrons originally present within the bond, and which would you expect to lose them?

The chlorine atom would gain them, while the hydrogen atom would lose them. What would be the nature of the ions that are formed in each case?

The chlorine would form the chloride anion, while the hydrogen would form the hydrogen cation. Draw a representation of this bond cleavage using the curly arrow notation introduced above.

$$H \text{———} Cl \;\rightleftharpoons\; H^+ \;+\; Cl^-$$

Note that only one arrow is used, and that it has a complete arrowhead, which represents the two electrons in the bond. It is not necessary to use two arrows, each with only half an arrowhead, because the two electrons are moving together. The tail of the arrow is on the bond that is about to break, and the head is on the chlorine atom, which is the destination of the electrons. Note that the Conservation of Charge is obeyed by the formation of the monocation and the monoanion. This type of bond cleavage is called heterolytic fission.

Draw the bond fission that would occur in chloromethane.

$$CH_3 \overset{\delta+}{\text{———}} Cl^{\delta-} \;\rightleftharpoons\; CH_3^+ \;+\; Cl^-$$

The most polarised bond is the C–Cl bond, with the chlorine being at the more negative end. If this bond should break, it is reasonable to suppose that it would break by continuing to polarise in the original direction, and so the electrons in the bond would go to the chlorine, and the carbon species that is left would be a cation.

Suggest what would be the nature of the bond fission that occurred in a water molecule.

$$H_2O \;\rightleftharpoons\; OH^- \;+\; H^+$$

$$\overset{\delta+}{H} \diagup \overset{\delta=}{O} \text{———} \overset{\delta+}{H} \;\rightleftharpoons\; HO^- \;+\; H^+$$

The O–H bonds are polarised towards the oxygen, and if this polarisation is continued until the bond is broken, then, after the heterolytic cleavage, a hydroxide anion and a hydrogen cation will result. Suggest why it would be unlikely that a hydride anion would be produced from the heterolytic cleavage of water.

Using the principle of the Conservation of Charge, if a heterolytic fission occurred that resulted in a hydride ion, then the other ion must be an OH^+ species. This is unlikely, because it would mean that the O–H bond had cleaved in the opposite direction to the original polarisation, and further, would produce a species that has only six valence electrons residing on the oxygen atom.

In summary, homolytic cleavage usually occurs with homoatomic bonds, or when the degree of polarisation is small. The result of such a cleavage is two radicals. Heterolytic cleavage occurs when the degree of polarisation is large, or when the resultant ionic species are particularly stable. The consequence of the two electrons in the original bond both going to one atom means that ions are formed, one of which has an extra lone pair and the other an empty orbital. By using curly arrows to indicate the movement of the electrons, you have written out the mechanism of these bond cleavages.

So far, we have only looked at single covalent bonds. In organic chemistry there are many compounds that contain double, or even triple, bonds between atoms. We will now examine some examples of these.

3.4 Isolated multiple bonds

The nitrogen atom has five electrons in its valence shell, so it needs to be involved in three covalent single bonds in order to obtain the stability that is conferred by having an octet of electrons. If two nitrogen atoms combine with each other so as to form a diatomic molecule, then each atom provides the three extra electrons that are needed by the other to satisfy the octet.

Draw a dot and cross structure for the diatomic molecule of nitrogen.

$$^x_x N \; ^x_{\bullet} {}^x_{\bullet} {}^x_{\bullet} N^{\bullet}_{\bullet}$$

Each nitrogen has one lone pair and is involved in three bonds, each of which has two electrons. The second and third bonds in the dinitrogen molecule are different from the first bond, which, it will be recalled, is called a σ bond. These multiple bonds are called π bonds, and are often formed from two p atomic orbitals.

Calculate the charge that is resident on each nitrogen atom in a nitrogen molecule.

The charge on each is zero. There is a total of eight electrons around each nitrogen in the valence shell, which normally has five accommodated there, and each nitrogen atom has a half share in the six electrons that are involved in the three bonds between the atoms. This results in a zero net charge on each atom.

The oxygen atom, which has six electrons in its valence shell, requires two covalent single bonds in order to obtain the stability of the octet of electrons. Oxygen can combine with another oxygen atom so as to form a diatomic molecule, in which each atom provides the two extra electrons needed by the other to satisfy the octet requirement.

Draw a dot and cross structure for the diatomic molecule of oxygen.

$$^{x\,x}_{x\,x}O \; ^x_{\bullet} {}^x_{\bullet} O^{\bullet\,\bullet}_{\bullet\,\bullet}$$

Each oxygen has two lone pairs and is involved in two bonds, each of which has two electrons. Calculate the charge that is resident on each oxygen atom.

The charge on each is zero. There is a total of eight electrons around each oxygen in the valence shell, which normally has six, and each oxygen atom has a half share in the four electrons that are involved in the bonds between the atoms. This results in a zero net charge on each atom.

This proposed electronic structure for diatomic oxygen contains a double bond between the two atoms, which on the face of it, would suggest that the molecule ought to be quite unreactive. In fact, diatomic oxygen is quite reactive. Suggest a reason for this.

Generally, paired electrons are less reactive than unpaired electrons. In oxygen, under normal circumstances, one of the bonds between the atoms breaks and so forms a biradical species. Draw the mechanism for this fission.

$$\overset{x\ x}{\underset{x}{\overset{x}{\text{O}}}} = \overset{\bullet\bullet}{\text{O}} \colon \leftrightarrow \overset{x\ x}{\underset{x}{\overset{x}{\text{O}}}} - \overset{\bullet\bullet}{\underset{\bullet}{\text{O}}} \colon$$

The π bond undergoes homolytic fission, and so forms the biradical dioxygen. Radicals possess unpaired electrons, which would be more stable if they combined and became paired, thus they tend to be quite reactive. Hence, the biradical form of dioxygen is quite reactive as it has two radical parts.

In the molecule diimide, $HN = NH$, there is a double nitrogen/nitrogen bond. Draw the dot and cross structure for this molecule.

$$\underset{H\ ^{x}\qquad\quad H}{\overset{x\ x\quad\ \bullet\bullet}{\underset{\bullet}{\text{N}}\ ^{x\ x}_{\bullet\ \bullet}\ \underset{\bullet}{\text{N}}}}$$

Suggest the polarisation that is present between the central atoms of the three molecules that have just been discussed.

In the case of dinitrogen and dioxygen, the multiple bonds were homoatomic, and so there is no permanent polarisation across the central bond. In the case of diimide, exactly the same atoms appear on each side of the central bond, and so again there is no permanent charge separation. However, temporary and induced polarisations may occur in each of these cases as may occur in homoatomic single bonds.

Carbon may form double or triple bonds with itself. Draw a dot and cross representation for these species, namely $R_2C = CR_2$ and $R-C \equiv C-R$. Note the symbol R is used to indicate a general alkyl group.

$$\underset{R\ ^{\bullet}\qquad\quad ^{\bullet}R}{\overset{R\ ^{x}\qquad\quad ^{\bullet}R}{\underset{x}{\overset{\bullet}{\text{C}}}\ ^{x\ x}_{\bullet\ \bullet}\ \overset{\bullet}{\underset{\bullet}{\text{C}}}}} \qquad R\ ^{x}_{\bullet}\text{C}\ ^{x\ x\ x}_{\bullet\ \bullet\ \bullet}\ \text{C}\ ^{\bullet}_{\bullet}\ R$$

Again, there is no permanent polarisation of the multiple bond, and so each carbon is electrically neutral.

Carbon may also form multiple bonds with many other elements. It is this ability that is one of the reasons for the richness of the chemistry of carbon. One of the commonest heteroatomic multiple bond systems in which carbon partakes involves oxygen. Draw the dot and cross structure of this carbon/oxygen double bond system.

$$
\begin{array}{c}
\text{H} \\
\overset{\bullet}{\underset{x}{\text{C}}} \ \overset{x\ x}{\underset{\bullet\ \bullet}{}} \ \overset{\bullet\ \bullet}{\underset{\bullet}{\text{O}}} \\
\text{H}
\end{array}
$$

Calculate the charge on the carbon and the oxygen atoms, and also suggest whether there is any polarisation of this bond.

$$
\begin{array}{c}
\text{H} \\
\diagdown \\
\qquad \overset{\delta+}{\text{C}} \longrightarrow \overset{\delta-}{\text{O}} \\
\diagup \\
\text{H}
\end{array}
$$

There are no whole units of charges resident on either the carbon or oxygen atom. However, the bond between the carbon and oxygen has a permanent polarisation due to the difference in electronegativities, with the oxygen atom being negative and the carbon being positive. The sum of the separated charges is zero. The carbon/oxygen double bond system is called a carbonyl group.

Draw the mechanism for the heterolytic fission of the π bond within the carbonyl group.

$$
\begin{array}{c}
\text{H} \\
\diagdown \\
\qquad \text{C} =\!\!= \text{O} \\
\diagup \\
\text{H}
\end{array}
\quad \leftrightarrow \quad
\begin{array}{c}
\text{H} \\
\diagdown \\
\qquad \overset{+}{\text{C}} \longrightarrow \overset{-}{\text{O}} \\
\diagup \\
\text{H}
\end{array}
$$

Note that a single arrow with a complete arrowhead is used, with the tail on the bond that is breaking and the head on the oxygen atom. The result is an ionic species that has both positive and negative poles. These opposite charges reside on adjacent atoms. This heterolytic fission is very important in the chemistry of the carbonyl group.

The attraction of electrons along the multiple bond network is called the mesomeric effect: those atoms or groups that attract electrons exert a negative mesomeric effect, and vice versa. The symbol is M, and it is used with the appropriate plus or minus sign.

Draw a dot and cross representation of the cyanide group. Then suggest the direction of polarisation, if any, along this bond system.

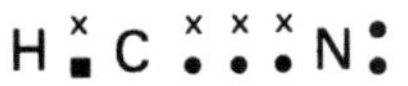

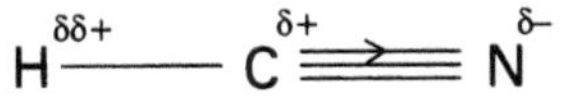

$$H \overset{x}{\underset{\bullet}{}} C \;\; \overset{x \;\; x \;\; x}{\underset{\bullet \;\; \bullet \;\; \bullet}{} } N \overset{\bullet}{\underset{\bullet}{}} \qquad\qquad H \overset{\delta\delta+}{\text{———}} C \overset{\delta+}{\underset{}{\Longrightarrow}} N^{\delta-}$$

The bond system is slightly polarised towards the nitrogen, with each atom bearing only a small partial unit of electric charge.

Draw the dot and cross structure of the imine group, $R_2C=NX$, and again suggest the direction of polarisation, if any. Note the symbol X is used to indicate a group that exerts a negative inductive effect, i.e. one that attracts electrons.

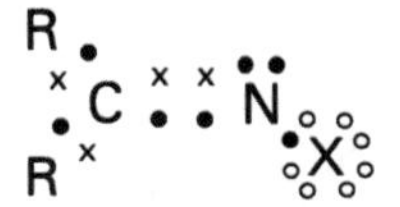

$$\underset{R}{\overset{R}{\underset{x}{\overset{\bullet}{}}}} \; C \; \overset{x \;\; x}{\underset{\bullet \;\; \bullet}{}} \; N \;\; \underset{X}{}$$

Again, the bond system is slightly polarised towards the nitrogen, but this time the degree of polarisation depends on the nature of X. If X is a strongly electronegative element or group, then the degree of polarisation is higher.

In a general compound that contains a carbonyl group, $R_2C=O$, the oxygen has two lone pairs, either of which may readily be protonated. Write an equation for the protonation of the carbonyl group, and draw a dot and cross structure for the product.

$$R_2C=O + H^+ \rightleftharpoons R_2C=OH^+$$

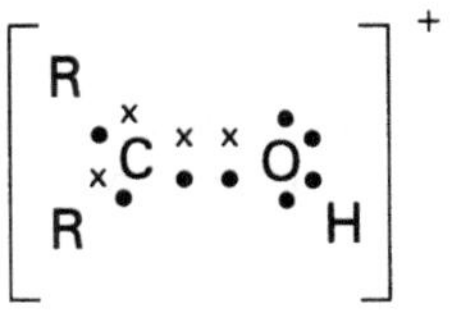

$$\left[\underset{R}{\overset{R}{}} \; C \; \overset{x \;\; x}{\underset{\bullet \;\; \bullet}{}} \; O \;\; \underset{H}{} \right]^+$$

Calculate the charge on the carbon and oxygen atoms of the product.

The carbon is still neutral, while the oxygen now carries a charge of plus one. This is to be expected as the carbon is involved in four covalent bonds, while the oxygen is involved in three.

As the oxygen now carries a positive charge, it attracts electrons even more strongly than it did when it was neutral. Has the inductive effect of the oxygen along the single bond system increased or decreased?

It has increased. What about the mesomeric effect along the double bond system?

This has increased as well. Draw the result of the heterolytic fission of the double bond.

$$\underset{R}{\overset{R}{\diagdown}} C = \overset{+}{O}H \quad \leftrightarrow \quad \underset{R}{\overset{R}{\diagdown}} \overset{+}{C} \!\!-\!\! OH$$

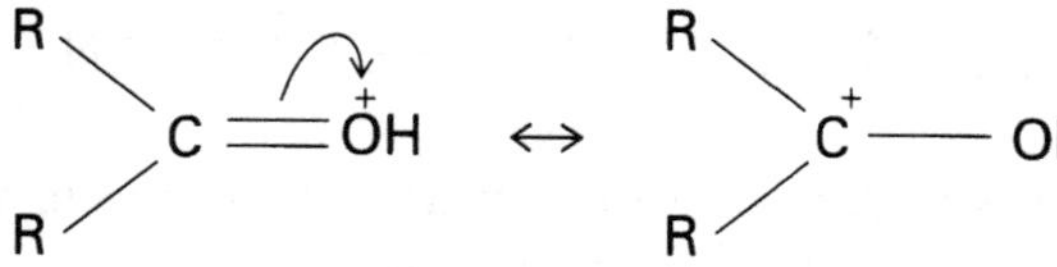

The positive charge now resides on the carbon atom.

If the nitrogen in an imine bond is protonated, draw the structure of the protonated product, and the product after the carbon/nitrogen double bond has been broken heterolytically.

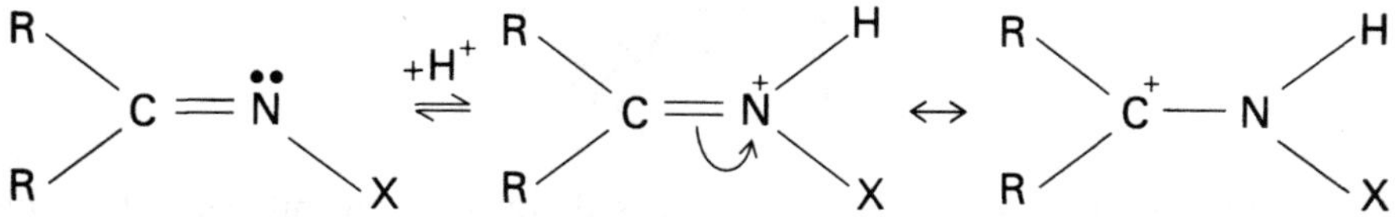

$$\underset{R}{\overset{R}{\diagdown}} C = \overset{..}{N} \underset{X}{\diagup} \quad \overset{+H^+}{\rightleftharpoons} \quad \underset{R}{\overset{R}{\diagdown}} C = \overset{+}{N} \underset{X}{\overset{H}{\diagup}} \quad \leftrightarrow \quad \underset{R}{\overset{R}{\diagdown}} \overset{+}{C} \!\!-\!\! N \underset{X}{\overset{H}{\diagup}}$$

Initially, the nitrogen carried the positive charge and then, after the heterolytic fission of the double bond, the carbon bears the positive charge.

In summary and simplifying slightly, the inductive effect operates along the single bond, or σ, system, and results in partial charge separation. In contrast, the mesomeric effect operates along the multiple bond, or π, system, and results in whole units of charge being separated.

3.5 Conjugated multiple bonds

When multiple bonds alternate with single bonds, the whole arrangement is called a conjugated system. It has certain properties that are of great importance in organic chemistry.

In the molecule $R_2C=CH-CH=CR_2$, there are two carbon/carbon double bonds separated by a carbon/carbon single bond. Draw the result of homolytically cleaving one of the carbon/carbon double bonds.

$$R_2C = CH - CH = CR_2 \quad \leftrightarrow \quad R_2C = CH - \overset{\bullet}{C}H - \overset{\bullet}{C}R_2$$

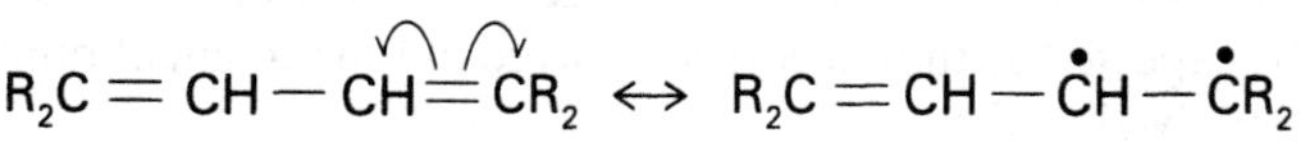

Draw the result of breaking the other double bond in the same manner.

$$R_2C = CH - \overset{\bullet}{C}H - \overset{\bullet}{C}R_2 \quad \leftrightarrow \quad R_2\overset{\bullet}{C} - \overset{\bullet}{C}H - \overset{\bullet}{C}H - \overset{\bullet}{C}R_2$$

Now, reform the double bond, not between the terminal atoms, but rather between the central atoms.

$$R_2\overset{\bullet}{C} - \overset{\bullet}{C}H \overset{\frown}{} \overset{\bullet}{C}H - \overset{\bullet}{C}R_2 \;\leftrightarrow\; R_2\overset{\bullet}{C} - CH = CH - \overset{\bullet}{C}R_2$$

The result is a biradical, but the unpaired electrons are in the 1,4 positions, or terminal positions, and not the 1,2 or adjacent positions. The numbers refer to the carbon atoms in the chain. Try to draw a single mechanistic step that will result in this 1,4-biradical being formed from the original molecule with the two double bonds.

$$R_2C = CH - CH = CR_2 \;\leftrightarrow\; R_2\overset{\bullet}{C} - CH = CH - \overset{\bullet}{C}R_2$$

Now consider a heteroatomic conjugated system, e.g. $R_2C=CH–CH=O$. Draw the product that results from the heterolytic cleavage of the carbonyl bond.

$$R_2C = CH - CH = O \;\leftrightarrow\; R_2C = CH - \overset{+}{C}H - \overset{-}{O}$$

The carbonyl carbon atom will exert a strong inductive effect, because it has been made slightly positive due to the $-M$ effect of the carbonyl oxygen. More importantly though, it will also exert a strong mesomeric effect on the carbon/carbon double bond. We saw earlier that it is through the π system that whole units of charge may be separated. Draw the result of the heterolytic cleavage of the carbon/carbon bond, and then the subsequent formation of a new carbon/carbon double bond between the central carbon atoms.

$$R_2C = CH - \overset{+}{C}H - \overset{-}{O} \;\leftrightarrow\; R_2\overset{+}{C} - CH - \overset{+}{C}H - \overset{-}{O}$$

$$\leftrightarrow\; R_2\overset{+}{C} - CH = CH - \overset{-}{O}$$

Notice that the terminal carbon/carbon bond cleaves in the direction that places the newly formed carbanion next to the existing carbonium ion. If it cleaved in the opposite sense, then two positive charges would be adjacent to each other, which would be unstable. Hence, by using the mesomeric effect, the charge has been moved a further two atoms down the chain in a very efficient manner, i.e. retaining the full unit of the charge. This is in contrast with the inductive effect, which, it will be remembered, only transmitted a partial charge to the

47

adjacent atom, and after that the effect became even smaller with increasing distance.

In the triene system, i.e. one with three double bonds, e.g. $R_2C=CH-CH=CH-CH=CR_2$, the same type of redistribution of the electrons in the double bonds might occur. Draw a mechanism involving the redistribution of the electrons present in all the double bonds.

$$\leftrightarrow R_2\overset{\bullet}{C} - CH = CH - CH = CH - \overset{\bullet}{C}R_2$$

Now let us consider the cyclic version of this linear triene, i.e. 1,3,5-cyclohexatriene. This cyclic version has fundamentally different electronic characteristics from those found in the linear version. Draw a mechanism showing the possible redistribution of the electrons present in this compound.

This time, as a result of the cyclic nature of this molecule, it is possible to draw three new double bonds, and not just two. This arises, because what were the two unpaired electrons that were at the ends of the chain, are now adjacent to each other and so may interact to form the third bond. Superficially, the new molecule looks the same as the old one, except that the double bonds have moved one position around the ring. However, even though this looks like a small difference, in fact it belies the fact that the cyclic version of the triene has completely different properties from its linear cousin. This cyclic triene molecule is benzene. Draw a representation for the time-averaged redistribution of the electrons present in these three double bonds within this cyclic molecule.

In the time-averaged molecule, every carbon/carbon bond has both single and double bond character, and each bond is the same length in this symmetrical structure. This cyclic system is called an aromatic ring. The representation of benzene that shows each bond as an individual single or double is called a canonical structure and represents a hypothetical structure that does not exist in practice. The representation that indicates every bond as half-way between a single and double is called the resonance hybrid and is closer to the true picture of the molecule. This time-averaged redistribution of the electrons is called delocalisation. An aromatic ring is often represented by a regular hexagon with a circle inside it. This is convenient when the aromatic ring itself is not involved in the reaction under consideration and its presence is only incidental to the electron movements that are occurring elsewhere in the molecule. However, when drawing mechanisms that involve the interaction of electrons within the aromatic ring, a canonical structure should be used in order to keep track of the electrons that originate from each bond.

Benzene is the archetypal example of a compound that displays aromatic properties. Aromatic compounds are characterised by a special stability over and above that which would be expected as a result of the delocalisation of the double bonds in a linear system. Typically, this extra stability is associated with the closed loop of six electrons, the aromatic sextet, as occurs in benzene itself. However, larger and smaller loops are possible. So long as there are $(4n + 2)\pi$ electrons (where n is an integer from zero, upwards) present in (at least three) adjacent p sub-orbitals that form a closed circuit, then the resultant molecule will be aromatic. It is also possible for heteroatoms to form part of the cyclic structure, and for the structure to be charged. Furthermore, aromatic compounds, in contrast to unsaturated compounds, tend to undergo substitution reactions more readily than addition reactions. This is because it is usually thermodynamically favourable to preserve the aromatic stability rather than release the energy contained in the double bonds.

Draw one canonical form for each of the following aromatic compounds: a three-membered carbocyclic cation with 2 electrons ($n = 0$, 3 p sub-orbitals); a five-membered carbocyclic anion with 6 electrons ($n = 1$, 5 p sub-orbitals); and a six-membered neutral ring containing one nitrogen atom and five carbon atoms ($n = 1$, 6 p sub-orbitals).

These molecules are cyclopropenyl cation, cyclopentadienyl anion, and pyridine respectively. Notice that in the case of pyridine, the lone pair on the nitrogen is not involved in the aromatic sextet, but rather projects from the nitrogen in the same plane as the ring, instead of being orthogonal to it, which is what would be required if it were to become involved in the aromatic ring of p sub-orbitals.

Reverting again to a linear conjugated system, let us now examine in slightly more detail what happens when a heteroatom is introduced to the system. After the redistribution of the π electrons in the $R_2C=CH-CH=O$ molecule, the carbonyl group ends up with more electrons than it had initially, i.e. it is an electron sink. There are many other groups that have a similar property.

Suggest a redistribution of the double bonds in the molecule $R_2C=CH-NO_2$ that would enrich the nitro group with electrons.

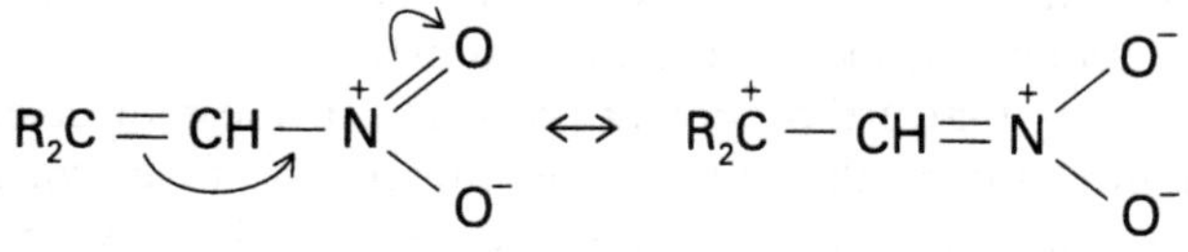

$$R_2C = CH - \overset{+}{N} \overset{\displaystyle O}{\diagdown} \quad \longleftrightarrow \quad R_2\overset{+}{C} - CH = \overset{+}{N} \overset{\displaystyle O^-}{\diagup}$$

Now, both oxygen atoms in the nitro group bear a negative charge, and so this group has acted as an electron sink in this system.

Suggest a mechanism whereby the cyano group may act as an electron sink.

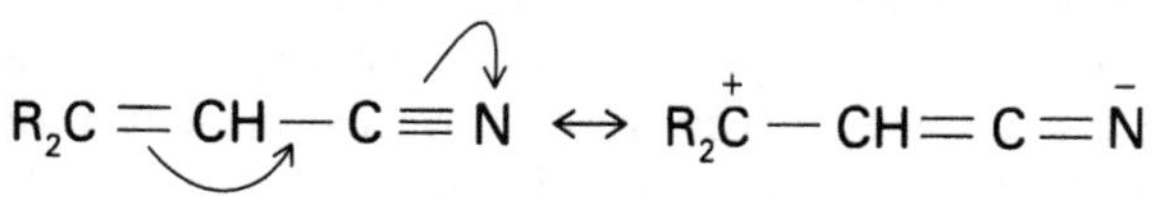

$$R_2C = CH - C \equiv N \quad \longleftrightarrow \quad R_2\overset{+}{C} - CH = C = \overset{-}{N}$$

In general, groups that have multiply bonded heteroatoms may act as electron sinks, because the multiple bond may cleave heterolytically and place the extra electron pair on the heteroatom.

If a group with a lone pair of electrons is placed at the other end of the chain, then it may donate its electron pair into the conjugated system. Such a group may, for example, be an amino group. Suggest a possible redistribution of the π electrons in the molecule $H_2N-CR=CH-NO_2$.

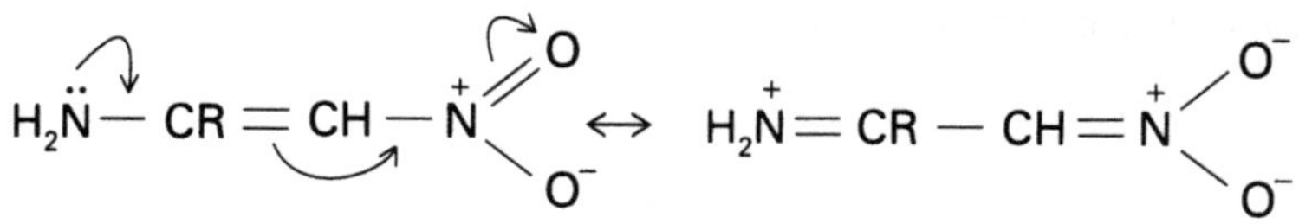

$$H_2\overset{..}{N} - CR = CH - \overset{+}{N} \overset{\displaystyle O}{\diagdown} \quad \longleftrightarrow \quad H_2\overset{+}{N} = CR - CH = \overset{+}{N} \overset{\displaystyle O^-}{\diagup}$$

Here, the amino group has provided the extra electrons, i.e. it has acted as an electron source. A hydroxyl group or halogen atom may act in a similar manner to the amino group. If an electron source and an electron sink are conjugated together, then a permanent dipole may be formed as a result of the mesomeric redistribution of the π electrons. This type of scheme can be characterised as a 'push-me, pull-you' system, and is often instrumental in initiating reactions.

So far we have considered the distribution of the electrons that occurs in a molecule, and any resultant redistribution that might take place. We have either assumed, or ignored, the shape of the molecule under consideration. In the next chapter, we will study how to determine the shape of a molecule from first principles.

3.6 Summary of bond polarisation and fission

A homoatomic bond exists between identical atoms. In such a case the bond is unpolarised, because the electrons are equally shared, and so there is no permanent charge separation.

However, random fluctuations can lead to a temporary charge separation, which is called the London effect. Temporary charge separations can also be induced by the approach of a fixed charge. The resultant charged species is called a dipole. The small amount of charge that is separated is indicated by a $\delta+$ or $\delta-$ sign as is appropriate. The sum of the δ charges is zero, while the sum of the magnitudes of these small charges is less than one.

A heteroatomic bond joins two atoms of two different elements. The electronegativity of an element reflects the affinity of that element for electrons: the higher it is, the greater the affinity. Accordingly, the element with the higher electronegativity will bear a permanent, partial, negative charge, while its neighbour will bear an equal positive charge. Thus, there will be a permanent dipole. This is indicated by an arrowhead on the bond joining the atoms, pointing towards the negative end, e.g. $H^{\delta+} \longrightarrow F^{\delta-}$. This effect is called induction, and is indicated by $\pm I$. The inductive effect may be transmitted along a carbon chain, but its effect very rapidly becomes negligible.

For most purposes the induced polarisation of the H–C bond can be ignored, while in H–O, H–F and H–Cl bonds it is significant. One consequence of this polarisation is that lone pairs can interact with the $\delta+$ hydrogens by forming a hydrogen bond, which is a weak Coulombic interaction. This bond can be either intermolecular or intramolecular.

When a carbon atom is bonded to oxygen, nitrogen or chlorine, it forms the positive end of the bond. In contrast, if it is bonded to a metal, it forms the negative end of the bond.

The induced polarisations within a molecule can be summed together as vectors, to give the overall dipolar motion of the molecule. For some molecules, like methane and boron trichloride, the result is zero; while for others, like water and ammonia, the result is non-zero.

Homoatomic bonds tend to prefer homolytic bond fission, which results in radicals being formed; while heteroatomic bonds often undergo heterolytic bond fission, which results in ions being formed.

The cleavage of bonds can be represented by curly arrows. The tail indicates the source, or start, of the electron movement, while the head indicates the destination. Half an arrowhead is used to represent one electron, while a complete arrowhead represents a pair of electrons moving together. The source is usually electron rich, e.g. a multiple bond, or a lone pair; while the destination is electron poor, e.g. a positive charge, empty orbital or electronegative atom.

Symmetrical multiple bonds do not have any permanent polarisation. However, unsymmetrical ones, like carbonyl, $C=O$, and imine, $C=N$, are permanently polarised. These bonds can easily break heterolytically to give charged intermediates, in which opposite charges reside on adjacent atoms.

The attraction of electrons along a multiple bond system is called the mesomeric effect, and is given the symbol $\pm M$. The mesomeric effect of an oxygen in a carbonyl group can be increased by protonating one of the oxygen lone pairs: $R_2C=OH^+$. This protonation facilitates the heterolytic fission of the $C=O$ double bond to result in the formation of the carbonium ion R_2C^+-OH.

The inductive effect operates along the single bond, or σ, system, and results in partial charge separation. In contrast, the mesomeric effect operates along the multiple bond, or π, system, and results in whole units of charge separation.

When multiple bonds alternate with single bonds, the whole arrangement is called a conjugated system. Whole units of charge can be transmitted along a conjugated system by the redistribution of the electrons present in the double and single bonds as a result of the mesomeric effect. In contrast, the inductive effect only transmits a partial charge ineffectively along a σ system.

A six-membered cyclic conjugated triene system is called an aromatic ring. The individual single and double bonds are shown in individual canonical structures, while the resonance hybrid indicates the delocalisation that takes place in reality.

Conjugated systems that have an electron sink at one end and an electron source at the other, usually have a permanent dipole that is established by the mesomeric effect. Furthermore, such systems are often instrumental in initiating reactions. Generally, groups that contain multiply bonded heteroatoms act as electron sinks, e.g. carbonyl, nitro and cyano groups, because a π bond can break and so place a negative charge on a heteroatom (i.e. not a carbon or a hydrogen atom). Conversely, groups that have lone pairs of electrons tend to act as electron sources, e.g. amino and hydroxyl groups.

4

Shapes of Molecules

4.1 Introduction

So far, we have only considered the electronic properties of the atoms within a molecule. We will now turn our attention to the spatial arrangement of those atoms within the molecule.

There are two main types of bonding, ionic and covalent. Ionic bonding is characterised by the non-directional nature of the Coulombic attractions between ions, i.e. the electrostatic force radiating from the central ion is felt equally in all directions. The main factor that influences the structure of the crystal lattice is the relative sizes of the cations and anions, because this affects how the ions will pack together within the lattice.

In simple ionic compounds, each ion only occupies one type of environment, with all the ions of the same type having exactly the same geometric relationship to all the other ions in the crystal lattice. In more complicated ionic compounds, it is possible for ions of one species to occupy one of a limited number of environments, but this is the exception rather than the rule at this level.

Covalent bonds are different from ionic bonds, in that they are directional in nature. Furthermore, each covalent bond has a particular length. The consequence of this is that when one atom is covalently bonded to another atom, the relative position in space of these two atoms is fixed. This means that in a molecule held together by covalent bonds, each and every atom has a defined, and predictable, geometric relationship to every other atom within the molecule. Furthermore, in the case of covalent compounds, it is now the norm for every atom to be considered individually in respect of its geometric relationship to every other constituent atom.

We will now look more closely at certain types of covalent bonding before examining some of the more common geometric possibilities that occur in organic molecules.

4.2 Types of bonding

Each covalent bond is formed from a molecular orbital that may accommodate a maximum of two electrons. Molecular orbitals are formed from atomic orbitals, each of which may also accommodate a maximum of only two electrons. The number of molecular orbitals created in the final molecule is the same as the original number of atomic orbitals present in the combining atoms.

For example, in each hydrogen atom there is one atomic orbital, which holds a single 1s electron. When two such atoms combine to form the hydrogen molecule, two molecular orbitals are formed from the two atomic orbitals of the original hydrogen atoms. One molecular orbital encompasses the two hydrogen nuclei. This is the bonding orbital, because it allows the electrons that are accommodated within it to reside between the two positive nuclei. Thus, there is some electron density between the nuclei, and this acts as an attractive force that keeps the two nuclei together, and so forms a stable molecule. The other molecular orbital is the anti-bonding orbital. Its shape is such that there is very little electron density between the two nuclei. As a result, there is nothing to override the natural repulsive effect exerted by the two nuclei on each other. The shapes of each of these orbitals are shown below.

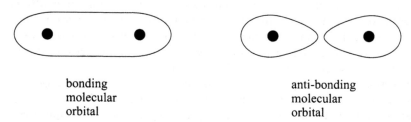

bonding
molecular
orbital

anti-bonding
molecular
orbital

For every bonding molecular orbital that is formed, there is an anti-bonding molecular orbital formed as well. In certain circumstances, non-bonding orbitals may be formed. If an anti-bonding orbital is occupied it destabilises the molecule, hence its name. With respect to molecular bonding orbitals of single bonds, there are two important points about the related anti-bonding orbitals: first, they project in approximately the opposite direction in space to the bonding orbitals to which they are related; and secondly, they are usually empty in an otherwise stable molecule. Similar issues relate to anti-bonding orbitals of multiple bonds; however, the location of these orbitals relative to their related bonding orbitals is more complicated than is the case with the anti-bonding orbitals associated with single bonds.

The reason why we must consider anti-bonding orbitals at all is because electrons must move from one orbital to another. If the electrons are to remain attached to an atom or molecule they cannot just float about in space without being contained in some orbital. As we saw above, each orbital may only accommodate two electrons, so if an electron-rich species is going to attack another species that already has a full complement of electrons, there needs to be present an empty orbital that is able to accommodate the incoming electrons before the electrons that were originally on the centre being attacked have departed and made room for the new electrons. The empty orbital is often an anti-bonding orbital. Most importantly, the geometry of the anti-bonding orbital will often determine the line of approach for the incoming reagent, and so will impose certain restrictions on the way a mechanism must proceed. This in turn will affect the stereochemistry of the product.

The shapes of organic molecules may be readily accounted for by the electron pair repulsion theory, which, in summary, states that any given electron pair will repel all other electron pairs. Thus, an electron pair will adopt a position in space that is furthest from all the other electron pairs. There is, however, one important caveat. If two or more electron pairs are directed in the same direction in space, e.g. the two pairs of electrons in a double bond or the three pairs of electrons in a triple bond, then they are treated as if they were only one pair for the purpose of determining the geometry of the resultant molecule. Furthermore, to a first approximation, the geometries adopted are regular in shape.

Accordingly, given the above guidelines, suggest the shape that will be adopted by a triatomic molecule of general formula AB_2 that only has electrons in the two molecular orbitals required to bond the B atoms to the central A atom, i.e. there are only two pairs of bonding electrons, and no other lone pairs of electrons to consider.

$$B \text{———} A \text{———} B$$

The geometry will be linear, i.e. the bond angle is 180°. Even though this is rather rare in organic molecules, there are important classes of compounds that have this geometry, e.g. alkynes and cyanides.

Now suggest the geometry of the molecule that has the general formula AB_3, which has only three bonding pairs of electrons, and no lone pairs.

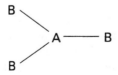

This geometry is called trigonal planar, and the bond angle is 120°. This geometry is important in all compounds that contain a doubly bonded carbon, or a carbonium ion.

Now suggest the geometry of the molecule that has the general formula AB_4, i.e. there are only four bonding pairs of electrons.

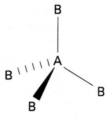

This adopts a regular tetrahedron, and is the first structure that requires the third dimension, i.e. it is not a flat structure. The bond angle is about 109°. This is the fundamental building block of organic molecules.

Now suggest the geometry of the molecule that has the general formula AB_5.

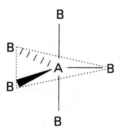

This structure is called a trigonal bipyramid. There are two different bond angles in this structure. Between the atoms in the trigonal plane, the angle is 120°, while between the axial and equatorial atoms the angle is only 90°, i.e. a right angle. This is an important shape in the transition of one saturated carbon centre to another.

We will now look at some of these geometries in a little more detail, starting with that of the tetrahedron, i.e. AB_4 compounds, because this is the most important shape in organic chemistry.

4.3 Tetrahedral geometry

This is the commonest geometry that is encountered in organic molecules, because this is the shape that is adopted by a saturated carbon, i.e. one that is bonded to four other groups by four single bonds. Write down the electronic configuration of the carbon atom in the ground state, i.e. the most stable state.

$1s^2$, $2s^2$, $2p_x^1$, $2p_y^1$

There are four electrons in the valence shell of carbon, so in order to achieve an octet it requires four more. The shape of the s atomic orbital is spherical, while the shape of the p orbital is like two spheres that just touch at the nucleus. Each of the p orbitals is orthogonal, that is at right angles, to each of the others. (More strictly, there is no shape to an orbital, as it is just a wave function; however, that is of little use to most organic chemists. Thus, a shape is allocated to each orbital that approximates to a volume in space around each nucleus where the electron in question is most likely to be found.) If a carbon atom were bonded only by means of the p orbitals, what would be the bond angle?

As each p orbital is orthogonal to the others, the bond angle must be 90°. In practice this is not observed; instead the bond angle is about 109°. In order to explain this observation, it is hypothesised that one electron from the valence shell, i.e. one of the 2s electrons, is promoted to the empty $2p_z$ orbital to give an electron configuration of $2s^1$, $2p_x{}^1$, $2p_y{}^1$, $2p_z{}^1$. Then all four atomic orbitals are mixed to give four equal hybrid orbitals called sp^3, i.e. formed from one s orbital and three p orbitals. This is the essence of hybridisation theory.

  If these four hybrid orbitals are the same, suggest what are their relative energy levels to one another, and also the directions in space that they occupy with respect to one another.

As they are the same, then the energy levels must be equal, i.e. they are degenerate. As they are equal in energy and also as they have the same shape, each one must occupy a similar position in space relative to each of the others, and so each one points towards a different corner of a regular tetrahedron. Thus, the bond angle formed from an sp^3 hybridised carbon is approximately 109°, which is the angle described between one corner of a regular tetrahedron, its centre, and any one of the three remaining corners.

  These four hybrid orbitals may then combine with four other atomic orbitals to form four bonding and four anti-bonding molecular orbitals. Thus, suggest the shape of the tetrahydride of carbon. Draw it using a stereochemical projection.

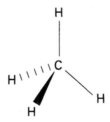

The tetrahydride of carbon is called methane and is a regular tetrahedron.

The molecular orbitals that have been formed between the sp^3 hybrid atomic orbitals and the atomic orbitals of the hydrogen atoms all have spherical symmetry about the axis connecting the atoms that are bonded together. In other words, there is no nodal plane that passes through both of the bonded nuclei. The molecular orbital that corresponds to such a bond is called a σ bond. A representation of a bonding σ orbital formed between an sp^3 carbon and an s type atomic orbital of a hydrogen atom is shown below. This is a normal carbon/hydrogen single bond.

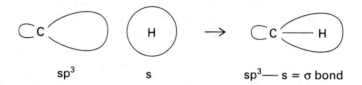

Notice that the lettering used for atomic orbitals is the lower case Roman alphabet, while for molecular orbitals the Greek alphabet is used.

If one of the hydrogen atoms is replaced by another methyl group, $-CH_3$, the resultant molecule is called ethane. What will be the shape of this molecule? Again draw this using a stereochemical projection.

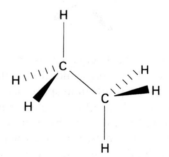

Around each carbon the bonds are still arranged in a tetrahedral fashion. Now redraw this using the sawhorse projection.

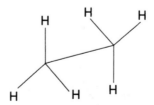

Now that the three-dimensional structure of the molecule is important, the dot and cross notation, which was so useful when counting electrons around each atom, is of less utility. First, it is rather cumbersome; and secondly, it does not give the spatial information that we now require. It is still useful, however, in ensuring that no electrons have been accidentally omitted.

In the nitrogen atom, the same hybridisation of the atomic orbitals occurs and so four sp^3 hybrid orbitals are formed. These are distributed around the central atom in the same tetrahedral manner as in carbon. Thus, suggest the shape of the ammonia molecule.

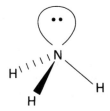

This shape is often called a trigonal pyramid because that is the shape that is mapped out by the hydrogen and nitrogen atoms. However, the true geometry is more closely related to the tetrahedron, the fourth corner of which is occupied by the lone pair. The nitrogen is bonded to three hydrogens, but this only accounts for six of the eight electrons that surround the nitrogen. The other two are in the lone pair. The lone pair has similar spatial requirements as the pair of electrons in each of the bonding orbitals, and so exerts a similar repulsive force on the three bonding orbitals as do the latter on each other. Hence, even though, initially, this might look like an AB_3 geometry because there are only three hydrogen atoms joined to the central nitrogen, it is really an AB_4 geometry, because of the presence of the lone pair. Put another way, ammonia is isoelectronic with methane, i.e. it has the same distribution of electrons, even though there are differences in the nuclear arrangement.

Suggest the shape of the dihydride of oxygen, i.e. water.

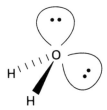

This molecule appears bent if one only looks at the atoms. The true geometry, however, is based on a tetrahedron, with two corners occupied by the hydrogen atoms and the other two occupied by the lone pairs. So again it is based on the AB_4 geometry, and not the AB_2 geometry as it might at first appear; this is because of the presence of the two lone pairs.

The lone pairs are held slightly closer to the central atom than the electrons that are involved in the bonding orbitals. Suggest a reason for this.

Simply, in a bonding orbital there are two nuclei trying to share the electrons, while in a lone pair the attraction of the nucleus comes only from one side. Thus, naturally, in a lone pair, the electron pair is drawn closer to the nucleus with which it is associated.

If the lone pairs are held slightly closer to the nucleus than the bonding pair, what are the consequences for the bond angles between the hydrogen atoms in the methane, ammonia and water molecules?

In methane, the HCH bond angle is that found in a perfectly regular tetrahedron, namely about 109°. However, in ammonia, the HNH angle is reduced slightly, because the lone pair, being closer to the nucleus, repels the bonding electrons slightly more strongly than they do each other, and so forces the bonding electron orbitals together. The resultant bond angle is about 107°. In water the effect is greater, because there are now two lone pairs, and so the HOH angle is about 105°.

For molecules that have lone pairs there is the possibility of protonation. In ammonia, the protonated species is the ammonium ion. Draw the shape of this ion.

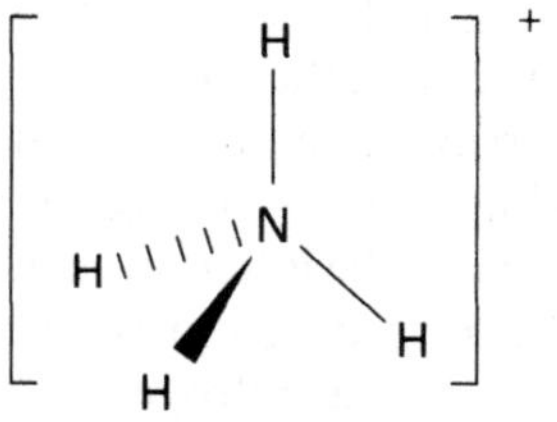

This molecule is a perfectly regular tetrahedron.

Suggest what will be the shape of the hydrazine molecule, N_2H_4, and its mono-protonated cation, $N_2H_5^+$.

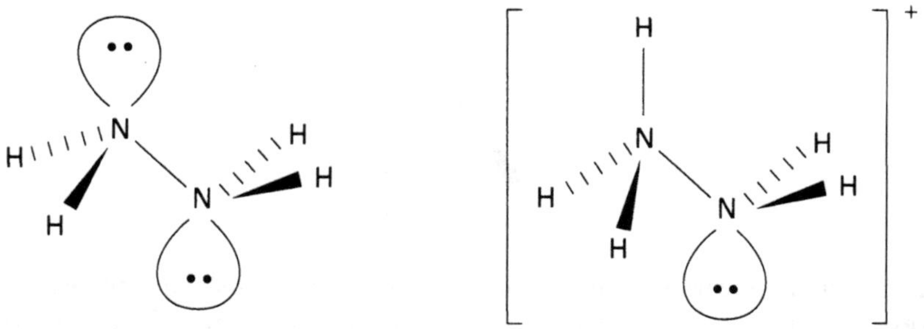

Hydrazine is similar to ethane, but with a lone pair replacing a hydrogen atom on each of the central atoms. There is free rotation around the nitrogen/nitrogen single bond. In the protonated form, the geometry around each nitrogen atom is still based on the tetrahedron.

What is the shape of the molecule that results from the mono-protonation of water, i.e. the hydroxonium ion?

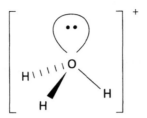

This molecule is similar in shape to the ammonia molecule, and so is still based on the tetrahedron.

What is the shape of the molecule which results from the deprotonation of water, i.e. the hydroxide anion?

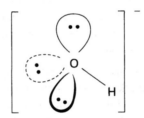

This ion, only having two atoms, must be linear, but the three lone pairs project towards the corners of tetrahedron, which is centred on the oxygen anion. Thus, the tetrahedral shape still forms the basic template for the orientation of the lone pairs. The position in space of these lone pairs is important, because that will determine the direction of approach of a protonating reagent.

Methane may be deprotonated under extreme conditions to give the methyl carbanion, CH_3^-. What is the shape of this ion?

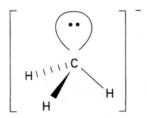

Again this ion is based on a tetrahedron, with three of the corners occupied by the hydrogen atoms and the last one occupied by the lone pair.

When one of the sp^3 hybrid orbitals of a carbon combines with an s type atomic orbital of a hydrogen, two molecular orbitals must be formed: one of which is the bonding, while the other forms the anti-bonding orbital. The bonding orbital extends from the carbon towards the hydrogen atom, which is placed at the corner of a tetrahedron. In which direction do you think the related anti-bonding orbital is pointed?

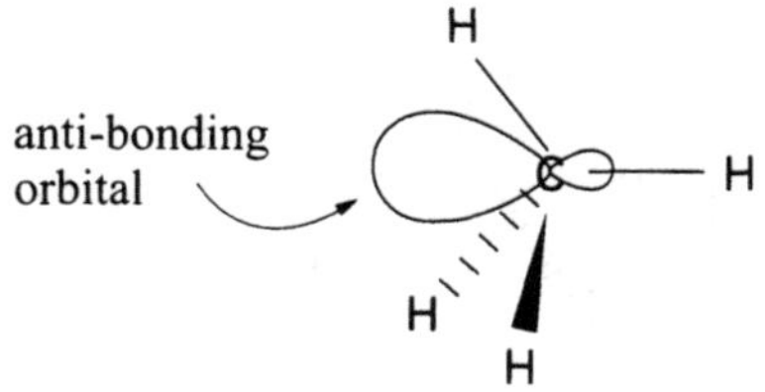

This anti-bonding orbital points in the opposite direction from the bonding one, and so is directed between the three remaining carbon/hydrogen bonds. The same holds true for the other three anti-bonding orbitals. An anti-bonding molecular orbital is indicated by a post-scribed asterisk after the usual symbol for the bonding orbital, e.g. σ^*. If a carbon centre that has a full complement of eight electrons is attacked by an electron-rich species, from which direction do you think this species will approach?

Electrons can only move from orbital to orbital, and as each orbital can only contain two electrons, the attacking species cannot place more electrons in an already full bonding orbital. Thus, an incoming electron-rich species must approach along the line of the empty anti-bonding orbital, into which it may then transfer an electron or an electron pair. Hence, the direction not only of the bonding orbitals, but also of the anti-bonding orbitals, is very important in understanding mechanistic organic chemistry.

The common feature of all of the examples examined in this section is that there are four directions in which the electron pairs, be they bonding orbitals or lone pairs, project from the centre in question. This results in a tetrahedral geometry.

We will now examine the AB_3 geometry.

4.4 Trigonal planar geometry

The hybridisation theory may be used to account for the direction of the bonds in an atom that is involved in three single bonds only, and which does not have any lone pairs. If the s orbital and only two of the three p orbitals are hybridised, leaving the p_z orbital untouched, then three hybrid orbitals are formed that are called sp^2 orbitals.

If these three orbitals are degenerate, what configuration relative to each other will they adopt?

These three orbitals will form a trigonal planar shape, with a bond angle of 120°. This leaves the last p orbital to project above and below the plane formed by the three hybrid orbitals, i.e. it is orthogonal to the plane. Thus, suggest the shape of the molecule boron trichloride, BCl_3.

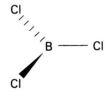

In the isolated molecule of boron trichloride, there are only three electron pairs around the boron, and so they adopt the trigonal planar geometry. The last orbital, p_z, is left unused and so is an empty orbital. Draw a diagram of the BCl_3 molecule, and incorporate into that diagram the unused p_z orbital.

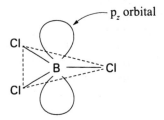

Now draw the shape of the methyl carbonium cation, CH_3^+.

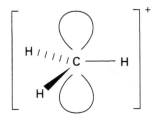

Again this is a trigonal planar species, with an empty orbital that projects above and below the plane of the C–H bonds.

The trichloromethyl carbanion, CCl_3^-, which is formed by the deprotonation of trichloromethane, $CHCl_3$, may, under favourable conditions, lose a chloride anion. Write down the balanced equation for the loss of the chloride anion from CCl_3^-.

$$CCl_3^- \rightleftharpoons {:}CCl_2 + Cl^-$$

The ${:}CCl_2$ must be neutral to ensure that the principle of Conservation of Charge is obeyed. Draw the dot and cross structure of the CCl_3^- ion and so deduce the electronic configuration of the :CCl₂ species, and hence suggest the shape of this molecule.

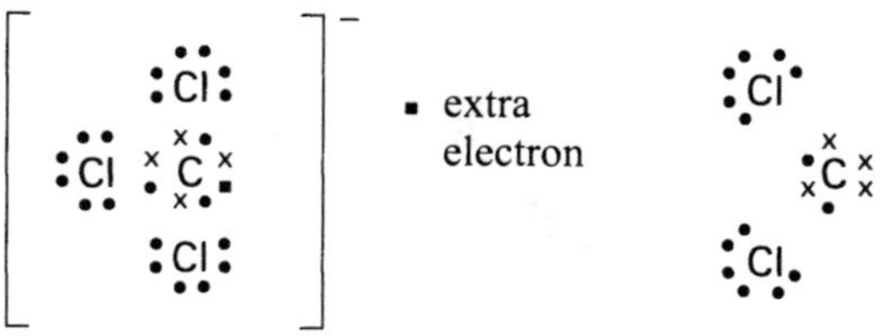

$$\left[\begin{array}{c} \vdots\ddot{Cl}\vdots \\ \vdots\ddot{Cl} \times \overset{\times\bullet}{\underset{\times\bullet}{C}} \times \vdots \\ \vdots\ddot{Cl}\vdots \end{array} \right]^{-}$$

■ extra electron

The value of the dot and cross structure here is to help to ensure that all the electrons are accounted for before attempting to deduce the structure. The electron pair repulsion theory gives the result that the :CCl₂ molecule has a trigonal planar shape. There are two C–Cl single bonds and also (formally) a lone pair. It will be recalled that this carbon species is called a carbene. It is a most interesting species, because it has both (formally) a lone pair and an empty orbital, and yet is electrically neutral. In fact, this species does not have a fully formed lone pair, because that would contravene Hund's rule, which (it will be recalled) states that electrons prefer to half fill all the available orbitals before pairing-up with each other to form fully occupied orbitals. Thus, a carbene is both electron deficient and a diradical, which accounts for its reactivity.

Earlier, we looked at the BCl₃ molecule and observed that it adopted a trigonal planar shape. Suggest what will be the structure of aluminium trichloride, AlCl₃, in the gas phase.

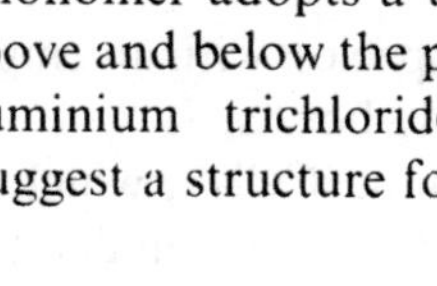

The aluminium trichloride monomer adopts a trigonal planar shape, with an empty orbital that extends above and below the plane in which the Al–Cl bonds are located. However, aluminium trichloride forms a dimer at room temperature and pressure. Suggest a structure for this dimer.

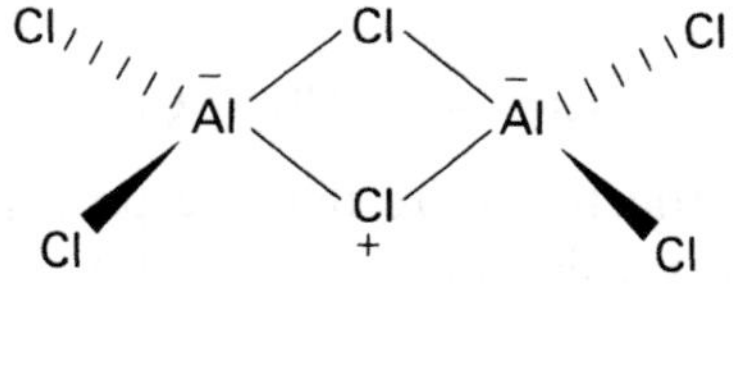

One of the lone pairs from a chlorine atom forms a dative bond with the aluminium atom of a second molecule. Similarly, one of the chlorines from the second molecule forms a dative bond with the aluminium of the first molecule. The chlorine lone pair was able to attack the aluminium atom, because of the presence of the empty orbital. Now each aluminium atom has four single bonds around it, and so adopts a tetrahedral configuration. Notice that the non-bridging chlorine atoms now all lie in the same plane, which is orthogonal to the plane that contains the bridging chlorines.

If instead of considering group III elements such as boron or aluminium, or electron depleted carbon species such as carbonium cations, one considers electrically neutral carbon species that are hybridised into the sp^2 configuration, then a further complication arises. For example, if two carbons atoms are hybridised into the sp^2 configuration, then each one will have a spare p_z orbital remaining, which will be occupied by an unpaired electron. A normal σ type carbon/carbon bond may be formed by using one of the sp^2 hybrid orbitals from each carbon atom, i.e. forming an sp^2/sp^2 σ bond. Furthermore, the remaining two sp^2 orbitals on each carbon may be used to bond to two hydrogen atoms to form four more σ type carbon/hydrogen bonds, i.e. forming four sp^2/s σ bonds. Draw the ethene molecule, representing the five σ type bonds that have been discussed so far as single lines, and also drawing out the remaining p_z orbital on each carbon.

$$p_z - p_z = \pi \text{ bond}$$

The σ bond framework gives the trigonal planar structure to each end of the ethene molecule. However, this is only part of the story, for the two p_z orbitals may interact by overlapping side on, i.e. by overlapping each lobe, one above and the other below the plane of the molecule. This is the second carbon/carbon bond in ethene. This second bond formed between the two p_z atomic orbitals, gives rise to the π bond molecular orbital. Notice that because the π bond is co-linear with the C–C σ bond, the geometry adopted is related to the AB_3 system. Notice also that in the π type of bond there is now a nodal plane that transects the two nuclei. This nodal plane is characteristic of a π bond.

If one end of the ethene molecule, comprising a carbon atom and two hydrogen atoms, were rotated by 90° relative to the other end, what would be the result?

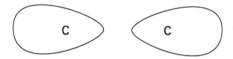

Now the p_z orbitals of each carbon atom cannot overlap successfully, and so it is not possible to form the second bond. So, by rotating around the C–C axis, the π bond has been broken. This cleavage would require energy. Accordingly, rotation about an axis that has a π bond requires energy, i.e. is hindered. It is the fact that rotation is hindered that ensures that one end of the molecule is co-planar with the other end, i.e. the whole molecule is flat.

Suggest the orientation of the σ* anti-bonding orbital that is formed along with the formation of the σ bonding orbital between the two carbons.

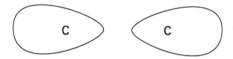

This empty, anti-bonding, orbital projects out, away from the carbon/carbon bond, between the remaining two hydrogen atoms. The π bond also has an anti-bonding orbital associated with it, which is designated π*. Suggest where this may be located in space with respect to the bonding orbital.

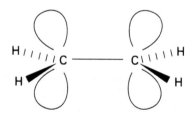

This empty orbital projects away from the double bond, but unlike the σ* anti-bonding orbital, it is above and below the plane of the atoms. This makes it more approachable to an incoming attacking species.

Carbon may form double bonds with other atoms apart from itself, such as oxygen and nitrogen. Suggest the shape of the general ketone, $R^1R^2C{=}O$, where R is the symbol for a general alkyl group, and the superscripts indicate that the two alkyl groups can be different, while $C{=}O$ represents the carbonyl group.

Notice that the three groups are in the same plane. Draw in the position of the lone pairs on the oxygen, and on both the carbon and oxygen atoms draw in the anti-bonding molecular orbital that is associated with the π molecular bonding orbital.

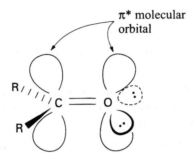

The lone pairs are in the same plane as the other groups, while the anti-bonding orbital projects above and below this plane and away from the lone pairs.

The carbon/oxygen double bond is polarised by both the inductive and mesomeric effects of the oxygen. Draw a diagram that shows the inductive effect.

Draw another diagram showing the mesomeric effect.

Now, suggest the shape of the general imine, $R_2C=NX$.

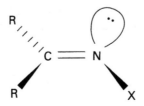

If the two R groups are different, then the X group may be on the same side as one or the other. This gives rise to the possibility of geometric isomers due to the lack of free rotation around the carbon/nitrogen double bond.

Thus, in summary, the trigonal planar geometry can result from having just three bonding electron pairs around the central atom, as in BCl_3 or CH_3^+, or having an sp^2 hybridised atom with an associated π bond, as in $H_2C=CH_2$. The common feature is that, in each case, there are only three directions in which pairs of electrons project from the atom under consideration.

4.5 Linear geometry

If only one p orbital and the s orbital are hybridised together, then two sp hybrid orbitals are formed. How will these two hybrid orbitals be orientated with respect to each other?

The bond angle will be 180°, so that they will be diametrically opposed to each other. There are also two remaining p orbitals, the p_z and the p_y orbitals. What orientation will these two orbitals adopt?

They will be orthogonal to each other, and also to the two sp hybrid orbitals. If two atoms were hybridised into the sp configuration, then the remaining p orbitals could be empty as in beryllium dichloride, $BeCl_2$; or alternatively, they could form two π bonds, each orthogonal to the other, but co-linear with the axis described by the sp orbitals. Draw the structure of the ethyne molecule, in which two carbons are joined by one σ bond and two π bonds, indicating the σ bonds with a line and drawing out the p orbitals that will form the π bonds.

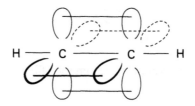

The ethyne molecule is linear. There are two π bonds that restrict rotation, but in this case it is not noticeable, because of the cylindrical symmetry around the σ bond.

If the ethyne molecule is deprotonated it gives rise to an anion, C_2H^-. Draw the shape of this anion, and in particular indicate the direction of the lone pair.

$$\left[\text{H} - \text{C} \equiv \text{C} \bigcirc : \right]^-$$

The lone pair projects directly away from the triple bond, and so the whole molecule is still linear. If this anion is used as a reagent, it is possible for it to attack atomic centres that are quite sterically hindered, because it is long and thin. For this reason it is sometimes called a 'Heineken' reagent, as it can reach the parts that other reagents cannot reach!

Note that the sp* orbital projects in the opposite direction to the sp orbital to which it is related, but this is in exactly the same direction as the other sp orbital. Thus, the sp* orbital is not readily accessible to accept incoming electrons. Furthermore, the π* anti-bonding orbital associated with one of the π bonding orbitals is co-orientated with the other π bonding orbital, and vice versa. Hence, this is a reason why triply bonded carbons are not readily attacked by electron rich species, because none of the anti-bonding orbitals are accessible to incoming nucleophiles. Another reason is that the high density of electrons present in the two π molecular orbitals tends to repel the approach of any further electrons. For these reasons, triply bonded carbons tend always to do the attacking, rather than being attacked.

The cyano group is another group in which there is a triple bond, this time between carbon and nitrogen. Draw the shape of the related anion, CN^-, indicating the lone pairs.

$$\left[\overset{..}{\bigodot} C \equiv N \bigodot \overset{..}{} \right]^-$$

Two atoms must form a linear unit. Furthermore, both lone pairs project directly out from the triple bond. Thus, suggest the shape of the protonated form of the anion, namely the molecule HCN.

$$H - C \equiv N \bigodot \overset{..}{}$$

The anion is protonated on the carbon end to give rise to a linear molecule, HCN, which is called hydrogen cyanide or hydrocyanic acid. A reagent that has lone pairs on two or more different atoms with which it may attack another compound is called ambident. The atom that attacks depends upon the nature of the atomic centre that it is being attacked. In the case of protonation, the hard, less easily polarised, atom reacts with the proton, which is itself a hard species. This type of interaction will be studied in more detail in Chapter 7.

There is another, rather unusual, functional group that is related to the cyanide group. It contains a carbon atom that is only doubly bonded to a nitrogen atom, but which is not involved in any further bonding. This is the isocyano group, RNC. Draw the dot and cross structure of this grouping.

$$R \, \overset{..}{\underset{\blacksquare}{N}} \, \overset{x \; x}{\underset{..}{}} \, \overset{x}{\underset{x}{C}}$$

The nitrogen atom is involved in one single and one double bond. Furthermore, there is a lone pair on that atom. What orientations would you expect these three groups to adopt around the central nitrogen atom?

The shape is based on the trigonal planar geometry, AB_3: with one B group being the lone pair; another, the R group; and the last, the doubly bonded carbon atom. However, in reality this functional group is linear, and not bent at an angle around the nitrogen atom. Suggest a redistribution of the electrons that would account for this observation.

$$R - \overset{..}{N} = C \quad \leftrightarrow \quad R - \overset{+}{N} \equiv \overset{-}{C}$$

If the nitrogen donates its lone pair into the empty orbital of the carbon atom and so forms a dative bond, then there is now a triple bond between the carbon and nitrogen atoms. What is the shape of the group with this electronic distribution, and in what direction does the lone pair on the carbon atom now project?

$$R - \overset{+}{N} \equiv \overset{-}{C} \quad :$$

The group is now co-linear with the R group and the lone pair on the carbon projects directly away from the carbon/nitrogen triple bond. This is the favoured electronic configuration of the isocyano functionality.

Thus in summary, the common factor in the molecules that display linear geometry is that there are only two directions in which the electron pairs project. This can be achieved by having only two bonding molecular orbitals, as in $BeCl_2$; or in a species that has sp hybridisation and two π bonds, as in alkynes or cyanides.

4.6 Further examples

In this section, we look at a few examples that develop the basic ideas that have been introduced in the previous sections. When confronted with a molecule of unknown structure it is often useful to build towards it from segments that are well known, and using hypothetical intermediates to help.

The first example is quite simple, but serves to introduce the idea. The structure of water has already been discussed. You will recall that it is a bent triatomic molecule based on a tetrahedron, with the two lone pairs of the oxygen pointed towards the unoccupied corners of the tetrahedron. Given this structure, what is the shape of the closely related compound, hydrogen peroxide, H_2O_2? In this case, we will start with diatomic oxygen, which must be linear, because there are only two atoms in it. Write down its electronic configuration.

Each oxygen has a full octet, so if an electron is to be added then at least one of the oxygen/oxygen bonds must be broken. Write down a mechanism whereby one of the oxygen/oxygen bonds is broken homolytically, and an electron is added to form the superoxide anion, O_2^-.

Adding another electron forms the peroxide anion, O_2^{2-}. Write down the electron configuration of this anion.

In this anion the lone pairs are arranged around each oxygen atom in a tetrahedral manner. Draw the shape of this ion.

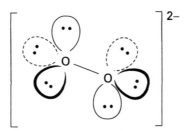

If two protons are added, one to each oxygen, then we form the desired molecule, H_2O_2. Now draw the shape of this molecule.

71

This is a bent molecule. Hydrogen peroxide is a relatively easy example, but it illustrates the methodology that can be used to find the structure of more complicated compounds. The next example follows the same pathway, but is slightly more difficult.

The structure of ammonia, NH_3, has been discussed previously; however, the structure of hydrogen azide, HN_3, is less obvious. In this case, we will start with diatomic nitrogen. Write down the dot and cross structure of diatomic nitrogen.

$$\overset{x}{\underset{x}{}} N \overset{x \; x \; x}{\underset{\bullet \; \bullet \; \bullet}{}} N \overset{\bullet}{\underset{\bullet}{}}$$

There is a triple bond between the two nitrogen atoms, which is composed of one σ bond and two orthogonal π bonds. Furthermore, there is a lone pair on each nitrogen that points away from the multiple bond. Each nitrogen has a full octet. Now form the hypothetical anion N_2^-, by the addition of an electron. Write down the mechanistic pathway for the production of this intermediate.

$$:N \equiv N: \; \leftrightarrow \; :N = N: \; \overset{\beta^-}{\rightarrow} \; \left[:\overset{\bullet\bullet}{N} = \overset{\bullet\bullet}{N}\bullet \right]^-$$

First, one of the π bonds must be broken to give the diradical; and then, an electron is added to one of the nitrogen atoms. This nitrogen now has two lone pairs. Draw the structure of this anion.

$$\left[\quad N = N \quad \right]^-$$

If the unpaired electron on the nitrogen of this anion now forms a covalent bond with one of the unpaired electrons on a nitrogen atom, we form the N_3^- anion, which is the deprotonated form of the molecule in which we are interested. Write down the mechanism for this reaction, indicating the lone pairs on the reagents and products.

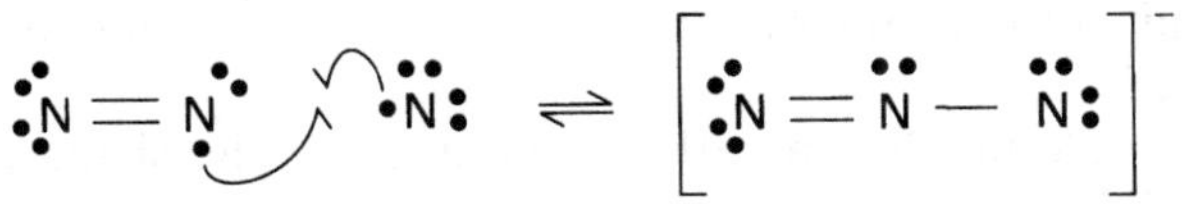

$$\ddot{N} = N \quad \cdot \ddot{N} : \quad \rightleftharpoons \quad \left[\ddot{N} = \ddot{N} - \ddot{N} : \right]^{-}$$

At the moment the ion is unsymmetrical. One terminal nitrogen has two lone pairs and is doubly bonded to the central atom, and so it bears a negative charge. The other terminal nitrogen has two lone pairs, but only one covalent bond; thus, it is electron deficient, but neutral. Suggest a redistribution of the electrons so as to make this anion symmetrical.

$$\bar{N} = \ddot{N} \quad N \quad \leftrightarrow \quad \bar{N} = \overset{+}{N} = \bar{N}$$

Now each terminal nitrogen is doubly bonded and carries a negative charge, while the central atom is doubly bonded to each of the two terminal nitrogen atoms, and so it bears a single positive charge. Suggest a shape for this molecule.

$$\left[:\bar{N} = \overset{+}{N} = \bar{N}: \right]^{-}$$

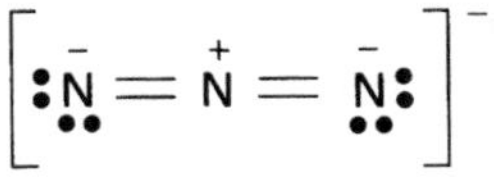

It is linear. If you have not already done so, indicate on this anion the orientations of the lone pairs.

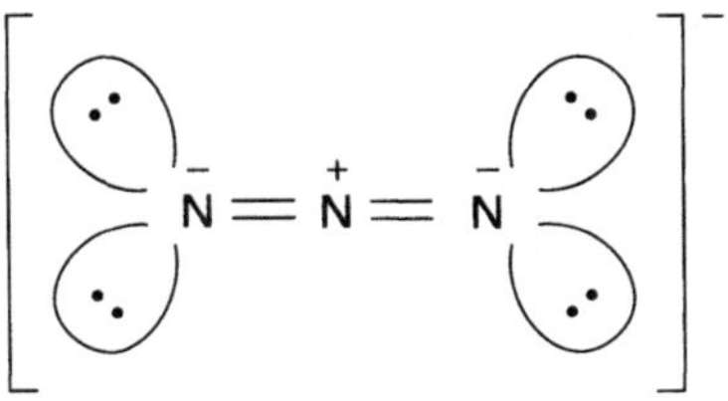

$$\left[\overset{\cdot\cdot}{\underset{\cdot\cdot}{}} \bar{N} = \overset{+}{N} = \bar{N} \overset{\cdot\cdot}{\underset{\cdot\cdot}{}} \right]^{-}$$

Each of the two lone pairs on the terminal nitrogens adopts a trigonal planar configuration, which are co-planar with each other, i.e. the molecule is flat. If this anion is now protonated, hydrogen azide is formed, HN_3. Suggest a shape for this molecule.

$$\underset{H}{\overset{\overset{+}{N} = \overset{-}{N}}{N = N = \bar{N}}}$$

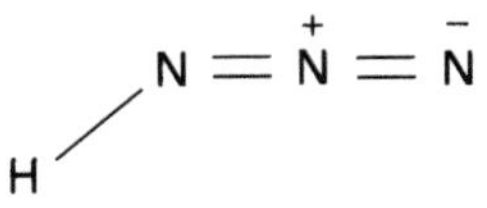

One of the lone pairs of the anion forms a dative bond with the proton. Thus, the final molecule is bent: the three nitrogen atoms are in a straight line, while the hydrogen is off to one side. So, by systematically building up the molecule from smaller units of known structure, the geometry of an unknown molecule has been deduced.

The next example involves a carbonyl derivative. The shape of a general carbonyl containing compound, $R^1R^2C=O$, has been discussed above. However, if one of the general alkyl groups is replaced by an electronegative group, such as a hydroxyl group, then the properties of the carbonyl group change. Write down the general formula for such compounds.

Now suggest some possibilities for X, and name the resulting compounds.

For example, when X is the group indicated below on the left, then the corresponding name is found to the right.

OH	carboxylic acid
Cl	acid chloride
OR	ester
NH_2	primary amide
NHR	secondary amide
NR_2	tertiary amide
OCOR	acid anhydride

Draw the structure of the carboxylic acid.

$$R - C \overset{\displaystyle O}{\underset{\displaystyle O-H}{}}$$

Write down a balanced equation for the deprotonation of this molecule using a general base, B, as the reagent.

$$RCOOH + B \rightleftharpoons RCOO^- + BH^+$$

Draw the structure of the resultant anion, indicating the lone pairs on the oxygen atoms.

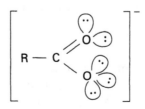

The carbonyl group is an electron sink and the oxygen anion is a source of electrons, and so the latter may donate electrons to the former. Suggest a possible movement of electrons that may occur.

In this new species, the bonding to the oxygen atoms is reversed, otherwise the molecule is the same. In reality, the electronic configuration is a mixture of these two, hypothetical, canonical structures. The electrons are said to be delocalised over the three atoms to give rise to the resonance hybrid of the carboxylate anion.

This resonance interaction affects the geometry that is adopted by some of the derivatives of the carboxylic acids. Draw the structure of a general carboxylic acid ester.

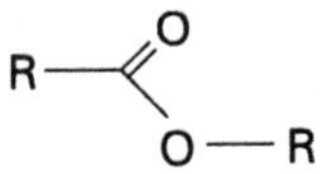

There is a single bond between the carbonyl carbon and the bridging oxygen bond, and so the groups may rotate with respect to each other around this bond, which in turn gives rise to many different conformers. However, there are two conformers of particular interest. Draw the two conformers of a general ester that have all the groups in one plane.

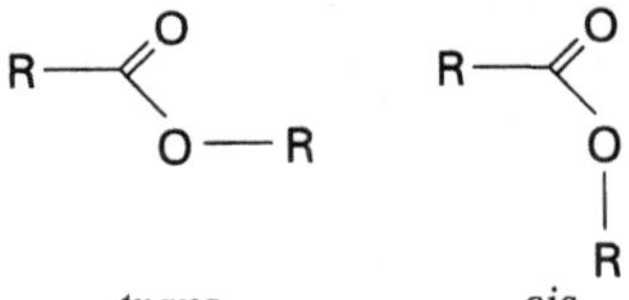

In essence, these are the *cis* and *trans* conformers of the ester, as the ester alkyl group is either adjacent to the carbonyl alkyl group or is opposite it. Suggest which one is the favoured conformer.

The *trans* conformer is the favoured one. Suggest two reasons that might account for this observation.

The first is that there is less steric interference between the two alkyl groups if they are in the *trans* conformation. The other reason is that, while in the *trans* position, the lone pairs on the oxygen may interact with the carbonyl group. Draw the mechanism for this interaction.

75

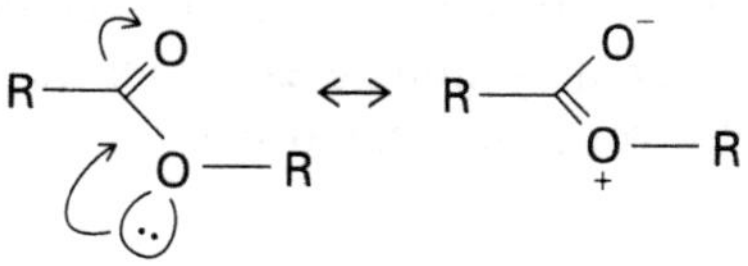

As a result of this interaction, the *trans* conformer is favoured. In a secondary amide the same considerations may apply. Hence, suggest the favoured conformer of a general secondary amide.

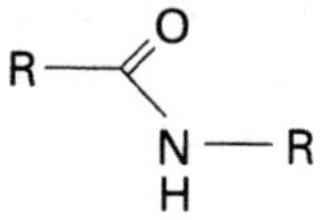

In biological systems this is important, because small energy differences often determine the overall geometry of interacting species.

In the next chapter, the stabilisation of charged species will be examined, and one of the most important mechanisms available involves the delocalisation of a lone pair in a manner that is similar to the stabilisation of the *trans* conformer of an ester or amide functionality.

4.7 Summary of shapes of molecules

There are two main types of bonding, ionic and covalent. Ionic bonds are characterised by their non-directional nature. The relative sizes of the component ions is the principal factor in determining the structure of the crystal lattice. In contrast, covalent bonds are directional in nature, and of a certain length. Usually in a simple crystal, every ion occupies exactly the same position as does every other ion of the same type. However, in covalent molecules, this is not the case. Thus, every atom in a covalent molecule must be considered individually with respect to its spatial relationship to all the other component atoms.

The electrons in a molecule occupy molecular orbitals. The number of molecular orbitals produced is the same as the original number of atomic orbitals. Each molecular orbital can hold a maximum of two electrons. An equal number of bonding and anti-bonding molecular orbitals are formed from the atomic orbitals. Occasionally, non-bonding molecular orbitals are produced. Anti-bonding orbitals are important, because the electrons from the attacking species often move into these orbitals in the process of forming the new compound. Thus, the orientation of the anti-bonding orbital will often determine the direction of attack, especially when the species being attacked has no empty bonding orbitals.

The shapes of organic molecules are determined by the electron pair repulsion theory, which states that any given electron pair in an orbital will repel all other electron pairs. There is one caveat, namely if two or more electron pairs are co-linear, then they act, for these purposes, as if they were one electron pair. Thus, an electron pair will adopt a position in space that is furthest from all the other electron pairs. To a first approximation, the geometries adopted are regular in shape. Thus, the AB_2 system is linear; the AB_3 system is trigonal planar; the AB_4 system is tetrahedral; and the AB_5 system is trigonal bipyramid.

The hybridisation theory states that atomic orbitals can be mixed, i.e. hybridised, to form new atomic orbitals that are degenerate, i.e. equal in energy and orientation. Thus, for example, one of the 2s electrons in the ground state of carbon is promoted to give the electron configuration of $2s^1$, $2p_x^1$, $2p_y^1$, $2p_z^1$ in the valence shell. These four atomic orbitals can then be hybridised to form four identical sp^3 hybrid atomic orbitals. These hybrid orbitals project towards the corners of a regular tetrahedron. This is the shape adopted by the AB_4 system, which includes carbon species such as CH_4 and CH_3^-, nitrogen species such as NH_3 and NH_4^+, and oxygen species such as H_2O and H_3O^+, i.e. they are all isoelectronic with each other. Any vertex unoccupied by an atomic centre is occupied by a lone pair. They all have in common the fact that there are four directions in space in which the electron pairs project.

The hybridisation between one s orbital and two p orbitals results in three sp^2 hybrid orbitals, which form a trigonal planar shape. In this scenario, the third p orbital is left unhybridised, and is at right angles to the plane described by the three sp^2 orbitals. Examples of this geometry include BCl_3, CH_3^+, $:CCl_2$ and doubly bonded carbon and nitrogen species such as $R_2C=O$, $R_2C=CR_2$ and $R_2C=NR$. They all have in common that there are three directions in space in which the electron pairs project.

The hybridisation between one s orbital and one p orbital results in two sp hybrid orbitals, which form a linear shape. In this scenario, the second and third p orbitals are left unhybridised, and are orthogonal to each other, but co-linear with the axis described by the sp orbitals. Examples of this geometry include $BeCl_2$, $RC\equiv N$, $RC\equiv CR$ and $R-N^+\equiv C^-$. They all have in common that there are two directions in space in which the electron pairs project.

When an atomic orbital combines with another atomic orbital to form a molecular orbital, the bonding and anti-bonding molecular orbitals that are formed project in different directions in space. When an electron-rich species attacks such a bond, it approaches along the line of the anti-bonding orbital, so that the attacking electrons can move into the empty anti-bonding orbital. Hence, the direction not only of the bonding orbitals, but also of the anti-bonding orbitals, is very important in understanding mechanistic organic chemistry. An anti-bonding orbital is indicated by a raised postscripted asterisk, e.g. σ^*.

If the molecular orbital has cylindrical symmetry about the axis connecting the atoms that are bonded together, i.e. there is not a nodal plane that passes through both of the bonded nuclei, then it is called a σ bond. An example is a bond formed between an sp^3 hybridised atomic orbital and an s atomic orbital, as in a carbon/hydrogen single bond. In contrast, a molecular orbital that has a nodal plane that passes through both of the bonded nuclei is called a π bond. An example is a bond formed between two p_z atomic orbitals in the second bond in a carbon/carbon double bond. Rotation around a double bond is difficult, because the overlap between the p_z orbitals would be lost.

If faced with deducing the structure of an unknown molecule, it is often helpful to work from a known structure. Thus, for example, the structure of HN_3 can be deduced by starting with an N_2 molecule, and proceeding through the N_2^- anion and the N_3^- anion, by the addition of an electron and a nitrogen atom respectively, before finally adding a proton to give the desired product.

5

Stabilisation of Charged Species

5.1 Introduction

When two molecules react with each other, new bonds are made while others are usually broken. At the half-way point between the making and breaking of any given bond there is a transitory stage between the original compounds and the products that result from the bond reorganisation. This transitory stage is called a transition state and represents a species that is very short lived. It does not have an independent existence and may not be isolated like a true intermediate in a reaction pathway. In all transition states, the distribution of charge is different from that found in the original molecules. This is because of the movement of the electrons that has occurred in the bonds that are breaking and forming.

Having passed through the transition state, the result of this initial bond reorganisation may not be the final product that is to be produced from the reaction, but instead it may be a true intermediate. Such intermediates may be neutral, but more often they bear an electrical charge. These intermediates may exist long enough to undergo bond rotation or other bond flexing movements, which may affect the eventual stereochemistry of the product. It may also be possible to isolate or trap these intermediates.

In order to facilitate the reaction, the charges that have been created by the reagents within the transition states and intermediates may be stabilised. It is the mechanisms that are available for such stabilisation that are covered in this chapter.

The two main forms of charge stabilisation have already been introduced in the context of polarised bonds, i.e. the inductive and mesomeric effects. Other mechanisms exist, and these may be of overriding importance when they are applicable. However, for most reactions it is the effects of inductive and mesomeric groups that have most influence upon the stability of the charged species, and so we shall start with an investigation of these mechanisms.

5.2 Inductive effects

5.2.1 *Charged hydrocarbon species*

The simplest, commonly encountered, charged hydrocarbon species is the methyl cation, CH_3^+. Draw a dot and cross structure of this ion, and suggest a shape for it.

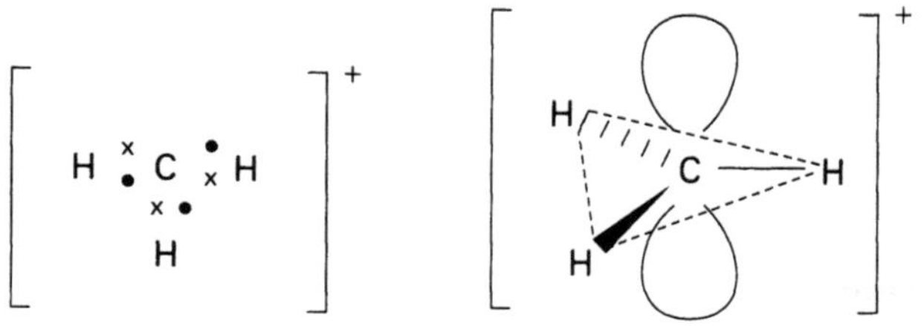

There are six electrons around the carbon in three single σ bonds. There is an empty p_z orbital. The structure adopted is trigonal planar, with the empty orbital at right angles to this plane. Suggest the direction of polarisation along these three single σ bonds. Indicate this polarisation on a diagram of the charged species.

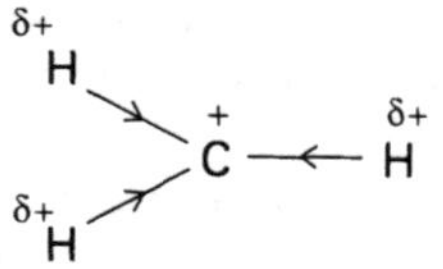

The central carbon bears the single positive charge, and hence its effective electronegativity is increased, i.e. it attracts electrons more strongly. It exerts this attractive force upon the electrons that are within the adjacent carbon/ hydrogen bonds. This reduces the charge that the carbon is carrying, and in effect places a $\delta+$ charge on each of the three hydrogens. This distributes the charge, and so reduces the charge density on the carbon. So strictly, the charge present on the central carbon is not a single plus one, but rather plus one minus three δ pluses, for the total charge present on the ion must sum to plus one.

Suggest the shape of the ethyl cation, $CH_3CH_2^+$, and indicate the charge polarisation that is present.

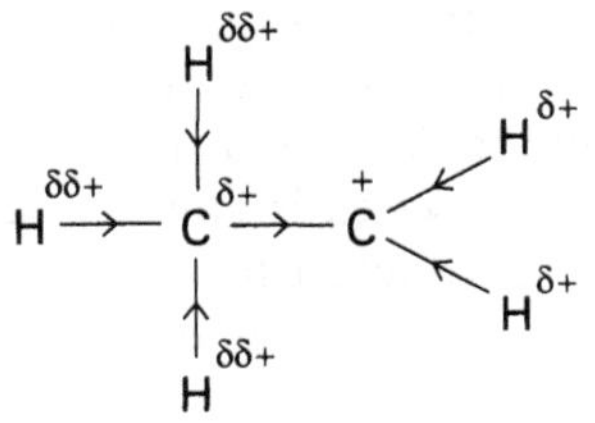

$$\overset{\delta\delta+}{H} \longrightarrow \overset{\underset{\displaystyle\overset{\uparrow}{H}}{\overset{\displaystyle\overset{H}{\big\downarrow}}{\underset{\delta\delta+}{C}}}}{} \overset{\delta+}{\longrightarrow} \overset{+}{C} \overset{\delta+}{\underset{\delta+}{\big\langle}} \begin{smallmatrix} H \\ H \end{smallmatrix}$$

The positive carbon adopts the trigonal planar configuration, while the other carbon is in the normal tetrahedral configuration. The carbon bearing the positive charge may inductively draw towards itself the electrons within the two σ bonds to the hydrogens, and so place a δ+ charge on each of them. It may furthermore draw towards itself the electrons within the σ bond to the second carbon atom, which in turn may attract the electrons in the three carbon/hydrogen bonds of the terminal methyl group and so spread out the positive charge over another three atoms. This lowers still further the charge density on the original charged carbon, and so stabilises it even further. This species is called a primary carbonium ion, because there is one carbon attached to the carbon that bears the charge.

In the propyl cation with the charge upon a terminal carbon, $CH_3CH_2CH_2^+$, suggest the polarisation that is present and indicate this on a diagram of the molecule.

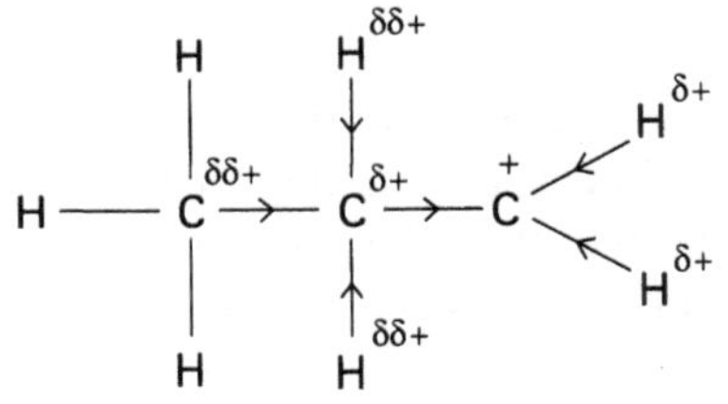

Here the addition of the extra methyl group is of less assistance in the stabilisation of the charge, because the transmission of the charge by the inductive effect becomes less efficient with distance. For most purposes, the effect can be ignored after the second σ bond from the charged centre.

Now consider the other possible propyl cation, namely the one with the charge upon the central carbon, $CH_3CH^+CH_3$. In this case, suggest the polarisation that is present, and indicate this on a diagram of the molecule.

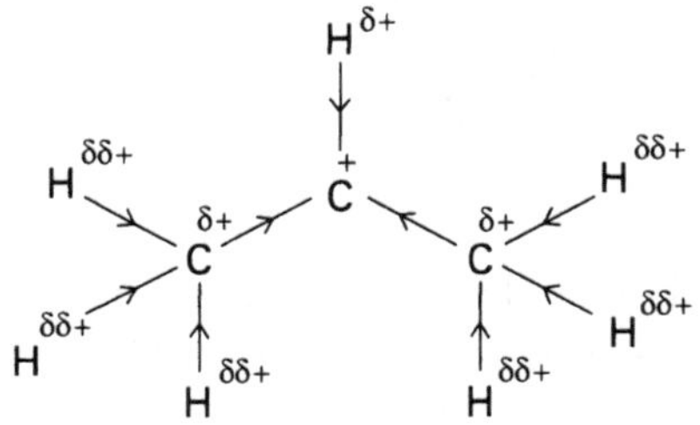

In this case, there is a further diminution of the charge upon the central carbon, because the extra atoms are that much closer, i.e. only two bonds away, and the electrons within their σ bonds may be effectively drawn towards the charge by induction. This species is called a secondary carbonium ion, because there are two carbons attached to the carbon that bears the charge. Suggest the order of stability of the four cations that have been discussed so far.

$$CH_3CH^+CH_3 > CH_3CH_2CH_2^- > CH_3CH_2^+ > CH_3^+$$

Suggest where the *t*-butyl cation, $(CH_3)_3C^+$, should be placed in the above list of stability of the cations. This species is called a tertiary carbonium ion, because there are now three carbons attached to the carbon that bears the charge.

It would be the most stable cation, because it has the largest number of bonds within two bond lengths of the charge.

Earlier we mentioned the shape of the methyl cation; suggest the shape of the *t*-butyl cation, $(CH_3)_3C^+$.

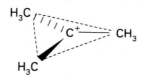

This molecule is trigonal planar around the central charged carbon. It is observed experimentally that carbonium ions are more stable if it is possible for them to adopt a planar configuration. One reason for this is that it allows for the maximum separation of the substituents, and so reduces the steric crowding to a minimum. Hence, carbonium ions at a bridgehead are less stable than would be expected if the considerations discussed earlier were the only factors involved in determining their stability.

Hydrocarbon species may also bear a net negative charge, i.e. they can exist as anions. Draw the methyl anion, CH_3^-.

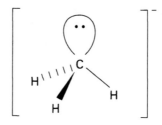

This species is tetrahedral in shape as there are eight electrons around the carbon in three σ single covalent bonds to the hydrogens plus one lone pair. Suggest the direction of polarisation within the bonds in this species.

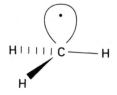

The carbon atom is electron rich and this now decreases its electronegativity, which was initially greater than that of the hydrogen atoms. Now, however, electrons are being pushed towards the hydrogen atoms. Yet each one already has a full complement of two electrons accommodated around it, and so the negative charge is not effectively transmitted to the neighbouring hydrogen atoms. The overall result is that the anion is destabilised.

In the ethyl anion, $CH_3CH_2^-$, suggest the shape of this charged species and also indicate what might be the polarisation of the bonds. Comment on the degree of stabilisation relative to the methyl anion.

Each carbon is tetrahedral: one has three carbon/hydrogen σ single bonds, while the other has only two such bonds and a lone pair. This ion is even less stable than the methyl anion, because there are even more hydrogens that would prefer to donate electrons to the carbon that carries the negative charge.

Suggest the relative order of stability of the following anions: $CH_3CH_2^-$, CH_3^-, $CH_3CH^{(-)}CH_3$ and $CH_3CH_2CH_2^-$.

$$CH_3^- > CH_3CH_2^- > CH_3CH_2CH_2^- > CH_3CH^{(-)}CH_3$$

The relative stability of the hydrocarbon anions is the opposite of that proposed for the related cations.

The last hydrocarbon species that we shall consider is the free radical. Suggest a shape for the methyl radical, $CH_3^{\cdot}$.

The overall shape is a flattened tetrahedron. The central carbon has six electrons in three covalent bonds to the three hydrogen atoms, and also an unpaired electron that occupies the p_z orbital. However, because there is only one electron in that orbital, it does not have the same repulsive effect as a pair of electrons. As a consequence the carbon/hydrogen bonds are not forced together as much as they are in, for example, the methyl anion, CH_3^-, which adopts an almost regular tetrahedron shape. Suggest the direction and degree of polarisation that is present in the methyl radical.

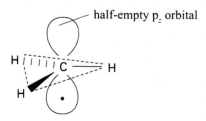

The carbon is electrically neutral, but it does have a greater electronegativity than hydrogen, and so it attracts slightly the electrons within the carbon/hydrogen bonds. Now suggest the relative order of stability of the primary, secondary and tertiary carbon radicals.

tertiary > secondary > primary

The order is the same as that of the carbonium ions, because they share the same electronic property in that electrons are attracted towards the carbon that bears the unpaired electron from the adjacent hydrogen atoms. Draw the methyl radical using the stereochemical notation, indicating the half-empty p_z orbital.

half-empty p_z orbital

There are two forms of the methyl radical, one with the unpaired electron above the plane of the other atoms, and the other with it below. In the methyl cation, all the atoms are in the same plane with one lobe of the p_z orbital above this plane and the other below. It is observed that the carbon radical species can invert its configuration very readily. Presumably, it passes through a shape that is closely related to that adopted by the methyl cation as the unpaired electron goes from above to below the plane of the atoms, and vice versa.

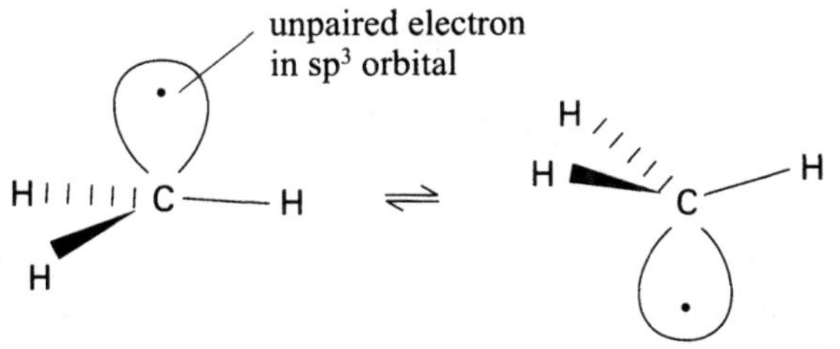

unpaired electron
in sp³ orbital

5.2.2 Charged non-hydrocarbon species

Electronegative atoms draw negative charge towards them. Indicate the direction of the polarisation along a carbon/chlorine σ bond.

$$\overset{\delta+}{C} \longrightarrow \overset{\delta-}{Cl}$$

The chlorine exerts a negative inductive effect, and so pulls the electrons within the σ bond towards itself. This in turn makes the carbon atom slightly positive.

If one of the hydrogen atoms within the methyl cation is replaced by a chlorine atom to give the chloromethyl cation, CH_2Cl^+, would this ion be more or less stable than the corresponding methyl cation, CH_3^+?

It would be less stable, because the chlorine atom would make the carbon atom even more positive and so destabilise it.

Now, consider the related anion in which a hydrogen atom has been substituted by a chlorine atom, i.e. the chloromethyl anion, CH_2Cl^-: would this substituted anion be more or less stable than a methyl anion?

In this case, the chloro anion would be more stable than the related hydrocarbon anion, because the electronegative chlorine would decrease the charge that is present on the carbon. Accordingly, suggest the order of stability of the following anions: $CHCl_2^-$, CH_2Cl^- and CCl_3^-.

$$CCl_3^- > CHCl_2^- > CH_2Cl^-$$

The greater the number of electronegative atoms that are joined to the carbon that bears the extra electron, the greater the degree of stabilisation of the resultant anion. The trihalogenomethyl carbanion is, in fact, so stable, that it acts as a good leaving group in the haloform reaction.

5.3 Mesomeric effects

5.3.1 Charged hydrocarbon species

The mesomeric effect operates through the π bond system. Draw the dot and cross structure of the cation, $CH_2{=}CH{-}CH_2^+$.

$$\left[\begin{array}{c} H \underset{\times}{} \quad \overset{H}{\underset{\times}{}} \quad \underset{\times}{} H \\ \overset{\bullet}{\underset{\times}{C}} \overset{\bullet}{\underset{\times}{}} \overset{\bullet}{\underset{\times}{C}} \overset{\bullet}{\underset{\times}{}} \overset{\bullet}{\underset{\times}{C}} \\ H \qquad\qquad H \end{array} \right]^{+}$$

The moiety that has the cationic centre in the carbonium ion adjacent to a carbon/carbon double bond is called an allylic cation. The diagram represents one of the canonical structures for this cation.

Draw the stereochemical diagram of this allylic cation indicating both the p_z orbitals that are involved in the π double bond and also the empty p_z orbital on the carbon that bears the positive charge.

p_z/p_z interaction to yield π bond

empty p_z orbital

The p_z orbitals that are involved in the carbon/carbon double bond are parallel to the empty p_z orbital, while all the carbon and hydrogen σ bonds are in the same plane as each other, which is orthogonal to the p_z orbitals. This means that the electrons within the π double bond may easily be delocalised into the empty p_z orbital.

Write the mechanism for the redistribution of the electrons within the π bond and p_z orbital system so as to give the other canonical structure of this allylic cation.

Draw a single structure that corresponds to a hypothetical combination of these two canonical structures.

Each terminal carbon is now joined to the central carbon by a bond that is half-way between a single and double bond. Furthermore, each terminal carbon bears half a unit of positive charge. Thus, the charge has very effectively been spread out over two carbons, i.e. delocalised. The structure with the intermediate bonds closely represents the actual electronic distribution that is present in the ion. This structure is called the resonance hybrid.

If a further single and double bond unit is added to this system, we obtain the cation with five carbons, $CH_2=CH-CH=CH-CH_2^+$. Draw the three canonical structures of this cation.

The addition of another $CH_2=CH-$ unit is said to form the vinylogue of the original compound. Write down the mechanism for the mesomeric transmission of the positive charge along this five carbon chain.

It is readily seen that the charge is very efficiently transmitted along a chain that possesses alternate double and single bonds, i.e. a system that is conjugated. In such a system, the degree of delocalisation is greater, and so the amount of stabilisation is greater.

There is one major restriction in this type of stabilisation, namely that the redistribution of the electrons must not result in a double bond at a bridgehead: that is, at a junction point between three ring systems. This is Bredt's rule.

Earlier, we mentioned the shape of various carbocations. Remind yourself of the shape of the *t*-butyl cation, $(CH_3)_3C^+$, and then draw this cation.

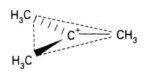

The cation is trigonal planar. Now redraw this cation using the stereochemical notation, showing the position of the empty p_z orbital and also one of the methyl groups with one of its carbon/hydrogen bonds aligned parallel to this empty orbital.

The electrons within the carbon/hydrogen bond are capable of being delocalised into the empty p_z orbital. Draw the mechanism for this, and hence the resultant products.

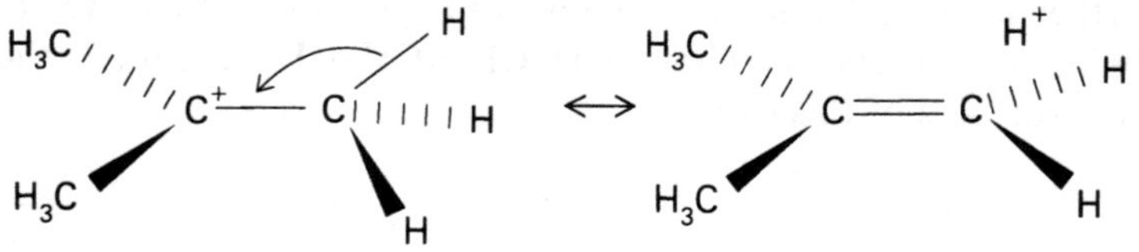

The unsaturated product is 2-methylpropene; and, in addition there is a bare proton positioned above it. This alternative structure represents another canonical structure of this cation. As there are nine possible hydrogens that could be involved in this redistribution, this results in a total of ten canonical structures for this particular cation. Hence, there is great scope for the delocalisation of the positive charge. This mechanism has been suggested as a possible way in which the hydrogens that are adjacent to a positive charge may help to stabilise it. It is called hyperconjugation (or σ-conjugation) and, as may be seen from the required geometry, it most readily occurs when the carbonium ion adopts a planar conformation. This is another reason why carbonium ions are more stable if they are able to adopt a flat shape.

5.3.2 Charged non-hydrocarbon species

It is often found that a carbon that is bearing a positive charge is bonded to a heteroatom that has a lone pair of electrons, e.g. in the R_2C^+–OH system.

Draw a stereochemical diagram for this cation, indicating the empty p_z orbital on the carbon atom and the lone pairs on the oxygen atom.

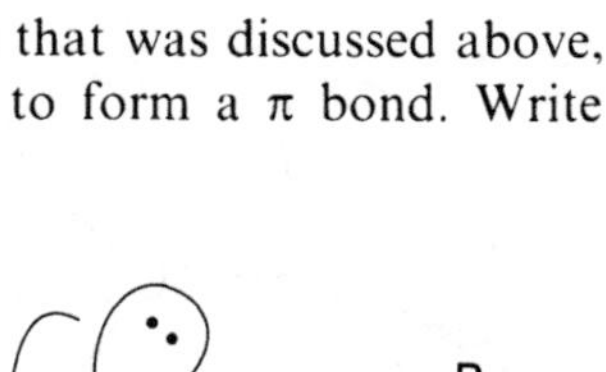

Like the hyperconjugation that was discussed above, the lone pair may move into the empty p_z orbital to form a π bond. Write the mechanism for this reaction.

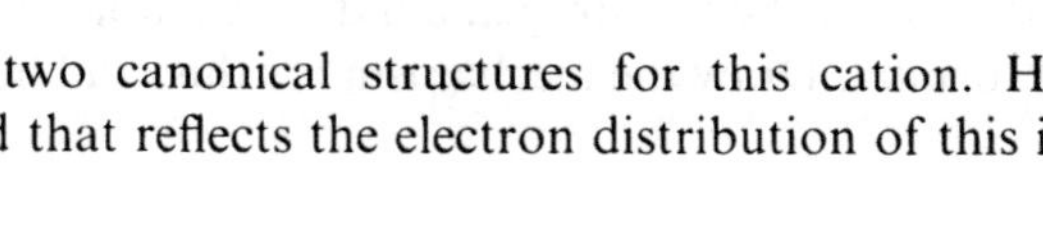

Now there are two canonical structures for this cation. Hence, draw the resonance hybrid that reflects the electron distribution of this ion in reality.

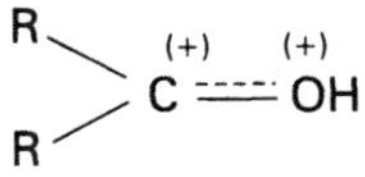

$$R_2C^{(+)} \text{---} \overset{(+)}{OH}$$

In this diagram, the extra bond is half formed, with the charge shared between the carbon and the oxygen.

A similar stabilisation may occur in the nitrogen analogue, $R_2C^+-NH_2$. Draw the mechanism that represents this stabilisation.

$$R_2\overset{+}{C} - \overset{..}{N}H_2 \longleftrightarrow R_2C = \overset{+}{N}H_2$$

Here the lone pair on the nitrogen is donated into the empty p_z orbital.

So far, we have looked at the stabilisation of a positive charge using the mesomeric effect, but it is equally possible to invoke this mechanism to account for the stabilisation of a negative charge. Instead of having a group that has an extra lone pair, which may be used to form a double bond, what is needed is a group that has a multiple bond that might cleave. In this case, one part of the multiple bond breaks heterolytically, and so places the extra charge on an electronegative element.

The first example of this type is concerned with the stabilisation of a negative charge on an oxygen in the carboxylate anion, RCO_2^-. Draw the dot and cross structure of this anion.

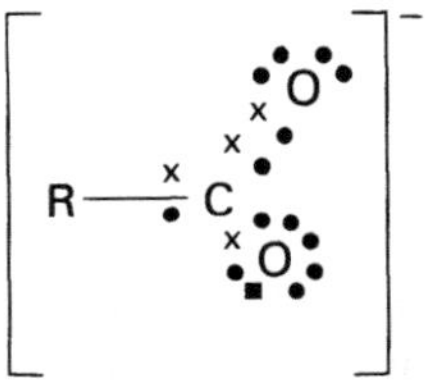

If the carbon/oxygen double bond breaks heterolytically, placing the negative charge upon the oxygen, then the carboxylate carbon will be left with a positive charge. Draw the mechanism for this initial step.

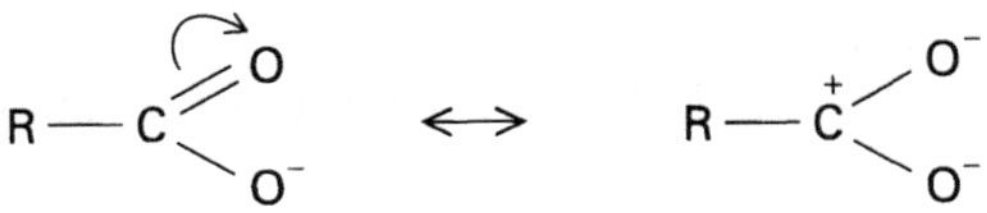

This species is symmetrical about the charged carbon. Now draw the second part of the mechanism in which the oxygen that originally bore the negative charge donates this to the carboxylate carbon, and in so doing, forms a new bond.

$$R-\overset{+}{C}\overset{O^-}{\underset{O^-}{\Big\langle}} \quad\longleftrightarrow\quad R-C\overset{O^-}{\underset{O}{\Big\langle}}$$

Here the original arrangement of the carbon/oxygen single and double bonds has been swapped over. Usually these two steps are joined together. Draw a mechanism that shows both parts on one diagram.

$$R-C\overset{O}{\underset{O^-}{\Big\langle}} \quad\longleftrightarrow\quad R-C\overset{O^-}{\underset{O}{\Big\langle}}$$

Now draw the resonance hybrid of the carboxylate anion that is a combination of these three canonical structures.

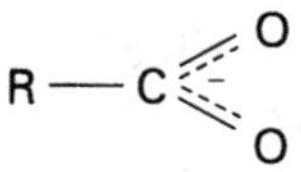

$$R-C\overset{O}{\underset{O}{\Big\langle}}$$

If, instead of the extra charge coming from the oxygen in the carboxylate anion, it comes from a carbon that is attached to the carboxylic acid group, then the same principles would apply. Thus, for example, in the anion, $R_2C^{(-)}-CO_2H$, write a mechanism that results in the delocalisation of the negative charge, which originally resides on the carbon, onto one of the carboxylic acid oxygens.

$$R_2\bar{C}-C\overset{O}{\underset{OH}{\Big\langle}} \quad\longleftrightarrow\quad R_2C=C\overset{O^-}{\underset{OH}{\Big\langle}}$$

Here the carbon/oxygen double bond has been broken, which places the negative charge on that oxygen, and a new carbon/carbon double bond has been formed.

Returning to the carboxylate anion, instead of the charge being relocated onto the other oxygen of the carboxylate anion, it could potentially be placed upon the carbon that is joined to the carboxylate group. In this case, the carboxylate group will cleave itself from the rest of the molecule. In order for this to occur it must be possible for the negative charge to be stabilised on the carbon atom in some manner.

Write a mechanism for the cleavage of the $R_3C-CO_2^-$ anionic species.

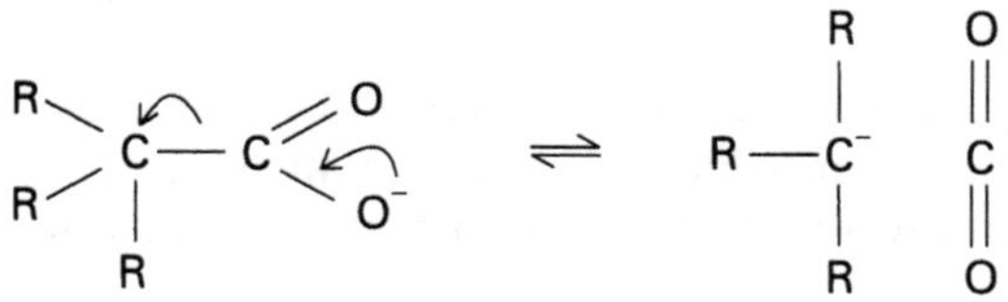

$$\underset{R}{\overset{R}{\Big\rangle}}\underset{|}{\overset{}{C}}-C\overset{O}{\underset{O^-}{\Big\langle}} \quad\rightleftharpoons\quad R-\underset{R}{\overset{R}{\underset{|}{\overset{|}{C^-}}}}\quad\underset{O}{\overset{O}{\underset{||}{\overset{||}{C}}}}$$

The grouping that originally formed the carboxylate anion now leaves as a neutral molecule of carbon dioxide. This by-product usually escapes from the reaction mixture, and so may not re-react to re-form the original compound. If the negative charge could be accommodated elsewhere, then the carboxylate group could be used as an effective leaving group. This possibility will be examined in more detail in Chapter 16, when, in particular, fragmentation reactions that occur by anionic cleavage are considered in some detail.

The multiple bond that is broken need not be a double bond; instead it may be a triple bond, which can react so as to form a double bond while accommodating a negative charge. For example, the cyano group, as in $NCR_2C-CO_2^-$, is capable of stabilising a negative charge. Write a mechanism for this.

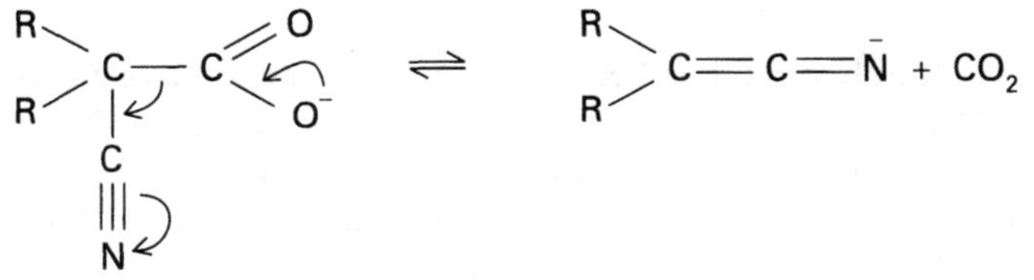

In this case, a carbon/carbon double bond is formed that is directly adjacent to a carbon/nitrogen double bond with the terminal nitrogen bearing the extra charge.

The stabilisation of negative charges that were initially formed at a carbon atom, plays a very important part in organic chemistry. One of the commonest groups that performs this function is the carbonyl group. Write down the mechanism whereby a negative charge, that was originally resident on a carbon atom, may be delocalised onto an adjacent carbonyl group.

If this anion is now protonated on the oxygen, a neutral species is formed. Write down the structure of this compound.

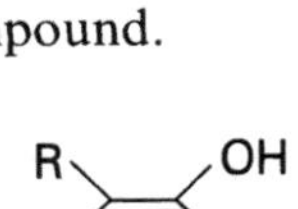

This type of compound is called an enol, and the reorganisation of the carbon/oxygen double bond in the carbonyl group to the carbon/carbon double bond in the enol compound is called the keto/enol tautomerism. Notice, also, that this is accompanied by a hydrogen atom changing place, from the carbon to the oxygen. Tautomers are different from canonical structures in that the former represent real compounds that may be isolated (at least in principle), while the latter represent only hypothetical structures that help in the understanding of the electron distribution.

The nitro group, $-NO_2$, also performs the function of stabilising a negative charge on an adjacent carbon atom. It does this by carrying out a similar electron redistribution to the keto/enol tautomerism. Write a mechanism for the delocalisation of a negative charge that is initially placed upon a carbon atom that is adjacent to a nitro group, $R_2C^{(-)}-NO_2$.

Here again a double bond has been broken; this time a nitrogen/oxygen bond, and a new carbon/nitrogen double bond has been formed in its place. This is called the nitro/aci tautomerism, and again each tautomer may be isolated, because each tautomer represents a real compound and not just a hypothetical model. These tautomers are shown below.

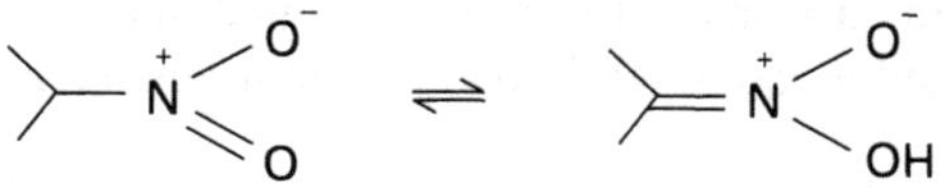

nitro   aci

This concludes these sections on inductive and mesomeric effects. It is important to remember that these two mechanisms are by far the most common that affect the stability of a carbanion or carbonium ion. However, there are a number of other mechanisms that, when they are relevant, have a particularly powerful influence. We shall now consider these in turn.

5.4 Degree of s orbital character

A lone pair of electrons orbits only one nucleus. The nucleus carries a positive charge, while the lone pair is a negative charge that is separated from the nucleus. Would the degree of stabilisation of the electrons within the lone pair increase or decrease as the average distance between the lone-pair electrons and the nucleus increases?

As the charge separation increases, the stabilisation would decrease, because the Coulombic interaction decreases.

For any given quantum level, is the s sub-orbital closer to the nucleus than the p sub-orbitals?

The s sub-orbital is closer to the nucleus. Thus, place the following hybrid orbitals, sp^3, sp and sp^2, in order of increasing average distance from the nucleus.

$$sp < sp^2 < sp^3$$

With increasing p sub-orbital content, the hybrid atomic orbital becomes increasingly distant from the nucleus. Accordingly, suggest the order of increasing stability for a lone pair in these hybrid orbitals.

$$sp > sp^2 > sp^3$$

This order reflects the ease with which a proton may be removed from a carbon hybridised in any one of the above ways, i.e. this reflects the acidity of such a hydrogen.

Now suggest the order of stability of the various dicarbon anions, $C_2H_5^-$, $C_2H_3^-$ and C_2H^-.

$$C_2H^- > C_2H_3^- > C_2H_5^-$$

The ethynyl anion is by far the most stable of the three, and may be formed under quite mild conditions by the addition of a suitable base. To remove the proton from a carbon that is hybridised in the sp^3 format requires a very strong base indeed, such as butyllithium.

Suggest the order for the stabilities of the corresponding carbonium ions, $C_2H_5^+$, $C_2H_3^+$ and C_2H^+.

$$C_2H_5^+ \gg C_2H_3^+ > C_2H^+$$

Only the ethyl carbonium ion is produced under normal conditions. The vinyl carbonium ion, $C_2H_3^+$, is only produced under extreme conditions and if you need to invoke it in a mechanism you are either wrong, or suggesting a rather advanced mechanism.

5.5 d orbital involvement

Third row elements, such as phosphorus and sulphur, have empty d sub-orbitals into which electrons may be placed from an adjacent carbon that bears a negative charge. These elements are not bound by the Octet rule and may accommodate more than eight electrons in their valence shell. Thus, these elements may act as an electron sink.

Suggest a possible mechanism for the stabilisation of the sulphoxide carbanion, $RSO_2CR_2^-$, that utilises this property of the sulphur atom to expand its octet.

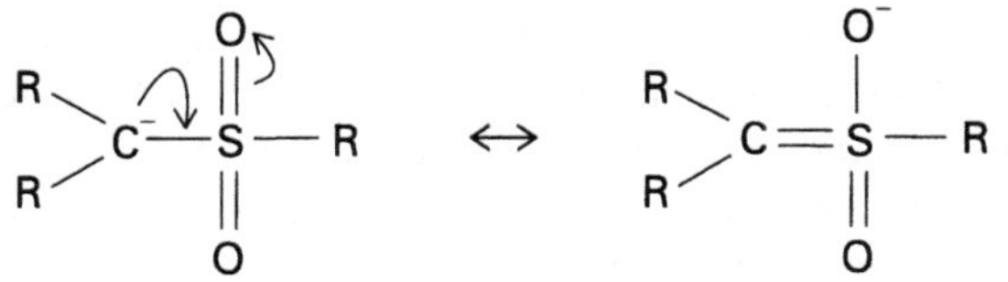

Here a sulphur/oxygen double bond is broken and replaced by a sulphur/carbon double bond.

Phosphorus exhibits similar properties in that it may expand its octet. In the Wittig reaction, the reagent is a phosphorus/carbon compound in which the carbon bears a negative charge next to a phosphorus, which is carrying a positive charge. Such an arrangement is called an ylid. Draw the structure for the phosphorus ylid, $Ph_3P^+-CR_2^-$, indicating the lone pair and the position of the charges.

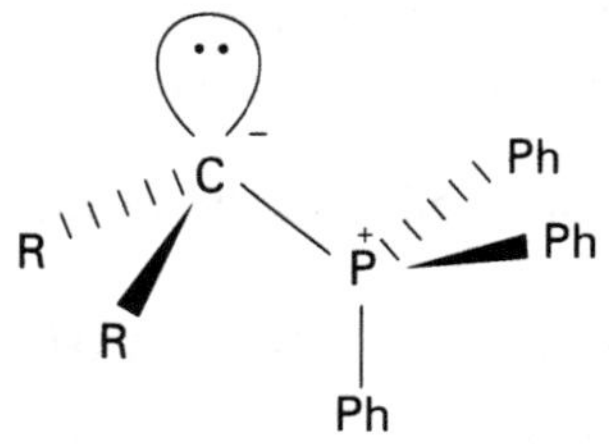

This is only one of the possible canonical structures, but this is the one that is used when involving the ylid in a reaction, because it allows the movement of electrons to be followed easily.

Suggest another canonical structure for this compound that involves the stabilisation of the negative charge by the empty d orbitals of phosphorus.

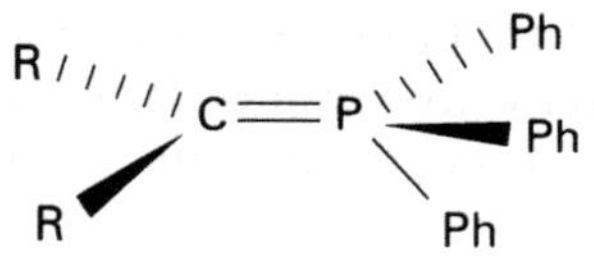

Now the phosphorus has expanded its valency to five and accommodates ten electrons around its nucleus in the valence shell. Later on in this reaction sequence, the phosphorus forms an oxygen containing compound with the formula $Ph_3P=O$, which again has a valency of five. It is the propensity of phosphorus to form this latter compound that is one of the driving forces behind the Wittig reaction.

5.6 Aromatic character

It has been observed that cyclic conjugated systems are particularly stable. In general, this applies to systems that have $(4n + 2)\pi$ electrons in a continuous closed loop of p orbitals, where n varies from zero upwards in steps of one unit. This is Hückel's rule, and the property is called aromaticity.

Draw the two major canonical structures of the cylcohexatriene compound, where n equals one, i.e. where there are six π electrons.

This molecule is benzene. Benzene is often represented by a regular hexagon with a circle inside to indicate the delocalisation, but this is not very useful when the electrons in bonds within a benzene ring are being utilised in a reaction pathway. When this occurs it is more useful to indicate each single and double bond separately, because it is then easier to account for all the electrons.

It is not necessary for there to be six atoms in a ring in order for a compound to be aromatic. For example, how many π electrons are there in the cyclopentadiene molecule?

There are four π electrons in this compound. However, at present, there is not a continuous circuit of p orbitals around the ring, because one of the carbon atoms in the ring is sp^3 hybridised, i.e. the methylene group, $-CH_2-$. This group prevents delocalisation occurring all around the ring. However, if a hydrogen cation is removed from the methylene group to form an anion, then the now deprotonated carbon may re-hybridise to give an sp^2 orbital and an occupied p_z orbital. This results in a continuous ring of p orbitals that are so orientated as to allow ready delocalisation from one to the other. Draw this anion, i.e. the cyclopentadienyl anion, $C_5H_5^-$, showing the p orbitals of the double bonds and the newly formed p orbital that is occupied by the lone pair.

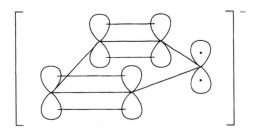

This system now has a continuous ring of five p orbitals (centred on the five carbon atoms forming the ring), which contains a total of six electrons, two from each of the two double bonds and two more from the lone pair. Thus, this system is also aromatic.

The cyclopentadienyl anion is sometimes drawn as a regular pentagon, inside which is circle, which in turn contains a negative sign:

The circle is used to indicate the delocalised nature of the bonding, and in particular the aromatic character.

So far, we have looked at two species in which n equals one. There is another species in which n equals one. In this case, a cation is formed from the cycloheptatriene, i.e. the cycloheptatrienyl cation, $C_7H_7^+$. Suggest how this may be formed and the resultant structure.

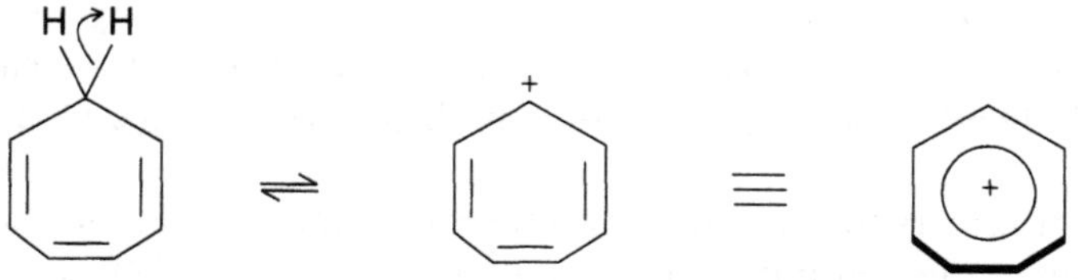

In the cycloheptatriene molecule there are three double bonds and hence six p orbitals; if a hydride anion is removed from the only methylene unit and that carbon is re-hybridised from sp^3 to sp^2 so as to create an empty p orbital, this completes the circuit of p orbitals around the cycloheptane ring. Then there is a continuous series of p orbitals that contain six π electrons, which gives rise to an aromatic system.

If a five-membered ring is fused with a seven-membered ring to form a 5/7 ring system, and if, in the larger ring there are three double bonds, while in the smaller ring there are two double bonds, suggest what, if any, redistribution of electrons may occur so as to form a dipole and the direction of any such dipole that might be formed.

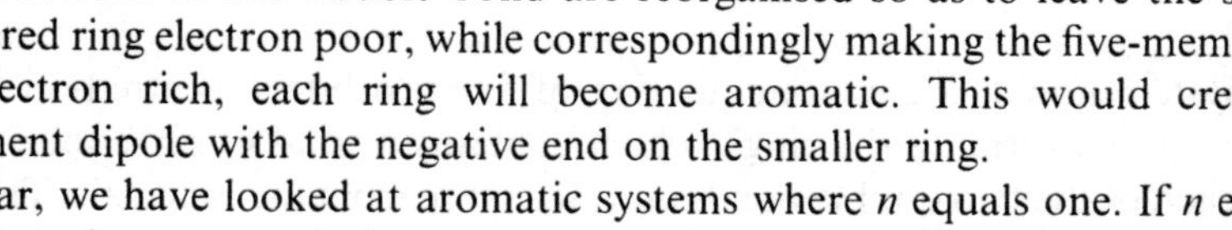

If the electrons in one double bond are reorganised so as to leave the seven-membered ring electron poor, while correspondingly making the five-membered ring electron rich, each ring will become aromatic. This would create a permanent dipole with the negative end on the smaller ring.

So far, we have looked at aromatic systems where n equals one. If n equals zero, then there can only be two π electrons. Suggest a system in which this arrangement may exist.

If you are having problems, start by thinking about the formation of the two aromatic ions illustrated above. In those cases, the aim was to obtain six π electrons in a continuous ring of p orbitals. When forming the anion, there were only four electrons initially in two double bonds and the extra two electrons were supplied by the removal of a hydrogen cation from a methylene unit. In the case of the cation, there were already six π electrons, but there was a need for an extra p orbital to complete the ring. This extra orbital had to be empty, and so a hydride anion was removed from a methylene unit.

When *n* equals zero, only two π electrons are required. It would seem logical to achieve this by removing a hydride anion from a methylene unit that is adjacent to a double bond containing the two π electrons required. Thus, draw the mechanism for the removal of a hydride anion from cyclopropene to give the cyclopropenyl cation, $C_3H_3^+$.

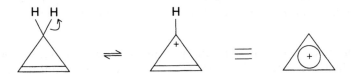

Here there are two π electrons in three p orbitals that form a continuous ring, and so the system is aromatic. Notice that, for the case of *n* equals zero, there is no aromatic system that is anionic in nature, because for this to happen there must be one fewer p orbitals than there are π electrons, which in this case would require only one p orbital, i.e. no ring.

If, instead of $(4n + 2)$π electrons in a continuous ring of p orbitals, there are merely $(4n)$π electrons, then the system is destabilised instead of being stabilised. This is called anti-aromaticity. Suggest an ion that might be formed from cyclopentadiene that would be anti-aromatic in nature.

In this cation, there are four π electrons in five p orbitals, and so this corresponds to *n* equals one. This results in the destabilisation of the ring, and so the formation of this ion is disfavoured.

Now suggest a neutral cyclic species in which *n* equals two.

In this case, there are eight π electrons in eight p orbitals that are contiguous. The system shows non-aromatic character, e.g. it is not flat and the single and double bonds are distinguished by being different lengths. If it were planar, it would be anti-aromatic, and this is one of the reasons why it is not planar.

5.7 Hydrogen bonding

In the hydrogen chloride molecule there is a permanent dipole. Draw this molecule showing the direction of this dipole.

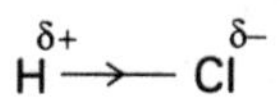

$$\overset{\delta+}{H} \longrightarrow \overset{\delta-}{Cl}$$

The chlorine has a small negative charge, while the hydrogen carries a corresponding positive charge. Suggest which atom has the higher charge density.

Each atom must carry the same charge, but the chlorine atom is much larger than the hydrogen atom. This means that the hydrogen atom has the higher charge density. In fact, because hydrogen is so small, the charge density is relatively high, so that it is capable of interacting with a negative charge. Suggest a source that is electron rich in the hydrogen chloride molecule.

The lone pairs of the chlorine atom are electron rich and so may become involved in an electrostatic interaction with the δ positive centre of the polarised hydrogen atom.
   This Coulombic interaction is called a hydrogen bond, and is a source of stabilisation in a variety of molecules, in particular in those that have an atom bearing a negative charge that is close to a polarised hydrogen atom. For example, such a configuration is present in *ortho*-hydroxybenzoic acid after the carboxylic acid group has been deprotonated. Draw the resultant anion.

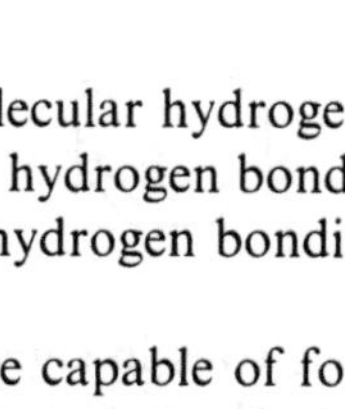

In this case there is an intramolecular hydrogen bond, while in the case of the hydrogen chlorine molecule the hydrogen bond was intermolecular. In the case of *ortho*-hydroxybenzoic acid, hydrogen bonding is also present in the neutral acid, but to a lesser extent.
   In dicarboxylic acids that are capable of forming intramolecular hydrogen bonds, it is observed that the first proton is removed more easily than would be predicted by just considering the inductive effect of the second carboxylic acid. Suggest the structure that *cis*-butenedioic acid adopts after one deprotonation.

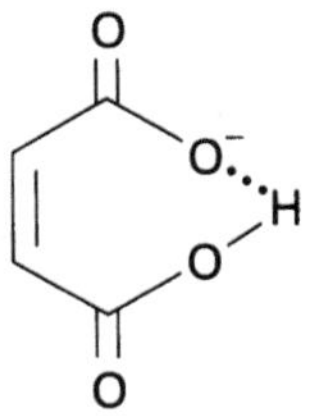

Here, a seven-membered ring is formed by the interaction between the deprotonated carboxylate group and the polarised hydrogen of the second carboxylic acid group. Suggest the form of hydrogen bonding that may exist in the mono-deprotonated anion of *trans*-butenedioic acid.

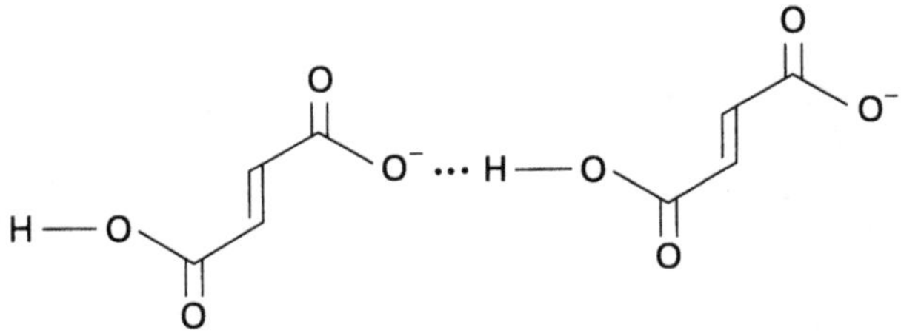

Note that in this case, only intermolecular hydrogen bonding is possible, because of the rigid nature of the central carbon/carbon double bond. The difference in the stability of the intermolecular versus the intra-molecular hydrogen bond is reflected in the difference in the acidities of the first hydrogen to be removed in each case. The *trans* isomer has a dissociation constant of 3.02, while for the *cis* isomer it is 1.92, i.e. the latter is a stronger acid.

Now suggest the relative ease of removal of the second carboxylic hydrogen from each of these two species.

In each case it is more difficult to remove the second hydrogen than it was to remove the first. This reflects the fact that it is harder to remove a positive proton from an anion than it is from a neutral species. Furthermore, the second hydrogen in the *trans* isomer is more acidic than the equivalent hydrogen in the *cis* isomer, because it is harder to remove the proton from the seven-membered ring than it is from an open chain form of hydrogen bonding.

5.8   Steric effects

The concept of intramolecular hydrogen bonding may be taken a little further by looking at some further steric considerations. For example, draw a possible structure of 2,4,6-trinitroaniline that shows the intramolecular bonding.

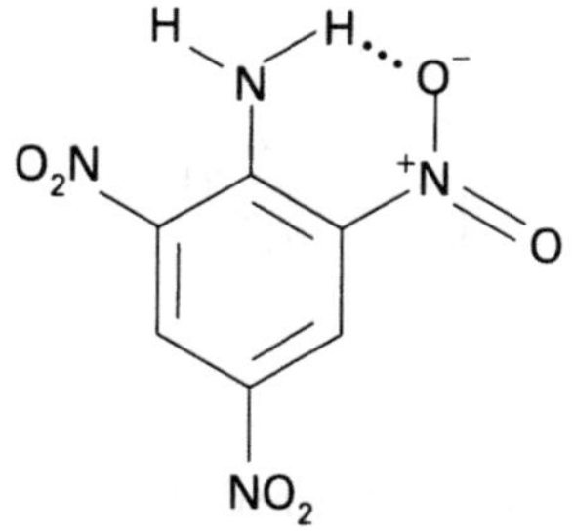

Now write down another canonical form of this molecule that still has the intramolecular hydrogen bonding present.

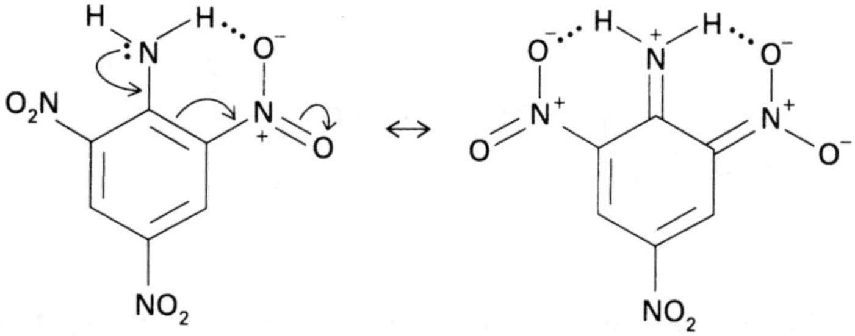

In this species, the lone pair of the nitrogen is involved in the aromatic ring and so can no longer act as a proton acceptor.

Now consider the related molecule 2,4,6-trinitro-*N*,*N*-dimethylaniline. Draw this molecule, and in particular indicate the hydrogen atoms on the methyl groups.

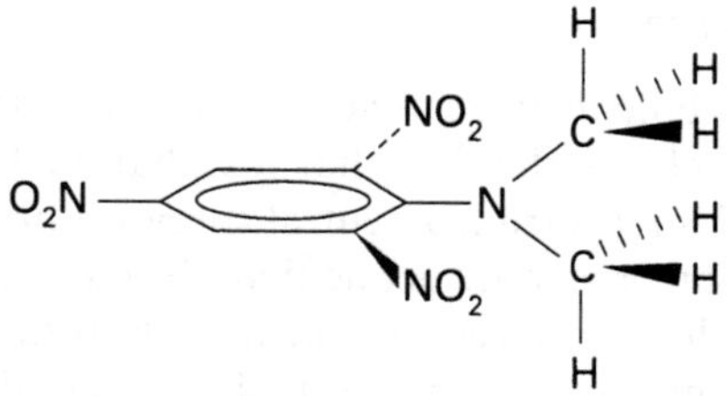

Once the hydrogen atoms are indicated on the methyl groups, it becomes abundantly clear that there is not enough room for the *N*,*N*-dimethylnitrogen group to lie co-planar with the benzene ring. Rather, this group must now be orthogonal to the ring. As a consequence, the lone pair on the nitrogen would also be orthogonal to the delocalised π system and so it can no longer interact with the aromatic ring. Bearing this in mind, suggest which of the two aniline derivatives is the more basic.

2,4,6-Trinitro-N,N-dimethylaniline is about 40 000 times more basic than 2,4,6-trinitroaniline, because of the impossibility of the lone pair on the nitrogen being delocalised into the aromatic ring, and so it is free to act as a base and react with a proton.

In the next chapter, some thermodynamic and kinetic effects will be considered. We have already seen an example of a thermodynamic effect, namely when the difference between intra- and inter-molecular hydrogen bonding was discussed. In the next chapter the reason why intramolecular hydrogen bonding is favoured over intermolecular hydrogen bonding will be examined.

5.9 Summary of stabilisation of charged species

The state that occurs mid-way along the bond-breaking/bond-forming route is called the transition state, and represents a species that is very short lived. In all transition states, the distribution of charge is different from that found in the original molecules. In order to facilitate the reaction, the charges that have been created by the reagents within the transition state may be stabilised.

The inductive effect results from the difference in the electronegativities of the atoms under consideration. It results in the placement of small, δ, charges on the atoms that were originally adjacent to the charged atom. It operates through the σ bonds, and its effect is felt for only about two bonds, after which it may, for most purposes, be ignored. Both carbonium cations and carbanions can be stabilised by this effect. For carbonium ions, the order of stability is: tertiary > secondary > primary. The order is similar for alkyl radicals. Furthermore, carbonium ions prefer to adopt a planar configuration, and thus they only exist at a bridgehead under extreme conditions. For carbanions, the order of stability is: primary > secondary > tertiary. In contrast to carbonium ions and alkyl radicals, the inductive effect cannot effectively stabilise the negative charge in a hydrocarbon carbanion. However, electronegative substituents, such as chlorine, which destabilise carbonium ions, act to stabilise carbanions. The effect of such substituents is cumulative.

The mesomeric effect operates through the π bond system. It is an efficacious method of transmitting whole units of charge, and so confers stability by effectively delocalising the charge over several atoms. Bredt's rule states that a double bond cannot be formed at a bridgehead. Hyperconjugation (or σ-conjugation) is a type of mesomeric stabilisation that involves the participation of the electron pair of a carbon/hydrogen σ bond. It accounts for the increased stability of carbonium ions that are adjacent to carbon/hydrogen bonds. Further, the mesomeric effect accounts for the stabilisation of a carbonium ion that has an adjacent heteroatom bearing a lone pair, e.g. as in the R_2C^+–OH or R_2C^+–NH_2 systems. Moreover, the mesomeric effect may be invoked to account for the stabilisation of a negative charge in such species as: RCO_2^-, $R_2C^{(-)}$–CO_2H, NCR_2C–CO_2^- and enolates, such as $R_2C^{(-)}$–$C(=O)R$. This mechanism underlies the keto/enol tautomerism. Tautomers are different from canonical structures in that the former represent real compounds that may be isolated, while the latter represent only hypothetical structures that help in the understanding of the electron distribution.

The greater the degree of s character in an orbital that contains a lone pair, the more stable it is. Thus, the order of stability of the following dicarbon anions is: $C_2H^- > C_2H_3^- > C_2H_5^-$. Conversely, increased s character in the unoccupied orbital decreases the stability of a carbonium ion. Thus, the order of stability of the following carbonium ions is: $C_2H_5^+ \gg C_2H_3^+ > C_2H^+$.

Third row elements may act as an electron sink, because they are not bound by the Octet rule. This is because such atoms have available d orbitals that allow the atoms to hold more than eight electrons in their valance shells.

Cyclic conjugated systems that have $(4n + 2)\pi$ electrons in a continuous closed loop of p orbitals are particularly stable. This is Hückel's rule, and the property is called aromaticity. When $n = 1$, examples include: the benzene molecule, C_6H_6; the cyclopentadienyl anion, $C_5H_5^-$; and the cycloheptatrienyl cation, $C_7H_7^+$. When $n = 0$, an example includes the cyclopropenyl cation, $C_3H_3^+$. If there are $(4n)\pi$ electrons in a contiguous ring of p orbitals, the system is destabilised, and is called anti-aromatic, e.g. $C_5H_5^+$. Cyclooctatetraene, C_8H_8, displays non-aromatic properties in that it is not planar and there are discrete single and double bonds.

The Coulombic interaction between a δ positive hydrogen atom and an electron rich atom such as a chlorine atom, is called a hydrogen bond. Such bonds may be intermolecular or intramolecular, e.g. in hydrogen chloride or deprotonated *ortho*-hydroxybenzoic acid respectively. Hydrogen bonding often affects the relative ease with which the hydrogen cations are removed from a poly-acid.

Steric factors, such as whether or not a group can align itself so that the lone pair can interact with another system, are often important. For example, they account for the increase in basicity of 2,4,6-trinitro-N,N-dimethylaniline over 2,4,6-trinitroaniline.

6

Thermodynamic and Kinetic Effects

6.1 Introduction

The interplay between thermodynamic and kinetic effects is very important, and as such it is possible to spend vast amounts of time and effort on it. This inevitably leads to complexity, which can be rather daunting and overbearing at first sight. However, thankfully, only a few basic ideas are actually needed to guide you through most of the situations that will arise at this level.

The principal guideline is that reversible reactions are governed by thermodynamic considerations, while irreversible reactions are controlled by kinetic factors. After we have looked at a few general principles, we will examine these two topics in a little more detail. Finally, in this chapter, we will make a few general comments on catalysts, with particular reference to acid/base catalysis.

In any general reaction between A and B to give C and D, the reagent A must physically approach the substrate B so that the electrons within each molecule may interact; this results in some bonds being broken and re-formed so as to produce the new molecules C and D. This process may be represented on a diagram called a reaction co-ordinate diagram.

On such a diagram, the ordinate, the y-axis, represents the total free energy of the system; the abscissa, the x-axis, represents how far the reaction has proceeded from the starting materials, A and B, to give the products, C and D. For a simple, one-step, reaction in which the products are more stable than the starting materials, the curve that represents the reaction pathway finishes lower than it started, i.e. the right hand side is lower than the left.

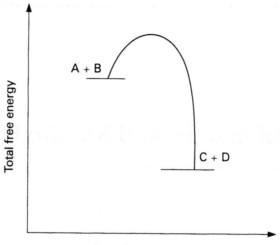

The peak of the curve represents the transition state between the starting materials and the products. The difference between the top of the peak and the starting materials represents the amount of energy that is needed to start the reaction, i.e. the activation energy. The difference between the starting level, which corresponds to the sum of the free energies of starting materials, and the finishing level, which corresponds to the sum of the free energies of the products, represents the overall change in free energy during the reaction; this is the change in the Gibbs free energy, ΔG. If the entropy factors, i.e. those factors that are concerned with the disorder within the system, are similar on each side of the reaction, then the free energy corresponds approximately to the enthalpy of the reaction, i.e. the difference between the bond energies and such like of the starting materials and the products. If the products have less energy stored in them than the starting materials, then energy will be released as the reaction proceeds. In such a case, the reaction is said to be exothermic. If energy is absorbed as the reaction proceeds, then the reaction is said to be endothermic. The less the energy that is contained within the species under consideration, then the lower down the ordinate axis will it be placed.

For a two-step reaction, the reaction co-ordinate diagram is as follows. In this case, there is an energy well, i.e. there is a peak either side of it. The species that is caught in the energy well is called an intermediate and may be isolated. The presence of the peaks on either side of the intermediate indicates that energy is required for both the forward or the reverse reaction to proceed from this intermediate species, i.e. there is an activation energy barrier that the intermediate must overcome before it may revert either to the starting materials or proceed to the final products. This is indicative of a real species, which can be isolated. This is in contrast to a transition state, which occupies the cusp of the

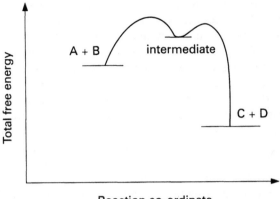

Reaction co-ordinate

curve, and so cannot be isolated, because it can spontaneously revert to either the starting materials or proceed to the products.

There are two simple rules of thumb that can be usefully employed, even though the second is not strictly true, but rather serves as a guide. The first states that the larger the energy difference between the starting materials and the products, the further the equilibrium will lie to the side that has the lower energy. The second states that the greater the difference in energies between the starting materials and the products, then the quicker equilibrium will be reached, because the activation energy barrier tends to be smaller.

We will now look at some thermodynamic considerations in a little more detail.

6.2 Thermodynamic effects

6.2.1 *General considerations*

For a reaction to proceed spontaneously the change in Gibbs free energy, ΔG, must be negative. Deduce whether this means that the free energy of the products must be lower or higher than that of the starting materials for the reaction to proceed in the forward direction.

For the change in the free energy to be negative, the free energy of the products must be lower than that of the starting materials.

The change in the Gibbs free energy comprises two parts, namely, the change in the enthalpy, ΔH, and the change in the entropy, ΔS, of the system, and these terms are related by the following equation, where T is the temperature in Kelvin:

$$\Delta G = \Delta H - T\Delta S$$

The enthalpy refers to the bond energies and physical state of the components: fewer and weaker bonds increase the enthalpy of the system, as does converting a solid to a liquid. Thus, what is favoured enthalpically is many strong bonds in a solid compound. This arises because, slightly confusingly, a strong bond has a more negative bond energy than a weaker bond, i.e. more energy is released upon its formation. This means that it is positioned lower down the ordinate on a reaction co-ordinate diagram, which reflects its greater stability.

The entropy refers to the amount of disorder: the greater the disorder of the species present in the system, the higher the entropy of that system. Thus, what is favoured entropically is the gaseous state over the solid state. If the starting materials and the products have approximately the same amount of disorder, then the reaction is controlled by the changes in the enthalpy of the system, which broadly follows the changes in the bond energies associated with the starting materials and those of the products.

There is another complicating factor, namely the activation barrier that must be surmounted before a reaction can proceed. Even if a reaction is exothermic, i.e. the products are positioned lower down the ordinate axis than the starting materials, in order to produce them the starting materials must be provided with the appropriate activation energy. Clearly at room temperature, there is a certain amount of thermal energy present. If this is sufficient for the reaction under consideration, then the reaction occurs spontaneously at room temperature. However, more commonly, some extra energy is needed in order to overcome this barrier. This is the reason why many potentially exothermic reactions, such as the combustion of coal in air, do not occur immediately, but rather require a little push, e.g. a fire lighter. Yet once going, the reaction produces sufficient excess energy so that unreacted components may undergo the reaction.

We will now examine how the enthalpy and entropy effects can interact in a simple reaction, before we examine each in a little more detail. Write down a general chemical equation for the hydrolysis of alkyl halides. We will look at the entropy effects first, because these are often simpler.

$$RX + OH^- \rightleftharpoons ROH + X^-$$

To a first approximation, the same sorts of species exist in the same relative ratios on each side of this equation, i.e. there is one anion and one neutral molecule on each side of the equation. Thus, it may be assumed that the entropy difference between the starting materials and the products is small, and so may be ignored.

If we now look at the alkyl halide in a little more detail, suggest what will be the major differences between an alkyl chloride and an alkyl iodide.

There are two major differences. The first is that in the alkyl chloride the C–X bond is more polarised than in the alkyl iodide; the second is that the C–Cl bond is much stronger than the C–I bond, which means that the C–Cl bond has a more negative bond energy than the C–I bond.

Let us assume initially that the degree of polarisation is the dominant effect upon the rate of hydrolysis. If this were the case, would an alkyl chloride hydrolyse faster than an alkyl iodide, or vice versa?

With this assumption that the rate of hydrolysis is determined by the degree of polarisation of the carbon/halogen bond, then the alkyl chloride would be hydrolysed faster than the alkyl iodide. This hypothesis may be justified by saying that the more highly polarised the bond is initially, the closer it is to breaking heterolytically, and so the more readily it will actually break into the resultant ions.

In contrast, let us assume it is the strength of the C–X bond that is the dominant factor in determining the rate of hydrolysis of the alkyl halide. In this alternative hypothesis, in which we are considering the enthalpy effects, what would be the relative rates of hydrolysis?

Now the opposite result is achieved, namely, that the compound with the weaker bond would hydrolyse faster, i.e. the alkyl iodide. This result comes from the rule of thumb that was given earlier: namely, that the larger the difference between the energies of the starting material and the product, the faster the rate, because the activation energy tends to be lower. In this case, the alkyl iodide is less stable than the alkyl chloride, because it has the weaker C–X bond. Thus, the iodo compound would be higher up the ordinate axis on the reaction co-ordinate diagram. In each case, though, the product is the same, i.e. the alcohol. Thus, there is a larger energy difference for reaction with the iodo compound, and so that species reacts faster.

In practice, it is observed that the alkyl iodide hydrolyses faster, and so it may be assumed that the strength of the C–X bond has more influence on the rate than does the degree of polarisation of the bond.

This highlights the important point that even though a particular hypothesis can be justified, as was the case when it was suggested that the rate of hydrolysis was determined by the degree of polarisation, this does not mean that either the justification or the hypothesis is correct. One must always test a hypothesis against the observed facts. This illustrates the pre-eminence of experimental results.

6.2.2 *Enthalpy effects*

Let us now study a few more examples in which enthalpy has a large effect. By examining the difference in the bond strengths of related isomers, it may be possible to identify which isomer is more stable. For example, a ketone such as propanone may form an enol under basic conditions. The structures of both isomers are shown below.

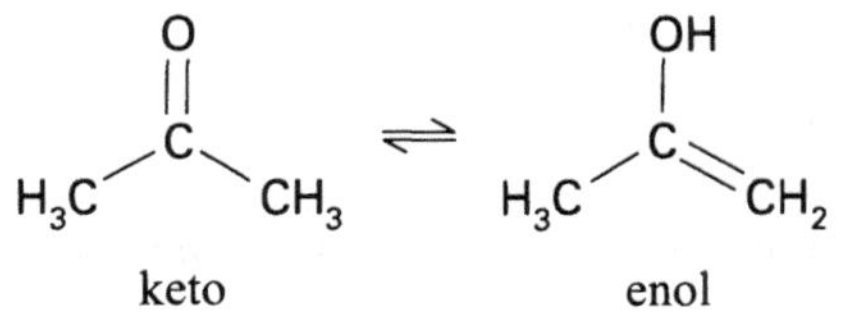

keto   enol

This isomeric relationship is called tautomerism, which in this particular case is an example of the keto/enol variety. Now list those bonds that are different between the two compounds.

In the ketone there is a carbon/oxygen double bond, three C–H bonds from an sp^3 carbon and one carbon/carbon single bond. In the enol tautomer, in contrast, there are two C–H bonds from an sp^2 carbon, a carbon/carbon double bond, a carbon/oxygen single bond and an O–H bond. This may be simplified by assuming that a C–H bond from an sp^3 carbon is about as stable as a C–H bond from an sp^2 carbon. Now one is looking at the difference between one carbon/oxygen double bond plus one C–C bond and plus one C–H bond for the ketone, and one carbon/carbon double bond plus one C–O bond and plus one O–H bond for the enol tautomer.

Every type of bond requires a different amount of energy to separate the component parts. This value is called the bond energy and it is measured in $kJ\,mol^{-1}$. Using the values given below, calculate the difference in the bond energies of the keto and enol isomer of the carbonyl compound.

O–H	$-460\ kJ\ mol^{-1}$
C–H	$-420\ kJ\ mol^{-1}$
C=O	$-760\ kJ\ mol^{-1}$
C=C	$-630\ kJ\ mol^{-1}$
C–O	$-380\ kJ\ mol^{-1}$
C–C	$-360\ kJ\ mol^{-1}$

The figures for the ketone are $(-760\,kJ\,mol^{-1}) + (-360\,kJ\,mol^{-1}) + (-420\,kJ\,mol^{-1})$, which equals $-1540\,kJ\,mol^{-1}$; while for the enol tautomer the figures are $(-630\,kJ\,mol^{-1}) + (-380\,kJ\,mol^{-1}) + (-460\,kJ\,mol^{-1})$, which equals $-1470\,kJ\,mol^{-1}$. Hence, the enol is less stable by about $70\,kJ\,mol^{-1}$. Note, however, that in the enol tautomer there is also the possibility of hydrogen bonding, which could contribute some enthalpic stabilisation, but it may also increase the order of the system and so decrease the entropy, which would be a destabilising factor; thus, on balance, the effect of the hydrogen bonding can be ignored to a first approximation.

Another common tautomeric system involves the nitro/aci isomeric compounds, and is illustrated below.

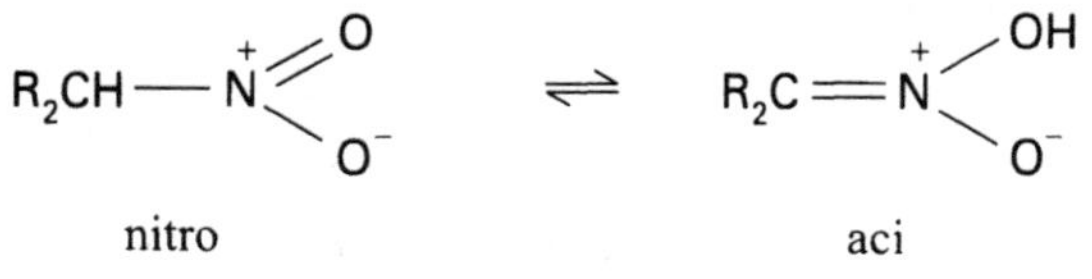

nitro   aci

Given the following values for the relevant bond energies, calculate the relative stability of the nitro/aci tautomers:

C–H	-420 kJ mol^{-1}
O–H	-460 kJ mol^{-1}
C–N	-315 kJ mol^{-1}
C=N	-610 kJ mol^{-1}
N–O	-210 kJ mol^{-1}
N=O	-630 kJ mol^{-1}

The figures for the nitro tautomer are $(-315 \text{ kJ mol}^{-1}) + (-630 \text{ kJ mol}^{-1}) + (-420 \text{ kJ mol}^{-1})$, which equals $-1365 \text{ kJ mol}^{-1}$, while for the aci isomer they are $(-610 \text{ kJ mol}^{-1}) + (-210 \text{ kJ mol}^{-1}) + (-460 \text{ kJ mol}^{-1})$, which gives $-1280 \text{ kJ mol}^{-1}$. So in this case, there is a larger difference in the stability of the two tautomers than in the case of the keto/enol tautomerism. In each of these two tautomeric systems, both isomers actually exist, and may be isolated as identifiable compounds. However, the difference in stability is reflected by the fact that, at equilibrium, the nitro/aci ratio is much greater than the keto/enol ratio.

Lastly, it is interesting to look at the relative bond energies for the single and multiple bonds of carbon with another carbon atom, oxygen atom or nitrogen atom. The figures are:

C–C	-360 kJ mol^{-1}
C=C	-630 kJ mol^{-1}
C≡C	-840 kJ mol^{-1}
C–N	-315 kJ mol^{-1}
C=N	-610 kJ mol^{-1}
C≡N	-860 kJ mol^{-1}
C–O	-380 kJ mol^{-1}
C=O	-760 kJ mol^{-1}

It may be seen that, all other things being equal in the carbon/oxygen system, there is little difference between a carbon atom forming one carbon/oxygen double bond or two carbon/oxygen single bonds. However, for the carbon/ nitrogen and the carbon/carbon systems the situation is different, i.e. a corresponding calculation does not produce such a balanced result. By using the above numbers, suggest what are the relative stabilities in the carbon/nitrogen system.

There is little difference between the double and single bond systems. However, if a triple carbon/nitrogen bond is being broken, this will release about 250 kJ mol^{-1}, but a further 295 kJ mol^{-1} would be released if the reaction proceeded to the carbon/nitrogen single bond stage. Using the rule of thumb mentioned above, namely that the greater the difference in energies between the starting materials and the products the faster the reaction, then the reaction from the double bond to the single bonds should be faster than that from the triple to the double bond. Thus, one can conclude that if a triple carbon/nitrogen bond is broken, then it is unlikely that the reaction would stop at the double bond stage, but rather that it would continue to the compound containing single carbon/nitrogen bonds.

By using the values for the carbon/carbon system, suggest the relative stabilities of those compounds.

Here it is apparent that both the multiple bond systems are less stable than the single bond configuration. This is reflected in the fact that addition reactions to carbon/carbon multiple bond systems are generally very facile.

6.2.3 *Entropy effects*

Now we shall turn our attention to the other thermodynamic parameter, namely ΔS. Entropy is a measure of the disorder of the system: the greater the disorder, the more the system is favoured. In other words, chaos is the favoured state of things, and to introduce order requires energy. For the three phases of matter, suggest which is the most disordered and which is the least.

gas $\gg$ liquid $>$ solid

If the phase of the products is different from that of the starting materials, then there will be a large entropy factor that must be taken into consideration. For example, if a reaction produces three molecules from only two molecules of starting materials then the disorder of the system is greatly increased. Similarly, if the products are gaseous, when the reagents were solids or liquids, then the entropy would be increased.

A different situation occurs where a charged species is formed from an uncharged precursor in a polar solvent. Suggest whether the entropy would increase or decrease in such a situation. Try to justify your answer.

The entropy would decrease because the charged species would order the solvent molecules around it, either with hydrogen bonding if the solvent is protic, i.e. had acidic protons, or otherwise by electrostatic interactions.

Further, the accompanying change in entropy may be used to account for the fact that intramolecular hydrogen bonding is favoured over intermolecular hydrogen bonding. Suggest an explanation for this.

In an intramolecular hydrogen bond only the motions internal to the molecule are being hindered. However, in an intermolecular hydrogen bond, order is imposed on other molecules, and this greatly reduces the entropy of the system. Hence, intermolecular hydrogen bonding is disfavoured when compared to intramolecular hydrogen bonding.

From the equation that was given at the beginning of this section, it may be seen that temperature has an effect on the entropy factor. As the temperature is raised, is the effect of entropy greater or smaller?

The change in the Gibbs free energy is directly related to the product of the temperature and entropy, and hence as the temperature is raised the effect of entropy becomes larger. This may be used to advantage, as a reaction may be controlled by altering the temperature at which it is performed, and so increasing or decreasing the influence of the change of entropy.

In a reaction mixture where every step is reversible, so long as sufficient time is allowed, the final ratio of all the possible products will be determined by the relative thermodynamic stabilities of each of the compounds. Such a reaction is said to be under thermodynamic control. However, if for any reason a step is irreversible, or is in essence irreversible, because the reverse reaction is very slow with respect to the length of time that is being allowed for the reaction, then the ratio of the possible products is no longer determined by thermodynamic considerations alone, but also by the speed at which each reaction takes place, i.e. the kinetic factors. A reaction in which the product distribution is determined by the rate of formation of the various products is said to be under kinetic control. We shall now examine some of these factors.

6.3 Kinetic effects

6.3.1 Forcing reactions

As with thermodynamic factors, it is possible to spend a great deal of time and effort in understanding the kinetic factors that affect a reaction. In particular, it is possible to use some very elaborate mathematical models, which are very elegant, but tend to be rather off-putting to the uninitiated. However, there are several very useful guidelines that may be used to advantage without any need to have a detailed understanding of the underlying mathematics.

In many reactions, it is possible for the products to react in such a manner that the starting materials can be re-formed. This is the case when a general ester is hydrolysed under acid conditions, e.g. the hydrolysis of ethyl ethanoate:

$$CH_3CO_2CH_2CH_3 + H_2O + H_3O^+ \rightleftharpoons CH_3CO_2H + CH_3CH_2OH + H_3O^+$$

Once the carboxylic acid and the alcohol are formed they are capable of re-reacting so as to re-form the ester. Now consider the situation when the ethyl ester group is substituted by a *t*-butyl ester group. Write down the equation for the hydrolysis of this new ester under acid conditions.

$$CH_3CO_2C(CH_3)_3 + H_2O + H_3O^+ \rightleftharpoons CH_3CO_2H + (CH_3)_3COH + H_3O^+$$

This looks like it would suffer from the same problem. However, under certain acidic conditions the *t*-butyl carbonium ion is produced instead of the alcohol.

This carbonium ion readily loses a proton. Write down the equation for this deprotonation.

$$(CH_3)_3C^+ \rightleftharpoons H^+ + (CH_3)_2C=CH_2$$

The product is 2-methylpropylene, which is a gas under normal laboratory conditions. Suggest what will happen to this product if the reaction vessel is not sealed.

Assuming that the reaction is being carried out in a liquid solvent, then the gaseous product, namely 2-methylpropylene, will bubble off from the mixture. Suggest what will be the consequence of the loss of the 2-methylpropylene.

If the 2-methylpropylene is no longer in the reaction mixture then it cannot under any circumstances react with the other product, i.e. the carboxylic acid, to re-form the starting materials. This means that the reaction will be driven over to form more of the product. Suggest a thermodynamic reason why the equilibrium lies over to the product side.

There is a large increase in the entropy of the system due to the formation of the gaseous product, and this favourably affects the position of the equilibrium towards the product. So there are two good reasons that both favour the formation of the desired product when a gaseous co-product is formed in a reaction.

This principle is also illustrated in the following example. If 1-aminobenzoic acid is treated with sodium nitrite, the diazo intermediate may be formed. This compound may then lose a proton from the carboxylic acid group to form an ion that contains both a positive and a negative charge. The skeletal formula of this compound is shown below.

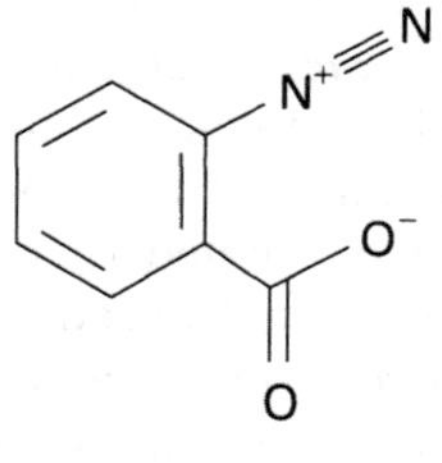

Suggest a mechanism whereby a molecule of nitrogen and a molecule of carbon dioxide are formed.

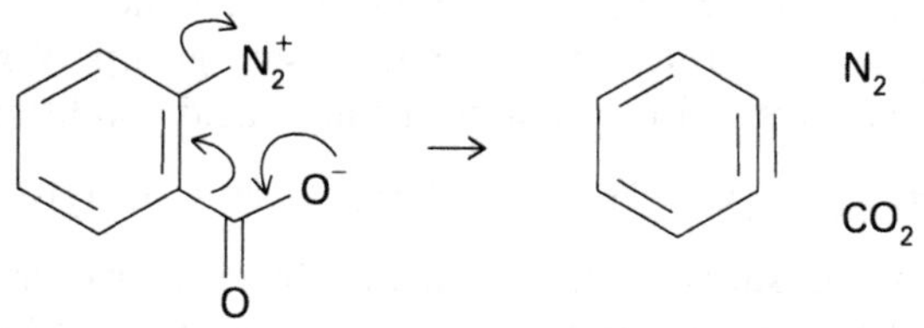

Here, the electron rich carboxylate group donates a pair of electrons to the aromatic ring, while the electron poor diazo group reclaims a pair. This leaves three neutral molecules, two of which are gases that bubble off from the reaction mixture, and one hydrocarbon with the molecular formula C_6H_4, which contains a triple bond in place of one of the double bonds of the original aromatic ring. This molecule is called benzyne, and is rather unstable and very reactive. The fact that two gaseous molecules were also formed in this reaction helps drives the reaction over to produce this unusual product.

6.3.2 *Ring formation*

Another useful concept is that certain sizes of ring are more easily formed than others. Let us first consider the formation of a six-membered ring. Draw out a six-carbon aliphatic chain in such a manner that the ends are close together. Remember that in an aliphatic carbon the hybridisation is sp^3, and so the CCC bond angle is about 109°.

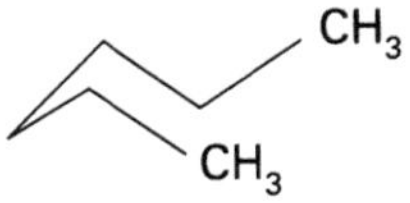

If the hexane chain adopts a shape that is similar to the chair conformation of cyclohexane, it is observed that the terminal carbons are within a normal single bond length of each other. Thus, they are close enough to react with each other to form a bond and so close the ring.

Now let us look at the formation of a five-membered ring from an aliphatic pentane molecule. Write down a conformation that will allow the terminal carbons to be close to each other.

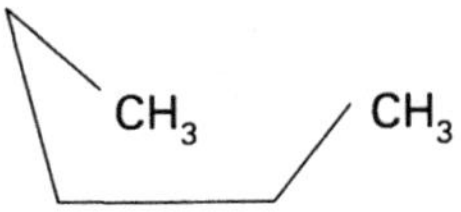

Here, if the carbon chain adopts a similar shape to that of cyclopentane, then the terminal carbons would be close enough to react together.

Now let us look at the formation of a four-membered ring from an aliphatic butane chain. Suggest what conformer may be adopted in order to minimise the distance between the terminal carbons.

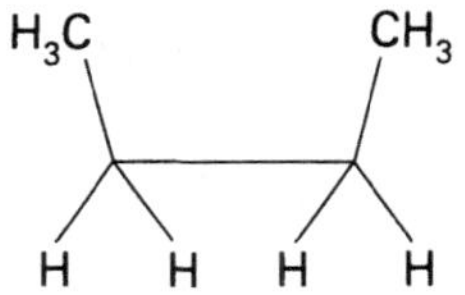

In this case, there is no conformation that brings the terminal carbons close together without deforming the natural bond angles. Suggest what will be the consequence of this for the ease of formation of a four-membered ring by the closure of a butyl chain.

As the ends of a four-membered chain cannot get close together it means that they cannot react together easily. This means that four-membered rings are hard to form by this method, i.e. the interaction of appropriate groups on the ends of the chain.

The tetrahedral sp^3 hybridisation is very common in second row elements. However, in the third row elements there is the possibility of using one or more of the d sub-orbitals that are now available. We used the electron pair repulsion theory in Chapter 4 to predict the shape of molecules of general formula AB_n, where n varied from two to five. If n equals five or six, what do you think will be the shape of such molecules?

When n equals five, as we have already seen, the shape adopted is a trigonal bipyramid; when n equals six, it is a square bipyramid, or an octahedron. What is the angle between an axial and an equatorial bond in either of these geometries?

In both cases the axial/equatorial angle is $90°$. Now suggest two third row elements, one of which can have a valency of five and the other can have a valency of six.

Phosphorus may have a valency of five, while sulphur may have a valency of six. When they do so, they adopt a trigonal bipyramidal or an octahedral geometry respectively.

So, now we have an element that may have a bond angle that is naturally $90°$, and so would not be under any strain if it formed part of a four-membered ring. This is demonstrated in the Wittig reaction in which an intermediate species has a four-membered ring containing a phosphorus/oxygen bond along one side. So, one way around the problem of forming a four-membered ring is to use a third row element in one corner of the ring.

Let us now look at the formation of a three-membered ring. Draw a conformation in which the terminal carbons of a propane molecule are close together.

$$CH_3 \qquad CH_3$$

Here, the terminal carbons are approximately within one bond length of each other and so may in theory react to close the ring. However, what would be the internal bond angle in such a compound?

The internal bond angle would be 60°, which is a lot less than the normal angle of 109° for an sp³ hybridised carbon atom. What does this suggest about the stability of such a ring?

Even though three-membered rings may be fairly easily formed, they do have a tendency to re-open, because of the great strain imposed on the bonds being forced to adopt such a small bond angle. This is revealed in their chemistry; for example, under normal conditions is it not possible to add hydrogen directly to a carbon/carbon single bond in an alicyclic compound like cyclohexane. In contrast, in cyclopropane the ring readily opens up to form the aliphatic compound.

6.3.3 *Abstraction of protons*

Another general guideline involves the reactivity of hydrogens attached to carbon atoms in environments that have different accessibility. The first step in many organic reactions often involves the abstraction of a proton from the starting material so as to form a carbanion. This, of course, requires that there is an available hydrogen to be attacked. Suggest which type of hydrogens in the butane molecule would most likely be attacked.

On a purely statistical basis there are more hydrogens on the terminal methyl groups, i.e. six, than there are on the methylene units, which only have four in total. So, if all other things are equal, the hydrogens on the terminal methyl groups would be attacked in preference to those on the methylene carbons. Note also that, in this case, the effect is further enhanced because the resultant anion would be more stable if the charge were on the terminal carbon rather than on one of the middle carbons.

The idea of accessibility may be taken further than a mere statistical effect. If there are two hydrogens that are equal in all respects except that one of them is partly hidden by some other part of the molecule, suggest what would be the difference in reactivity of these two hydrogens when treated with either a base that was small in size or one that was large.

If the base is very small, such as the hydroxide anion, then it will be able to approach, and so react with, a hydrogen in almost any position, even if it is partly obscured by another part of the molecule. The converse is also true: thus, if the base is very large, i.e. a hindered base which has a large steric demand, then such a base would be able to react only with hydrogens that are accessible. Hence, a degree of selectivity may be introduced by choosing the size of the base that will be used to abstract the proton.

So far, we have only considered a single step reaction sequence. In many organic reactions there are many individual steps between the starting materials and the final products, with many separate intermediates. The principles that we have discussed with respect to a single step apply equally to each individual step of a multi-step reaction pathway. The overall rate of such a pathway is determined by the rate of the slowest step, which is called the rate determining (or rate limiting) step.

We will now examine briefly some aspects of catalysis, and see how that can affect the rate of a reaction by affecting the rate determining step.

6.4 Catalysis

6.4.1 *General considerations*

A catalyst is a component of the reaction mixture that affects the rate of the reaction while not being consumed itself. It may speed up or slow down the rate of the reaction; however, normally only those catalysts that increase the rate of a reaction are considered to be useful. Even though a catalyst is not consumed in the reaction, it may become involved, and change its oxidation state for example, so long as by the end of the reaction any changes that have occurred to it have been reversed so that there is no net change in its chemical composition. The physical state of the catalyst may however be permanently changed.

If a catalyst has a different physical state from the reaction mixture that it is catalysing, it is said to be a heterogeneous catalyst. An example of this would be a metal catalyst, such as platinum, which catalyses the hydrogenation of gaseous alkenes. If, however, the catalyst has the same physical state, it is said to be a homogeneous catalyst, for example, the solvated proton in an acid catalysed reaction that is occurring in solution.

Many catalysts are only required in trace amounts in order to affect the rate of the reaction. Some, however, are required in similar quantities to the reagents, in which case they are called stoichiometric catalysts. This often indicates that the catalyst is forming a complex with one of the reagents, and it is this complex that is the attacking species in the reaction. An example of this is the Lewis acid catalysis of the Friedel–Crafts reaction.

In thermodynamic terms what controls the rate of a reaction?

The rate of a reaction is related to the activation energy barrier that exists between the starting materials and the transition state. On the reaction co-ordinate diagram, this is represented by the difference between the starting materials and the peak of the reaction co-ordinate curve. If the activation barrier is large, how will this affect the rate of the reaction?

The larger the activation energy barrier, the slower the rate of that reaction. Suggest how a catalyst might affect the rate of a reaction.

The simplest way a catalyst might affect the rate is by lowering the activation energy barrier, and this will result in a faster rate of reaction. This is usually achieved by the catalyst providing an alternative reaction pathway that involves the catalyst. The activation energy for this alternative route must obviously be smaller than the normal route in order to increase the overall rate of the reaction. This alternative route will pass through a different transition state. However, the final products of a catalysed reaction are the same as in the uncatalysed reaction. Bearing this in mind, what effect has a catalyst on the equilibrium position of the reaction?

The equilibrium position of any given reaction is determined by the difference in energy between the starting materials and the products. The catalyst does not affect the stability of the starting materials or products involved in the reaction, and so cannot have any effect on the equilibrium position of a reaction. This means that a catalyst cannot be used to make a reaction proceed that would otherwise be thermodynamically impossible. A catalyst may only alter the rate of a reaction that is otherwise possible.

We shall now look at a particular type of homogeneous catalyst that is very important in organic chemistry, namely acid/base catalysis.

6.4.2 *Acid/base catalysis*

The simplest example of acid/base catalysis is when the rate of a reaction is found to be directly proportional to the conjugate acid of the solvent, $[SH^+]$. If water is the solvent, then the rate is proportional to the concentration of the hydroxonium ions, $[H_3O^+]$. This is called specific acid catalysis, because the rate is affected only by the concentration of the specific acid species, SH^+. The rate is unaffected by the concentration of any other acid, i.e. any other proton donor, which is in the reaction solution, so long as the concentration of SH^+ ions remains constant. Specific acid catalysis is characteristic of reactions where there is a rapid, reversible protonation of the substrate before the rate determining step. This type of catalysis occurs in the acid catalysed hydrolysis of acetals, $R_2C(OR)_2$.

If the rate of a reaction is found to be dependent not only on the concentration of the conjugated acid of the solvent, but also on the concentration of any other acid species that might be present, then the reaction is said to be general acid catalysed. In such a case, the catalysis proceeds by way of proton donors in general and is not just limited to (in the case of water as the solvent) hydroxonium ions. General acid catalysis is characteristic of reactions in which the protonation of the substrate is the rate limiting step. An example is the acid catalysed hydrolysis of ortho-esters, $RC(OR)_3$.

The same distinctions may be drawn for base catalysed reactions. In the reverse of the aldol condensation reaction, the rate of the reaction is found to be dependent only upon the concentration of hydroxide ions, and so this is an example of specific base catalysis. This is indicative of rapid reversible deprotonation of the substrate before the rate limiting step.

In general base catalysis, any base may affect the rate of the reaction. Again by analogy with the case of general acid catalysis, this type of reaction is characterised by the fact that there is a slow deprotonation of the substrate that precedes the fast subsequent reactions to the products. An example of this type of reaction is the base catalysed bromination of a ketone.

In the next chapter, we will examine acid/base characteristics in more detail.

6.5 Summary of thermodynamic and kinetic effects

At this level, a simple approach will suffice for most occasions. Thus, in brief, reversible reactions are governed by thermodynamic considerations, while irreversible reactions are controlled by kinetic factors.

The course of a reaction may be followed by plotting the change in the Gibbs free energy of the system against how far the reaction has proceeded. Such a plot is called a reaction co-ordinate diagram. In a simple, one-step, reaction, the peak represents the transition state between the starting materials and the products. The difference between the starting point and the peak is the activation energy, and the difference between the starting point and the finishing point represents the energy that is released or absorbed by the reaction, i.e. whether the reaction is exothermic or endothermic. This difference is the change in the Gibbs free energy, ΔG, of the reaction. In a two-step reaction, an intermediate may be formed, which occupies an energy well, and which has activation energy barriers to both the forward and backward reactions.

Generally, the larger the energy difference between the starting materials and the products, the further the equilibrium will lie to the side that has the lower energy. Also, the greater the difference in energies, the faster equilibrium is reached, because the activation energy barrier tends to be smaller.

For a reaction to proceed spontaneously the change in Gibbs free energy, ΔG, must be negative, assuming that there is sufficient energy in the system to overcome the activation barrier. The change in Gibbs free energy is related to the change in the enthalpy, ΔH, and the change in the entropy, ΔS, of the system by the equation: $\Delta G = \Delta H - T\Delta S$. The enthalpy refers to the bond energies and such like, while the entropy refers to the amount of disorder.

Slightly confusingly, a strong bond has a more negative bond energy than a weaker bond, i.e. more energy is released upon its formation. By examining the bond energies present in related isomers, it is possible to determine which is the more stable, e.g. for the keto/enol tautomers, it is found that the ketone is more stable than the enol by about $70 \, kJ \, mol^{-1}$. Further, the relative stabilities of the C–C, C=C and C≡C; C–N, C=N and C≡N; and C–O and C=O systems can be deduced by considering the energy differences between each stage. In the carbon/oxygen system, there is no great preference for either the single or double bonded configuration; while in the carbon/nitrogen system, it is unlikely that if a triple bond were to be broken it would stop at the double bond;

rather, it would proceed to the single bonded configuration. The carbon/carbon system has a tendency to go from multiple bonds to single bonds.

The disorder of a system is increased if the system produces gaseous products, when the reagents were solids or liquids, or the number of components increases. Conversely, the production of a charged species in a polar solvent reduces the entropy, because it imposes order on the solvent molecules. The effect of a change in entropy is enhanced by raising the temperature. This can be used to control which particular reaction is favoured.

A reaction can be forced to completion by ensuring that the products cannot react so as to re-form the starting materials, e.g. the use of a *t*-butyl group as a leaving group, because it readily forms the gaseous product 2-methylpropylene, which can escape from the reaction. Similarly, an unfavourable, but desired, product can be formed if the co-products are particularly favourable, e.g. the formation of benzyne from 1-aminobenzoic acid, which is facilitated by the simultaneous production of the gaseous co-products, nitrogen and carbon dioxide.

Five- and six-membered carbon rings are easy to form, while four-membered rings are more difficult. This is because the terminal atoms of a four-unit carbon chain cannot readily get close enough to each other to allow them to interact to form the ring, while in the other cases the terminal atoms can readily get close enough. This is a consequence of the stereochemical constraints that flow from the geometry of the sp^3 hybridisation. Accordingly, by invoking a different geometry that may be found, for example, in third row elements, it becomes easier to form a four-membered ring, e.g. in the Wittig reaction. Three-membered rings are easy to form, because the terminal carbons are in close proximity. However, such rings are also easy to break, because of the strain imposed by the acute carbon/carbon bond angle.

Hydrogens that are accessible are likely to be abstracted first. Note also any statistical weighting. Selectivity can be introduced by using a hindered base, which has a large steric demand, so that it is only able to abstract readily approachable hydrogens.

The overall rate of such a multi-step reaction is determined by the rate of the slowest step, which is called the rate determining (or rate limiting) step.

A catalyst is a component of the reaction mixture that affects the rate of the reaction, while not being consumed itself. A catalyst does not affect the equilibrium position of a reaction. It may be heterogeneous or homogeneous, depending on whether it is in a different or similar physical state to the reaction mixture. Catalysts can be present in trace or stoichiometric amounts. A catalyst increases the rate of a reaction by lowering the activation energy barrier of the main reaction.

Specific acid catalysis is where the rate is affected only by the concentration of the conjugated acid of the solvent, e.g. H_3O^+ ions in aqueous reactions. An example would be the acid catalysed hydrolysis of acetals, $R_2C(OR)_2$. It is characteristic of reactions in which there is a rapid, reversible protonation of the substrate before the rate determining step. In a general acid catalysed reaction, the rate is affected by the concentration of any acid species that is present, e.g. the acid catalysed hydrolysis of ortho-esters, $RC(OR)_3$. In such a case, the catalysis proceeds by way of proton donors in general and is not just limited to conjugated acid of the solvent. It is characteristic of reactions in which the protonation of the substrate is the rate limiting step.

The same distinctions may be drawn for base catalysed reactions. Thus, for example, the reverse of the aldol condensation reaction is dependent only on the concentration of hydroxide ions, i.e. it is a specific base catalysed reaction; while the base catalysed bromination of a ketone is an example of a general base catalysed reaction.

Acid/Base Characteristics

7.1 Introduction

All the properties that have been discussed so far, for example the shape of a molecule, have been inherent in the molecule in question. This means that they have not required the presence of another molecule for those properties to be revealed. However, the acid/base properties of a molecule do depend on the presence of other molecules in the reaction mixture. For example, nitric acid is usually considered to be an acid, but this is because we normally encounter nitric acid when it is dissolved in water. When nitric acid is dissolved in concentrated sulphuric acid it acts as a base. So, to display the acid/base properties of a compound, one needs the presence of a molecule with the opposite property.

Many mechanisms in organic chemistry start with an acid/base reaction. This may be just a simple Brönsted–Lowry protonation of a hydroxyl group, which results in the activation of a C–OH bond; or it may be a Lewis acid/base reaction as, for example, when aluminium trichloride complexes with a halogenoalkane in the first step of the Friedel–Crafts reaction. In each case, the initial intermediate usually reacts further and leads to the desired product. In inorganic chemistry, the acid/base reaction may be all that is of interest, e.g. the treatment of a carbonate with an acid to liberate carbon dioxide. However, it is unusual in organic chemistry for the acid/base reaction to be an end in itself. It is for this reason that acid/base characteristics are normally considered as a property of the molecule, similar to the nucleophilic and electrophilic properties to which they are closely related, rather than as a fundamental reaction type as is the case in inorganic chemistry.

We shall now look at some different definitions as to what constitutes an acid or a base, and then investigate some of the reactions that reveal these properties.

7.2 Definitions of acids and bases

The first thing is to define what we mean by an acid or a base. This is not straightforward as there are several definitions, each of which has its uses. We shall look at the four principal definitions in turn.

7.2.1 Arrhenius

Historically, this is the original definition of what is meant by an acid or a base. It is also the one with which you are most likely to be familiar. An acid is a hydrogen cation donor, while a base is a hydroxide anion donor.

Identify the acid and base in the following reaction:

$$HCl_{(aq)} + NaOH_{(aq)} \rightarrow NaCl_{(aq)} + H_2O$$

The hydrogen chloride is the acid because it donates the proton, i.e. the hydrogen cation; while the sodium hydroxide is the base because it donates the hydroxide anion. The above reaction is the classic: 'acid plus base gives salt plus water'. This definition is of limited applicability, because the base is restricted to those compounds that can produce a hydroxide ion, and so this effectively restricts a base to being a metal hydroxide. The solvent must be water, which, while it may be convenient for the inorganic chemist whose reactions are usually performed in aqueous solutions, is not very common for the organic chemist who normally use non-aqueous solvents. Thus, we will not consider this definition further.

7.2.2 Brönsted–Lowry

Acids and bases are defined as proton donors and acceptors respectively. This now includes a much wider array of compounds as possible bases. In the following reaction, which is the acid and which is the base?

$$HCl_{(g)} + NH_{3(g)} \rightleftharpoons NH_4{}^+Cl^-{}_{(s)}$$

Note that the reagents are gases, and that the product is a solid, i.e. the reaction is not being performed in solution. The hydrogen chloride is still considered to be the acid, because it donates a proton. The ammonia molecule is now considered to be the base, because it accepts a proton in this reaction. Why would the ammonia molecule not be considered a base under the Arrhenius definition?

The ammonia molecule does not produce a hydroxide ion. What would have been the position if the reaction had been carried out in the aqueous phase?

In the aqueous phase, the ammonia molecule would have reacted with a water molecule to form an ammonium ion and a hydroxide ion. Write an equation for this reaction.

$$NH_{3(aq)} + H_2O \rightleftharpoons NH_4^+{}_{(aq)} + OH^-{}_{(aq)}$$

If ammonia had been allowed to react with hydrogen chloride in the aqueous phase, then it would have fallen within the Arrhenius definition of an acid/base reaction. Notice that it is the water molecule that is central to the Arrhenius definition, and this is one of the main reasons why it is of little use in conventional organic chemistry. Similarly, because the Brönsted–Lowry definition does not depend on water, but only upon the presence of two compounds of complementary properties, it is this definition that is of more use to organic chemists.

Now consider the following reaction, performed in an organic solvent, and identify which is the acid and which is the base:

$$RCO_2H + RNH_2 \rightleftharpoons RCO_2^- + RNH_3^+$$

The carboxylic acid, RCO_2H, is the acid; while the amine, RNH_2, is the base. Now consider the reverse reaction, and identify the acid and the base:

$$RCO_2^- + RNH_3^+ \rightleftharpoons RCO_2H + RNH_2$$

Now the carboxylate anion, RCO_2^-, is the base; while the alkylammonium cation, RNH_3^+, is the acid. Notice that the carboxylic acid after deprotonation formed a base, i.e. the carboxylate anion. This is called its conjugate base. Suggest the name for the alkylammonium cation, RNH_3^+, which is formed after the protonation of the amine, RNH_2.

The alkylammonium cation is called the conjugate acid of the amine. Write down the general equation for a Brönsted–Lowry acid/base reaction, labelling the acid and the base, and the related conjugate base and acid.

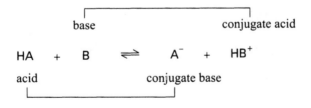

The related conjugate acid and base are always formed in a Brönsted–Lowry acid/base reaction.

On rare occasions the acid/base reaction is the sole purpose of the reaction. For example, what will be the product when the benzoate anion, $C_6H_5CO_2^-$, dissolved in an aqueous solution, is treated with hydrochloric acid? Write an equation for this reaction.

$$C_6H_5CO_2^-{}_{(aq)} + HCl_{(aq)} \rightleftharpoons C_6H_5CO_2H_{(s)} + Cl^-{}_{(aq)}$$

The protonated form of the benzoate anion is benzoic acid. Suggest whether it would be more or less soluble in the aqueous solution than its related conjugate base.

The benzoic acid, which is neutral, is less soluble than the related anion. In fact, it is so insoluble that it precipitates out from the solution as white flaky crystals. This great difference in solubility may be used to advantage to purify benzoic acid. The benzoate anion can be extracted into an aqueous bicarbonate solution, leaving behind any neutral organic impurities. Then, after acidification, the free acid may be extracted into an organic solvent. This second extraction allows the acid to be separated from impurities that are soluble in water.

The same principle may be used to form the free amino compound from its conjugate acid, this time by using a base. Write down the equation for the reaction between the hydrochloride of a general amine, $RNH_3^+Cl^-$, and a general base, B.

$$B + RNH_3^+Cl^- \rightleftharpoons BH^+ + Cl^- + RNH_2$$

Some organic molecules are stored as the related salt, because the salts are more stable and also because they may be easier to purify in that form. An example is the hydrochloride of glycine, $CH_2(CO_2H)NH_3^+Cl^-$, which is a white crystalline solid. Write down the equation for when this compound reacts with a base, B.

$$B + CH_2(CO_2H)NH_3^+Cl^- \rightleftharpoons BH^+ + Cl^- + CH_2(CO_2H)NH_2$$

This reaction liberates the free amino acid, glycine. If glycine is allowed to stand in solution, it reacts with itself to form a cyclic dehydration product called diketopiperazine:

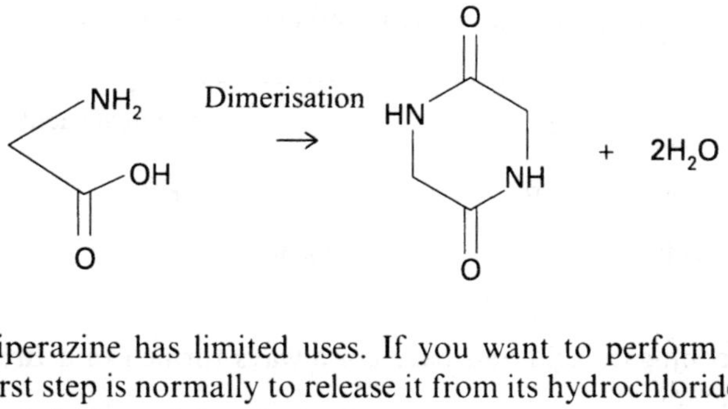

This diketopiperazine has limited uses. If you want to perform reactions on glycine, the first step is normally to release it from its hydrochloride and use the free glycine as it is formed freshly *in situ*.

So far we have mainly looked at organic molecules, or their ionic derivatives, reacting with an acid or base in an aqueous environment. Before we look further at the acid/base reactions of organic molecules in a non-aqueous solvent, it is instructive to look more closely at the reaction of a general acid, HA, dissolved in water and its reaction to form the hydroxonium ion, H_3O^+. Write down the equation for the reaction of this general acid with water, labelling the acid and base and the related conjugate base and acid.

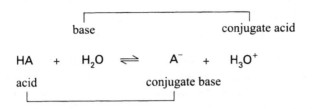

From this equation, an expression for the equilibrium constant, K, may be obtained. Write down this expression.

$$K = ([A^-][H_3O^+])/([HA][H_2O])$$

The concentration of the water remains almost constant, so rearranging the equation and defining K_a as equal to $K[H_2O]$, we arrive at the following expression:

$$[H_3O^+] = K_a[HA]/[A^-]$$

Now take the negative logarithm to the base ten of each side:

$$pH = pK_a + \log_{10}([A^-]/[HA])$$

This expression now contains the pH of the solution and also the term pK_a. The pK_a term is often used to indicate the strength of any particular acid. For a strong acid is this value high or low?

The stronger the acid, the lower the value of the pK_a term. These values may even become negative for very strong acids. However, very nearly all organic acids have pK_a values that are more positive than about plus two. In the equation that defined pK_a, water was used as the complementary molecule to the acid so that the latter could display its acidic properties. (The subscript 'a' refers to the fact that this equilibrium constant is concerned with acidic properties, and not to the fact that the reaction occurs in an aqueous environment. Hence, the opposite term, which is used to describe the strength of bases, has the subscript 'b'.) However, we could have used any other complementary molecule, such as a general alcohol, ROH. Write down the acid/base equation for the reaction between an alcohol and a general carboxylic acid, RCO_2H.

$$ROH + RCO_2H \rightleftharpoons ROH_2^+ + RCO_2^-$$

The stronger the acid, RCO_2H, the more of the protonated form of the alcohol, ROH_2^+, will be formed. This will be related to the pK_a value of the acid, but remember that the latter is defined for the aqueous system. This even excludes D_2O from being the solvent, let alone a general alcohol, ROH. Thus, the pK_a value as calculated for an aqueous system is only a rough guide to the strength of an acid in an organic solvent.

A general alcohol is protic, which means that it may be a donor of hydrogen ions. So far, we have only seen it acting as an acceptor of hydrogen ions; now write down the equation for the reaction of an alcohol acting as a proton donor to a general primary amine.

$$ROH + RNH_2 \rightleftharpoons RO^- + RNH_3^+$$

The degree to which the alcohol is deprotonated is dependent upon the strength of the amine as a base, which in an aqueous system is indicated by the pK_b value.

Write down the chemical equation of the reaction of a general base, B, reacting with water to form the hydroxide anion, OH^-, and so deduce the expression for pK_b.

$$B + H_2O \rightleftharpoons BH^+ + OH^-$$
$$K_b = [BH^+][OH^-]/[B]$$
$$pOH = pK_b + \log_{10}([BH^+]/[B])$$

At room temperature and pressure, pOH is related to pH by the expression:

$$pOH + pH = 14$$

Hence, the pH of the solution may be found. Notice that the pK_b of a base, which is a measure of its relative strength as a base, is only defined with water as the solvent, and so even though these values may be of interest in organic solvents, they are not strictly applicable and only act as a guide to the situation that is present in organic solvents.

We have now looked at reactions that involve a protic solvent such as an alcohol. However, there are many organic solvents that are not protic, such as benzene or hexane. In these solvents, there are no hydrogens joined to heteroatoms like oxygen or nitrogen, but only ones that are involved in carbon/hydrogen bonds. These bonds do not break easily, and so under normal conditions do not cleave to give hydrogen ions.

There are, however, some compounds that are usually non-protic solvents, but under extreme conditions may lose a hydrogen cation, for example trichloromethane. Suggest what would be the result of losing a hydrogen cation from this molecule.

$$CHCl_3 + B \rightleftharpoons BH^+ + CCl_3^-$$

This anion is stabilised by having three $-I$ groups attached to the carbon that bears the negative charge. This species may lose a chloride ion to form the dichlorocarbene, which itself may react further.

Another solvent that, under extreme conditions, may lose a proton is tetrahydrofuran, THF. Suggest which hydrogen is lost.

The hydrogen that is lost is the one which is closest to the oxygen. This anion may react further. Suggest a mechanism for the possible redistribution of electrons.

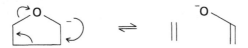

This reaction is driven by the fact that once the ethene is formed, it bubbles off from the reaction mixture and so the ethanal enolate anion is unable to undergo the reverse reaction. This decomposition has practical consequences. It means that THF cannot be used as the solvent when a very strong base is to be employed, unless precautions are taken to avoid the above reaction. The commonest precaution is to perform the reaction at low temperature, e.g. $-78°C$, when the deprotonation of the THF is very slow.

The acid/base property of a molecule is a thermodynamic property and as such it is affected by temperature. This may be illustrated by the autolysis of water, in which two molecules of water react with each other, one acting as a base while the other acts as an acid. Write down the autolytic reaction of water.

$$2H_2O \rightleftharpoons H_3O^+ + OH^-$$

What is the pH of a sample of pure water at standard temperature and pressure?

The pH of pure water at standard temperature and pressure is 7. Given that this reaction is endothermic, what would be the effect on the pH of this sample when the temperature is increased?

According to Le Châtelier's principle, as the temperature rises, an endothermic reaction will proceed further to the right as written, which in this case, means that more hydrogen ions will be produced and so the pH will fall. Note, however, that at all temperatures the concentration of hydrogen cations still remains equal to the concentration of hydroxide anions, even though the solution has become more acidic, as indicated by the drop in the pH value for the solution. This highlights the point that pH is only a measure of the hydrogen ion concentration, and that there is nothing particularly special about 7 being the value that is quoted for a neutral solution. Thus, a solution could be neutral, i.e. have the same concentration of hydrogen cations as hydroxide anions, but have a pH of 5: it would just be a rather hot solution.

The essence of thermodynamic reactions is that they are reversible. If a very strong base is allowed to react with a molecule that has a removable proton, and the resultant conjugate base is much weaker, then the reaction becomes in effect irreversible, and so kinetic factors might become important. What factors will affect which proton is removed when the acid/base reaction is controlled by kinetic considerations?

There are two principal driving forces that will determine which proton is removed. The first is: which hydrogen is most weakly bonded? The second is:

which hydrogen is more accessible to the base? The first is a reflection of which hydrogen is the most acidic in thermodynamic terms. The second reflects the steric demands of the base that is being used.

If the base that is used is large, and so has a large steric demand, would it be more likely to remove a primary or tertiary proton, assuming that they each had the same pK_a?

The sterically hindered base would be more likely to remove the primary proton, because it would not be able to reach the crowded tertiary proton. Bearing these factors in mind, which proton is most likely to be removed when $CH_3COCH_2CO_2Et$ is treated with a general base?

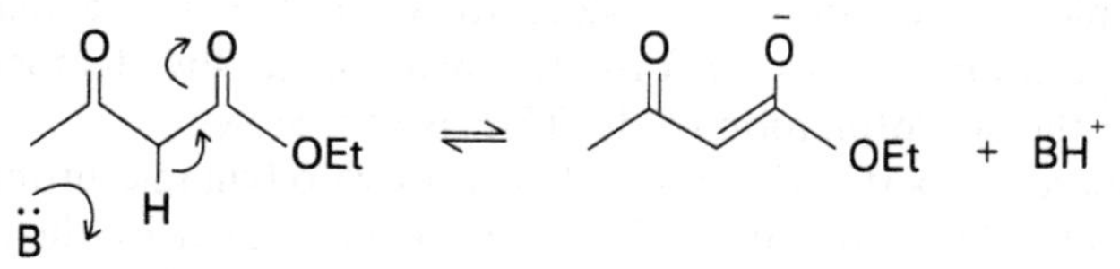

The middle proton is removed regardless of whether the base employed is NaOEt or lithium diisopropylamide, LDA, because that proton is the most acidic and, even though the protons on the methyl group are the most accessible, the methylene protons are still accessible enough for both bases to abstract them freely. However, if the reagent is modified slightly, to become $CH_3COCHRCO_2Et$, so that now the most acidic proton is attached to a tertiary carbon, then there is a difference in how this modified molecule will react with NaOEt or LDA. Suggest what is this difference.

The OEt^- anion may still gain access to the tertiary proton, which is the most acidic, and so remove it. However, LDA has a much larger steric demand and so cannot approach a tertiary proton without incurring some steric hindrance. Thus, it attacks the next most acidic proton that is readily available, which in this case is one of the terminal methyl protons. Draw the two anions that result from the removal of the two different protons.

Notice that in both cases, the resultant anions are stabilised by a carbonyl group: in one case by only one, and in the other by two. The essence of acid/base reactions that are controlled by kinetic considerations is that the reaction is irreversible, or at least effectively so on the time scale under consideration. For each example given above, write down the conjugate acid of the reacting base that is formed after the deprotonation of the initial reagent.

$$EtO^- + H^+ \rightleftharpoons EtOH$$
$$Pr_2N^- + H^+ \rightleftharpoons Pr_2NH$$

Both these conjugate acids are very weak acids. The weaker the resultant conjugate acid, the less likely it will react to re-form the original reagent, because of the greater activation barrier to the reverse reaction. Hence, the starting material is denied another chance to form the thermodynamically favoured product. Thus, a thermodynamically disfavoured product can be produced if it is formed more rapidly and, once formed, cannot revert to the starting materials.

Charge stabilisation of the anion, which is formed after deprotonation, is still important, but to a lesser extent than when the acid/base reaction is reversible. If butyllithium is used as a base, then very unreactive hydrogens may be removed, which result in unstabilised anions. However, the conjugate acid is butane, which is a very, very weak acid; also, as it is a gas, it leaves the reaction vessel. Thus, if butyllithium is used, the deprotonation becomes effectively irreversible.

7.2.3 Lewis

Acids and bases are defined as electron pair acceptors and donors respectively. This now includes a much wider array of compounds. An example is the reaction between boron trichloride and ammonia to give the adduct, H_3NBCl_3. Suggest which is the acid and which is the base in this example.

$$H_3N: \quad BCl_3 \quad \rightleftharpoons \quad H_3\overset{+}{N} \longrightarrow \overset{-}{B}Cl_3$$

base acid   adduct

The ammonia is the base, while the boron trichloride is the acid, because it accepts the lone pair from the ammonia. This definition is of great use to the organic chemist in, for example, the Friedel–Crafts reaction. Here, the reagent, an alkyl halide, reacts with an aromatic compound such as toluene, in the presence of a catalyst such as aluminium trichloride. There must be approximately the same amount of the catalyst as there is alkyl halide, and not just trace amounts of the catalyst, i.e. the aluminium trichloride is acting in this reaction as a stoichiometric catalyst. The reason becomes apparent when the mechanism is studied. Suggest what may be the reaction between the alkyl halide and the aluminium trichloride.

$$R\overset{..}{X} \quad AlCl_3 \quad \rightleftharpoons \quad R - \overset{+}{X} \longrightarrow \overset{-}{A}lCl_3$$

Here, the alkyl halide acts as a base, while the aluminium trichloride acts as the electron pair acceptor.

The adduct formed between the alkyl halide and the aluminium trichloride may react further by fragmenting. Suggest how this may happen.

$$R \overset{+}{\frown} \overset{-}{X} \longrightarrow \overset{-}{AlCl_3} \;\rightleftharpoons\; R^+ \;+\; [AlCl_3X]^-$$

This is a Lewis acid/base reaction in reverse, in that the adduct is fragmenting to form an acid and a base; however, they are different from those that originally formed the adduct. The carbonium ion may rearrange and so react via a different carbon atom than the one to which the halogen was originally attached. This rearrangement is called the Wagner–Meerwein rearrangement, and is a complicating factor in the Friedel–Crafts reaction. Both of these reactions are explored in more detail in Part II of this book.

7.2.4 Cady–Elsey

This definition is principally concerned with ionic solvents. A few important organic species may be synthesised in the superacid solvents to which this definition has most application.

In this definition an acid is defined as a species that raises the concentration of the cation of the solvent, while a base raises the concentration of the anion of the solvent. This definition is of use when dealing with autolytic solvents such as liquid ammonia. Write down the autolytic reaction of ammonia, and so identify the solvent anion and cation.

$$2NH_3 \rightleftharpoons NH_2^- + NH_4^+$$

The anion is an amide ion, while the cation is an ammonium ion. Suggest what would be an acid in this solvent.

Ammonium chloride would be an acid as it would raise the concentration of the cation, namely the ammonium ion. Now suggest what would act as a base in this solvent.

Sodium amide would act as a base as it would raise the concentration of the anion, namely the amide ion. This definition may be used for water as well. Suggest what is the autolytic reaction for water, and then what would be a potential acid and base using this definition.

$$2H_2O \rightleftharpoons OH^- + H_3O^+$$

In this case, hydrochloric acid for example, would act as an acid, because it raises the concentration of hydroxonium ions. In contrast, sodium hydroxide would act as a base, because the anion is the hydroxide ion. This scheme of events, you will recall, is similar to the original definition with which we started, namely the Arrhenius definition.

The main use of this definition is for solvents that are capable of forming both anions and cations by autolysis. There are other examples, such as liquid antimony pentafluoride, that do not depend upon the transfer of a proton to effect the ionisation of the solvent. Suggest the autolytic reaction for this solvent.

$$2SbF_5 \rightleftharpoons SbF_4^+ + SbF_6^-$$

Here the anion is formed by the addition of a fluoride ion, while correspondingly the cation is formed by the loss of the fluoride ion. Suggest in what manner hydrogen fluoride will act in this solvent.

$$HF + SbF_5 \rightleftharpoons H^+ + SbF_6^-$$

In this solvent, hydrogen fluoride acts as a base, because it raises the concentration of the anion of the solvent. This is rather different to the reaction of HF in water, where is acts as an acid.

Liquid antimony pentafluoride has been used as a solvent to study the properties of carbonium ions, because such ions are relatively stable in this solvent.

Another occasion when this definition is of some importance to an organic chemist occurs in the Birch reduction, which is performed in liquid ammonia. Sodium metal is dissolved in liquid ammonia to give solvated electrons, which may be represented, rather simplistically, as NH_3/β^-. This species may decompose to give NH_2^- and a hydrogen atom. This nascent hydrogen may then react in a rather different manner to that in which the solvated electron would have done, and so potentially can introduce an unwanted side reaction.

We will now examine electrophilic and nucleophilic properties, to which the acid/base properties mentioned above are closely related.

7.3 Electrophilic and nucleophilic properties

A species that is attracted to electrons is called an electrophile. It is usually electron deficient in that it has an empty orbital. Suggest a molecule that has already been considered in which one element has an empty orbital.

Molecules such as boron trihalide or aluminium trihalide have an empty orbital on the central atom, because in each case there are only three electron pairs around it. Electrophiles may have a full complement of electrons if they also bear a full positive charge. Suggest an example that falls into this category.

Such species would include the ammonium cation or the hydroxonium cation. These species will attract electrons Coulombically. Some species are both electron deficient and positively charged and so, not surprisingly, are very electrophilic. Suggest an example of such a species.

The commonest example would be the carbonium cation. There is another category in which the electrophilic species has a full complement of electrons yet does not bear a full positive charge, but still attracts electrons. Suggest an example that fits into this group.

An example would be a hydrogen atom that is joined to an element that has a high electronegativity, such as chlorine, nitrogen or oxygen. In such a case the electronegative element draws electrons away from the hydrogen and so, even though the hydrogen does not bear a full positive charge, nor has it a completely empty orbital, it is electron deficient and so may attract other electrons. Suggest a situation in which such an electron deficient hydrogen does attract electrons.

An example is when a hydrogen is involved in a hydrogen bond, such as that existing between two water molecules.

The opposite property to electrophilicity is called nucleophilicity. This is when a species is attracted to areas that are electron deficient. This property may be displayed in a number of different ways. One of the commonest ways is for the nucleophile to have an excess of electrons, which may be either in the form of a lone pair on a neutral species, or it may be as a full negative charge on an anion. Suggest an example of each type.

The ammonia molecule has a lone pair, but is electrically neutral; while the alkoxide anion has a lone pair and a full negative charge. The excess of electrons need not arise from having a lone pair, but may be found in a multiple covalent bond. Suggest an example that would fall into this category.

A molecule like ethene, which has a carbon/carbon double bond, has a centre of high electron density in the π bond. Thus, nucleophiles characteristically have a high electron density. Would they be attracted to a positive or negative centre?

An area of high electron density would be attracted to a positive centre. Ammonia, which has a lone pair, and so a high density of electrons, is attracted towards a positive centre. Draw the reaction between the ammonia molecule and a general positive centre, A^+.

$$H_3N: \quad A^+ \quad \rightleftharpoons \quad H_3\overset{+}{N} \longrightarrow A$$

This is the interaction of a nucleophile with an electrophile. However, this type of reaction is already familiar. Replacing the general electrophile, A^+, with a proton highlights this similarity with a general acid/base reaction. Write down the reaction of the ammonia with the proton, labelling each species as is appropriate with the terms nucleophile, electrophile, acid and base.

$$H_3N: \quad\quad H^+ \quad\quad \rightleftharpoons \quad\quad \overset{+}{N}H_4$$

nucleophile	electrophile	adduct
base	acid	

The nucleophile, the ammonia molecule, is also acting as a base, while the electrophile, the proton, is also acting as an acid. So a base may also be thought of as a nucleophile, because both are electron rich species and seek positive centres. The above reaction between the ammonia and the proton may be classified as a Brönsted–Lowry acid/base reaction. It may also be classed as a Lewis acid/base reaction, depending on whether one views the ammonia as a proton acceptor or as a donor of a lone pair of electrons.

Thus, there is a close relationship between nucleophiles and electrophiles on the one hand, and acids and bases on the other. We will now develop this idea a little further.

In certain circumstances the hydroxide anion may be classified more appropriately as either a base or a nucleophile. Suggest reactions in which one description is more suitable than the other.

as a base: $\quad OH^- + H^+ \rightleftharpoons H_2O$

as a nucleophile: $\quad HO^- \quad \overset{+}{C}R_3 \quad \rightleftharpoons \quad HO \text{------} CR_3$

The hydroxide anion may act as a base when it reacts with a proton to form a water molecule. This is a typical acid/base reaction under the Brönsted–Lowry definition. However, it may also act as a nucleophile when it is attracted towards a positive carbon centre. There is, however, an important difference between the positive centre of the proton and that of the carbonium ion. The difference lies in their relative charge densities: has the proton or the carbonium ion the greater charge density?

The hydrogen cation has a much higher charge density than the carbonium ion. A charged centre with a high charge density generates a very high potential gradient around itself. Thus, would a charged centre with a high charge density be better at polarising another charged centre than one with low charge density?

A centre with a high charge density would be better at polarising another charge centre. A centre with a high charge density is called hard, while one with a low charge density is called soft. Suggest a charged species that has a low negative charge density, and so would be classified as soft. Also, suggest a neutral species that has a low density of excess electrons.

The iodide anion has a low charge density, as does the carbon/carbon double bond. Both of these would be classified as soft nucleophiles.

Species with the opposite property are the electrophiles and acids. Suggest a charged electrophilic species that has a high charge density.

A common example would be the proton. Now suggest a species that is an electrophile, but with a low charge density.

The bromine molecule is an example of an electrophilic species that has a low charge density. In this case, the covalent bond between the bromine atoms may be polarised by being in close proximity to a charged centre. Any dipole that is induced would only be small, and so the corresponding charge density must be low.

133

Nucleophiles with a high charge density, such as hydroxide ions, tend to react more readily with electrophiles with a high charge density, such as protons. Thus, hard nucleophiles tend to react with hard electrophiles. The reaction of a hydroxide ion with a proton is also typical of a Brönsted–Lowry acid/base reaction.

Soft nucleophiles tend to react more readily with soft electrophiles. Suggest an example of this type of reaction.

For example, the reaction between the soft nucleophile, e.g. an alkene, and a soft electrophile, such as a bromine molecule, is shown in the following reaction.

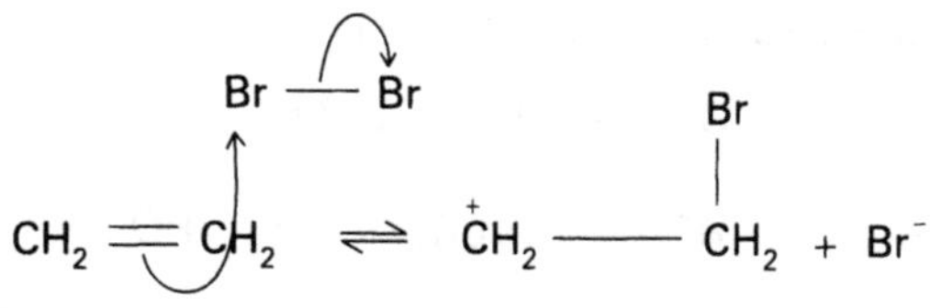

$$CH_2 = CH_2 \;\rightleftharpoons\; \overset{+}{C}H_2 \text{——} CH_2 \;+\; Br^-$$

This general rule of thumb is of assistance when a reagent like the hydroxide anion has the possibility of reacting with either a proton or a carbonium ion. While both of these are considered to be hard, the charge density on the proton is obviously greater, and so the hydroxide ion reacts preferentially with it, rather than the slightly softer carbonium ion. So commonly, a hydroxide ion will abstract a proton rather than attacking a carbon centre.

Finally, in this chapter we will look at some external factors that might influence the course of the reaction.

7.4 External factors

There are a number of other factors that will affect how a species reacts in a particular set of conditions.

One of the most easily appreciated is when an acid and its conjugate base bear a different charge, for example when ethanoic acid, CH_3CO_2H, loses a proton to give rise to the ethanoate anion. Write down the equation for this reaction, clearly showing the change in the charge of the organic component.

$$CH_3CO_2H \;\rightleftharpoons\; CH_3CO_2^- + H^+$$

The neutral carboxylic acid becomes the carboxylate anion, which bears a single negative charge. This charged species will interact with the solvent molecules in a different manner from that of the uncharged species. If the ethanoic acid was originally dissolved in ethanol, CH_3CH_2OH, suggest how the solvent/solute interaction changes as the ethanoic acid loses its proton.

ethanoic acid/ethanol interaction   ethanoate anion/ethanol interaction

The uncharged ethanoic acid interacts by means of a weak hydrogen bond between the carbonyl oxygen and the hydrogen of the hydroxyl group of the ethanol. However, the carboxylate anion interacts more strongly, because of the full charge that now resides on the carboxylate group, and so attracts the hydrogen on the hydroxyl group more strongly.

This demonstrates that there is a differential solvation of the conjugate acid and base. This differential solvation will naturally affect the relative solubilities of the conjugate acid and base, as illustrated by the earlier example of sodium benzoate and its conjugate acid, benzoic acid.

Furthermore, the difference between a conjugate base and the acid is a full unit of charge. This often has the effect of altering which resonance structure is favoured. In 2,6-dimethyl aniline, the amino group is free to rotate, and so the lone pair is capable of delocalisation into the aromatic ring; hence, this compound is not very basic in nature. Once it is protonated, delocalisation into the ring is no longer possible. However, in the N,N-dimethyl derivative, the dialkyl group may only position itself orthogonally to the aromatic ring, and so the lone pair on the nitrogen is no longer capable of being incorporated into the aromatic ring. It is now capable of acting as a base and reacting with a proton more readily. Thus, the *ortho*-methyl groups prevent the mesomeric delocalisation of the lone pair, and so there is a great difference in the basicity and nucleophilicity of the nitrogen group.

On a slightly different note, changing a solvent may have a great effect on the course of a reaction, because of the changes that thereby might occur in the solvation of the different solutes. For example, the change from water to ethanol will shift a reaction from substitution to elimination, because there is a greater charge separation in the substitution reaction that is favoured by the more polar nature of the water. Without this extra stabilisation, the elimination reaction is preferred when performed in ethanol.

Moreover, the actual presence of the solvent itself, and its ability to interact with the solutes, can have a fundamental effect upon the acid/base characteristics of the solutes under consideration. Thus, for example, the *t*-butoxide anion is a stronger base than the methoxide anion in solution in the corresponding alcohols. However, in the gas phase, the reverse is true. Conventionally, the explanation given for the more usually encountered solution result, is that the presence of more electron donating methyl groups in the *t*-butoxide anion increases the charge density on the oxygen, and so makes it more basic. However, as this explanation involves an intrinsic property of the *t*-butoxide anion, it ought to be equally true in the gas phase as it is in solution. Yet, the finding that in the gas phase *t*-butoxide anion is a weaker base than methoxide means that this conventional explanation must be wrong. Rather, the inherent property that determines the basicity of these anions in the gas phase relates to the number of groups over which the negative charge can be dispersed. In the *t*-butoxide anion there are more groups than in the methoxide anion, and so the former is more stable, and therefore less basic. The solution result can be explained readily by the fact that methoxide anion can interact far more strongly with the solvent than can the larger *t*-butoxide anion. Thus, in solution the methoxide anion is stabilised to a greater extent, and so becomes less basic than the *t*-butoxide anion. Thus, one must be wary of relying only upon relative results gleaned from experiments conducted in solution in attempting to deduce intrinsic properties of the molecule under consideration. Often, the effect of solvation will mask the property purportedly being studied.

That brings to an end the last introductory chapter on the basic principles of mechanistic organic chemistry. Now that we have studied these tools of the trade, we will apply them in helping us understand the main types of reaction mechanism.

7.5 Summary of acid/base characteristics

To reveal the acid/base properties of a molecule, the presence of another molecule with the opposite properties is required. In the Arrhenius definition, an acid is a hydrogen cation donor, while a base is a hydroxide anion donor, e.g. H^+ and OH^- respectively. Furthermore, this definition requires the solvent to be water. In the Brönsted–Lowry definition an acid is a proton donor, while a base is a proton acceptor, e.g. H^+ and NH_3 respectively. There is no requirement for water to be the solvent. In the Brönsted–Lowry definition, each component is related to its conjugate opposite component, e.g. the acid, RCO_2H, is related to its conjugate base, RCO_2^-. In the Lewis definition, acids and bases are defined as electron pair acceptors and donors respectively, e.g. BCl_3 and NH_3 respectively. In the Cady–Elsey definition, an acid is a species that raises the concentration of the cation of the solvent, while a base is defined as a species that raises the concentration of the anion of the solvent, e.g. NH_4^+ and NH_2^- respectively in liquid ammonia.

Many solvents undergo autolysis, i.e. one component acts as the acid, while the other acts as a base.

The term pK_a is often used to indicate the strength of an acid. It is strictly only applicable to aqueous solutions. Similarly, the term pK_b indicates the strength of a base. However, these terms can be used to estimate the relative strengths of acids and bases when used in organic solvents. Protic solvents are ones that potentially can donate a proton, e.g. ROH. Even non-protic solvents can often lose a proton under extreme conditions, e.g. $CHCl_3$ can lose a proton to form, eventually, the dichlorocarbene, $:CCl_2$.

Sterically hindered bases can be used to remove selectively only an accessible hydrogen, even if it is not the most acidic.

A species that is attracted to electrons is called an electrophile. It is usually electron deficient in that it has an empty orbital or bears a positive charge. Conversely, when a species is attracted to areas that are electron deficient it is called a nucleophile. Such species are electron rich, e.g. have a lone pair of electrons, or a multiple bond.

Electrophiles are akin to acids, while nucleophiles are similar to bases. A centre with a high charge density is called hard, while one with a low charge density is called soft. Nucleophiles with a high charge density, such as hydroxide ions, tend to react more readily with electrophiles with a high charge density, such as protons. Thus, hard nucleophiles tend to react with hard electrophiles. Conversely, soft nucleophiles tend to react with soft electrophiles.

Charged and uncharged species tend to interact with the solvent differently, i.e. there is differential solvation. This can also be achieved by changing the solvent in which the reaction is performed, e.g. by changing the solvent from water to ethanol; elimination is thereby favoured over substitution.

PART II

Mechanisms

Introduction to Mechanisms

8.1 Preliminary considerations

In inorganic chemistry, even though there are a large number of reactions that each element can undergo, it is possible to divide all the reactions into four general reaction types, namely: redox, acid/base, precipitation and complexation. It is even possible to reduce these four categories to three, if one considers that complexation reactions form the first stage along the path towards precipitation. In fact, a further simplification is possible, because complexation reactions may also be seen as a type of acid/base reaction. This is particularly true when one uses the Lewis definition of what comprises an acid and base. Thus, from this very brief analysis, it is clear that the vast array of inorganic reactions may be ordered into a small number of fundamental reaction types. It is this ready division that greatly assists in the understanding of inorganic chemistry, and makes the subject less daunting to the student.

In contrast, in organic chemistry there seems to be an enormously large number of reactions that are possible. Furthermore, many of these, particularly at first sight, seem to be unique and seemingly unrelated to any other example. This gives the initial impression that organic chemistry is incapable of being readily rationalised into a small number of principal reaction types like inorganic chemistry. Yet, thankfully, this first impression is wrong, and it is possible to organise this multitude of reactions into a small number of fundamental reaction types, namely: substitution, addition, elimination, rearrangement and redox reactions. Of these reaction types, the first three are fairly straightforward, while the latter two are generally more complex.

In a substitution reaction, one group is replaced by another. Usually a heteroatom or group, i.e. not a hydrogen atom or hydrocarbon moiety, is replaced by another heteroatom or group. In an addition reaction, the adduct, which is the molecule that is going to be added to the substrate, is broken into approximately equal parts, and these parts then add across a multiple bond, which is commonly a carbon/carbon or carbon/oxygen double bond. Elimination reactions are broadly overall the reverse of addition reactions. Thus, in an elimination reaction usually a compound is formed that has more multiple bonds than did the starting material, i.e. there is a greater degree of unsaturation in the final product than in the starting materials.

As well as those reactions that fall neatly into one or other of the three simple reaction types, there are some reactions where an addition reaction is followed by an elimination reaction, or vice versa, resulting in an overall substitution of a group. These reactions will be considered separately in this book under the heading of sequential addition/elimination reactions.

More complicated still, in a rearrangement reaction, the backbone of the molecule is broken and reformed in a different configuration. Often, this involves the breakage of a carbon/carbon bond, but on some occasions it may involve the cleavage of a carbon/oxygen or carbon/nitrogen bond. Lastly, there are redox reactions. On many occasions redox reactions may be considered conveniently under one of the first four headings. However, for some other reactions, it is more convenient to consider them separately, because some of the detailed pathways of the mechanisms do not readily fall within any of the other categories.

In addition to the potential divisions mentioned above, a secondary classification may be employed that utilises the divergent electronic natures of the species that attack the starting material, e.g. nucleophilic, electrophilic or free radical.

Nucleophilic reagents are electron rich, usually because they have a lone pair or possess a multiple bond, and so they have two electrons with which to form a new bond with the species being attacked. Nucleophiles may be neutral or bear a negative charge. They tend to attack centres of low electron density, such as empty orbitals or positive charges.

In contrast, electrophilic reagents are the opposite of nucleophilic ones. They are often positively charged, or at least are electron poor by virtue of having an empty orbital, i.e. an atomic orbital in which there are no electrons. Electrophiles generally attack centres of high electron density, such as double bonds.

Free radicals are molecules that have an odd number of electrons and so possess an unpaired electron. Unlike the nucleophilic and electrophilic species, which are often charged, free radicals are usually neutral, although on occasions charged radical species may exist. Radicals are capable of attacking unpolarised covalent bonds such as carbon/hydrogen bonds, which are otherwise usually inert to either nucleophilic or electrophilic reagents.

There is a further way in which reactions may be sub-divided, namely into pericyclic and non-pericyclic reactions. Pericyclic reactions usually involve the simultaneous movement of more than one pair of electrons in any given reaction step. In contrast, in non-pericyclic reactions, usually either one or two electrons move sequentially in each step of the reaction. Pericyclic reactions may be further divided into three principal types, namely: electrocyclic, sigmatropic and cycloaddition.

In this book, the divisions based upon the five principal reaction types and the nature of the attacking species will be used. Yet, however logical the above divisions may appear, there are other possible divisions. For example, many textbooks divide the presentation of the reactions by reference to the dominant carbon functional moiety, e.g. aliphatic, aromatic or carbonyl. This approach is of value if one wishes to highlight the differences between the ways in which compounds that contain these moieties react with different reagents. However, underlying the differences there are many fundamental similarities in the manner in which these different moieties react, and it is these similarities that this book endeavours to highlight. Once these parallels have been understood, the differences may be appreciated more easily. Accordingly, this is the approach that is adopted in this book.

8.2 Summary of introduction to mechanisms

Inorganic reactions can be divided into four categories, namely: redox, acid/base, precipitation and complexation. By considering complexation reactions as the first stage towards precipitation, and further by considering complexation reactions as a type of acid/base reaction, it is possible to reduce this number to two fundamental categories for inorganic reactions.

Organic reaction mechanisms can similarly be reduced to a small number of fundamental reaction types, namely: substitution, addition, elimination, rearrangement and redox reactions.

A secondary classification may be employed that utilises the divergent electronic natures of the species that attack the substrate, e.g. nucleophilic, electrophilic or free radical.

Reactions in which more than one pair of electrons move simultaneously in a concerted manner, rather than sequentially, are called pericyclic reactions. These can be divided into three principal types, namely: electrocyclic, sigmatropic and cycloaddition.

Many textbooks divide the presentation of the reactions by reference to the dominant carbon functional moiety, e.g. aliphatic, aromatic or carbonyl. This approach is of value if one wishes to highlight the differences between the ways in which compounds that contain these moieties react with different reagents. However, by dividing the reactions into the five principal reaction types and the nature of the attacking species as first suggested, one highlights the fundamental similarities in the manner in which these different moieties react. Once these parallels have been understood, the differences may be appreciated more readily.

9

Nucleophilic Substitution Reactions

9.1 Introduction

Substitution reactions are one of the most important types of reactions available to the synthetic organic chemist. In its simplest form, a functional group, X, is replaced by a different functional group, Y. This overall reaction may be represented by the following equation, in which R represents a general carbon moiety.

$$RX + Y \rightleftharpoons RY + X$$

Y may be a nucleophile, an electrophile or a free radical. These terms may be used to distinguish the different mechanistic pathways that are characterised by the involvement of these differing attacking species.

The substitution of one functional group for another is called a functional group interchange (or interconversion), FGI. Usually a heteroatom or group, i.e. not a hydrogen atom or hydrocarbon moiety, is replaced by another heteroatom or group. Thus, for example, a hydroxyl group may be substituted for a bromine atom. However, it is possible for a heteroatomic functional group to be replaced by one containing carbon atoms, e.g. a cyano or alkyl group. In this way, the carbon chain of the original molecule can be extended, and so a simple precursor can be elaborated into a more complex molecule. This is the basis of organic synthesis.

The commonest type of substitution reaction involves the attack by a nucleophile, Nuc^-, on the substrate, with the concomitant departure of a different nucleophile in order to ensure the overall conservation of charge. Reactions that involve an initial attack by either an electrophile or a free radical are less prevalent and will be discussed in the following chapters.

9.2 Substitution at a saturated carbon

9.2.1 Introduction

In a nucleophilic reaction at a saturated carbon centre, the incoming nucleophile acts as a soft nucleophile; that is, it attacks the relatively soft electrophilic centre of a $\delta+$ charged carbon atom; relative that is to the other possible sites of attack, namely, for example, the harder proton. In a later chapter, we will study a class of reactions in which the incoming nucleophilic agent acts as a hard nucleophile; that is, as a Brönsted–Lowry base. In such a case, the nucleophile removes, as the first step in the reaction pathway, a proton from the reacting carbon species. This latter type of reaction results in an elimination reaction occurring. In practice, no reagent is either completely soft or completely hard in nature, and so a mixture of substitution and elimination reactions nearly always occurs in any one reaction mixture. However, various factors may favour one route over another, and in this chapter and the next two, we shall concentrate upon substitution reactions before looking at elimination reactions in Chapter 14.

In a carbon centre that is involved in four covalent single bonds, how many electrons are there around the carbon atom? Demonstrate your answer by focusing on the carbon atom in a molecule of methane, CH_4, and drawing a dot and cross structure for it.

$$
\begin{array}{c}
\text{H} \\
\overset{x\ \bullet}{\underset{\bullet\ x}{\text{H}\ \overset{\bullet}{\underset{x}{\text{C}}}\ \text{H}}} \\[2pt]
\text{H}
\end{array}
$$

The answer is, of course, eight: a full octet of electrons is accommodated around the carbon atom in this molecule. What is the maximum number of electrons that a carbon atom may normally accommodate in its outermost, that is its valence, shell of orbitals?

As carbon is a second row element, the maximum number of electrons that it may normally accommodate in its valence shell is eight.

Now let us turn our attention to the electronic characteristics of the incoming nucleophile. Give two examples of typical nucleophiles.

The incoming nucleophile is electron rich, and so your examples may have included such anionic species as hydroxide, halide, hydride or cyanide anions; or may have included such neutral species as ammonia or water molecules. Note that in all cases, whether the species in question was anionic or neutral, there was always a lone pair of electrons. Nucleophiles are (almost) never positive species, because the positive charge on a cation indicates that it is electron poor, which is the opposite electrical property to that of a typical nucleophile.

In a simple nucleophilic substitution reaction, a carbon atom that already has eight electrons is approached by an electron rich nucleophile. This would lead to a situation in which there were more than eight electrons around the carbon atom. This simple exercise of electron counting highlights the central problem with substitution reactions: how does a carbon atom that already has its full complement of electrons manage to substitute one group for another without expanding the octet of electrons by occupying orbitals in the next quantum level?

It is immediately possible to identify one mechanistic pathway that cannot be correct, namely the simple addition of the incoming nucleophile to give an adduct, which then subsequently disassociates to give the desired products. This cannot be a permissible pathway, because, in forming the adduct, ten electrons would reside on the carbon atom. If you are in any doubt about this, count the number of electrons around the central carbon atom in the resultant anion in the following hypothetical reaction.

$$CH_4 + H^- \rightleftharpoons CH_5^-$$

The methane carbon had eight electrons to start with, and then with the addition of the hydride ion, two more are added to give ten in all. This would require the extra two electrons to occupy the third quantum level, which would require a large amount of energy. Thus, even though this is possible, it is not usual. Hence, this mechanism does not happen under normal circumstances. It is however important to note that given the right conditions, it could happen if enough energy were supplied to the system.

There are two solutions to the problem posed above that are found to occur in practice. Suggest what they may be.

The first solution proposes that the group that is to be substituted departs from the carbon species, taking two electrons with it. In this case, the remaining carbon species now has only six electrons, and so may readily accept the two electrons that the incoming nucleophile has to offer. This mechanism is called unimolecular substitution, and may be written in general terms as follows.

1. $CH_3X \rightleftharpoons CH_3^+ + X^-$
2. $CH_3^+ + Y^- \rightleftharpoons CH_3Y$

This type of reaction occurs in two discrete steps. There is a carbonium ion intermediate, which exists as a discrete species. This intermediate can, on occasions, be isolated.

In the second mechanism, the leaving group departs simultaneously with the incoming nucleophile approaching. Thus, at any one time the central carbon atom sees only eight electrons, and so does not violate the Octet rule. This mechanism is called bimolecular substitution, and may be represented in the following manner.

$$CH_3X + Y^- \rightleftharpoons CH_3Y + X^-$$

147

This reaction occurs in only one step, and so there is no discrete intermediate. In contrast, there is a transition state that cannot be isolated, but only hypothesised.

9.2.2 *Unimolecular substitution*

Originally the difference between unimolecular and bimolecular substitution reactions was deduced from kinetic studies on a wide range of reagents. It was observed that for some reactions the overall rate of substitution depended only upon the concentration of the substrate, i.e. the species undergoing substitution, and that the rate was independent of the concentration of the attacking species, i.e. the nucleophile. The reaction, therefore, is a first order reaction; that is, the sum of the indices of the concentration of the reagents in the rate equation equals one. These reactions are called unimolecular nucleophilic substitution reactions, and are given the label S_N1.

In this mechanism, it is proposed that the first step consists of the heterolytic fission of the bond that joins the carbon atom with the group that will be substituted. In the molecule $(CH_3)_3CBr$, identify the polarised bond, and then indicate the direction in which it is polarised.

$$\overset{\delta+}{>}\!\!-\!\!\!\longrightarrow\overset{\delta-}{Br}$$

Only the C–Br bond is polarised to any significant extent, and it is polarised towards the bromine atom. Hence, it is likely that this is the bond that will be broken most readily. Write down the first step of the S_N1 reaction in which the carbon/heteroatom bond is broken, using curly arrows to indicate the movement of electrons. Identify the products.

$$\overset{\delta+}{>}\!\!-\!\!\!\longrightarrow\overset{\delta-}{Br} \quad \rightleftharpoons \quad \overset{+}{>}\!\!-\!\!\!- \;\; + \;\; Br^-$$

The neutral starting compound has been split into two parts: the first is an electron-deficient carbonium ion, while the second is an electron-rich leaving group, namely the bromide anion. The sum of the charges of the products is zero, which equals that of the starting material.

Only the concentration of the starting material affects the overall rate of substitution in this type of reaction. From this it may be deduced that only the starting material is required in the rate determining step, i.e. the slowest step of the reaction pathway. This explains why the reaction is called unimolecular, because only one molecule is required in the rate determining step. Whatever happens thereafter must occur more quickly than this initial step. Thus, the subsequent formation of the final product, by the attack of the incoming nucleophile on the carbonium ion, must be fast.

There are a number of factors that may affect the rate of this type of reaction. Some of these factors are inherent in the molecule itself, while others are external and so relate to the environment in which the reaction is performed.

The most obvious internal factor is the bond strength of C–X. Suggest whether a strong or a weak heteroatomic bond would increase the rate of an S_N1 reaction, and why.

A weak C–X bond will increase the rate, because it will break more easily and so form the carbonium ion more easily.

Another internal factor that is related to the leaving group is the stability of the leaving group after cleavage. Suggest how the stability of the leaving group may affect the rate, and give some examples of good and bad leaving groups.

Generally, good leaving groups are those that form soft nucleophiles: such as the bromo or iodo substituents that form bromide or iodide anions respectively; or those that form weak Brönsted–Lowry conjugate bases, e.g. those groups that leave to form anions of strong acids such as tosylate ($p\text{-}CH_3C_6H_4SO_3^-$), or neutral molecules, such as water and amines. Soft nucleophiles are good leaving groups, because they are easily polarised and so the heterolytic fission of the C–X bond may take place more readily. Furthermore, in general, it is advantageous if the leaving group is readily solvated, because this means that there is some favourable energy of solvation, which is then available to compensate for the energy required to break the C–X bond initially. In fact, in solution, the effect of the solvent must not be overlooked, because it often has a determining effect upon which reaction pathway is favoured.

It is possible to form the unstable phenyl cation, $C_6H_5^+$, from the diazonium salt, $C_6H_5N_2^+$, because of the great propensity of the diazo group, N_2^+, to form a very stable leaving group. Suggest the mechanism for this reaction, and also suggest what is the identity of the stable leaving group.

$$C_6H_5 \overset{\frown}{\underset{}{—}} \overset{+}{N} \equiv N \;\rightarrow\; C_6H_5^+ + N_2$$

The diazo group reorganises the electrons associated with it so as to form the dinitrogen molecule, which is the stable leaving group, and the phenyl cation. The nitrogen molecule is a very thermodynamically stable species, and is also a gaseous product; it thereby greatly increases the degree of entropy for the forward reaction. The phenyl cation, which is a useful synthetic intermediate, cannot be readily formed except by using this decomposition, which is the reason why the diazonium salts of benzene derivatives are so important in the synthesis of aromatic compounds.

149

Turning our attention from the leaving group to the carbon moiety, it is observed that the structure of the resulting carbonium ion has a large effect upon the ease of substitution.

Suggest the relative stability of the carbonium ions, CH_3^+, $C_2H_5^+$, $(CH_3)_2CH^+$ and $(CH_3)_3C^+$, and then suggest which related parent species would undergo substitution fastest by the SN1 mechanism, and why?

$$(CH_3)_3C^+ > (CH_3)_2CH^+ > C_2H_5^+ > CH_3^+$$

The tertiary carbonium ion is more stable than the secondary carbonium ion, which in turn is more stable than the two other carbonium ions. Consequently, tertiary substituted compounds undergo SN1 substitution much faster than secondary or primary ones. This is because there is a smaller energy of activation required to form the tertiary carbonium ion as it is more stable, and so its rate of formation is faster. Notice also that the greater the degree of substitution in the starting material, i.e. the larger the number of alkyl groups on the carbon atom that is to bear the positive charge, then the greater the release of steric strain that will be experienced in going from an sp^3 configuration to an sp^2 configuration.

Two further examples may be used to illustrate the dependence of the rate of SN1 substitution on the stability of the resultant carbonium ion intermediate. First, suggest what are the relative rates of substitution between CH_3CH_2Br and $PhCH_2Br$; and second, the relative rates of substitution between $(CH_3)_3CBr$ and 1-bromonorbornane, the latter having a bicyclic structure with the bromine attached to the bridgehead carbon.

$$PhCH_2Br > CH_3CH_2Br$$

$$(CH_3)_3CBr > \text{1-bromonorbornane}$$

In the first case, both bromides are primary, and so SN1 substitution would be expected to be slow. This is true for bromoethane. However, for the benzyl derivative, it is possible for the charge on the resultant carbonium ion to be delocalised into the benzene ring, and this will greatly increase its stability, and so greatly increase the rate of substitution by the SN1 mechanism. In the second case, even though both compounds are tertiary, the norbornyl derivative is almost inert to SN1 substitution, because the required carbonium ion is unable to attain the planar conformation that is needed for it to be stabilised, in particular by hyperconjugation. Thus, as the necessary carbonium ion intermediate is thermodynamically unfavourable, the preceding activation energy is high, and so the rate of substitution by the SN1 pathway is slow.

Other factors that affect the stability of the carbonium ion intermediate will also affect the rate of SN1 substitution. In particular, the effect of inductive or mesomeric groups that influence the electronic distribution around the charged carbon have a great effect. This may be illustrated by studying the observed rates of substitution in a series of related deuteriated and undeuteriated compounds. Thus, suggest whether $(CH_3)_3CBr$ or $CD_3(CH_3)_2CBr$ would be substituted faster and why?

The SN1 mechanism proposes a carbonium ion as the intermediate, and this cation may be stabilised by +I or +M groups. The alkyl groups are +I, and so help to stabilise the positive charge. However, the alkyl groups may also be involved in hyperconjugation, in which the adjacent hydrogens donate the electron pair in their hydrogen/carbon σ bond to help stabilise the positive charge. The involvement of these hydrogen/carbon bonds may be deduced from the observation that when the protium atoms are replaced by deuterium atoms, the overall rate of substitution is reduced. This is because deuterium atoms are less effective at stabilising the positive charge. This effect is an example of the secondary isotope effect, i.e. an isotopic substitution that has an effect on the rate of the reaction, even though that isotope is not directly involved in the bond being broken or formed.

Apart from these inherent properties, which affect the rate of SN1 substitution, there are several external factors that may have an effect. First, it is preferable to have a weak incoming nucleophile. This is because, if it were strong, then it might attack the carbon centre that is to form the carbonium ion, and so change the mechanistic pathway from unimolecular to bimolecular. Hence, a weak nucleophile favours an SN1 substitution route, because it does not promote an SN2 route. There are, however, other important factors concerning the solvent. Suggest whether a solvent of high or low polarity would favour the SN1 reaction.

The result of the first step in an SN1 reaction is to produce a pair of ions. If these ions can be solvated, then this will favour the forward reaction because of the favourable energy of solvation. Furthermore, as both a cation and an anion are produced, it is best to have a solvent that is capable of solvating both species of ion. Thus, solvents of high polarity generally favour SN1 substitution.

If the ionic concentration of the reaction mixture is increased by the addition of other ions, it is found that the overall rate of substitution is increased. This is called the salt effect. However, if the anion of the leaving group is added to the reaction mixture, then, in contrast, it is found that the rate decreases. Suggest a reason for this latter observation.

This effect is a variation of Le Châtelier's principle, and is called the common ion (or mass-law) effect. The first step of the SN1 reaction is to produce the leaving group, and if there is already a large concentration of this present, the forward reaction is disfavoured as it would produce more of that species.

The interplay of all these different factors affects the overall rate for any specific substitution reaction. However, given any two compounds, it is usually possible to predict which will react the faster by balancing the various factors against each other.

Apart from the speed with which a reaction occurs, it is also of interest to follow the stereochemical consequences of a reaction. This may be studied by observing the result of a group departing from an asymmetrical carbon atom. What is the structure of the carbonium ion that results from the chiral precursor, $CH_3CH_2(CH_3)CHBr$, and hence what is the stereochemical consequence of any substitution reaction that involves this carbonium ion as an intermediate?

Like other carbonium ions, it is planar. The incoming nucleophile may then attack with equal ease from either above or below the carbonium ion, and so form equal amounts of each enantiomer of the product. Thus, the stereochemical consequence will be a complete loss of the steric information that was contained in the original compound, i.e. there will be complete racemisation.

Suggest what would be the stereochemical consequences if the intermediate carbonium ion only existed for a very short period of time before it was converted to the product.

The shorter the half life of the carbonium ion, the less racemisation there will be, and the greater the amount of inversion of configuration will occur. Perfect racemisation will only result when there is an exactly equal opportunity for attack from each side of the carbonium ion. This point may be illustrated as follows.

While the leaving group occupies the space to one side of the carbonium ion, it blocks any potential approach by an incoming nucleophile from that side. Thus, during the period that the leaving group remains close to the carbonium ion, the nucleophile must attack from the other side, and this will lead to the inversion of the configuration at the chiral centre. This protection only operates for the short period of time during which the leaving group departs, and so only has an effect on the resultant stereochemistry if the carbonium ion is very short lived.

If we take this one step further, and the incoming nucleophile approaches at the same time as the leaving group departs, we arrive at a situation in which there is no carbonium ion intermediate. In this case, the substitution reaction proceeds in one step and has now become an S_N2 reaction.

9.2.3 *Bimolecular substitution*

The other major type of nucleophilic substitution reaction is bimolecular, and is given the label S_N2. In this case, the rate of the reaction is observed to be dependent upon both the concentration of the species being substituted, i.e. the substrate, and also upon the incoming nucleophile. The reaction, therefore, is a second order reaction; that is, the sum of the indices of the concentration of the reagents in the rate equation equals two. The order of a reaction is an experimentally observed number and need not be an integer.

The mechanism for the SN2 reaction occurs in only one step, in contrast to the two steps involved in the SN1 reaction. The incoming nucleophile arrives at the same time as the leaving group departs, i.e. there is simultaneous bond cleavage and bond formation. Any reaction step that involves two molecules is said to have a molecularity of two. The molecularity is a theoretical concept and depends upon the details of the mechanistic step being proposed. By its very nature, it must be a whole number as it represents the number of molecules involved in that step.

The rate of an SN2 reaction may be affected by many different factors, which are similar to those that affect the rate of an SN1 reaction. These points may be illustrated by examining in detail the approach and simultaneous departure of the incoming nucleophile and leaving group respectively.

First, draw a diagram to illustrate the initial approach of a nucleophile, Nuc^-, to a general substrate, R_3CX, where X^- will eventually become the leaving group. Consider the direction from which the incoming nucleophile will approach.

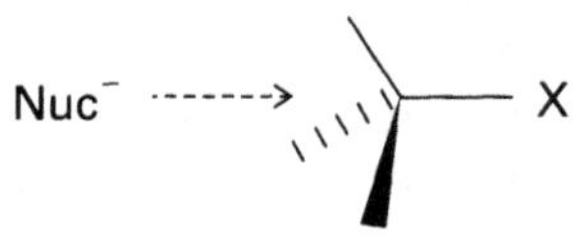

It seems reasonable to suppose that the incoming nucleophile will approach the substrate from a direction that is the opposite to the one in which the C–X bond is to be broken. Thus, the attack is from the opposite side to that from which the leaving group will eventually depart. This ensures that the incoming nucleophile does not interfere with the subsequent departure of the leaving group as the anion, X^-.

Second, consider the steric factors that operate to hinder or facilitate this approach. Illustrate your answer by comparing the attack upon *n*-butylbromide and 1-bromo-2,2-dimethylpropane (*neo*-pentylbromide).

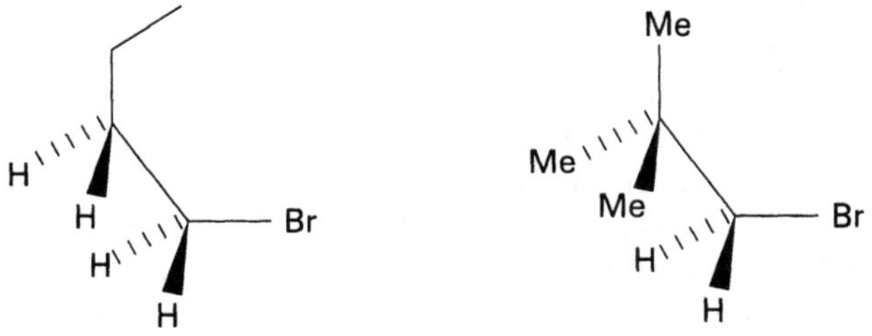

n-butylbromide   *neo*-pentylbromide

In both cases, the bromide ion is the leaving group and is attached to a primary carbon atom. In the case of bromobutane, the carbon chain is off to one side and does not interfere with the approach of the nucleophile. However, in contrast, in the case of *neo*-pentylbromide, there is the large *t*-butyl group that obstructs the approach of the incoming nucleophile. This steric hindrance decreases the rate of SN2 substitution.

Now consider the electronic interactions and the changes in electron distribution that occur as the nucleophile approaches the substrate.

$$\text{Nuc}^- \cdots\cdots \overset{\delta^+}{>}\!\!-\!\!-\overset{\delta^-}{\text{LG}} \qquad \text{LG = leaving group}$$

The C–X bond is polarised towards the leaving group, and so the carbon atom bears a $\delta+$ charge. The incoming nucleophile is usually negatively charged, but always electron rich, and so there is a Coulombic attraction between these two species. At first sight, it would be reasonable to suppose that the rate of SN2 substitution increases with the degree of polarisation of the C–X bond. This would mean that fluoro compounds would be substituted faster than iodo compounds, under otherwise similar conditions. In fact, the opposite is found to be the case when the reaction occurs in solution. (In passing, it should be noted that in the gas phase, the relative rates of reactions are often different from those found in solution, because of the solvent effects that are present in the latter.)

In order to understand why this order is observed in solution, we must consider the next stage in the approach of the incoming nucleophile. Draw a diagram to illustrate the mid-point between the reagents and the products, i.e. the transition state, and consider the changes that have occurred in the electronic distribution of the bonds that are being broken and formed.

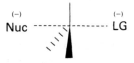

$$\overset{(-)}{\text{Nuc}} \text{-----} \overset{(-)}{\text{LG}}$$

In forming the transition state, the incoming nucleophile needs to form a partial bond with the carbon atom that is undergoing substitution. This carbon must not exceed its octet of electrons, and so the bond to the leaving group must be simultaneously partly broken. Thus, there is a movement of electrons within this bond towards the leaving group. A C–F bond is very strong, while a C–I bond is both weak and more easily polarised. In forming the transition state, it is preferable for the bond to the leaving group to be easily polarised, because this will facilitate the subsequent heterolytic fission of the C–X bond.

This highlights the difference between a bond that is highly polarised, e.g. carbon/fluorine, and one that, even though not highly polarised initially, may be

154

easily induced to become more so and eventually breaks heterolytically. An example of such an easily polarisable bond is carbon/iodine.

The same considerations apply to the incoming nucleophile; the more polarisable the nucleophile, the better it is as an attacking species in an Sn2 reaction. Bearing this point in mind, suggest why the overall rate of substitution of an alkyl fluoride by hydroxide anions is increased by the addition of iodide ions, and so suggest the intermediate compound that might be formed.

$$R\text{–}F + I^- \rightleftharpoons R\text{–}I + F^-$$

$$R\text{–}I + OH^- \rightleftharpoons ROH + I^-$$

The C–F bond is both strong and less readily polarised than a C–I bond, and the OH$^-$ ion, even though electron rich and capable of forming strong C–O bonds, is not easily polarised. Hence, both species are reluctant to undergo the required movement of electrons within the bonds to form the transition state. Iodide ions, however, are more easily polarised and so may readily attack a saturated δ+ carbon, and form the transition state, which may then yield the iodo compound, R–I, as an intermediate. This, in turn, may now be more readily attacked by the OH$^-$ anion as the C–I bond is readily polarised. The use of an iodide anion in this manner is an example of nucleophilic catalysis.

As the incoming nucleophile approaches the substrate, the electron density increases on the carbon atom that is to be substituted. Suggest what substituent groups on this carbon would reduce this build-up of negative charge.

Both −I and −M groups would help stabilise the potential build-up of negative charge on the carbon that is to be substituted, and so assist this mechanism. Furthermore, −I groups would increase the positive character on that carbon in the substrate, and so enhance the initial Coulombic attraction between it and the incoming nucleophile.

The effect of −I groups also partly accounts for the reason why Sn2 reactions are favoured at primary as opposed to tertiary carbon sites, because alkyl groups are slightly +I in nature. There are two other reasons that favour primary over tertiary carbons in Sn2 reactions: suggest what they may be.

The first reason is that sterically there is less hindrance encountered by the approaching nucleophile. The second reason is that primary carbons only very reluctantly form carbonium ions, and so the rate of any competing Sn1 reaction is greatly reduced, which means that the presence of any Sn2 reaction that occurs will be more readily noticed.

The last stage of the Sn2 reaction, in which the leaving group departs, is the mirror image of the first stage in which the nucleophile approached the substrate.

The factors that have been considered so far have concerned the inherent properties of the molecules involved in the substitution reaction. The nature of the solvent also has a great effect on the rate of an Sn2 reaction as it did with an Sn1 reaction. Suggest whether a solvent of high polarity would favour or disfavour the Sn2 pathway.

In the transition state, the charge that was originally concentrated on the attacking nucleophile and the carbon/leaving group bond is now spread out over the two partially formed bonds. So, there is a decrease in the charge density in going from the starting reagents to the transition state. Thus, if the solvent is polar, there would now be a smaller energy of solvation. Hence, solvents of high polarity disfavour S$_N$2 reactions due to the greater loss of solution energy in such cases.

There is a further property of the solvent that will affect the rate of S$_N$2 substitution, apart from just its polarity. In the reaction of the azide anion, N_3^-, on iodomethane, the rate of substitution is increased by a factor of 10 000 in changing the solvent from methanol to *N,N*-dimethylmethanamide (dimethyl formamide, DMF), even though the two solvents have approximately the same polarity. Suggest why this is so.

Methanol is a hydroxylic solvent, which means that it is capable of hydrogen bonding with its solutes, and so forms a solvent shell around the potential nucleophile. DMF, however, cannot hydrogen bond with the azide ion, but can only interact with it Coulombically. Furthermore, methanol can solvate both anions and cations, while DMF can only solvate cations. Thus, in effect, this means that the azide ion is now free of the solvent shell, and so may more readily attack the substrate.

So far, we have considered the rate of S$_N$2 substitution and factors that affect it. In so doing we have, however, already noted one of the most important points that affects the stereochemical consequences of this type of substitution reaction, namely that the incoming nucleophile approaches the substrate from the opposite side to which the leaving group departs. Suggest what will be the stereochemical consequence of this.

In a reaction that proceeds purely by the S$_N$2 pathway there ought to be complete inversion at the carbon site that undergoes substitution. This means, that if it were a chiral centre beforehand, it would afterwards have the opposite enantiomeric configuration. This is called the Walden inversion. Notice that it is an absolute requirement for the substituted carbon to have its configuration inverted. Hence, predict the susceptibility of a substituent at a bridgehead carbon to undergo S$_N$2 substitution, in for example 1-bromonorbornane.

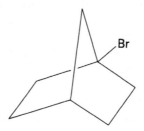

At such a bridgehead carbon, it is not possible for there to be either inversion of the carbon centre or for the incoming nucleophile to approach from the opposite side to the departing substituent. Hence, such a carbon is unable to be substituted effectively by this mechanistic pathway. You will recall that in the last section it was noted that such a carbon centre was inert to SN1 substitution, because of the impossibility of forming a flat carbonium ion. Thus, it will be appreciated that such a carbon will not readily undergo either SN1 or SN2 substitution, and so is generally inert to all types of nucleophilic substitution.

Notice also that in cyclopropyl derivatives, SN2 substitution does not occur (even though there is room for the approach of an incoming nucleophile), because the carbon being attacked cannot invert due to the restraint of the three-membered ring. Inversion would require the substituted carbon to go from one sp^3 configuration, with a normal bond angle of 109°, to another sp^3 configuration, via a trigonal bipyramid, which normally requires the bond angles for the groups in the trigonal plane to be 120°. The introduction of this extra bond strain is prohibitory to the inversion occurring.

There are occasions when the initial substrate is not sufficiently prone to attack by an incoming nucleophile and it needs activating in some manner. The commonest way is to make the potential leaving group a weaker base. This can be achieved by protonation so as to form the conjugate acid of the substrate. For example, a simple open chain ether is normally resistant to cleavage, and so usually requires protonation before it will cleave. This may be achieved by reacting the ether with hydrogen iodide. Write down the pathway for this cleavage, and identify the factors that have enabled this reaction to proceed.

$$HI + ROR \rightleftharpoons R_2OH^+ + I^-$$
$$R_2OH^+ + I^- \rightleftharpoons RI + ROH$$

The HI molecule protonates the ether oxygen. This increases the polarisation on the carbon that is to be substituted, and also changes the leaving group from the hard alkoxide ion to the softer alcohol molecule. Furthermore, the HI molecule provides a very good nucleophile in the form of an iodide ion that readily performs the SN2 reaction, which effects the cleavage of the ether. A substitution reaction that proceeds via the conjugate acid of the substrate is labelled SN1cA or SN2cA, or simply A1 or A2, depending upon whether the subsequent reaction follows a unimolecular or bimolecular pathway.

All of the reactions that we have studied so far have taken place between two molecules, i.e. they were intermolecular reactions. We will now turn our attention to the situation where one part of a molecule interacts with another part, i.e. there is an intramolecular reaction.

9.2.4 *Intramolecular substitution*

We have seen that primary and secondary halogenoalkanes when attacked by a hydroxide anion normally undergo inversion at the carbon centre at which substitution occurs, because they proceed via an SN2 pathway. In the case of 1-chloro-2-hydroxyalkanes, it is observed that when treated with a hydroxide anion to give the 1,2-diol, there has been retention at the carbon site that originally bore the chlorine atom. Suggest what may be the first step in the reaction, bearing in mind that hydroxide anions are hard nucleophiles.

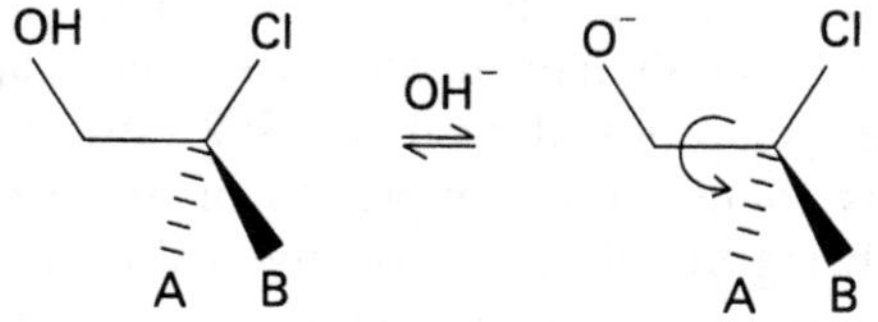

In this case, the hard hydroxide anion has removed the hard proton on the hydroxyl group to form an alkoxide ion in a simple Brönsted–Lowry acid/base reaction. Suggest what is the next step in this reaction, remembering that overall there is retention of the stereochemistry at the carbon bonded to the chlorine atom.

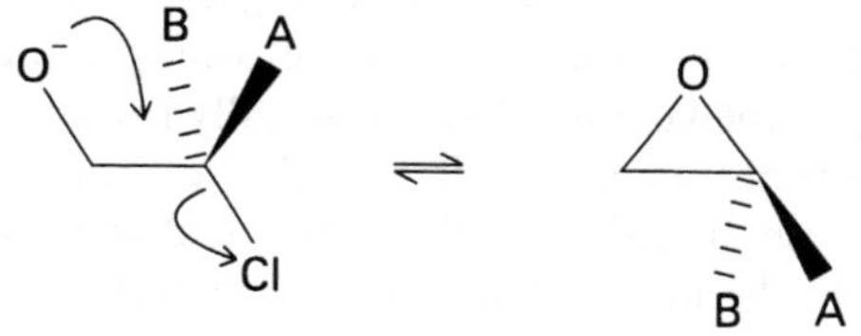

The alkoxide anion performs an internal attack upon the C–Cl bond. This proceeds in a similar manner to a normal SN2 reaction. Thus, there is an attack from the opposite side to the C–Cl bond, with the required inversion of configuration at that carbon, to produce the epoxide. Suggest what may follow in order to produce the 1,2-diol with overall retention at the carbon centre.

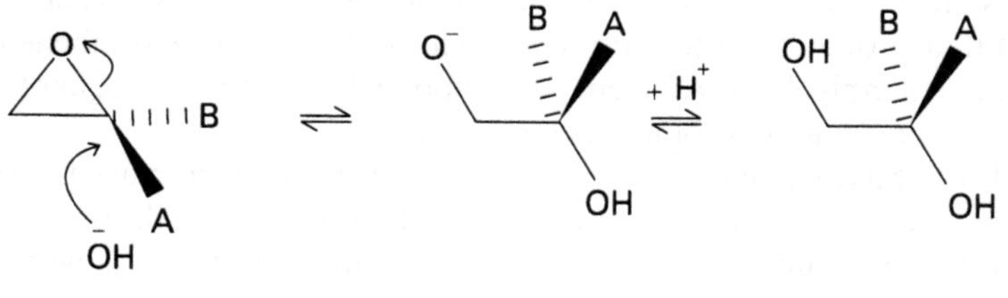

There is now an attack upon the epoxide by a hydroxide anion, which inverts the carbon centre for the second time, and this produces the monoanion of the diol, which is then protonated from the solvent to form the 1,2-diol. Notice how the two sequential inversions lead to overall retention of configuration.

This is an example of the neighbouring group effect, in which a nearby group that has an available lone pair, performs an internal SN2 reaction on the centre to be substituted. The neighbouring group is said to be lending anchimeric assistance in the reaction. In so doing, it forms a cyclic compound, which is then opened up by the original nucleophile to form the final product, but with overall retention of configuration at the carbon centre that has undergone substitution.

This may be illustrated by the reaction of α-bromocarboxylic acids, which undergo substitution of the bromine atom by hydroxide anions with retention of configuration when the concentration of the hydroxide anions is low. Suggest what is the mechanism for the substitution reaction at low and high concentrations of hydroxide anions.

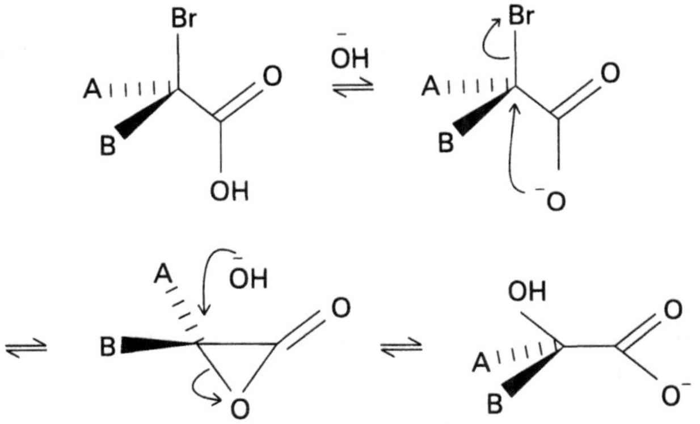

At low concentration, the reaction proceeds via an α-lactone; i.e. the carboxylate anion, which is generated by the hydroxide anion, acts as a neighbouring group and reacts internally. This lactone is then attacked by another hydroxide anion that opens up the ring to form the α-hydroxylcarboxylate anion, with overall retention of configuration at the α-carbon. At higher concentrations of hydroxide anions, the reaction proceeds directly by a normal intermolecular SN2 reaction with overall inversion at the α-carbon.

The neighbouring group effect is not limited to the internal reaction of an oxygen anion, but may also be effected by the lone pair on a sulphur or nitrogen atom. Furthermore, in the chapter on addition reactions to carbon/carbon multiple bonds, we will see a bromine atom acting as a neighbouring group when it forms a cyclic bromonium ion intermediate during the addition of a bromine molecule to an alkene.

The examples given so far have involved the formation of a three-membered ring. However, this effect is not limited to cases in which a three-membered ring may be formed. Suggest which other ring sizes may be observed.

Apart from three-membered rings, the two other common sizes are five and six. Four-membered rings are not observed in this type of mechanism. Three-, five- and six-membered rings may be easily formed, because with such ring sizes it is possible, without undue strain, to bring the ends within close proximity of each other and so allow them to interact.

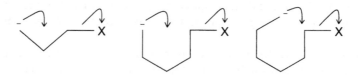

In the case of a four-membered ring, in which the normal bond angle of an sp^3 hybridised carbon is 109°, it is not possible for the ends of a four-carbon chain to be in close proximity, and so the molecule must be put under strain in order for the ends to become close enough in order to interact; there would therefore be a larger activation barrier for a reaction to proceed by this route.

Explain why the intramolecular substitution reaction occurs in preference to an intermolecular attack.

When two molecules react to form one molecule, there is a great reduction in the entropy of the system. There is not such a large decrease in entropy when a molecule reacts with itself, for in this case, there is only a loss of some free rotation within the molecule. Thus, intramolecular reactions are favoured entropically over intermolecular reactions. Hence, if the other factors are similar, intramolecular reactions are more facile than intermolecular reactions. Substitution reactions that occur via a neighbouring group mechanism are therefore often faster than those that occur by a simple intermolecular SN2 pathway.

When an alcohol is treated with thionyl chloride, SOCl$_2$, the rate equation is found to be dependent upon the concentration of both the substrate and the thionyl chloride. Yet it is observed that there is retention at the hydroxyl carbon. The rate of this reaction is found to increase with increasing polarity of the solvent. This observation, it will be recalled, is unusual for a normal SN2 reaction. With this simple, but important, piece of kinetic information, and a consideration of which atom bears the greatest positive charge, suggest what may be the first two steps of this reaction.

The sulphur atom is the most electron deficient atom, and so is prone to attack by the hydroxyl group to form a chlorosulphite intermediate. Notice that the C–O bond has not been broken in this step, and so there cannot have been any change in the configuration at this centre. This intermediate now fragments. Suggest how this may occur.

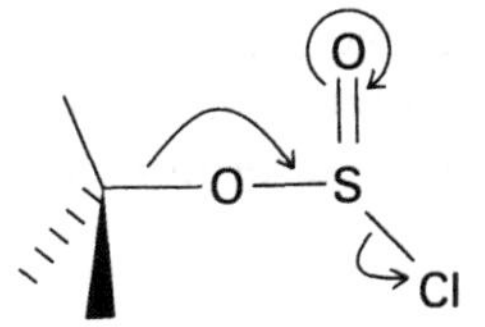

The C–O bond breaks to form the chlorosulphite anion. This further decomposes to form sulphur dioxide and a chloride ion. This is the slow step of the reaction.

Suggest what occurs next, and so provide an explanation as to why there is retention of configuration.

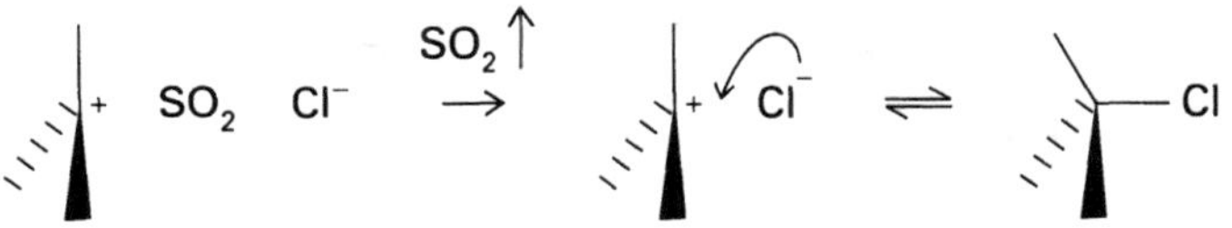

Once the sulphur dioxide has escaped, the chloride ion attacks the carbonium ion to form the substituted compound. The fragmentation of the chlorosulphite to give the sulphur dioxide and chloride ion occurs within a cage of solvent molecules. The sulphur dioxide escapes as a gaseous product, while the anion is still held in place by this solvent cage. Thus, the nucleophile is formed on the same side as the original C–O bond; it then attacks the carbonium from this side, and so results in retention of the configuration. The attack by the chloride anion occurs very quickly after the sulphur dioxide molecule is lost, and so there is no time for the solvent cage to collapse to allow the release of the anion. This reaction type is called internal nucleophilic substitution, S_{Ni}. Note that it is possible for the chlorosulphite intermediate to be attacked from the reverse side if there is another nucleophile present, and, in this case, the reaction would proceed with inversion of the carbon centre.

In this section, we have seen the formation of three-membered ring intermediates when a neighbouring group participates in a reaction. We will now look at some other reactions involving three-membered rings, but this time ones that exist in either the starting material or the product, rather than just as an intermediate.

9.2.5 *Formation and cleavage of three-membered rings*

One of the commonest three-membered ring systems is the three-membered cyclic ether, otherwise known as an epoxide, or oxirane. The first thing to appreciate is that a three-membered ring is under strain, because of the compression from the normal bond angle of 109° to about 60°. This strain is released when the cyclic compound forms the open chain compound. Unlike the cleavage of open chain ethers, which is quite difficult, the opening up of the epoxide ring is easy and may be achieved by water at neutral pH, even though this reaction is rather slow. The reaction is faster if either an acid or base catalyst is employed.

Write down the three possible hydrolysis mechanisms that may occur depending upon the pH of the solution. Pay particular attention to whether or not the substrate or intermediate is protonated, and also the nature of the attacking nucleophile.

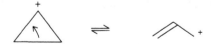

The formation of the vicinal diol, i.e. the 1,2-diol, may thus be achieved under a wide range of conditions.

Now, instead of studying a heterocyclic three-membered ring, we will examine a carbocyclic one, which has rather different properties. We saw earlier that a cyclopropane ring will not undergo an S$_N$2 reaction, because of the difficulty of the carbon inverting its configuration. The carbon atoms that constitute the ring are secondary and thus capable of forming carbonium ions. Such an ion, however, would be even more strained than the neutral molecule, because the carbonium ion normally adopts an sp^2 geometry, which has a bond angle of 120°. Suggest what would happen to a cyclopropyl cation if it were to be formed by some means.

One of the bonds in the three-membered ring may break and so form an allylic carbonium ion that is stabilised due to delocalisation. This ion may then react further to form the final substituted product.

This reaction may proceed in the reverse direction. For example, suggest the pathway whereby the allylic compound, 3-bromo-2-methylpropene, is converted to the 1-methylcyclopropyl cation.

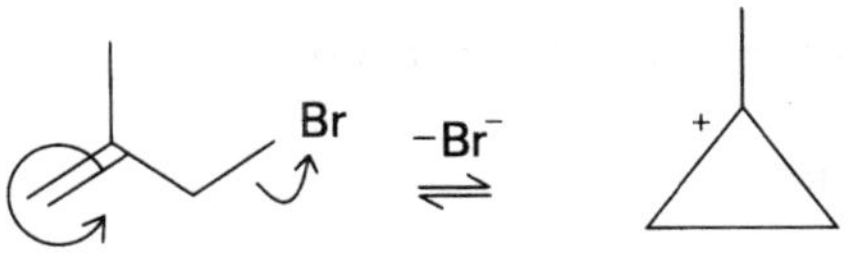

In this case, the product is a tertiary carbonium ion, which helps to stabilise an otherwise strained system. This is the methylcyclopropyl carbon skeleton.

An isomer of this carbonium ion has the primary carbon bearing the positive charge. This arrangement is called the cyclopropylmethyl carbon skeleton. If this carbonium ion is used, then it may rearrange to form the cyclobutyl carbonium ion. Suggest how this may occur.

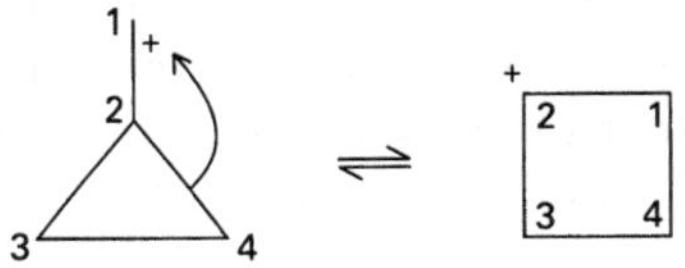

It is also possible to form the cyclopropylmethyl system when a 4-halo-genobutene system is attacked by a nucleophile. Suggest how this conversion may occur.

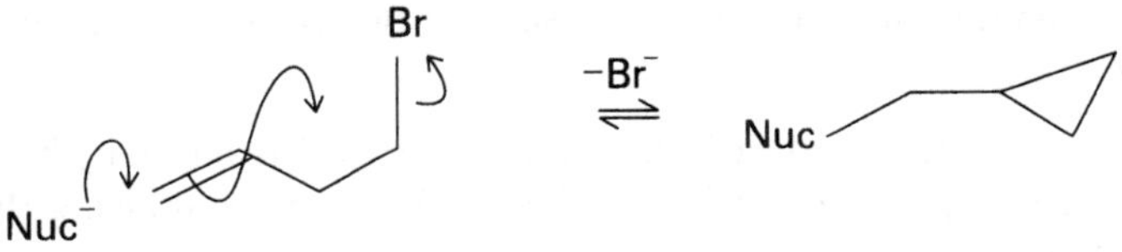

The 4-halogenobutene is a homoallylic compound, which means that the substituent is one methylene unit, $-CH_2-$, removed from the allylic position, which in turn is one methylene unit removed from the vinyl carbons. Instead of the incoming nucleophile attacking directly the carbon that bears the halogen atom, it attacks the unsaturated carbon. In this case, the π bond acts as an internal nucleophile to displace the halogen, forming the cyclopropyl ring at the same time.

As may be readily appreciated from the foregoing examples, substitution reactions on cyclopropyl systems and their derivatives may take some unexpected routes and lead to a variety of interesting products. Which pathway will actually be followed, or more realistically, what will be the spread of products, may be only determined by performing the experiment and identifying the products. The value of the mechanistic approach is that it will

163

give some indication of those synthetic reactions that might be worth pursuing, because there is at least a plausible pathway by which the desired products could be formed.

9.3 Substitution at an unsaturated carbon

9.3.1 SN1 mechanism

So far, we have looked at substitution reactions in which the leaving group was attached to a saturated carbon. We will now turn our attention to the situation when the leaving group departs from an unsaturated carbon and examine how this change modifies the mechanistic pathways that we have already studied.

When an alkene derivative, $CH_2=CHX$, forms a carbonium ion, consider whether this fragmentation would be more or less difficult than the related reaction if a saturated derivative had been used.

We already know from the chapter on the acidity of protons that in an sp^2 hybridised bond the electrons are held closer to the carbon than in an sp^3 hybridised bond. So, it is harder to break the C–X to form a carbonium ion, because it is, first, a stronger bond, and secondly, less polarised towards the leaving group.

It is found that a vinyl carbonium ion is slightly less stable than the corresponding alkyl carbonium ion. In particular, a secondary vinyl carbonium ion is less stable than a secondary alkyl carbonium ion.

Bearing all these factors in mind, it is apparent that vinyl carbons tend not to undergo the SN1 reaction unless conditions are so arranged as to stabilise the resultant carbonium ion intermediate. Suggest two possible ways in which this could be done.

It is possible to force a vinyl carbon to undergo an SN1 reaction by having a very good leaving group, such as triflate, $-OSO_2CF_3$, or by stabilising the positive charge with an α-aryl group.

Suggest what will be the shape of the intermediate vinyl carbonium ion, and then suggest what will be the stereochemical consequences of an SN1 reaction on a vinyl carbon.

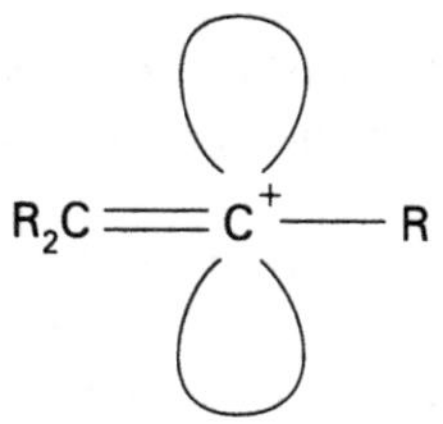

$$R_2C = \overset{+}{C} - R$$

When a vinyl carbonium ion is formed it is linear, and so the stereochemistry of the final product after the addition of the electrophile will be randomised.

In the case of an alkynyl halide, SN1 substitution is even more disfavoured than in the case of a vinyl halide, because it would result in a positive charge on an sp hybridised carbon, after having broken a bond that is stronger and less easily polarised than is the case of the vinyl halide.

9.3.2 SN2 mechanism

We will now consider the possibility of the SN2 reaction mechanism in unsaturated systems. First, draw a diagram of bromoethene, showing the π orbitals of the double bond. Then, consider the arrangement of the electrons in space and how they may interact with an incoming nucleophile.

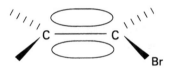

In the case of an alkene, there is a large electron cloud that is close to the line of approach of any potential nucleophile. This will exert a disfavourable Coulombic interaction and so inhibit the approach of the nucleophile.

Finally, consider the possibility of inversion at such an unsaturated carbon atom.

The transition state at a saturated carbon is a trigonal bipyramid, in which the three groups that are not involved in the substitution reaction orientate themselves in the trigonal plane, while the approaching and leaving groups interact along an axis that is orthogonal to this plane. The hypothetical transition state at an unsaturated carbon atom would be based upon an octahedron framework, with the approaching and leaving groups interacting along an axis that is orthogonal to a plane that contains the two lobes of the p orbital of the carbon being substituted, perpendicular to the two other σ bonds that are not involved in bond cleavage or formation. It is hypothetically possible for this arrangement to exist; however, it is observed that in an sp^2 hybridised carbon inversion of the configuration does not take place in practice.

Consider the case of a hypothetical SN2 reaction on an alkynyl halide, and predict whether such a molecule would react by this pathway.

An alkynyl halide cannot undergo an SN2 reaction at all, because, first, that would require the incoming nucleophile to approach through the carbon at the other end of the triple bond; second, there is no mechanism whereby the carbon could invert its configuration.

9.3.3 Tetrahedral mechanism

We have seen above that in a vinyl carbon, the SN1 is not particularly favoured, while the SN2 mechanism is highly unlikely.

Instead of performing the one step bimolecular SN2 reaction, alkenes react via two closely related bimolecular pathways. The first of these is called the tetrahedral mechanism and proceeds via a negatively charged intermediate. This mechanism is sometimes called the addition/elimination reaction, which is given the label Adn/E. This alternative name is unfortunate, because the other pathway is called the addition/elimination mechanism and proceeds via a readily detectable neutral intermediate. This latter mechanism will be considered in the chapter on sequential addition/elimination reactions. In this book, in an attempt to reduce the confusion, we will call the mechanism that proceeds via an anionic intermediate the tetrahedral mechanism, and reserve the name addition/elimination mechanism for the mechanism that proceeds via a neutral species.

Consider the general alkene, $ZCH=CHX$, and let it be attacked by a nucleophile, Y^-. Draw the mechanism for this attack, and suggest how the anionic intermediate may be stabilised.

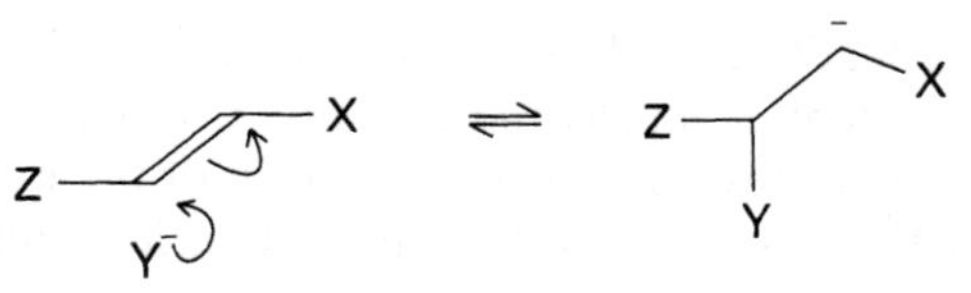

The adjacent carbon now bears a negative charge and this may be stabilised by either strongly $-I$ groups, or better still $-M$ groups such as carbonyl or cyano groups.

Suggest how this anionic intermediate may now react.

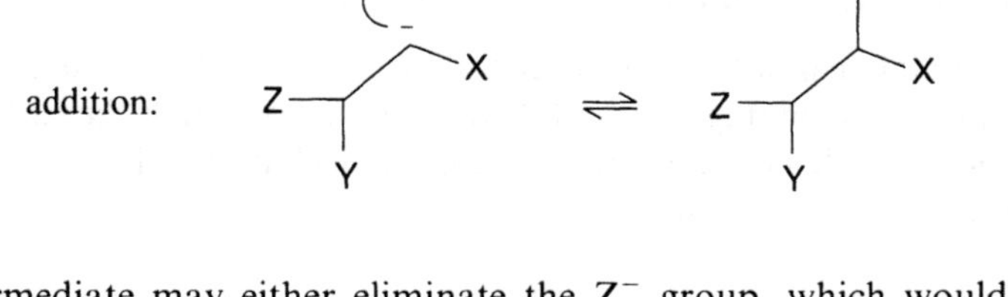

The intermediate may either eliminate the Z^- group, which would result in overall substitution, or it could add an electrophile, which would result in overall addition to the alkene bond. These two reactions usually compete.

166

9.3.4 *Ester hydrolysis*

In the two preceding subsections, we examined the possible substitution reactions which occurred at an unsaturated carbon that was multiply bonded to another unsaturated carbon. When, however, a carbon atom is doubly bonded to an oxygen atom, it forms a carbonyl group. Such compounds undergo a wide variety of reactions. In particular, when one of the substituents on the carbonyl carbon is an electron withdrawing group, substitution reactions occur very easily via the tetrahedral mechanism, with concomitant substitution of the electron withdrawing group for the incoming nucleophile. Furthermore, in the chapter on sequential addition/elimination reactions we will look at those reactions of the carbonyl group in which the carbonyl oxygen is itself substituted. These reactions again proceed via the tetrahedral mechanism, but often in these cases, there is an intermediate that may be isolated.

Returning to substitution reactions of the electron withdrawing group bonded to the carbonyl carbon, write down the structure of an ester, RCO_2R, and identify the electron withdrawing group that will undergo substitution.

$$
\begin{array}{c}
\overset{\delta-}{O} \\
\parallel \\
R \overset{\delta++}{\diagup} \overset{}{\diagdown} \overset{\delta-}{OR}
\end{array}
$$

The electron withdrawing group is the alkoxyl moiety.

Write down the overall reaction for the hydrolysis of an ester.

$$RCO_2R' + H_2O \rightleftharpoons RCO_2H + R'OH$$

On hydrolysis, the ester yields the carboxylic acid and the alcohol. This reaction proceeds under either acid or base conditions. The hydrolysis reaction may be either unimolecular or bimolecular. Furthermore, it is possible for either the bond between the carbonyl carbon and the alkoxyl oxygen to break (acyl cleavage); or for the bond between the alkoxyl oxygen and the alkoxyl carbon to break (alkyl cleavage). This gives rise to eight different possible reactions, which are distinguished by the following labels: A or B for acid or base conditions; 1 or 2 for a unimolecular or bimolecular pathway; and, AC or AL for the acyl or alkyl cleavage. The hydrolysis of an ester under basic conditions is also called saponification, because it is involved in the process whereby soap was originally made.

The two most important pathways that occur in practice are the AAC2 and the BAC2, both of which proceed via a tetrahedral intermediate. Write down these two pathways, paying particular attention to the nature of the attacking species and the protonation and deprotonation steps.

AAC2

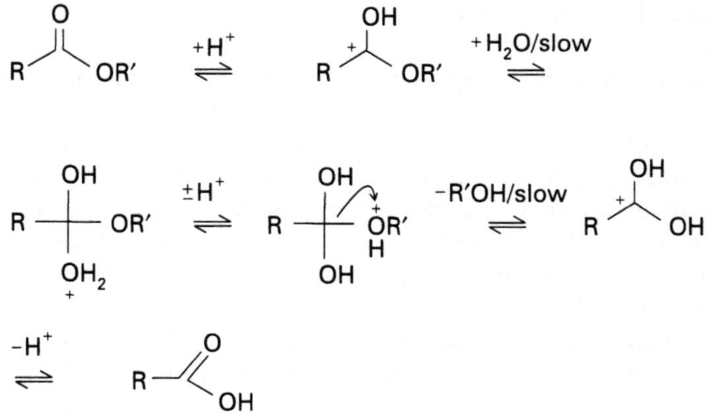

BAC2

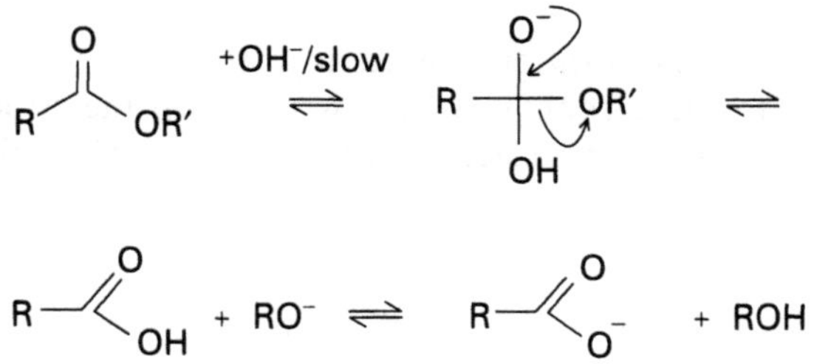

These are the normal pathways for acid or base hydrolysis of an ester. Notice that if the alcohol were chiral, then there would be retention of configuration and no rearrangement of the carbon chain.

The other six possible pathways only occur under special circumstances, if at all. We will look at the unimolecular reactions first, i.e. AAC1, AAL1, BAC1 and BAL1. In each case, these reactions proceed in a manner that is related to the SN1 reaction.

The initial part of the AAC1 reaction bears a resemblance to the SN1cA reaction. Thus, first the carbonyl oxygen, and then the alkoxyl oxygen are protonated. Write down these two steps and then suggest what the remaining steps are.

AAC1

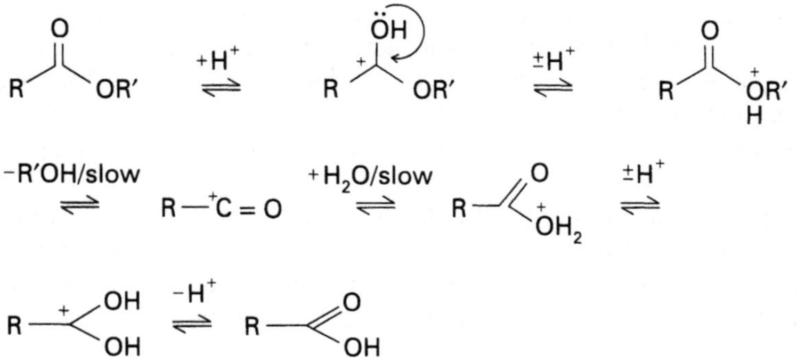

The cleavage of the acyl bond and the attack by the water molecule are both slow steps. This pathway is followed when R is very bulky, which hinders the approach by the nucleophile, and when there is an extremely strong acid in an ionising solvent.

In the A$_{AL}$1 mechanism, after the initial protonation of the carbonyl oxygen, the O–alkyl bond breaks to yield a carbonium ion. Write down this pathway and suggest how it proceeds. Suggest the characteristics of R′ that would favour this mechanism.

A$_{AL}$1

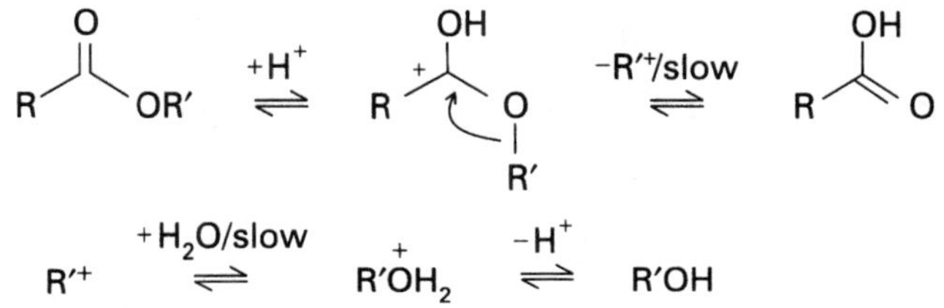

R' must be able to form a stable carbonium ion, and so a tertiary carbon would provide a suitable example. This is also the requirement in the related B$_{AL}$1 reaction pathway, which occurs only under very mild basic conditions or sometimes even neutral conditions. Write down this pathway. Only the final deprotonation of the conjugate acid of the alcohol requires a base in this case.

B$_{AL}$1

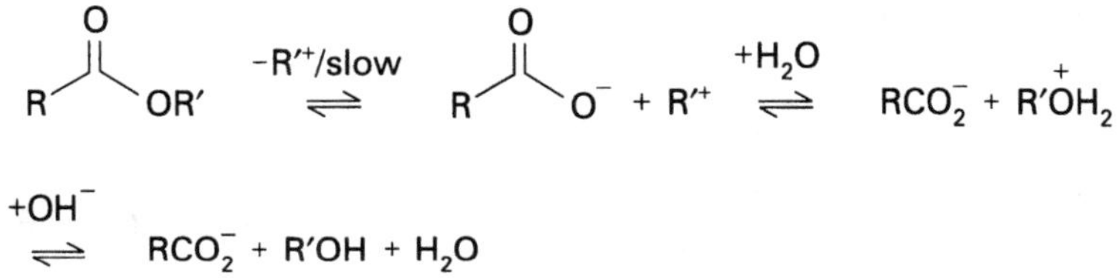

In all cases that involve *O*-alkyl cleavage, the proposed reaction proceeds via an S$_N$1 or S$_N$2 type substitution on the alkoxide carbon, while the leaving group is the acyloxy group, –OCOR, or its conjugate acid.

The last possible unimolecular pathway, B$_{AC}$1, is not observed in practice. Write down this pathway and then suggest a reason why this reaction mechanism does not occur.

B$_{AC}$1

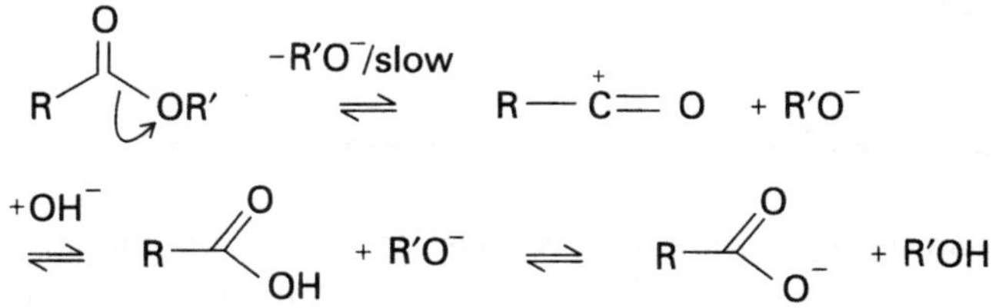

169

The pathway requires that an alkoxide anion leaves to form the acylium cation; however, the alkoxide anion is not a good enough leaving group to do that.

There are two remaining possible pathways. Both of these proceed via a bimolecular mechanism. The acid catalysed version, A$_{AL}$2, is similar to the S$_{N}$2cA pathway. This mechanism, like the one before, is not observed experimentally. Write down a route for this reaction, and also suggest why it is not observed.

A$_{AL}$2

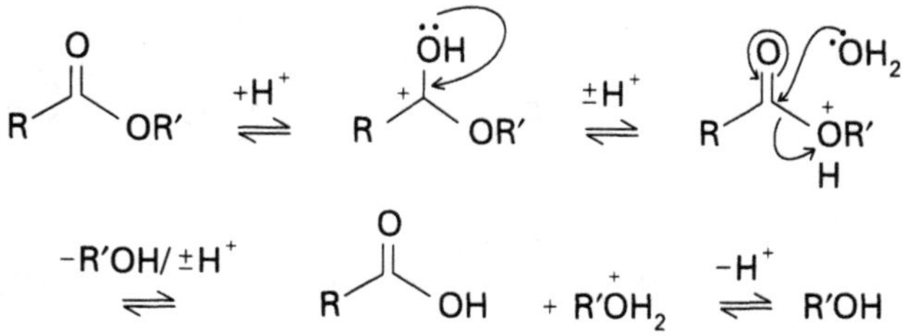

This reaction would require a water molecule to be a nucleophile in an S$_{N}$2 reaction, which it is rarely capable of being.

The last reaction type, B$_{AL}$2, requires an hydroxide anion to attack an alkyl carbon in preference to the carbonyl carbon. Write down this mechanism, and deduce what will be the stereochemical consequences on the configuration of the alcohol carbon.

B$_{AL}$2

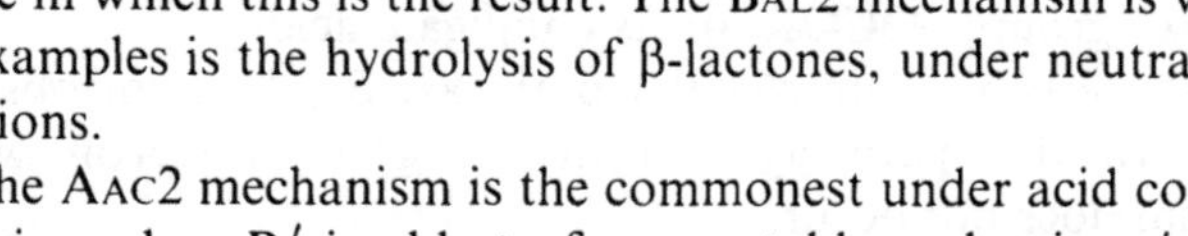

The alcohol carbon will be inverted, and so this pathway is easy to distinguish as it is the only route in which this is the result. The B$_{AL}$2 mechanism is very rare. One of the few examples is the hydrolysis of β-lactones, under neutral or very mild basic conditions.

In summary, the A$_{AC}$2 mechanism is the commonest under acid conditions, with A$_{AL}$1 occurring when R′ is able to form a stable carbonium ion; while under basic conditions B$_{AC}$2 operates in almost all cases. The other pathways only occur, if at all, in limited circumstances.

9.4 Summary of nucleophilic substitution reactions

In a substitution reaction one group is replaced by another group. Substitution reactions may be classified according to the nature of the attacking species, namely nucleophile, electrophile or free radical. The substitution of one functional group for another is called a functional group interchange or interconversion, FGI. By replacing non-carbon containing groups with ones that contain carbon, the original carbon skeleton can be augmented, thus permitting the synthesis of more complicated molecules. This is the basis of organic synthesis.

In a unimolecular substitution reaction, a group departs initially with a pair of electrons leaving an electron deficient carbonium ion intermediate, which is subsequently attacked by an incoming nucleophile. Note that this is a two step reaction, in which the intermediate can, on occasions, be isolated. In contrast, in a bimolecular substitution reaction, the leaving group departs simultaneously as the nucleophile approaches. In this case, the reaction occurs in one step, with only a transition state and no proper intermediate.

The rate of a unimolecular substitution depends only upon the concentration of the substrate, and not upon the concentration of the nucleophile. Such reactions are called unimolecular nucleophilic substitution reactions, SN1. The first step is a heterolytic cleavage of the carbon/heteroatom bond, to result in a carbonium ion.

The rate of substitution in an SN1 reaction is affected by both properties inherent to the substrate and external factors. Thus, considering inherent factors, the rate is increased by (a) a weaker C–X bond; (b) substituents that give rise to more stable leaving groups, e.g. in general, species that form on leaving weak Brönsted–Lowry conjugate bases or neutral molecules; (c) a more stable resultant carbonium ion, e.g. a tertiary over a secondary over a primary carbonium ion; (d) the possibility of delocalisation in the resultant carbonium ion; (e) a more planar resultant carbonium ion, i.e. not one at a bridgehead; and (f) the presence of +I or +M groups. For external factors, the rate is increased by (a) a solvent of high polarity; (b) the increase of the ionic concentration of the reaction mixture; and (c) the absence of common ions. Furthermore, the use of a weak nucleophile favours this pathway, because it is unlikely to attack the substrate itself, and so promote an alternative mechanism.

A unimolecular substitution reaction results in the stereochemical information contained at the substituted centre being lost, because the carbonium ion intermediate is planar and so can be attacked from above and below with equal ease. Thus, there will be racemisation. The longer the carbonium ion exists before it is attacked to form the product, the closer the resultant mixture will be to being perfectly racemised.

In contrast to a unimolecular nucleophilic substitution reaction, in a bimolecular nucleophilic substitution reaction, SN2, the rate of substitution depends upon the concentration of both the substrate and the attacking nucleophile. In such a reaction, there is simultaneous bond fission and formation. Both the order and the molecularity of the reaction are two.

The nucleophile must approach the substrate from the opposite direction from that in which the leaving group is departing. This results in inversion of configuration, namely the Walden inversion.

The rate of substitution in an SN2 reaction is affected by a number of factors that are similar, but usually opposite, to those that affect the rate in an SN1 reaction. An exception is that, like the situation found in SN1 reactions, the weaker the C–X bond the faster the rate. However, there are some additional factors that should be borne in mind, for example, steric crowding, which hinders the approach of the incoming nucleophile, and decreases the rate of substitution.

Iodide ion may be used as a nucleophilic catalyst, because it is both a good nucleophile and a good leaving group.

A substitution reaction that proceeds via the conjugate acid of the substrate is labelled SN1cA or SN2cA, or simply A1 or A2, depending upon whether the subsequent reaction follows a bimolecular or unimolecular pathway.

The neighbouring group effect can lead to a retention of configuration, due to two sequential inversions. The reaction proceeds via an intramolecular SN2 reaction to form a cyclic intermediate, which is then opened up by the attack of the nucleophile to form the final product. The intramolecular substitution reaction is favoured over the intermolecular substitution reaction due to the smaller decrease in entropy that such a route necessitates. This intramolecular attack can proceed via three-, five- or six-membered rings, but not four-membered rings.

The reaction between an alcohol and thionyl chloride proceeds via an internal nucleophilic substitution, SNi, in which retention is also observed at the substituted carbon.

Epoxides may open up under a wide range of pH conditions.

A cyclopropyl cation may open to give an allylic cation. This reaction is reversible.

SN1 substitution reactions occur less readily at unsaturated carbon centres than at saturated ones, because the resultant vinyl, or alkynyl, carbonium ion is less stable. However, in the vinyl carbonium ions at least, it is possible to encourage such a reaction by having a very good leaving group, such as triflate, or by stabilising the positive charge with an α-aryl group. The intermediate vinyl carbonium ion is linear, and so the resultant stereochemistry of the product is randomised. In contrast, direct SN2 reactions do not occur at either doubly or triply bonded carbon centres, because of the high electron density that disfavours the approach by a nucleophile and the problems associated with inversion of such an unsaturated centre.

However, alkenes do undergo substitution reactions via two closely related bimolecular pathways. The first of these is called the tetrahedral mechanism and proceeds via a negatively charged intermediate. The other pathway is called the addition/elimination mechanism and proceeds via a readily detectable neutral intermediate. In the tetrahedral mechanism, the nucleophile attacks one carbon of the alkene double bond, so the other carbon acquires a negative charge. This negative charge may be stabilised by either strongly $-I$ groups, or better still $-M$ groups such as carbonyl or cyano. The intermediate may either eliminate the electron withdrawing group, which would result in overall substitution, or it could add an electrophile, which would result in the overall addition to the alkene bond. These two reactions usually compete.

In carbonyl compounds that contain an electron withdrawing group attached to the carbonyl carbon, substitution of that electron withdrawing group readily occurs via the tetrahedral mechanism. The hydrolysis of an ester may occur in eight different ways, namely: the reaction may proceed under either acid or base conditions; be either unimolecular or bimolecular; and occur with either acyl or alkyl bond cleavage. These variations are distinguished using the labels A or B for acid or base conditions; 1 or 2 for unimolecular or bimolecular pathway; and AC or AL for acyl or alkyl cleavage. The hydrolysis of an ester under basic conditions is also called saponification. The two most important pathways that occur in practice are the AAC2 and the BAC2, both of which proceed via the tetrahedral mechanism. Notice that if the alcohol were chiral, there would be retention of configuration and no rearrangement of the carbon chain. The other six possible pathways only occur under special circumstances, if at all.

10

Electrophilic Substitution Reactions

10.1 Introduction

So far, we have looked at substitution reactions in which the attacking species has been a nucleophile. Now, we will look at reactions in which the attacking species has the opposite electrical properties, i.e. it is an electrophile.

In an electrophilic substitution reaction, the attacking species is an electron deficient species. Therefore, the leaving group must also be electron deficient, i.e. an electrofuge, in order to maintain the overall charge on the substrate. Write down the equation for the overall reaction of the electrophilic substitution of RX by an electrophile, E^+.

$$RX + E^+ \rightleftharpoons RE + X^+$$

It is obvious that the stability of the departing group as an electrofuge will be one of the principal factors that will affect how easy this pathway is.

In a nucleophilic substitution reaction, it was found that either the leaving group left before the incoming nucleophile arrived, or else the arrival and departure were simultaneous. These different pathways were required so that the quota of electrons around the carbon undergoing substitution did not exceed eight. In contrast, in the case of electrophilic substitution, because the incoming electrophile is electron deficient, it may attack the substrate to form an intermediate adduct, from which an electrofuge can leave in order to generate the substituted product. Write down the general equation for this two-step process.

$$RX + E^+ \rightleftharpoons [RXE]^+ \rightleftharpoons RE + X^+$$

This type of sequence may be characterised as being an addition/elimination process. The reverse of such a sequence is also possible. In this case, the reaction

175

would proceed via a two-step process that is comparable to the S_N1 reaction. Write down the general equation for the two stages of such a pathway for the substitution of RX by E^+.

$$\text{step 1:} \quad RX \rightleftharpoons R^- + X^+$$
$$\text{step 2:} \quad R^- + E^+ \rightleftharpoons RE$$

Such a heterolytic cleavage gives rise to an intermediate carbanion, and the stability of that species is important in determining the rate of this particular pathway.

In this chapter, we will first study the electrophilic substitution reactions that occur in aromatic systems, because they are well defined, and then we will look at those that occur in aliphatic systems.

10.2 Aromatic substitution

10.2.1 Arenium ion mechanism

In aliphatic systems that are substituted with an electron-withdrawing group, there is a $\delta+$ carbon centre that is readily accessible to an attack by an incoming nucleophile. Hence, nucleophilic substitution predominates in such systems. In contrast, in aromatic systems, it is immediately apparent that the electronic characteristics of the substrate are different, because there is a delocalised ring of electrons that exists above and below the aromatic ring. This area of high electron density naturally hinders the approach of an incoming negative species, but correspondingly attracts positive ones.

Thus, commonly, the attacking species in aromatic substitution reactions is an electrophile, which is usually either a positive ion or the positive end of a dipole. Similarly, the leaving group must lack an electron pair.

In nucleophilic substitution, the best leaving groups were those that could most readily accommodate an unshared pair of electrons. By analogy, suggest what will be the best leaving groups in an electrophilic substitution.

The best leaving groups in an electrophilic substitution reaction will be weak Lewis acids.

In aromatic chemistry, there is one electrophilic pathway that predominates. This is called the arenium ion mechanism. It occurs in three steps. The first is the formation of the attacking electrophile; the second is the attack by this electrophile on the aromatic ring; and the third is the departure of an electrofuge, which is usually a proton, with the assistance of a base. Write down the general equations for this sequence of reactions.

$$\text{step 1:} \quad Z\text{–}X \rightleftharpoons Z^- + X^+$$
$$\text{step 2:} \quad C_6H_5Y + X^+ \rightleftharpoons [C_6H_5YX]^+$$
$$\text{step 3:} \quad [C_6H_5YX]^+ \rightleftharpoons C_6H_5X + Y^+$$

The production of the attacking electrophile may occur in many different ways, and aromatic chemistry is often studied by reference to the formation of this species. This initial step is not usually rate limiting.

The remaining two steps proceed along very similar lines in most electrophilic aromatic substitution reactions. The attack by the electrophile is usually the rate limiting step. The cationic intermediate is called a Wheland intermediate, or σ-complex or an arenium ion, and can sometimes be isolated.

The stability of the Wheland intermediate will obviously have a large effect upon the overall rate of this reaction pathway. First, write down the two principal canonical forms of the benzene ring, C_6H_6, and then write down the resonance hybrid of benzene.

canonical structure   resonance hybrid

It is the extensive ability of benzene to form delocalised structures that accounts for the increased stability of the benzene system over the hypothetical cyclohexatrienyl system.

Now write down the mechanistic step that shows the attack of an electrophile on a benzene ring that gives rise to the Wheland intermediate.

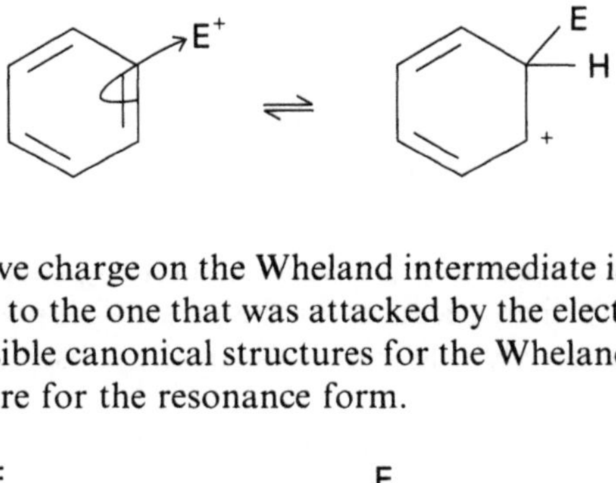

Note that the positive charge on the Wheland intermediate is initially placed on the carbon adjacent to the one that was attacked by the electrophile. Now write down the three possible canonical structures for the Wheland intermediate, and so suggest a structure for the resonance form.

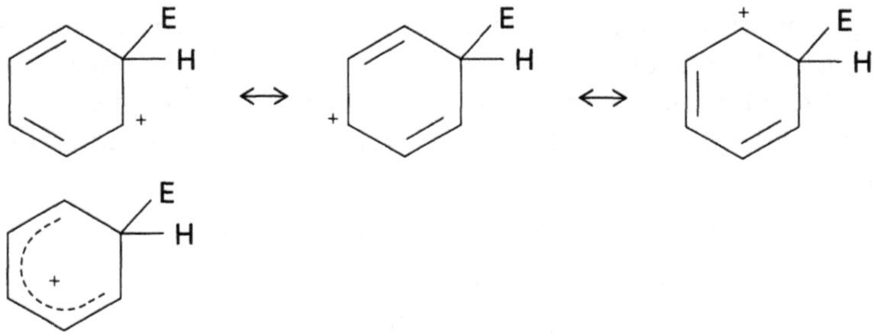

Even though there is still extensive delocalisation of the positive charge that has been introduced by the electrophile, it is limited to that which would have been available to a linear, non-cyclic system. The extra delocalising possibilities that were available to the benzene system, because it was cyclic, are no longer possible. This means that the cyclohexadienyl cation is significantly less stable than the initial benzene ring, because it has lost the aromatic stabilisation of the cyclic sextet of electrons.

The last step in this pathway is the loss of an electrofuge. If, as in this example, a proton is lost with the help of a base, then overall a substitution reaction has occurred. This is the commonest type of substitution reaction that occurs in aromatic chemistry. Write down this step and suggest what is the thermodynamic driving force.

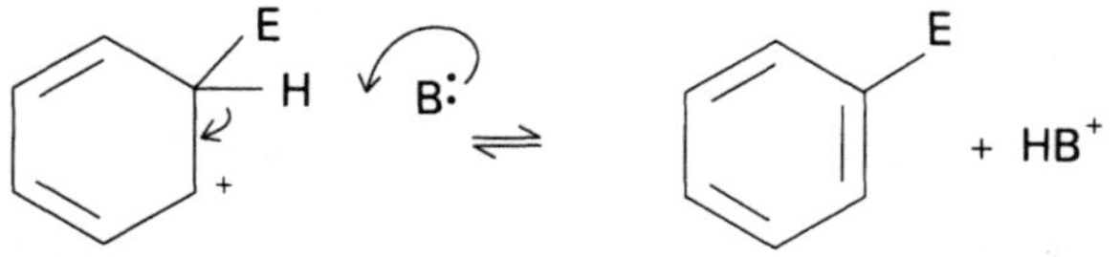

The restoration of the aromatic sextet, with the accompanying energy of delocalisation, provides the driving force for this step of the reaction.

It has been suggested that there is a step prior to the formation of the Wheland intermediate, which involves the complexation of the attacking electrophile with the π-system of the aromatic ring to form a π-complex, and which then converts into the σ-complex. The structure of such an intermediate may be represented as:

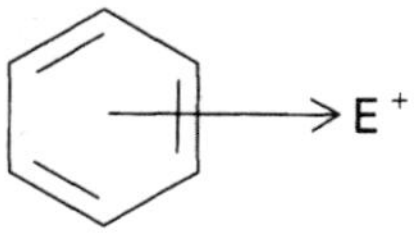

The involvement of this extra step is very difficult to prove in practice, and we will not consider it further, because, for the present purposes, it is sufficient to postulate that there is only one step between the substrate and the Wheland intermediate.

One of the commonest electrophilic substitution reactions of benzene is the simple nitration of the aromatic ring. In this reaction, the attacking electrophile is the nitronium ion, NO_2^+, which may be formed by the action of concentrated sulphuric acid on concentrated nitric acid. Bearing in mind that sulphuric acid is a stronger acid than nitric acid, write down the equation for the formation of this cation.

$$H_2SO_4 + HNO_3 \rightleftharpoons HSO_4^- + H_2NO_3^+$$
$$H_2NO_3^+ \rightleftharpoons H_2O + NO_2^+$$
$$H_2O + H_2SO_4 \rightleftharpoons H_3O^+ + HSO_4^-$$

Note that two molecules of sulphuric acid are consumed: the first to produce the protonated nitric acid that decomposes to produce the nitronium ion; and the second to protonate the water molecule, which is a side product formed from the decomposition of the protonated nitric acid.

Write down the second step of this reaction, which is the attack of the nitronium ion on the benzene ring, paying particular attention to the position of the positive charge on the intermediate.

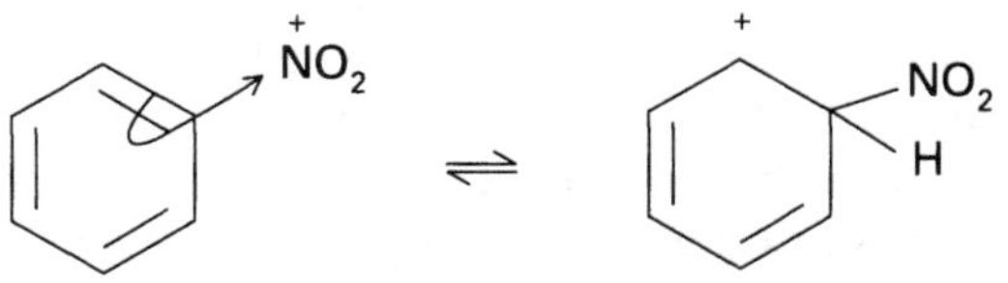

Notice that the positive charge initially resides on the carbon that is adjacent to the one that was attacked. The charge is, of course, in practice delocalised over the remaining conjugated double bond system.

Write down the last step in this reaction. Suggest the identity of the base that participates in this step.

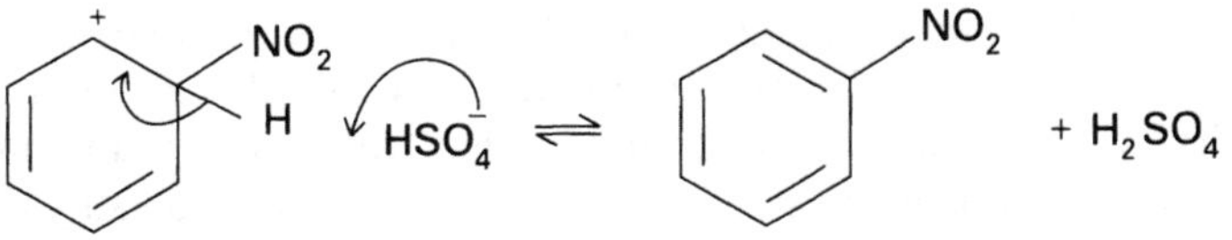

The hydrogen sulphite anion acts as a base to help remove the proton so as to re-form the aromatic ring to result in the nitrobenzene product.

There are various ways in which the nitronium ion may be formed, for example, directly from N_2O_5 in chloroform. The nitration of more complicated benzene derivatives can be controlled slightly by careful manipulation of the nitrating mixture.

Another common aromatic electrophilic substitution reaction is chlorination. In this case, the electrophile is produced by the action of chlorine on the Lewis acid, $AlCl_3$. Suggest how these two molecules may react together, and so suggest what is the nature of the electrophile.

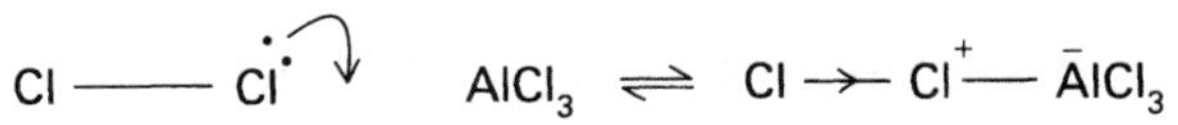

The $AlCl_3$, acting as a Lewis acid, complexes with the chlorine molecule, and this polarises the chlorine/chlorine bond so that the strongly $\delta+$ end may act as the electrophile in the attack on the benzene ring.

Write down the next step of this reaction.

The formation of the Wheland intermediate results in the heterolytic cleavage of the chlorine/chlorine bond to form the $AlCl_4^-$ complex, which later acts as the base to help remove the proton that is to be substituted.

Benzene rings may also be alkylated, thereby adding a carbon side chain to the ring. This is the Friedel–Crafts reaction, and is commonly performed by reacting an alkyl halide with the aromatic compound in the presence of a Lewis acid, such as $AlCl_3$. Suggest how the electrophile is formed.

The carbonium ion may also be formed from an alkene or alcohol. The carbonium ion formed from any of these starting materials is particularly prone to rearrangement reactions. These are called Wagner–Meerwein rearrangements, and severely limit the synthetic utility of this reaction to form simple alkyl substituted aromatic compounds. The tendency to rearrange may be reduced if the acyl derivative is used instead. This modification is called the Friedel–Crafts acylation reaction, and it has the further advantage that normally only monoacylation occurs, instead of the polyalkylation that happens using the simple Friedel–Crafts reaction.

A chloromethyl group may be added by the reaction of methanal and HCl with a $ZnCl_2$ catalyst. The overall reaction is called chloromethylation. However, the first step is hydroxyalkylation due to the reaction of the aldehyde with the benzene ring, and this is then followed by the acid catalysed substitution reaction of the hydroxyl group by the chlorine atom. Write down the steps in this reaction sequence, and so identify the role played by the Lewis acid.

$$HCl + ZnCl_2 \rightleftharpoons H^+ + ZnCl_3^-$$
$$H^+ + CH_2O \rightleftharpoons CH_2OH^+$$
$$CH_2OH^+ + C_6H_6 \rightleftharpoons [C_6H_6CH_2OH]^+$$
$$[C_6H_6CH_2OH]^+ \rightleftharpoons C_6H_5CH_2OH + H^+$$
$$C_6H_5CH_2OH + H^+ \rightleftharpoons C_6H_5CH_2OH_2^+$$
$$C_6H_5CH_2OH_2^+ + Cl^- \rightleftharpoons C_6H_5CH_2Cl + H_2O$$

The $ZnCl_2$ raises the acidity of the medium, and so causes an increase in the concentration of the CH_2OH^+ ions, which is the attacking reagent.

The examples given above are just a few of the many possible examples of different electrophiles being added to an aromatic system. In each case, the synthetic utility lies in the ease of producing the electrophile and the usefulness of the final product.

The substitution of pure benzene by an electrophile will result in the formation of a monosubstituted product, which is capable of undergoing further substitution reactions. When designing the strategy for the synthesis of an aromatic compound, there are two principal points that must be borne in mind, namely: first, the reactivity of the monosubstituted product compared with that of the original benzene; and second, the position on the aromatic ring where the second substitution reaction will take place. These two issues will now be examined, and it will be seen that they are, at least to some extent, dependent upon each other.

10.2.2 *Orientation and reactivity of monosubstituted benzene rings*

The rate of a subsequent substitution reaction may be either faster or slower than benzene, i.e. the ring may be activated or deactivated. Furthermore, the position of the second substitution may be at any one of three unoccupied positions. If the position of the first substituent is given the number 1, then the second substitution may occur at position 2, 3 or 4. The 5 and 6 positions just mirror the 3 and 2 positions respectively. These positions are often referred to by their old labels: *ortho*, *meta* and *para* respectively. It is found experimentally that the position of the second substitution is related to whether or not the ring has been activated or deactivated. That is why these two issues are considered together.

First, it is important to note that most aromatic electrophilic substitution reactions are under kinetic, and not thermodynamic, control. This is because most of the reactions are irreversible, and the remainder are usually stopped before equilibrium is reached. In a kinetically controlled reaction, the distribution of products (or product spread), i.e. the ratio of the various products formed, is determined not by the thermodynamic stabilities of the products, but by the activation energy barrier that controls the rate determining step. In a two-step reaction, it is a reasonable assumption that the transition state of the rate determining step is close in energy to that of the intermediate, which in this case is the Wheland intermediate; and so by invoking the Hammond postulate, one may assume that they have similar geometries.

The result of this analysis is that by looking at those factors that affect the stability of the Wheland intermediate, it is often possible to deduce which products will be favoured kinetically. Thus, if a particular Wheland intermediate is stabilised in comparison to its related intermediates, would that favour or disfavour the product that it precedes?

Normally, a smaller energy of activation needs to be surmounted in order to yield a more stable product in contrast to a less stable one. Hence, usually, the most stable intermediate is formed more quickly than other related intermediates. Once formed, the next step is fast, and so a more stable intermediate would lead to the preferential formation of the product that it preceded.

We will study the effect of inductive and mesomeric groups separately. First, for the general monosubstituted benzene ring, PhZ, substituted with a $+I$ group, there are three possible Wheland intermediates that may be formed by the addition of a general electrophile, E^+. Write down the *ortho*, *meta* and *para* substituted Wheland intermediates, and for each write out the three related canonical structures. Then predict the effect upon the stability of the intermediate of a group that has a positive or a negative inductive effect.

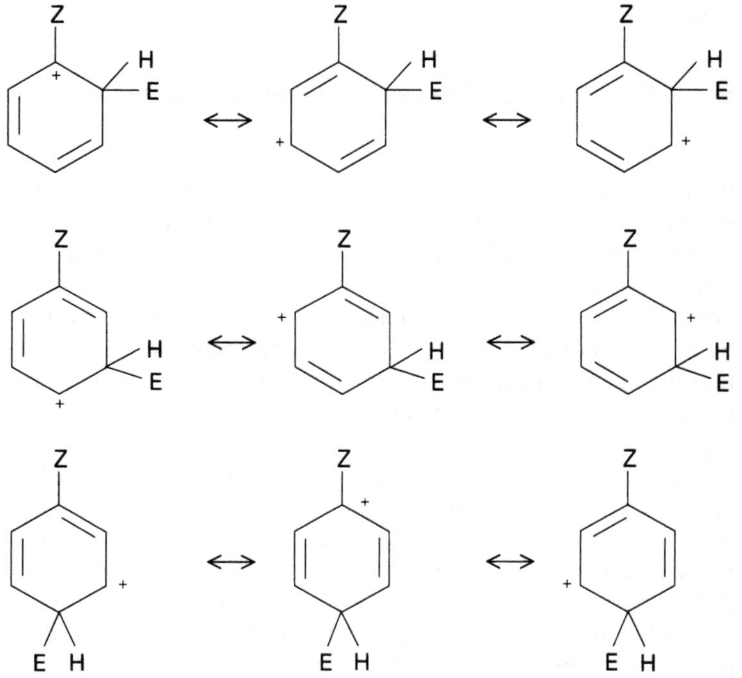

In all cases, the ring bears a positive charge. Thus, a $+I$ group will stabilise all the intermediates to some extent, while a $-I$ group will destabilise all of them to some extent. A further refinement may be made. The inductive effect diminishes with distance. In the *ortho* and *para* Wheland intermediates, one of the possible canonical structures places a positive charge upon the carbon atom that is joined to the inductive group. Thus, the inductive stabilisation of a $+I$ group will be greatest for the *ortho* and *para* substituted Wheland intermediates, and hence the disubstituted products with these orientations will be kinetically favoured.

Correspondingly, the reverse is true for $-I$ groups, i.e. they interact in an unfavourable manner with the positive charge. However, this unfavourable interaction is strongest when the positive charge is situated at the *ortho* and *para* positions rather than when it is at the meta position. Thus, in such a scenario, the *meta* position is favoured relative to the *ortho* and *para* positions by reason of it being less disfavoured. This analysis also shows why there is a link between the orientation and the rate of the second substitution of a monosubstituted aromatic ring.

Now we will turn our attention to the case of a mesomeric group, Z, which is capable of donating a lone pair of electrons into the ring. As before, there are three possible positions of attack for the incoming electrophile. In addition to the nine canonical structures that were considered for the inductive substituent, there is now an extra canonical structure for both the *ortho* and *para* substituted Wheland intermediates. Write down these two extra canonical forms, and suggest what effect they will have on the product distribution.

In the case of the *ortho* and *para* substituted Wheland intermediates, these extra canonical forms greatly increase the stability of the relevant intermediate, and so greatly favour the production of the related product.

From this analysis, it is possible to divide the various substituents into four groups. The first contains those moieties that possess an unshared pair of electrons capable of exerting a $+M$ effect on the ring. This would include such moieties as $-OR$, $-O^-$ and $-NR_2$. Such moieties activate the ring, i.e. substitution occurs faster in these derivatives than in pure benzene. Furthermore, they direct the incoming electrophile to the *ortho/para* positions.

The second group contains the halogens, which also direct the incoming electrophile to the *ortho/para* positions, but exert such a strong $-I$ effect that they withdraw sufficient electron density from the benzene ring to deactivate it.

The third group comprises those moieties that lack a lone pair of electrons and exert a $-I$ effect, e.g. $-NR_3^+$, $-NO_2$, $-CN$ and carbonyl derivatives. Each member of this group deactivates the ring and directs the incoming substituent to the *meta* position, because that is the least disfavoured position.

The fourth, and last, group has no lone pair of electrons and is $+I$ in effect. It consists of such moieties as alkyl and aryl groups, as well as negatively charged entities as carboxylate. This group activates the ring and directs the second substituent to the *ortho/para* positions. However, it should be noted that it is possible for alkyl groups to activate the ring further by hyperconjugation.

10.2.3 **Ortho/para ratio**

In the previous discussion we did not distinguish between the attack of the incoming electrophile at the *ortho* or *para* position. We will now do so. First, from a purely statistical reckoning, what would be the expected distribution between these two sites?

As there are two *ortho* protons for every one *para* proton, then from a statistical point of view the ratio between them ought to be 2 : 1.

It may be calculated that the charge density at the *ortho/meta/para* positions in the cation, $C_6H_7^+$, formed by protonating benzene, are in the ratio of 0.25:0.10:0.30 respectively. Using these figures, predict whether the amount of *para* substitution would be more than the 33% predicted on purely statistical grounds, or less.

In a Wheland intermediate, a disproportionate amount of the positive charge would reside at the *para* position. Hence, if there were a substituent that could stabilise that charge, it would have greater effect than a substituent at the *ortho* position; thus, the amount of *para* substitution would be increased over that predicted on purely statistical grounds. In line with this deduction, it is observed that *para*-quinonoid structures are more stable than those with *ortho*-quinonoid structures.

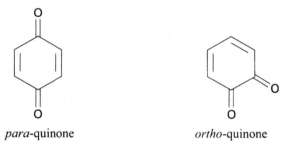

para-quinone   *ortho*-quinone

For groups that exert an inductive effect, it should be remembered that this effect diminishes quickly with distance, and so such groups will have a far greater influence at the *ortho* position rather than the *para* position. In the case of the halogens, predict whether an iodo or a fluoro substituted benzene would have the greater amount of *ortho* substitution.

Nitration of halogenobenzenes yields 12% of the *ortho* product for fluorobenzene, and 41% for the iodobenzene.

Apart from the purely electronic effects, there is the steric interaction that may occur between the incoming electrophile and the substituent already present. In the nitration of toluene and *t*-butylbenzene, suggest which compound experiences the higher degree of *ortho* substitution.

Toluene gives 58% *ortho* and 37% *para*, while *t*-butylbenzene gives 16% *ortho* and 73 % *para*. The *t*-butyl group physically blocks the entry of the incoming nitronium ion and so disfavours substitution at the *ortho* position. In addition, the arenium ion formed from the attachment of the electrophile to *t*-butylbenzene has a much greater disfavourable interaction between the bulky *t*-butyl group and the *ortho* nitro group. This disfavourable interaction raises the activation energy for the initial electrophilic attack, which is rate limiting, and so disfavours the formation of the *ortho* product.

The steric blocking of the *ortho* position may be achieved in another manner. Cyclohexaamylose is a compound whose structure is akin to an open cylinder. Molecules such as anisole, $C_6H_5OCH_3$, may be dissolved in a solution that contains cyclohexaamylose, and then the anisole molecules may enter the central cavity in the cyclohexaamylose. Thus, the ratio of the *ortho* to *para* chlorinated product formed from anisole when dissolved in a normal solvent to that formed when dissolved in a solution of cyclohexaamylose, increased from $1:1.48$ to $1:21.6$. Suggest a reason for this increased selectivity.

The association of the cyclohexaamylose and anisole molecules is similar to the situation found in an inclusion compound. The result is that the *ortho* positions are blocked by the sheath of the cyclohexaamylose molecule; while the *para* position is exposed at the bottom of the cylinder, and so can still be attacked by the chlorine electrophile.

Having started with the statistical distribution that favoured the *ortho* position by $2:1$ to the *para* position, we have seen a number of factors that favour the *para* position over the *ortho* position. There is, however, an important factor that may favour substitution at the *ortho* position. For example, in the nitration of methyl-[2-phenyl]-ethyl ether with N_2O_5, an unusually high proportion of *ortho* substituted product is produced. Suggest an explanation for this experimental observation.

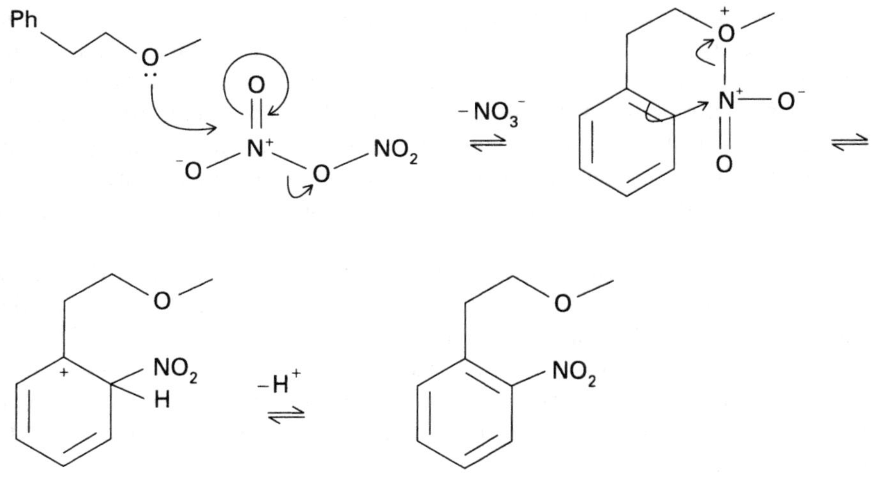

The complexation of the reagent with the lone pair of the ether oxygen, which then holds the attacking electrophile in close proximity to the *ortho* position, accounts for this high *ortho* yield.

10.2.4 *Multiple substitutions*

The substitution of an unsubstituted benzene ring by an electrophile to give a monosubstituted aromatic system proceeds by the arenium ion mechanism, which involves the Wheland intermediate. By studying the effects of inductive and mesomeric groups upon the Wheland intermediate, it is possible to predict the orientation of substitution that the second incoming electrophile would adopt. The simple electronic considerations did not distinguish between the *ortho* and *para* positions on the aromatic ring, and so the analysis was further refined. Now we will look at which position the third incoming electrophile will attack when there are already two substituents on the ring.

In the simplest case, the directing effects of the two groups reinforce each other. Suggest where the incoming electrophile will attack in a 1,3-dialkylbenzene, and in *p*-chlorobenzoic acid.

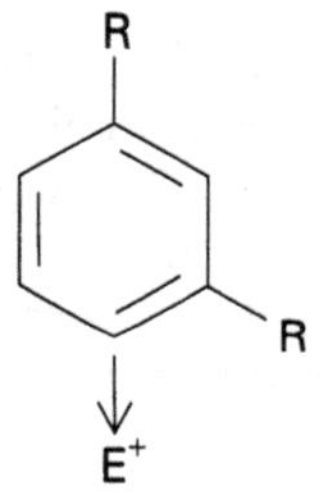

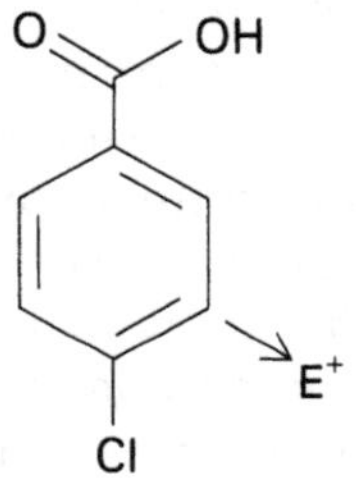

In the first case, the incoming electrophile will attack at the 4 position, because this is *ortho* to one alkyl group and *para* to the other. In the second case, the incoming electrophile will attack at the 3 position, because this is *meta* to the carboxyl group and *ortho* to the chloro group. (Strictly, in the last example, the favoured position of attack should be called the 2 position. This arises because the chlorine substituent has a higher priority than the carboxylic acid substituent when it comes to systematic naming. However, for the purposes of the above illustration, that refinement merely adds an unnecessary level of complexity.)

When the groups that are already present oppose each other in the directions in which they favour further substitution, it is more difficult to predict the position of attack of the third incoming electrophile. However, a few guidelines exist. First, activating groups have a larger influence than deactivating groups on the position of the incoming electrophile, with mesomeric groups exerting a greater influence than inductive groups. Hence, predict the position of attack in *o*-methylphenol.

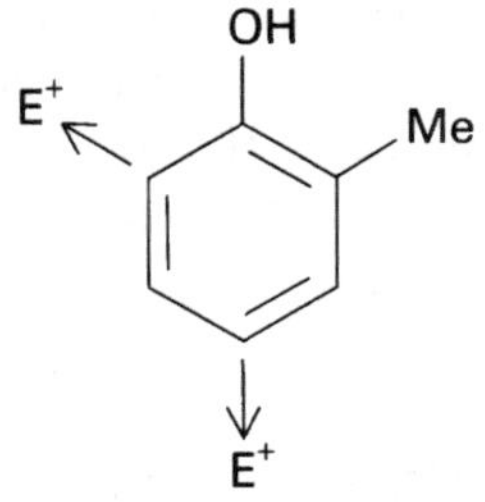

The hydroxyl group is more strongly activating than the methyl group as it is a +M group. Thus, the incoming electrophile will go *ortho* and *para* to the hydroxyl group, and the position of substitution will not be governed by the methyl group.

Another guideline is that the incoming group is unlikely to enter between two groups which are *meta* to each other. The reason for this is steric, and increases in importance as the size of the groups increases.

The last guideline is that where there is a *meta* directing group that is *meta* to an *ortho/para* directing group, the incoming group enters *ortho* to the *meta* group in preference to the *para* position. This is called the *ortho* effect. Hence, predict the position of chlorination of *m*-chloronitrobenzene.

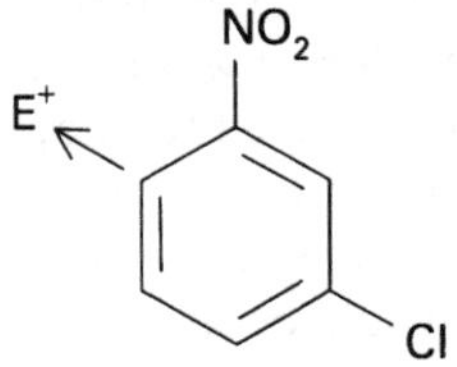

The incoming chlorine tends to attack *para* to the chlorine already present and *ortho* to the nitro group.

10.3 Aliphatic substitution

The mechanisms that occur in aliphatic electrophilic substitution reactions are less well defined than those that occur in aliphatic nucleophilic substitution and aromatic electrophilic reactions. There is still, however, the usual division between unimolecular and bimolecular pathways: the former consisting of only the S$_E$1 mechanism, while the latter consists of the S$_E$2 (front), S$_E$2 (back) and the S$_E$i mechanism.

The leaving group in all these cases is an electron deficient species, i.e. an electrofuge. One of the commonest types of electrofuges in aliphatic systems is a metal ion, as it is easily capable of bearing a positive charge. Thus, these types of mechanisms are often encountered in organometallic pathways. Generally, organometallic chemistry is studied only in outline at this level, so these mechanisms will be examined only briefly in this text. However, it is still useful to study the pathways because, even though this may be considered an advanced topic, much progress and understanding may be achieved by applying the basic ideas utilised in early chapters.

In the unimolecular mechanism, i.e. the S$_E$1 mechanism, the electrophilic substitution takes place in two steps that parallel those that occur in the S$_N$1 mechanism. Accordingly, write down the two steps for a general electrophilic substitution reaction between RX and E$^+$, and indicate which is the slow step.

Step 1: $R-X \rightleftharpoons R^- + X^+$ (slow)

Step 2: $R^- + E^+ \rightleftharpoons R-E$

The first step follows the expected first order kinetics. Suggest what type of substituents on the alkyl carbon would favour this reaction pathway, and then suggest what would be the effect of increasing the polarity of the solvent on this pathway.

The intermediate carbanion bears a negative charge and so $-M$ and $-I$ groups would help to stabilise it. Furthermore, as there is an increase in charge separation that accompanies the formation of the two charged intermediates, there is an increasingly favourable interaction with the solvent if it is polar. As a result, this route is favoured by increasing the polarity of the solvent. These conclusions mirror those observed in S$_N$1 reactions.

The factors that affect the rate of the reaction are easily predictable, namely ones that are similar to those that affect the rate of an S$_N$1 reaction. However, the stereochemical consequences of this pathway are slightly more difficult to predict. The simple carbanion would be expected to adopt a tetrahedral conformation, which is similar to that which a nitrogen atom adopts in a tertiary amine. Consider whether tertiary amines may be resolved into their enantiomers, and then by analogy, suggest what would be the stereochemical consequences of the S$_E$1 mechanism on the configuration at the substituted centre. Hence, suggest what would be the structure of a simple carbanion intermediate.

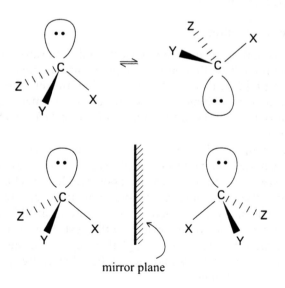

mirror plane

A simple tertiary amine cannot be resolved into its enantiomers at normal temperatures, because there is rapid interconversion between the two forms. This rapid interconversion is called the umbrella effect. A similar effect occurs in simple carbanions, and so they cannot be resolved either. This means that any stereochemical information at such a centre is lost.

Now consider the situation where a −M group is conjugated with the negative charge, which would result in the charge being delocalised onto that group. Also suggest what would be the stereochemical consequences upon a chiral centre at the carbanion carbon.

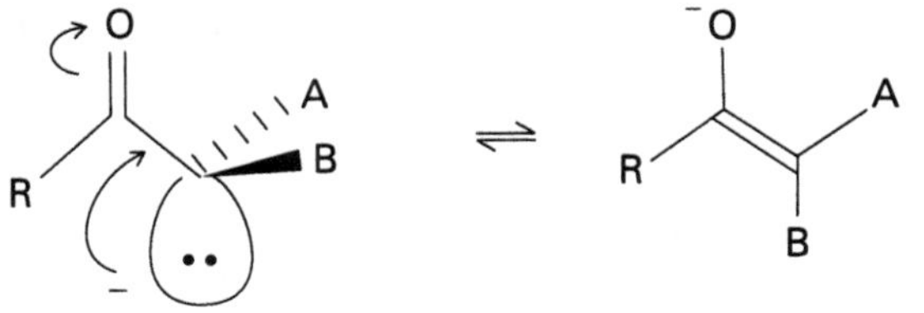

In this case, the anion would adopt a planar configuration. As a consequence of this, any chiral information at the α-carbon to the carbonyl would be lost on the delocalisation of the α-anion. Thus, in both cases, namely that of the simple carbanion and the stabilised carbanion, there will be racemisation at the carbanion centre. This result, though, depends for its analysis upon the anionic intermediate existing long enough, and being able to undergo the umbrella effect or to have the negative charge delocalised onto the −M group. In practice, the outcome is highly dependent upon the solvent, which may stabilise the charge and interact with the intermediate via the solvation shell, in which case retention or even inversion may be observed.

In the case of vinyl carbanions, which may maintain their configuration, retention is observed. Write down the reaction pathway for the electrophilic substitution of *cis*-2-bromo-but-2-ene when it is treated with lithium and then with carbon dioxide.

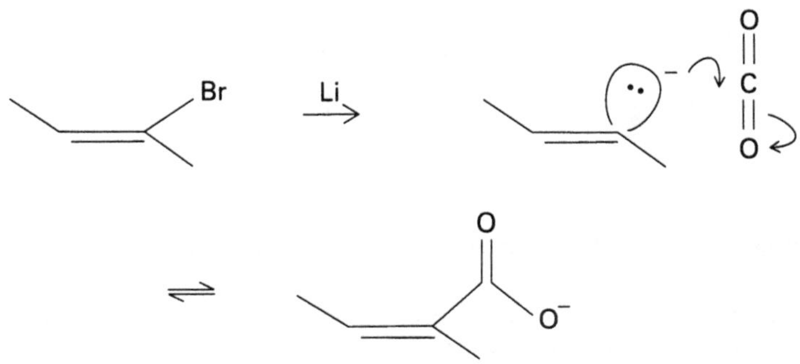

The lithium forms the carbanion, and this anion then reacts with retention of its configuration with carbon dioxide to form the carboxylic acid, called angelic acid, in about 70% yield. The *trans* isomer, tiglic acid, is only produced in about 5% yield.

We will now turn our attention to the bimolecular pathways. In the case of bimolecular nucleophilic attack, the incoming nucleophile contains an extra pair of electrons. In order to accommodate this pair of electrons, the leaving group has to depart at the same time. The lowest energy arrangement that allows this to occur has the incoming nucleophile arriving on the opposite side to that on which the leaving group departs.

In contrast, in the case of an incoming electrophile, there is no such build-up of electrons, and so it is possible for the attack to occur from either the front or the back of the substrate. Write down the pathway for each of these two possibilities.

S$_E$2 (front)

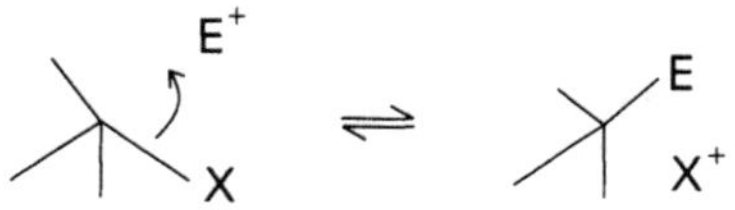

S$_E$2 (back)

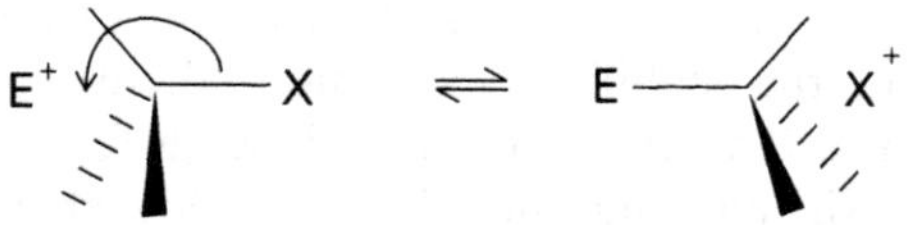

These two reactions are called S$_E$2 (front) and S$_E$2 (back). There is a third possibility. When the attacking species enters from the front, a part of it may assist in the removal of the leaving group by forming a bond with it at the same time as the new carbon/electrophile bond is being formed. The bond formation and bond cleavage is approximately simultaneous. Write down a representation of this pathway.

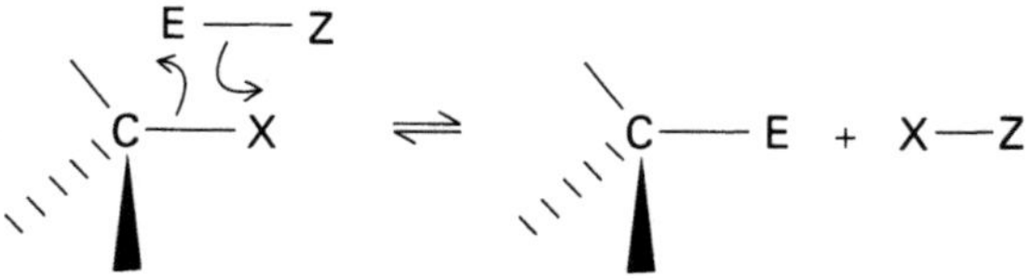

This mechanism is called the internal electrophilic substitution and is usually given the label S$_E$i. However, it sometimes goes under the label of S$_F$2 or S$_E$2 (cyclic).

The S$_E$2 (back) mechanism is different from the other two, because it results in inversion of the configuration of the carbon undergoing substitution. This has been observed; but more usually retention is observed, which would indicate that either the S$_E$2 (front) or the S$_E$i mechanism is operating. Suggest how it may be possible to distinguish between these two.

In the SE2 (front) mechanism there would be a greater degree of charge separation in the transition state than in the SEi mechanism. Thus, one would expect the rate of the former to be affected by both the polarity of the solvent and the addition of ions to the reaction mixture. The results of such experiments have been used to support the suggestion that one pathway is being favoured over the other in any particular case. However, the results from the same experiments have also been used to support the suggestion that further refinements are needed in order to understand the full details of the pathways of electrophilic substitution reactions. Thus, it has been suggested that, instead of the simultaneous formation and cleavage of bonds that occurs in SEi, there may be an alternative pathway, which is given the label SE2 (co-ord), or SEC.

The detailed study of electrophilic substitution reactions is beyond the ambit of this book. However, it is instructive to be aware that a number of different mechanisms may be suggested from an analysis of the experimental data, for it is a reminder that all the mechanisms that we are studying are merely hypotheses that fit the available observations gathered to date. If fresh data were forthcoming, which could not be explained by the established mechanisms, then modifications would have to be made to the suggested mechanisms, and so our understanding would develop.

10.4 Summary of electrophilic substitution reactions

In an electrophilic substitution, the attacking species is an electrophile, and the leaving group is an electrofuge. In contrast to nucleophilic reactions, an electrophilic substitution can occur by either the preliminary addition of the electrophile, or the elimination of an electrofuge, as well as the simultaneous approach and departure of the attacking and leaving groups.

In an aromatic system, the delocalised electrons in the aromatic ring discourage the approach of a nucleophile, but in contrast will encourage electrophiles. In aromatic chemistry, the arenium ion mechanism predominates. It occurs in three steps. The first is the formation of the attacking electrophile; the second is the attack by this electrophile on the aromatic ring; and finally, the departure of an electrofuge, which is usually an proton, with the assistance of a base. The cationic intermediate is called a Wheland intermediate, σ-complex or arenium ion, and can sometimes be isolated. The positive charge in the Wheland intermediate is notionally located on the adjacent carbon to the one that is attacked initially. The restoration of the aromatic sextet with the accompanying energy of delocalisation provides the driving force for the final step of the reaction.

The nitration of benzene follows this mechanism. The attacking species, the nitronium ion, NO_2^+, can be formed by the action of concentrated sulphuric acid on concentrated nitric acid.

The chlorination of the aromatic ring can be performed by reacting chlorine with aluminium chloride, which is a Lewis acid, to produce a complex in which one of the chlorine atoms that originated from the chlorine molecule bears a $\delta+$ charge and so can be more readily attacked by the electrons of the ring.

The alkylation of the aromatic ring is done using the Friedel–Crafts reaction. In this case the electrophile is produced by reacting an alkyl halide with a Lewis acid, such as $AlCl_3$. The resultant carbonium ion may also be formed from alkenes or alcohols. The carbonium ion formed in any of these ways is particularly prone to the Wagner–Meerwein rearrangement, and this severely limits the synthetic utility of this route. The tendency to rearrange may be circumvented if the acyl derivative is used instead. This modification is called the Friedel–Crafts acylation reaction, and it has the further advantage that normally only monoacylation occurs, instead of the polyalkylation that often happens using the simple Friedel–Crafts reaction.

A chloromethyl group may be added by the reaction of methanal and HCl with a $ZnCl_2$ catalyst. The overall reaction is called chloromethylation.

Aromatic electrophilic substitution reactions are under kinetic, and not thermodynamic, control. This is because many of the reactions are irreversible, and the remainder are usually stopped before equilibrium is reached.

Groups that exert a $-I$ effect will deactivate the ring to substitution and will favour *meta* substitution. Conversely, groups that exert a $+I$ effect will activate the ring, and favour *ortho* and *para* substitution. Similarly, groups that exert a $+M$ effect greatly activate the ring, and direct the new substituent to enter at the *ortho* and *para* positions. Halogeno substituents are exceptional, because they deactivate the ring (due to their strong $-I$ effect), but direct the second substituent to the *ortho/para* positions, because of their lone pairs. Activation of the ring means that substitution occurs faster than in benzene, and conversely deactivation means that the rate of substitution of the second group is slower than in benzene.

The *ortho/para* ratio is affected by a number of factors. Statistically, the *ortho* position should be favoured; however, due to the electronic distribution of the π electrons there is more substitution at the *para* position than would be expected on a purely statistical basis. Steric effects also favour attack at the *para* position.

When there are two substituents, the activating one, i.e. the $+M$ group, has the overriding influence as to the site at which the third substituent enters. Further, the third substituent is unlikely to enter between two groups that are *meta* to each other for steric reasons. Another guideline is the *ortho* effect. This states that where there is a *meta* directing group that is *meta* to an *ortho/para* directing group, the incoming group enters *ortho* to the *meta* group in preference to the *para* position.

The mechanisms that occur in aliphatic electrophilic substitution reactions are less well defined than those that occur in aliphatic nucleophilic substitution and aromatic electrophilic reactions. There is still, however, the usual division between unimolecular and bimolecular pathways: the former consisting of only the S_E1 mechanism, while the latter consists of the S_E2 (front), S_E2 (back) and the S_Ei mechanism. One of the commonest leaving groups is a metal ion, and so these pathways are often encountered in organometallic reactions.

The rate of SE1 reactions is affected by similar considerations as those that affect SN1 reactions. Further, generally any stereochemical information at the carbanion centre is lost, either due to the umbrella effect or due to the formation of a planar intermediate as a result of delocalisation to a −M substituent such as a carbonyl group. However, vinyl carbanions can maintain their configuration, and so retain their stereochemistry.

In the case of bimolecular aliphatic electrophilic substitution reactions, there are three principal pathways, namely SE2 (front), SE2 (back), and the internal electrophilic substitution, SEi (otherwise known as SF2 or SE2 (cyclic)). Inversion would occur with SE2 (back), while retention would occur with the other two. Experiments to distinguish between these latter two have indicated that the situation may be even more complicated, in that an alternative pathway, which is labelled either SE2 (co-ord), or SEC, may operate.

11

Radical Substitution Reactions

11.1 Introduction

So far, we have looked at substitution reactions in which the attacking species was either a nucleophile or an electrophile, i.e. a charged species possessing only paired electrons. There is a third possibility, namely that the attacking species is a radical, which has a single unpaired electron. Even though it is possible to have a radical that bears either a positive or negative charge, in this chapter we will only deal with those radical species that are electrically neutral, because these are far more common.

One of the most important features of free radical chemistry is that the reactions are not affected by the normal variations in reaction conditions, such as a change in the polarity of the solvent or the acid/base characteristics of the reagents, except insofar as these changes will favour or disfavour competing ionic reactions. This is because such factors are only relevant when dealing with species that interact Coulombically.

11.2 Photochlorination of methane

This reaction may be used to illustrate the mechanistic steps that occur in a radical substitution reaction. Its synthetic value is strictly limited, because, as we will see, there is little control over the product distribution. Furthermore, the reaction tends to proceed explosively, which limits its popularity.

The first step is to form the free radicals that will react further. In contrast to the heterolytic fission that is required for the production of polar intermediates, radicals are formed by homolytic fission of a covalent bond. This fission may be induced by light, heat, another radical, or even by a redox reaction. In the case of photochlorination, the fission is achieved by shining ultraviolet light at the reaction mixture. Write down the mechanism for the simple homolytic fission of a chlorine molecule.

$$Cl - Cl \rightleftharpoons Cl\cdot \quad Cl\cdot$$

The chlorine/chlorine single covalent bond is broken symmetrically to yield two radicals. In this first step, the overall number of radicals has increased. This is characteristic of an initiation step. Free radicals are, by their very nature, reactive species and so they tend to react quite quickly with any other species that happens to be in the vicinity, which, when the concentration of free radicals is low, will usually be a neutral molecule. The product of this second reaction step is another radical species. Many free radical reactions are characterised by the lack of free radical/free radical combination, and instead display a large number of reactions in which the initial free radical reacts with some other type of molecule in the reaction mixture.

A very common second step in radical substitution reactions is for the initial free radical to react with a hydrogen atom of the organic reagent, and, after abstracting this hydrogen atom, to produce a carbon radical. Write down the balanced equation for the reaction of a chlorine radical with a methane molecule, and then write down the mechanism.

$$CH_4 + Cl\cdot \rightleftharpoons CH_3\cdot + HCl$$

It is possible to write down an equation in which the initial chlorine radical attacks the methane molecule and, instead of forming HCl and the methyl radical, forms chloromethane directly and a hydrogen radical. In practice, only the first route is observed, and the reaction does not proceed by the second route. Suggest why this is the case.

In order to form the hydrogen radical and the chloromethane directly, the attacking radical would need to interact with the electrons of the central carbon atom. These electrons are masked to a certain extent by the groups that surround the central carbon atom. The observed pathway requires only that the attacking radical interacts with the electrons of the hydrogen atom, which are more accessible. There is no steric hindrance that inhibits this attack, and so it is more easily achieved.

The formation of the methyl radical maintains the number of radical species. This is characteristic of a propagation step. There are often many different propagation steps in any given reaction, whereas there is usually only one initiation step in which the original free radical is formed.

So far a hydrogen chloride molecule and a methyl free radical have been formed. Notice that the carbon radical was not formed originally, but only after the reaction with the more reactive chlorine radical, which in turn had been formed by the photoinduced homolytic fission. The hydrogen chloride takes no further part in the reaction. The methyl radical may react further with, for example, a chlorine molecule. Write down the equation and mechanism for this step.

$$CH_3\cdot + Cl\text{–}Cl \rightleftharpoons CH_3\text{–}Cl + Cl\cdot$$

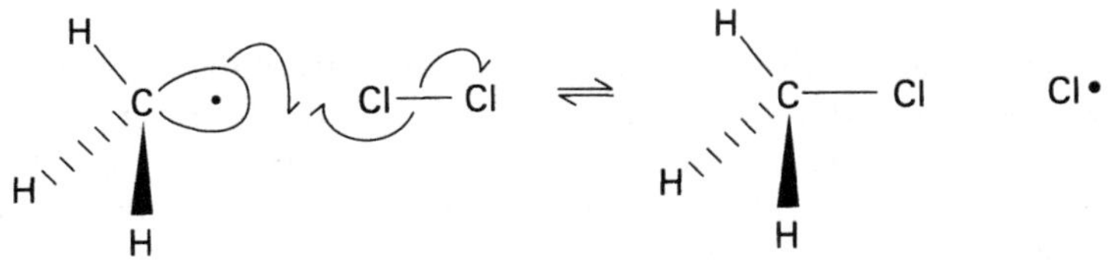

The number of free radicals remains constant in this reaction, and so this is another example of a propagation step. Here, one of the products, the chlorine radical, has been used before in the previous propagation step and so may be used to continue the reaction, i.e. propagate the reaction. The other product is chloromethane, which is the monosubstituted product. Write down a balanced equation for the production of this product from the starting materials.

$$CH_4 + Cl\text{–}Cl \rightleftharpoons CH_3Cl + HCl$$

For every molecule of chlorine consumed, only one chlorine atom ends up in the organic product, while the other combines with the hydrogen to form hydrogen chloride. Notice also, that a hydrogen has been substituted by a chlorine atom in the original methane molecule. Such a reaction is unusual in simple aliphatic compounds, because the hydrogen is normally very difficult to activate *per se*. It will be recalled, however, that a simple substitution of a hydrogen atom also occurred in the electrophilic substitution of an aromatic ring.

After the reaction between chlorine and methane has proceeded for a short time, there exists in the reaction mixture a significant concentration of chloromethane, and so it becomes likely that a chlorine radical could attack this monochlorinated species instead of a methane molecule. Write down the equations and mechanisms for such an attack.

$$CH_3Cl + Cl\cdot \rightleftharpoons CH_2Cl\cdot + HCl$$
$$CH_2Cl\cdot + Cl-Cl \rightleftharpoons CH_2Cl_2 + Cl\cdot$$

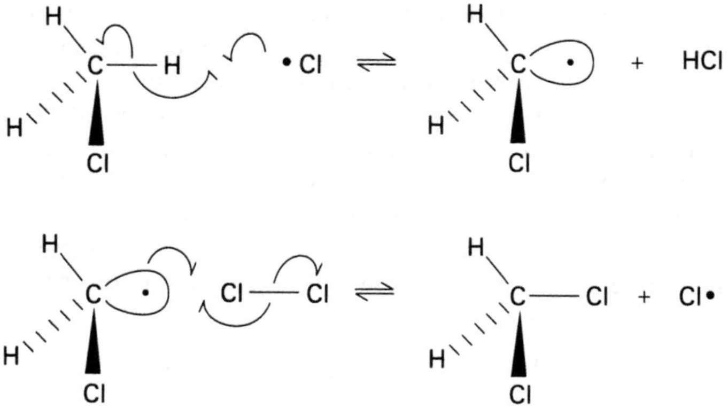

The chlorine radical has abstracted a hydrogen atom from the chloromethane to form the chloromethyl radical, and this may go on to react to form the disubstituted product. This, in turn, may react with more chlorine radicals and so on. It is this lack of control over which product is formed that makes this reaction not very useful synthetically.

There is a third type of reaction that may occur in a typical free radical mechanism. This is exemplified by the reaction of one radical with another. Write down some possible examples of this third type of reaction.

$$2Cl\cdot \rightleftharpoons Cl-Cl$$
$$2CH_3\cdot \rightleftharpoons H_3C-CH_3$$
$$CH_3\cdot + Cl\cdot \rightleftharpoons CH_3Cl$$

All of these reactions share the characteristic that the number of free radicals is reduced, and thus they are called termination steps, because they lead to the termination of the chain of propagation steps of the reaction.

Note that in several of the examples given above, the product of the termination reaction was either a starting material or one of the products formed by some other route, and so the existence of these termination reactions would not be apparent. However, the ethane molecule is not formed by any of the other routes, and so the presence of this side product in the reaction mixture indicates that these termination reactions might in fact occur. Generally, the presence of unusual side products is very useful in attempting to elucidate the pathway that any particular mechanism follows.

After having considered a specific example, namely the photochlorination of methane, we will now consider the general pathway for such reactions. A free radical reaction need only have an initiation step and a termination step. However, this would be most unusual, because normally one or more propagation steps interpose. The reason is because, as radicals are very reactive, the radical intermediates tend to react with the first species with which they come into contact, and, as usually the concentration of radicals is low, the most likely species that a radical will meet will be a non-radical. Hence, the presence of the propagation steps. There are four types of propagation steps, which may be represented as following:

1. $R\cdot + M \rightleftharpoons R\text{–}M\cdot$
2. $R\cdot + R'\text{–}H \rightleftharpoons R\text{–}H + R'\cdot$
3. $R\cdot \rightleftharpoons R'' + R'''\cdot$
4. $R\cdot \rightleftharpoons R''''\cdot$

In the first type, the radical combines with a neutral molecule to form a new, larger radical; in the second, one radical attacks a molecule and abstracts a hydrogen atom to form a new radical; in the third, there is a fragmentation reaction, which is similar to the reverse of the first type; while in the fourth, there is a rearrangement of the radical into another radical. Propagation steps like type two are given the label S$_H$2, which stands for bimolecular homolytic substitution reactions, while initiation reactions are called S$_H$1 reactions.

The reaction of one radical to form another and then another, and so on, by propagation steps is called a chain reaction. If the radicals are very reactive would there be more or less propagation steps in the chain reaction?

The more reactive the radicals, the more propagation steps occur before the chain reaction is brought to an end by a termination reaction. As with propagation reactions, there are often many possible termination reactions, some of which, as we have seen above, produce intermediates or starting materials of the reaction. This multiplicity of possible reactions makes radical reactions very difficult to study kinetically. One possible termination reaction consists of disproportionation. Suggest what would be the resultant products of the disproportionation termination reaction between two ethyl radicals, and further suggest a mechanism for this reaction.

$$\overset{\bullet}{C}H_2 - CH_2 - H \quad \bullet CH_2 - CH_3 \rightleftharpoons CH_2 = CH_2 + CH_3 - CH_3$$

Another possible termination reaction is the simple dimerisation of the radical. The large number of possible termination reactions adds to the complexity that already exists due to the many propagation steps that are available.

It was mentioned at the beginning of this section that many radical reactions are commenced by the addition of compounds that promote the formation of radicals. These compounds are called initiators, and often contain a weak bond that is easily broken, e.g. peroxides. Similarly, there are some species that scavenge free radicals and these are called inhibitors. Such species include molecular oxygen, nitric acid and benzoquinone. The addition of either initiators or inhibitors will greatly affect the rate of the radical reaction.

Now that we have studied the main type of reactions that occur in a typical radical reaction, we will study the reactivity and structure of radicals in more general terms.

11.3 Reactivity and structure

In the photochlorination reaction, the chlorine radical abstracted a hydrogen atom, and so produced a carbon radical. Notice that the possible alternative combination of a hydrogen radical and a chloromethane molecule was not produced.

In the case of methane, there was only one type of hydrogen that the chlorine radical could attack, but in a larger alkane there is often a choice. One of the principal factors that determines which hydrogen will be abstracted is the stability of the resultant carbon radical. Suggest what will be the order of stability of primary, secondary and tertiary carbon radicals.

tertiary > secondary > primary

Carbon radicals are like carbonium ions in that both are electron deficient species, i.e. the carbon atom does not bear a full octet of electrons. Hence, it is not unreasonable to assume that radicals will follow the same sequence as carbonium ions for stability; and this is found to be the case in practice.

In addition to the stability of the resultant radical affecting which hydrogen is abstracted, it is also observed that the order of preference for the abstraction of a hydrogen atom is related to the bond dissociation energy.

In the case of a general alkene, $RCH_2CH=CH_2$, suggest which hydrogen atom will be abstracted by an initiator, In·, and give reasons for your choice.

In this case, the strongest C–H bond is the vinyl carbon/hydrogen bond, and this bond is practically never broken. It is the allylic hydrogen that is abstracted, because this carbon/hydrogen bond is relatively weak. Furthermore, cleavage of this bond results in a radical that is stabilised by delocalisation.

In an arylalkane, such as ethyl benzene, which hydrogen will be attacked and why?

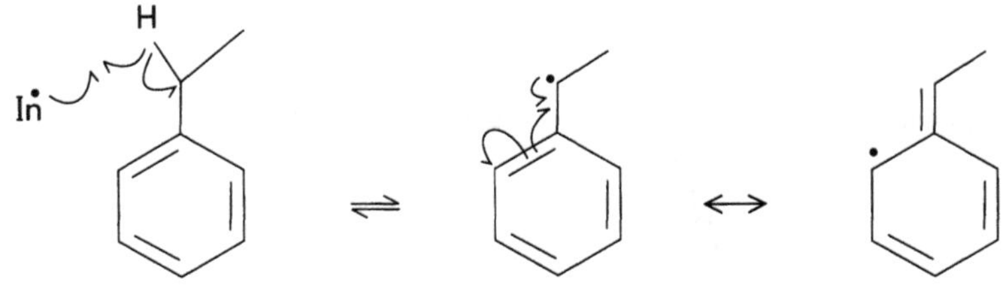

In nearly all cases where there is an alkyl group that is joined to an aromatic ring, the abstracted hydrogen will be on the α-carbon of the alkyl side chain, because such a bond is relatively weak and the resultant radical is stabilised by delocalisation of the unpaired electron around the aromatic ring. The aromatic ring is very rarely attacked to form an aryl radical. Thus, benzene reacts much more slowly to form the $C_6H_5\cdot$ radical. Suggest why this is so.

The aromatic C–H bond is relatively strong; also there is no possible delocalisation around the aromatic ring of the resultant unpaired electron.

   If a radical attacks an aromatic system, there is an tendency initially to form a simple adduct, rather than abstracting a hydrogen atom. Suggest what would be the structure of such an adduct, and also suggest why this is the favoured route.

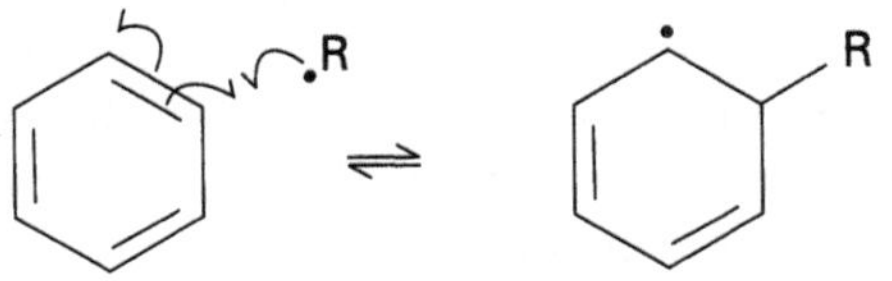

The adduct is similar to the one formed during an electrophilic aromatic substitution reaction. Even though there is the loss of the aromatic delocalisation, the resultant radical is itself delocalised to some extent. Once formed, such aromatic radicals may then react in a number of different ways. It is possible for two to dimerise or disproportionate. Alternatively, such a radical may undergo an overall substitution after the abstraction of a hydrogen atom, or even react further to result in overall addition. If the last option is followed, then often addition takes place at all three double bonds to give the hexasubstituted product, $C_6H_6X_6$.

   It is also found that both electron donating and electron withdrawing groups activate the ring. In all cases, substitution at the *ortho* position is favoured unless there is steric hindrance due to the original group, in which case the incoming radical will attack at the *para* position.

   The shape of the intermediate carbon radicals is important in determining the stereochemical consequences of radical substitution. For the simple alkyl radical, $CR_3\cdot$, suggest the various conformations that it could adopt, and suggest what would be the stereochemical consequences of these shapes.

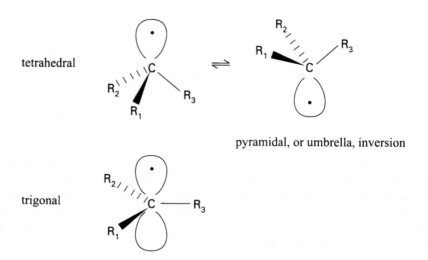

tetrahedral

pyramidal, or umbrella, inversion

trigonal

Such a substituted alkyl radical could adopt a tetrahedral arrangement in which the central carbon atom is sp^3 hybridised, or it could adopt an sp^2 arrangement that would result in a trigonal planar shape. In the former case, it would not be unreasonable to suppose that the radical would undergo umbrella inversion, rather like a tertiary amine, and so the original molecule would be racemised upon becoming a radical. The trigonal planar arrangement would also lead to a similar racemisation of the configuration. The exact shape of the carbon radical depends upon the substituents that are attached to the radical carbon. In the case of the methyl radical, it is almost planar, while in the trifluoromethyl radical it is essentially based upon a tetrahedron.

Suggest what will be the practical consequence of the tolerance of a radical species to its conformation; in particular, when attempts are made to form a radical at a bridgehead.

It is quite possible to form a radical at a bridgehead carbon, and sometimes this is the only way to perform any chemistry at such a position, for we have already seen that a bridgehead carbon is normally inert to attack by both the Sɴ1 and the Sɴ2 pathways.

It has already been mentioned that radicals are usually very reactive species. Different carbon radicals have different stabilities, which would lead one to presume that there would be some selectivity in the reaction; in fact, what is often observed is that there is no selectivity. This, of course, reduces the synthetic utility of these reactions.

However, if the reactivity of the attacking radical is reduced sufficiently, then some degree of selectivity may be introduced. An example of this is that the ratio of chlorination at a primary, secondary or tertiary carbon atom is 1:4.4:6.7; while for bromination the ratio is 1:80:1600.

Selectivity may also be achieved by altering the solvent in certain cases. Thus, chlorination of 2,3-dimethylbutane in aliphatic solvents yields the primary to tertiary product in the ratio of 3:2; while in an aromatic solvent the ratio is 1:9. Suggest an explanation for this result.

The chlorine radical forms a π complex with the aromatic solvent, which reduces its activity and so increases its selectivity.

Selectivity may also be achieved by having a very bulky radical that is unable to abstract certain protons. A similar point is observed with the use of sterically hindered bases to remove only primary hydrogens instead of the most acidic ones in the molecule. An example of a sterically hindered radical is that formed from *N*-chloro-di-*t*-butylamine, which abstracts primary hydrogens 1.7 times faster than tertiary ones.

11.4 Neighbouring group assistance

We have seen that photolytic halogenation usually leads to a wide variety of products with little or no selectivity. However, the bromination of alkyl bromides gives about 90 per cent substitution on the adjacent carbon. Furthermore, if the adjacent carbon is chiral then it retains its configuration. Suggest a mechanism that may account for these experimental observations.

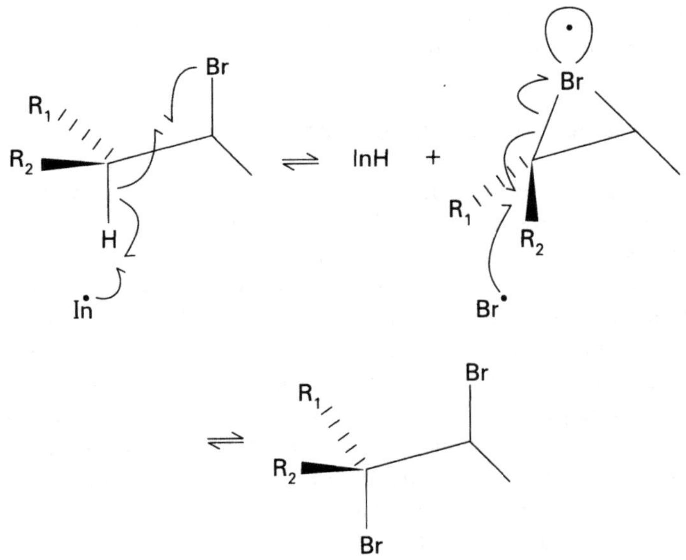

By invoking the assistance of the neighbouring bromine and forming a cyclic free radical intermediate, the retention of configuration and the selectivity of substitution is explained.

This cyclic free radical has obvious similarities to the involvement of a neighbouring group in S_N2 reactions. In those situations, the rate of the reaction was often increased, because of the favourable entropic factor. Normally in radical reactions, if the rate is increased, then the selectively goes down. In the case of *cis*-4-bromo-*t*-butylcyclohexane, both the rate and the selectivity of substitution by a bromine radical over the *trans* isomer is increased. Bearing in mind that the *t*-butyl group locks the conformation of the cyclohexane ring in such a manner that the *t*-butyl group is in an equatorial position, suggest the mechanism that is in operation.

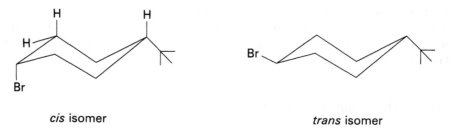

cis isomer   *trans* isomer

The *t*-butyl group must always be in the equatorial position, and so this determines the position of the smaller bromo substituent. In the *cis* isomer, the bromo substituent is in the axial position and so is capable of acting as a neighbouring species in the radical substitution reaction. This promotes substitution on the adjacent carbon, and also increases the rate of reaction. In the *trans* isomer, the bromine is in the equatorial position, and so cannot assist the substitution reaction in this way.

11.5 Further examples of radical reactions

The radicals that are formed in the initiation step are usually derived from molecules that have a weak covalent bond in them, which may easily be broken by light or heat. This was the case in the first example in which the weak chlorine/chlorine bond in a chlorine molecule was broken by the action of ultraviolet light. Light induced homolysis has advantages over thermal cleavage. For example, it may be used to break selectively some strong bonds that would not readily be cleaved by thermal means. Furthermore, if high temperatures are required to break a particularly strong bond, then this might also cause the rest of the molecule to be disrupted in an unselective manner. This advantage of light induced homolysis over thermal cleavage might be illustrated by the light induced cleavage of azoalkanes, $R-N=N-R$, to form alkyl radicals. Write down the mechanism for this cleavage.

$$R{-}N{=}N{-}R \;\; \overset{h\nu}{\rightleftharpoons} \;\; R\bullet \quad N{\equiv}N \quad \bullet R$$

As indicated earlier, heat may be used to form radicals derived from the fission of weak bonds. A very common example of this is the thermal decomposition of metal alkyls. For example, tetraethyl lead, $Pb(C_2H_5)_4$, readily gives ethyl radicals on heating. This type of compound was often used to prevent pre-ignition, or knocking or pinking, in petrol, which is when the fuel explodes rather than burns smoothly. The anti-knocking property derives from the fact that the ethyl radicals react with radicals produced in the combustion process and so terminate the chain reaction before it becomes explosive in nature. One of the problem with using thermal energy alone is that it is unselective, and so thermolysis often yields a variety of products.

In contrast to light and heat, it is possible to initiate a reaction using a compound that spontaneously forms radicals. For example, a very useful synthetic radical reaction is the bromination of an allylic carbon using *N*-bromosuccinimide, NBS. This reaction exemplifies the third manner in which radical reactions may be initiated. This reaction will not start unless there is a trace of a radical initiator, In·. This may be the result of some impurity, such as a small amount of a peroxide, which readily decomposes to form a radical. This initial radical, In·, then reacts with traces of bromine or HBr molecules to form bromine atoms. These then abstract the allylic hydrogen from the alkane, R–H. Write down the steps that have been considered so far.

Step 1a: In· + Br$_2$ $\rightleftharpoons$ In–Br + Br·

Step 1b: In· + HBr $\rightleftharpoons$ In–H + Br·

Step 2:   Br· + R–H $\rightleftharpoons$ HBr + R·

So far, the NBS has not been involved in the reaction. We know that HBr will react by adding across the double bond that is present in the allylic compound, but in this case that would be an unwanted side reaction. This is where the NBS comes into play. It reacts with the HBr to produce Br$_2$, which then reacts with the allylic radical to form the desired product. For the present, the only point about the structure of NBS you need to know is that it contains a weak nitrogen/bromine bond. Write down these steps.

Thus, the NBS acts so as to produce a very low, steady concentration of bromine molecules. The very low concentration accounts for the fact that there is very little addition of the bromine molecule across the double bond. The addition reaction of HBr is also kept to a minimum, because the HBr is consumed by the NBS. This allylic bromination is called the Wohl–Ziegler bromination.

Earlier in this section, it was mentioned that Pb(C$_2$H$_5$)$_4$ was used to ensure the smooth combustion of petrol by preventing knocking. Lead compounds are common reagents in radical reactions. An interesting example is the use of lead tetraacetate, Pb(OAc)$_4$, to cyclise alcohols that have a δ-hydrogen to form tetrahydrofurans. The four- and six-membered cyclic ethers are not formed (oxetanes and tetrahydropyrans respectively). Initially, an organolead compound is formed between the alcohol and the Pb(OAc)$_4$, which then cleaves either thermally or photolytically to give an oxygen heteroradical, which in turn abstracts the desired hydrogen. Write down the mechanism for these steps.

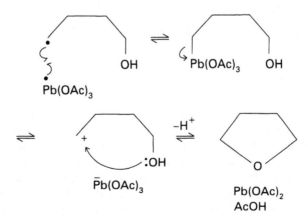

Notice that the desired cyclic ether cannot be formed directly by the oxygen radical attacking the δ-carbon, because that would mean the oxygen radical was attacking a quadrivalent carbon centre in preference to the univalent hydrogen, which is far more exposed.

At this stage of the reaction, there is a carbon radical and a lead radical. There is evidence that a carbonium ion is involved in this reaction, because cationic rearrangements may take place at the δ-carbon. Such rearrangements do not occur at radical centres. The carbon cation may be formed by combining the lead and carbon radical, and then heterolytically cleaving that same bond. Ring closure then takes place as would be expected. Write down these mechanistic steps.

However, it should be noted that even though this mechanism has been proposed by respected authors, it does not mean that it is above criticism. For example, the penultimate species proposed, the hydroxycarbonium ion, could readily isomerise so as to place the positive charge on another carbon atom. This just illustrates the point that very few mechanisms are in fact actually proved to occur in practice. Rather, any particular mechanism is suggested with a greater or lesser degree of confidence.

Lead compounds are often involved in radical reactions, because of the weakness of the carbon/lead bond, and the fact that lead has two accessible oxidation states. In the case of lead, though, because the oxidation states are plus two and plus four, the transition between them requires the addition or removal of two electrons. However, radical reactions usually require the involvement of only a single electron. Thus, one would expect transition metals that have readily occupied oxidation states that only differ by a single unit to facilitate radical reactions more easily than lead.

Not surprisingly, this is found to be the case. Thus, there are many reactions in which the $Cu(I)/Cu(II)$ couple plays a central role. For example, in the Sandmeyer reaction, an aromatic diazonium salt is converted to the corresponding chloride by the catalytic action of $Cu(I)Cl$. Suggest what will be the first step, which involves the diazonium ion, ArN_2^+, $Cu(I)Cl$ and a further chloride ion.

$$Ar-N_2^+X^- + CuX \rightleftharpoons Ar\cdot + N_2 + CuX_2$$

In this first step, the diazonium ion is reduced to form the phenyl radical. This step is helped by the fact that nitrogen is such a good leaving group. In the next step, $Cu(I)Cl$ is regenerated, and so it is a true catalyst for this reaction. Write down this reaction step.

$$Ar\cdot + CuX_2 \rightleftharpoons ArX + CuX$$

If this reaction is performed using copper and HCl, it is called the Gatterman reaction.

Another reaction in which the $Cu(I)/Cu(II)$ couple plays a central role is the Eglinton reaction. This time, though, a $Cu(II)$ salt is placed in the reaction mixture to start with. In this reaction, terminal alkynes, $R-C\equiv C-H$, are coupled in the presence of a $Cu(II)$ salt and a base such as pyridine. The reaction is performed in a vessel that is open to the air. Suggest a route for this reaction.

$$R-C\equiv C-H + pyr \rightleftharpoons R-C\equiv C^- + pyrH^+$$
$$R-C\equiv C^- + Cu^{2+} \rightleftharpoons R-C\equiv C\cdot + Cu^+$$
$$2R-C\equiv C\cdot \rightleftharpoons R-C\equiv C-C\equiv C-R$$
$$4Cu^+ + O_2 \rightleftharpoons 4Cu^{2+} + 2O^{2-}$$

Notice that the $Cu(I)$ is oxidised back to the $Cu(II)$ by atmospheric oxygen. Sometimes this reaction is carried out with a balloon of oxygen to help this process.

Another source of single electrons is the anode of an electrolytic cell. In the Kolbe reaction, an alkylcarboxylate salt, RCO_2^-, is decarboxylated and the resulting alkyl radicals couple to form the dialkyl product, R–R. Suggest the pathway that is followed by this reaction.

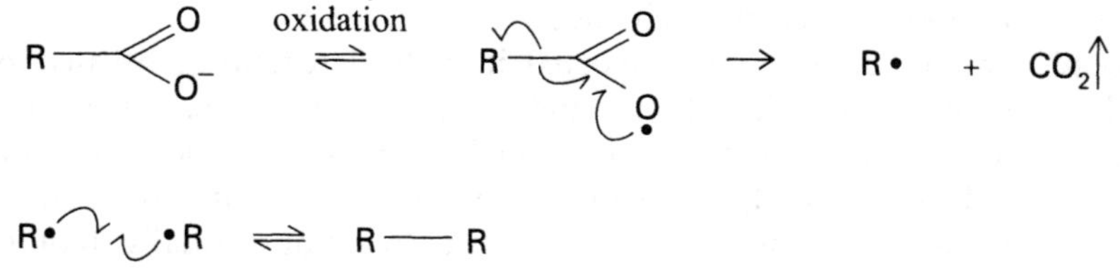

This is a good method of synthesising symmetrical alkanes. The Kolbe reaction is usually performed using the potassium or sodium salt of the carboxylic acid. If the silver salt is reacted with bromine, then decarboxylation occurs again, but this time the alkyl bromide is formed. This reaction is called the Hunsdiecker reaction. The first step involves the formation of an acyl hypohalite, RCO_2X. Suggest how this proceeds to the final product.

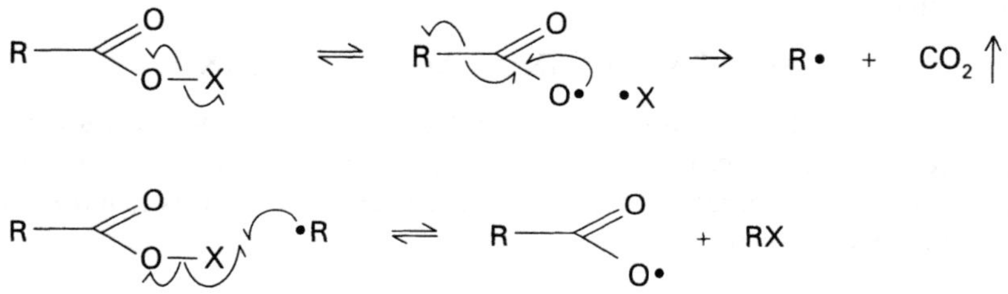

If iodine is used instead of bromine, then the exact ratio of the reagents determines which products are formed. From a starting mixture of 1:1, the alkyl halide is produced as above. However, if twice as much salt is used to the iodine, then an ester results, RCO_2R. Suggest how this may be formed.

The reaction proceeds like the Hunsdiecker reaction to yield the R–Br, which then reacts with the silver salt to give the ester. This version of the Hunsdiecker reaction is called the Simonini reaction.

A very important reaction of radicals is the formation of hydroperoxides, i.e. $RH + O_2 \rightleftharpoons R-O-O-H$. When this occurs by slow atmospheric oxidation, this is called auto-oxidation. Slow, in this context, means without combustion. The hydroperoxides that are formed often undergo further reactions, oxidising other parts of the molecule or different molecules altogether. The reason why this is so important is because this process is responsible for the hardening of paints, the perishing of rubber, the curing of varnishes and the process whereby fats become rancid. The reaction proceeds much faster in light, and usually requires a trace of an initiator, In·, because molecular oxygen is not reactive enough to abstract a hydrogen atom directly. Suggest what the route would be in the presence of such an initiator.

$$In\cdot + O_2 \rightleftharpoons In-O-O\cdot$$

$$In-O-O\cdot + RH \rightleftharpoons In-O-O-H + R\cdot$$

$$R\cdot + O_2 \rightleftharpoons R-O-O\cdot$$

$$R-O-O\cdot + RH \rightleftharpoons R-O-O-H + R\cdot$$

The initially formed In–O–O· is reactive enough to abstract the hydrogen from any weak C–H bond, such as a tertiary, allylic or benzylic bond. The auto-oxidation of ethers occurs at the α-carbon. However, the resultant compounds have a tendency to explode spontaneously. This is why ethers should not be exposed to sunlight, and preferably should be stored in dark bottles.

Radical reactions may be used to introduce functionality at a remote position from the original functional group. We have already seen an example of this in the formation of cyclic ethers. In that case, the ether linkage was introduced at the δ-carbon from the original hydroxyl group. This was an example of an intramolecular radical reaction. Another example is the Hofmann–Löffler reaction in which an *N*-alkyl-*N*-halogenoamine is heated under acidic conditions to give the *N*-alkyl-pyrrolidine or *N*-alkyl-piperidine product. Suggest a route that this reaction might follow.

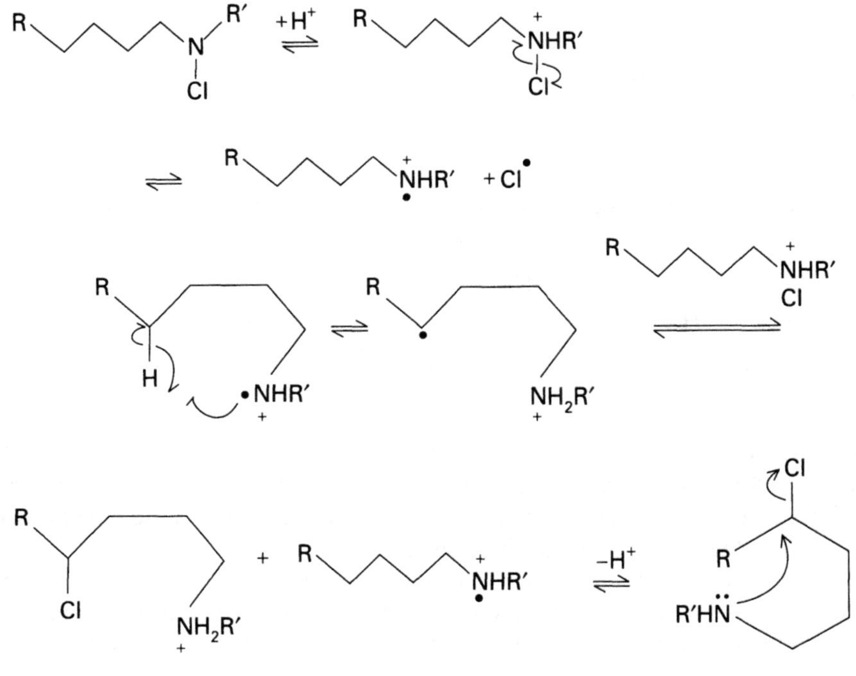

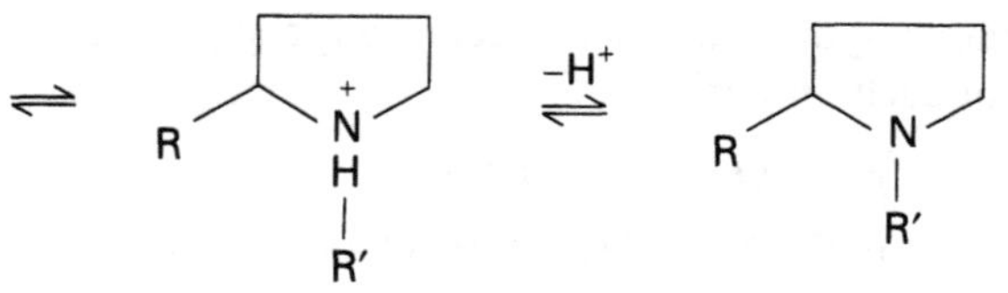

Protonation followed by homolytic cleavage of the N–Cl bond, produces a nitrogen radical that is capable of abstracting a hydrogen further down the alkyl chain. This carbon radical then combines with another molecule of the protonated starting material to form the rearranged halogenoamine, which then closes the ring via a nucleophilic substitution reaction.

A related reaction is the Barton reaction. In this case, a methyl group that is δ to a hydroxyl group is oxidised into an aldehyde functionality. The hydroxyl group is first converted into a nitrite ester by the action of nitrosyl chloride, NOCl, and after exposure to light the δ-methyl group undergoes nitrosation. Write down the pathway for these steps.

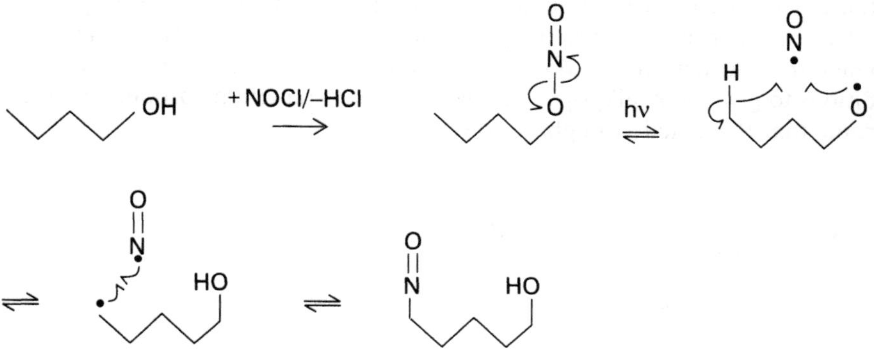

Suggest how this may now be converted to the desired aldehyde product.

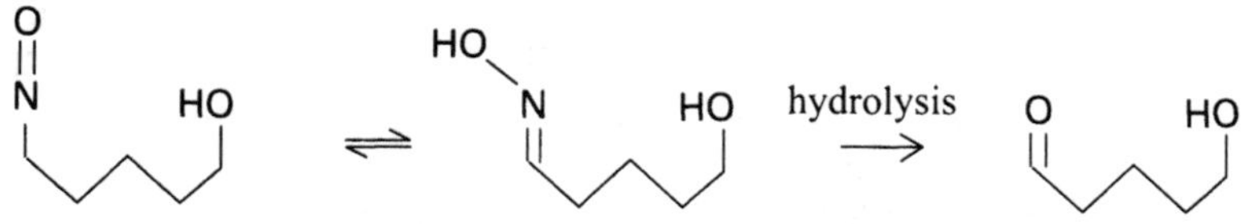

The nitroso compound tautomerises to give the oxime, which in turn may be hydrolysed to give the desired product. This reaction requires a six-membered cyclic transition state for the hydrogen abstraction, and so is quite specific as to where the aldehyde group will be formed.

That concludes this chapter on radical substitution reactions. Radical reactions are vastly underestimated in value, especially considering how important they appear to be in nature. For example, the double ring closure that results in the formation of the penicillin structure is thought to proceed by a radical mechanism. Yet one of the reasons why they are not studied in more detail is because they often proceed by less well defined mechanistic routes than do nucleophilic or electrophilic reaction pathways. A consequence of this is that often there is less control over the outcome. However, as illustrated by, for example, the Barton reaction, this is not always the case.

This chapter also brings to a close the study of substitution reactions. We will now take a look at addition reactions.

11.6 Summary of radical substitution reactions

Radicals have an unpaired electron and are usually electrically neutral. Accordingly, radical substitution reactions tend not to be affected by those factors, such as solvent polarity, that affect mechanisms involving charged species, such as nucleophilic or electrophilic substitution.

Radical substitution reactions involve an initiation step, in which the number of radicals increases; a number of propagation steps, which are also called S$_H$1 reactions, in which the number of radicals remains constant; and a number of termination steps, in which the number of radicals decreases. A very common propagation step is the abstraction of a hydrogen atom from a carbon/hydrogen bond, which is an example of a bimolecular homolytic substitution reaction, S$_H$2.

The reaction of one radical to form another and then another, and so on, by propagation steps is called a chain reaction. The more reactive the radicals involved, the greater the number of reactions in the chain before it is brought to an end by a termination reaction.

Initiators often trigger free radical reactions, because they easily form free radicals; while inhibitors consume free radicals and thus prevent the initiation or propagation of free radical reactions.

Tertiary radicals are more stable than their secondary isomers, which in turn are more stable than primary ones. The order of preference for the abstraction of hydrogen atoms increases with decreasing bond dissociation energy, coupled with any tendency to form a delocalised radical system.

In arylalkanes, the α-carbon on the alkyl side chain is invariably attacked. Aryl derivatives, if they undergo attack at all, usually give the addition product.

The stereochemistry at a radical centre is usually lost upon the formation of the radical. However, a radical is very tolerant of its exact stereochemistry, and so it is possible to form a radical at a bridgehead for example.

By reducing the reactivity of the attacking radical it is possible to introduce some degree of selectivity into a radical substitution reaction. Sometimes a judicious choice of solvent can also effect the degree of selectivity.

As in nucleophilic substitution, it is possible for a neighbouring group to assist in the radical substitution, in which case there is often retention of configuration and preferential substitution at the α-carbon at an increased rate.

211

Light or heat may be used to cleave a weak bond, e.g. the thermolytic cleavage of tetraethyl lead to give ethyl radicals; while generally, only light is used to cleave a stronger bond, e.g. in an azoalkane, $R-N=N-R$, because otherwise the large amount of thermal energy that would be needed would lead to the fission of other bonds. Allylic bromination may be achieved using NBS; this is called the Wohl–Ziegler bromination.

Radical reactions often involve weak organometallic bonds, such as C–Pb, as in the cyclisation of alcohols that have a δ-hydrogen to form tetrahydrofurans.

The Cu(I)/Cu(II) couple is utilised in the Sandmeyer reaction, in which an aromatic diazonium salt is converted into the corresponding aryl chloride by the catalytic action of Cu(I)Cl. There is a variation that uses copper and HCl, which is called the Gatterman reaction.

In the Eglinton reaction, terminal alkynes, $R-C\equiv C-H$, are coupled in the presence of a Cu(II) salt and a base such as pyridine to form dialkynes.

The single electron that is essential to a radical reaction may be provided by an electrolytic cell as in the Kolbe reaction. In this reaction, an alkylcarboxylate salt, RCO_2^-, is decarboxylated and the resulting alkyl radicals combine to form the dialkyl product, R–R. If the silver, rather than the sodium or potassium, salt is reacted with bromine, then the alkyl bromide is formed. This variation is called the Hunsdiecker reaction. If iodine is used instead of bromine, and twice as much salt is used to the iodine, then an ester results, RCO_2R. This is called the Simonini reaction.

Auto-oxidation is very important in many fields. It usually requires an initiator, because atmospheric oxygen is not reactive enough to form the hydroperoxide intermediate.

In the Hofmann–Löffler reaction an N-alkyl-N-haloamine is heated under acidic conditions to give the N-alkyl-pyrrolidine or N-alkyl-piperidine product. A related process is the Barton reaction, in which a methyl group that is δ to a hydroxyl group is oxidised into an aldehyde.

Addition Reactions to Carbon/Carbon Multiple Bonds

12.1 Introduction

Addition reactions represent one of the simplest type of organic mechanisms. However, there are many possible variations. They usually involve the addition of a small molecule to a substrate that has a multiple bond, which results in a product called an adduct. If the resultant adduct no longer has any more multiple bonds between the carbon atoms, then it is said to be saturated. Correspondingly, the original molecule, which contained one or more multiple bonds between carbon atoms, is said to be unsaturated. There are a number of different ways in which this general reaction type may occur. Thus, to start with, we will outline these various permutations and introduce some of the terms that will be used in the remainder of this chapter.

Write down a general equation for the reaction of a small symmetrical bimolecular species, A_2, with the symmetrical molecule $R_2C=CR_2$, and indicate which species is saturated, which is unsaturated and which is the adduct.

$$R_2C=CR_2 + A_2 \rightleftharpoons R_2AC-CAR_2$$

The alkene, $R_2C=CR_2$, is unsaturated, while the adduct, $R_2AC-CAR_2$, is saturated. This type of reaction is very common, e.g. the hydrogenation of an unsaturated carbon/carbon double bond to give the saturated carbon/carbon single bond. Write down the overall equation for the hydrogenation of ethene and identify the adduct.

$$H_2C=CH_2 + H_2 \rightleftharpoons H_3C-CH_3$$

The addition product is ethane. This type of reaction is fairly elementary, and so you are likely to have come across it before. This is not always the case, and as the examples become more complex and less commonplace, the value of a mechanistic approach increases, because you will be able to work out a likely reaction sequence and hence the likely products.

The addition of a small molecule such as hydrogen to an unsaturated molecule may occur in one step, i.e. once the hydrogen molecule has approached the alkene, they may react together to form directly the final product. Assuming that this is the case, it is not unreasonable to propose that there is a cyclic transition state. A consequence of this is that both hydrogen atoms would be added from the same side. This is called *syn*-addition.

Write down the overall reaction equation for the addition of a bromine molecule to ethene.

$$H_2C=CH_2 + Br_2 \rightleftharpoons BrH_2C-CH_2Br$$

The bromine/bromine bond is much weaker than the hydrogen/hydrogen bond, and so it is conceivable that, in this case, the bromine molecule breaks into two parts before the addition reaction is complete. If this were to happen, then the addition would occur in two steps, with each bromine species being added one at a time. In such a situation, it would be possible for the second bromine atom to add to the intermediate from either the same side as the first bromine species, i.e. *syn*-addition, or from the other side, in which case it is called *anti*-addition.

If the addition occurred stepwise, then there are three possible types of species that may attack the alkene. Suggest what these three species are.

$$E^+, Nuc^- \text{ and } R\cdot$$

The attacking species may be an electrophile, a nucleophile or a free radical.

If the attacking species is not symmetrical like a dihydrogen molecule or a bromine molecule, but instead is unsymmetrical, and if it is attacking an unsymmetrical alkene, then the addition may take place in either of two orientations. Illustrate these two orientations with the addition of AB to $X_2C=CY_2$.

$$AX_2C-CY_2B \text{ or } BX_2C-CY_2A$$

If it were possible to arrange the reaction conditions such that one orientation could be favoured over the other, then this would be useful synthetically.

If the reagent has more than one multiple bond that is capable of undergoing an addition reaction, or even if there is a multiple bond that is capable of reacting with more than one molecule, then that reagent is said to be polyunsaturated. Suggest an example of each type of polyunsaturated molecule.

1,3-butadiene, $H_2C=CH-CH=CH_2$, and ethyne, $H-C\equiv C-H$, respectively. In each case, and under the appropriate conditions, the molecule could react with two molecules of hydrogen. Write down all the possible open chain adducts that could result from the addition of just one molecule of hydrogen to (a) ethyne and (b) 1,3-butadiene.

In the case of ethyne there is only one possible product, i.e. ethene, $H_2C=CH_2$, which is the partial hydrogenated product. However, in the case of the conjugated system of 1,3-butadiene there are two possible products, i.e. 1-butene, $H_2C=CHCH_2CH_3$ or 2-butene, $H_3CCH=CHCH_3$, depending on whether the hydrogen molecule has been added across a double bond as depicted in the original molecule, or has added across the terminal carbons accompanied by a reorganisation of the residual unsaturation. The former is called 1,2-addition, while the latter is called 1,4-addition or conjugate addition.

Having introduced an number of different terms and variations on the general theme of addition reactions, we will now look in greater detail at these possibilities. To start with, we will study the simplest type of addition reaction, namely the catalytic hydrogenation of a simple alkene.

12.2 Cyclic addition

12.2.1 *Heterogeneous catalytic hydrogenation*

In the one step addition of two molecules to each other, it is reasonable to suggest that such an addition occurs via a cyclic transition state, with the new bonds being formed simultaneously with the old bonds being broken. It would also be reasonable to suggest that both parts of the molecule that is adding to the unsaturated molecule approach the double bond from the same side. This would be especially true for the addition of small molecules. This is called *syn*-addition.

The commonest example is the catalytic hydrogenation of an alkene. Write down a balanced equation for the hydrogenation of ethene.

$$H_2C=CH_2 + H_2 \rightleftharpoons H_3C-CH_3$$

The reaction is commonly catalysed heterogeneously by metals such as nickel, platinum or palladium. The alkene and the hydrogen are adsorbed onto the surface of the metal, and this results in the weakening of the bonds so that the species react with each other more easily. Some of the hydrogen will also be absorbed into the metal lattice, but this is of little use in effecting the reduction of the alkene.

Draw a mechanism that illustrates the movement of the electrons.

The *syn* nature of this addition may be more fully appreciated when a cycloalkene such as 1,2-dimethylcyclopentene is hydrogenated. Suggest what will be the geometry of the resultant alkane after hydrogenation with a nickel catalyst.

A further example may be provided by the partial hydrogenation of ethyne to ethene. In passing, it is important to note that one does not confuse partial hydrogenated with partly hydrogenated. The former means that all of the starting material has been hydrogenated to produce an intermediate that is capable of further hydrogenation; however, no starting material is left. The latter term means that only some of the starting material has been hydrogenated, and so some remains. A Lindlar catalyst is needed to effect this partial hydrogenation. This catalyst comprises palladium on calcium carbonate, partial poisoned with lead(II) oxide. Suggest what will be the stereochemistry of the product formed from the partial hydrogenation of but-2-yne using this catalyst.

The precise mechanism of catalytic heterogeneous hydrogenation is a matter of debate; however, the mechanism as proposed above agrees with the experimental observations, namely that *syn*-addition occurs.

12.2.2 *Other syn-addition reactions*

Another example of hydrogenation is when diimide, H–N=N–H, reacts with an alkene to form an alkane. There are two geometric isomers of diimide. Draw out both isomers.

trans   *cis*

Now write down the mechanism for the reduction of an alkene using the appropriate isomer of diimide.

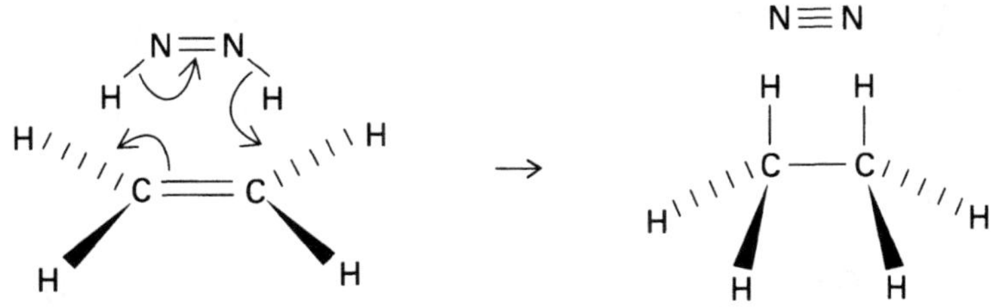

In this reaction, only the *cis* form of the diimide will react, because only this isomer can form the six-membered cyclic transition state. In this case, a molecule of hydrogen has been added across the alkene, but it is possible to add many other species across a carbon/carbon double bond, either directly, as in the case of the hydrogen molecule, or indirectly as in our next example.

The overall result of treating an alkene with osmium tetroxide is to add a molecule of hydrogen peroxide, H_2O_2. Write down a balanced equation for this overall reaction, bearing in mind that the initial intermediate is hydrolysed to a 1,2-diol.

$$H_2C{=}CH_2 + OsO_4 + 2H_2O \rightleftharpoons H_2COH{-}CH_2OH + H_2OsO_4$$

In this case, osmic acid, H_2OsO_4, is formed as a by-product. It is usually reoxidised to the osmium tetroxide by the use of hydrogen peroxide, which means that only a small amount of the toxic and expensive osmium compound is used.

Suggest a mechanism for the reaction between the osmium tetroxide and the alkene.

Again a concerted cyclic mechanism may be drawn, which in this case results in the formation of the cyclic osmic ester. If this compound is isolated, and then hydrolysed using ^{18}O labelled water, it is observed that there is no ^{18}O label in the resulting 1,2-diol. Suggest a mechanism that explains this result.

217

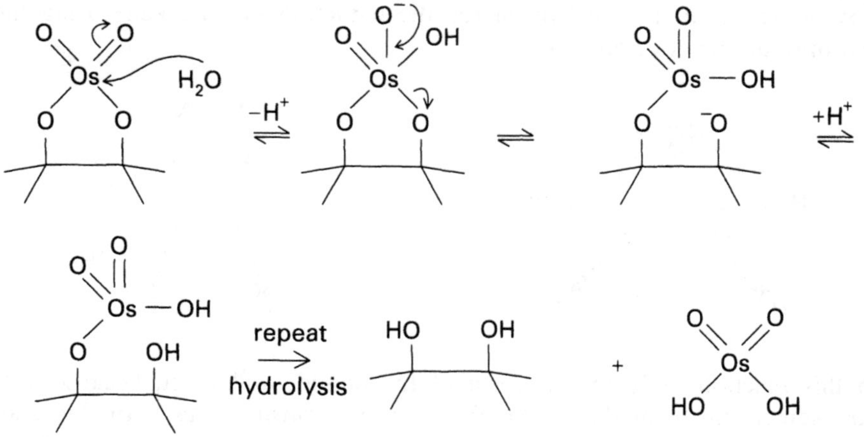

The result of the labelling experiment indicates that both oxygens in the 1,2-diol originate from the osmium tetroxide, and that the Os–O bonds are broken on hydrolysis. Suggest what is the stereochemical relationship between the two hydroxyl groups added in this manner.

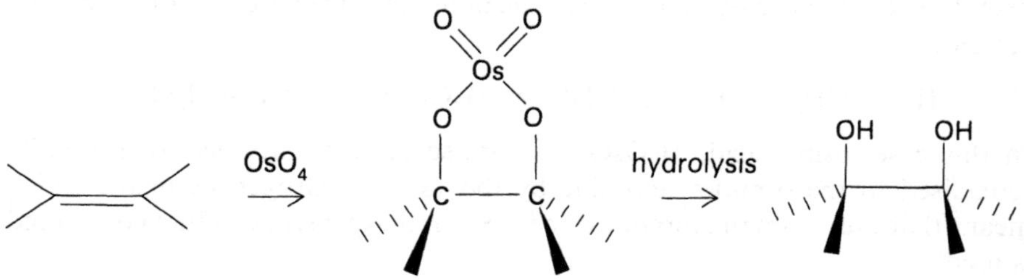

As both of the oxygens originate from the osmium tetroxide, which reacted via a cyclic five-membered ring, it means that they must be on the same side as each other, i.e. there was *syn*-addition.

So, in summary, this synthetic route gives rise to overall *syn*-addition of hydrogen peroxide.

12.2.3 Pericyclic reactions

Suggest the result of the addition of 1,3-butadiene across an alkene such as ethene.

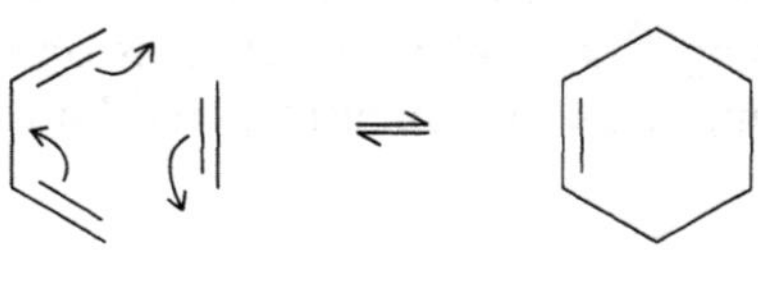

Here the two molecules have reacted together to give a cyclic adduct. This is the mechanism of the Diels–Alder reaction. Note that the 1,3-butadiene molecule may rotate around the central carbon/carbon bond, but that this rotation is not as easy as it would have been in the case of an isolated carbon/carbon single bond in, say, ethane. Suggest a reason for the slightly restricted nature of the rotation about this central bond in 1,3-butadiene.

It is possible to draw a canonical structure for this molecule that involves a separation of charges, but which in turn confers some double bond characteristics on the central part of the molecule. As a consequence, there are two identifiable conformers of 1,3-butadiene that result from this small energy barrier to free rotation. Suggest what are these two conformers.

cisoid   transoid

These two conformers are called the *cisoid* and *transoid* conformers respectively. Only the *cisoid* conformer will react in the Diels–Alder reaction indicated above. Suggest which conformer is the more stable, and why.

The *transoid* structure is more stable than the *cisoid*, because there is less unfavourable steric interaction between the groups located around the central bond. Accordingly, suggest whether cyclopentadiene would react more or less rapidly than the open chain 1,3-butadiene, and suggest a reason for any difference that might be observed.

The cyclopentadiene would react more rapidly, because for the open chain diene there is only a small concentration of the required *cisoid* conformer present in the reaction mixture. However, in the case of cyclopentadiene, all of the molecules are locked in a *cisoid* conformation.

Draw out the three geometric isomers of 1,4-diphenylbutadiene and then suggest which one will undergo the Diels–Alder reaction with a suitable reagent.

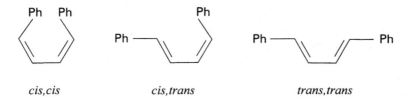

cis,cis   cis,trans   trans,trans

Only the *trans,trans* isomer will react, because in the other two isomers there is too much unfavourable steric interaction for the diene to adopt a planar conformation.

It is observed that the rate of the Diels–Alder reaction is increased by electron donating substituents on the diene, and by electron withdrawing groups on the alkene, or dienophile. This would suggest that the diene is acting as the nucleophile, while the dienophile is acting as the electrophile. Typical examples of a diene and dienophile are cyclopentadiene and maleic anhydride respectively. When these two compounds undergo the Diels–Alder reaction, it is possible for the reaction to proceed with two distinct orientations; suggest what is the geometry of each of these orientations.

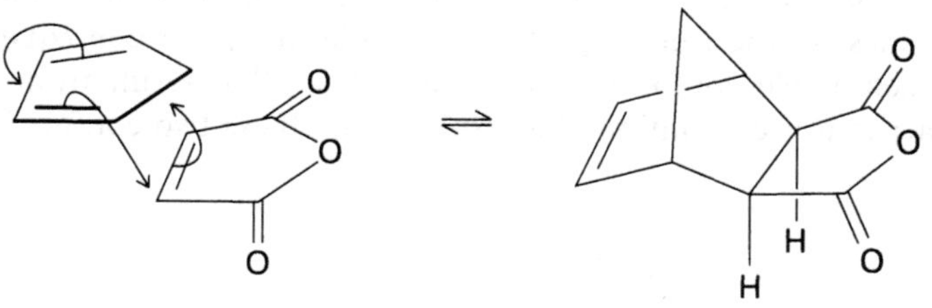

exo

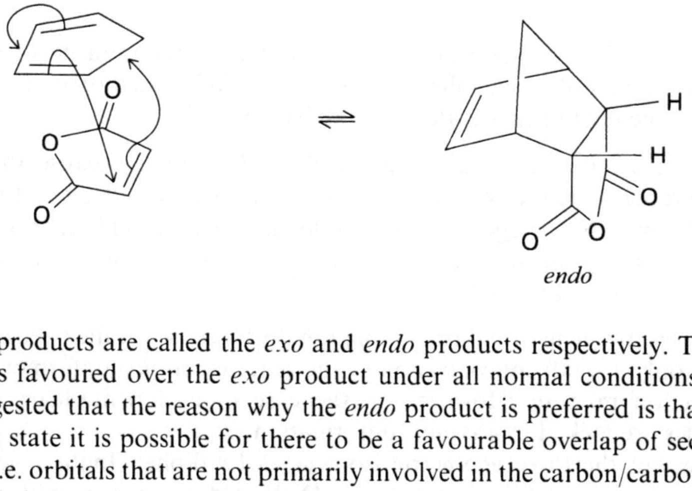

endo

The two products are called the *exo* and *endo* products respectively. The *endo* product is favoured over the *exo* product under all normal conditions. It has been suggested that the reason why the *endo* product is preferred is that in the transition state it is possible for there to be a favourable overlap of secondary orbitals, i.e. orbitals that are not primarily involved in the carbon/carbon bonds that are being made or broken. Notice that the result of adopting such a geometry during the transition state is that the addition to both the diene and the dienophile is *syn*, and that the stereochemistry of four adjacent carbon atoms is thereby controlled. This control over the stereochemistry of four carbon centres is one of the reasons why the Diels–Alder reaction is so important in synthetic chemistry.

It is usual to perform the Diels–Alder reaction with the type of reagents indicated above; however, it is possible to perform the reaction with reagents in which the electronic characteristics of the diene and dienophile are reversed.

Bearing this in mind, deduce the mechanism for the reaction between perchlorocyclopentadiene and cyclopentene.

This is an example of a Diels–Alder reaction in which the typical electronic characteristics have been reversed. Now suggest what would be the relative rates of reaction between the perchloro compound and (a) tetracyanoethylene, (b) maleic anhydride and (c) cyclopentene.

The perchloro compound reacts the fastest with cyclopentene, and only slowly with maleic anhydride, and not at all with tetracyanoethylene. The last is usually a very reactive reagent in Diels–Alder reactions operating under the normal electronic basis. However, under the reverse basis, what was favourable is now unfavourable, to such an extent that the reaction does not proceed to a noticeable extent.

In the hydrogenation reactions, two atoms of hydrogen were added across a double bond, i.e. the addition of H_2 occurred, while in the Diels–Alder reaction, two carbon/carbon bonds were formed, i.e. the addition of R_2 occurred. There is a reaction that falls between these two general reaction types, namely one new hydrogen/carbon bond is formed and one new carbon/carbon bond is formed, i.e. the addition of RH. This reaction is called the ene synthesis. Suggest the product of the addition between an alkene such as maleic anhydride and a reagent such as 3-phenylpropene.

It is a common feature of both the Diels–Alder reaction and the ene synthesis that the reaction is not affected by the presence of radicals or by changes in the polarity of the solvent, but only by the action of heat and light. The mechanisms suggested above have all been single step mechanisms that involved the concerted movement of electrons utilising cyclic transition states. In such a case, the bond formation occurs approximately simultaneously with the bond breakage, and so cannot be influenced by outside factors.

Suggest a possible reaction that might occur when two molecules of ethene are forced to react with each other.

It is found that it is very difficult to perform this reaction. It is suggested that the reason for this reluctance is that all these types of reaction are controlled by the symmetry, or phases, of the respective molecular orbitals involved, and in the case of the proposed dimerisation of ethene, the phases of the relevant orbitals are not compatible with ring closure. These reactions form part of a class called pericyclic reactions. There are three principal divisions, namely electrocyclic, cycloaddition and sigmatropic. The detailed study of symmetry controlled reactions is outside the scope of this book, and whenever a pericyclic reaction appears in this text, it will just be assumed that it is concerted without further consideration of the phase of the molecular orbitals involved.

We will now turn our attention from the concerted addition of a molecule that effectively occurs in a single step to reactions that occur in two discrete steps.

12.3 Electrophilic addition

12.3.1 *Addition of a symmetrical molecule*

The addition of a small symmetrical molecule, such as bromine, to a symmetrical substrate, such as ethene, is the simplest type of two step addition reaction. We will look at this reaction in some detail in order to highlight the factors involved.

First, write down an equation for the overall reaction, and name the reagents and the product.

$$H_2C{=}CH_2 + Br_2 \rightleftharpoons BrH_2C{-}CH_2Br$$

Ethene reacts with a bromine molecule to yield 1,2-dibromoethane. The first important point to notice is that the 1,2-dibromo product was formed. Write down the other possible positional isomer of this compound.

The other possible positional isomer is 1,1-dibromoethane, $H_3C{-}CHBr_2$. In this molecule, both of the bromine atoms are attached to the same carbon and not joined to adjacent ones. The presence or absence of isomers often gives very important clues as to the route that has been followed by the reagents from the starting materials to the eventual products. This is true not only of structural isomers, but also of stereoisomers. The 1,1-dibromo compound is called the *geminal* adduct, while the 1,2-dibromo compound is called the *vicinal* adduct. These terms are often abbreviated to *gem* and *vic*, respectively.

Draw out the molecular structure of ethene, and in particular show the electron orbitals that are associated with the carbon/carbon double bond. Also describe the geometry of this molecule.

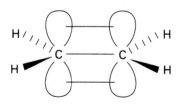

Ethene is a flat molecule, i.e. all six atoms are in a single plane; each carbon is in a trigonal configuration; and the π electrons are concentrated above and below the plane in which the atomic nuclei are situated. Given this arrangement, deduce what will be the charge distribution around the molecule, and hence suggest whether this molecule will be electrophilic or nucleophilic.

The π electrons, which are above and below the plane of the atoms, will be a centre of negative charge, and so this molecule is nucleophilic.

We will now turn our attention to the detailed reaction pathway for the bromination of an alkene. Suggest what will happen when a bromine molecule approaches the ethene molecule.

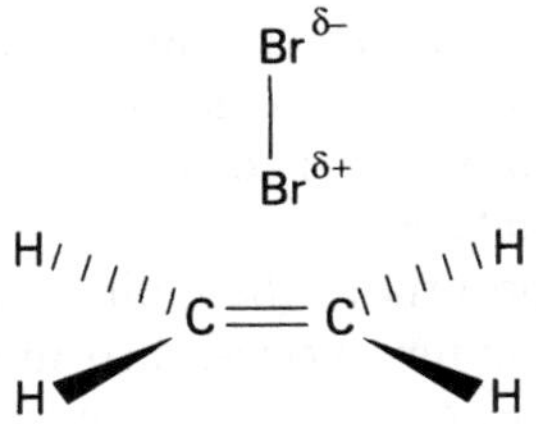

$$
\begin{array}{c}
Br^{\delta-} \\
| \\
Br^{\delta+}
\end{array}
$$

$$
H \cdots C = C \cdots H
$$

As the bromine molecule approaches the π electrons, they induce a temporary dipole in the electron cloud of the bromine. This is easily done, because the outer electrons on the bromine molecule are a long way from the two positive nuclei, and so are held less firmly than if they were more closely bound. In other words, the bromine molecule is a soft electrophile, because its electron cloud may be easily distorted. Note that all temporarily induced dipolar interactions are attractive in nature, and so facilitate the further approach of the reacting species.

Suggest what may happen as the bromine molecule is drawn closer to the ethene molecule.

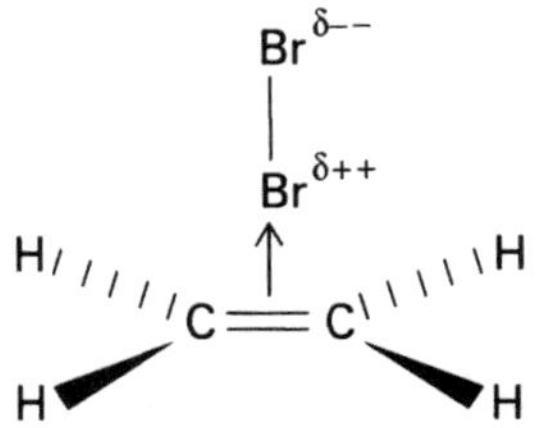

$$
\begin{array}{c}
Br^{\delta--} \\
| \\
Br^{\delta++}
\end{array}
$$

$$
H \cdots C = C \cdots H
$$

As the molecules get closer and closer, the interaction between the electrons in one molecule and the other becomes stronger and stronger. Eventually, there is a reorganisation of the old molecular orbitals centred on the original molecules into new molecular orbitals that are centred on a different arrangement of the atomic nuclei.

Suggest a likely movement of electrons, and indicate this by drawing arrows to show this movement of electrons. Also, suggest the nature and structure of the likely intermediates that will be formed along this reaction pathway.

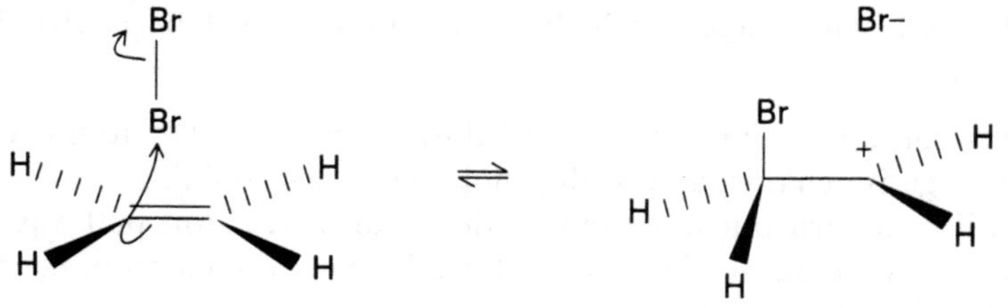

The electron-rich region of the π electrons cloud attacks the electropositive end of the bromine molecule. In so doing, the carbon/carbon double bond breaks, leaving one carbon with a positive charge. The bromine/bromine bond also breaks to form a bromide anion, thus conserving charge. Both of these bond cleavages are heterolytic in nature.

Indicate, using an appropriate three-dimensional representation, the structure of the bromocarbonium species; also indicate on the diagram, the bonds about which there is free rotation, and show the orientation of the empty p orbital that exists.

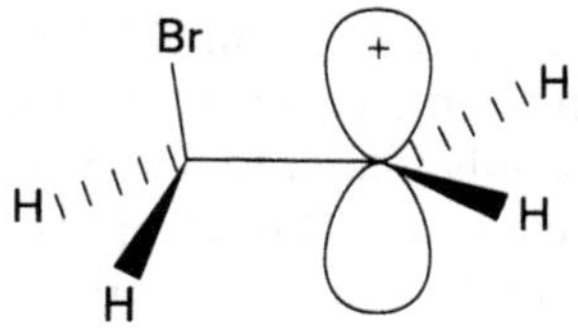

Notice that the three lone pairs that are on the bromine atom are quite close in space to the empty orbital that is on the carbon carrying the positive charge. Bearing this juxtaposition in mind, suggest a possible movement of electrons that might occur, and clearly indicate the position of the charge on the resultant intermediate.

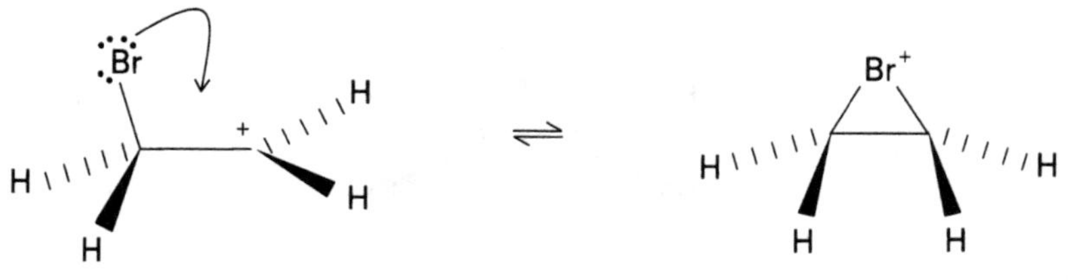

One of the lone pairs of electrons on the bromine atom is donated into the empty orbital and so forms a dative bond. As a result, the bromine atom becomes positively charged, because it was the donor of the electrons, while the carbon atom correspondingly becomes neutral.

Indicate the stereochemical structure of this three-membered species, and also indicate the direction of any polarisation that might exist along the bromine/carbon bonds.

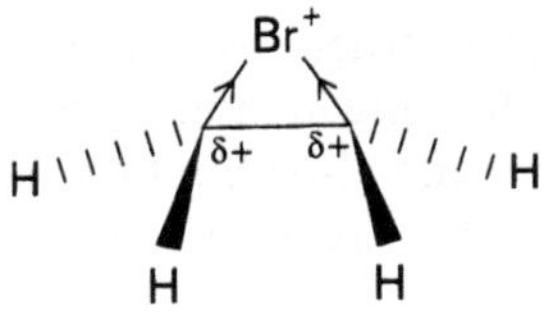

If the second bromine species, which is the bromide anion, reacts with a carbon atom, suggest the likely line of attack, and also show the electron movement that will occur as a result.

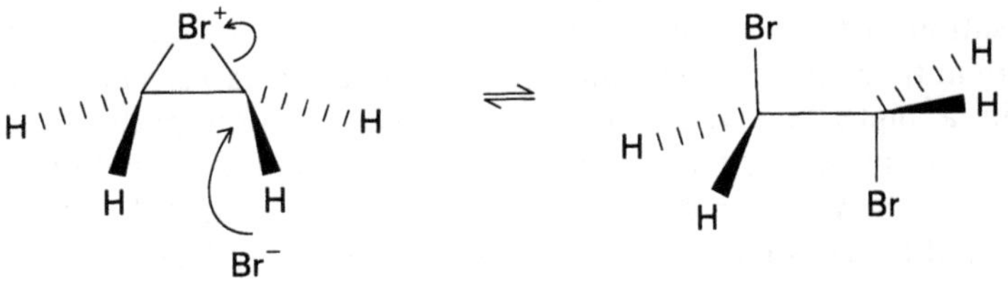

The bromide ion attacks the carbon from the opposite side to the bridged bromine atom, because the carbon/bromine antibonding orbital projects in this direction. This mode of addition is called *anti*-addition. This is clearly seen from the conformation of the final product before any rotation of the carbon/carbon bond has occurred. As the bromide ion attacks the $\delta+$ carbon, that carbon/ bromine bond breaks in order that the octet of electrons is not exceeded on the carbon. This mechanism explains why the final product is the 1,2-dibromo adduct and not the 1,1-dibromo adduct.

The pathway outlined above is a two step process with a positively charged intermediate. Suggest what may be the result of performing the addition reaction of bromine with ethene in the presence of chloride ions.

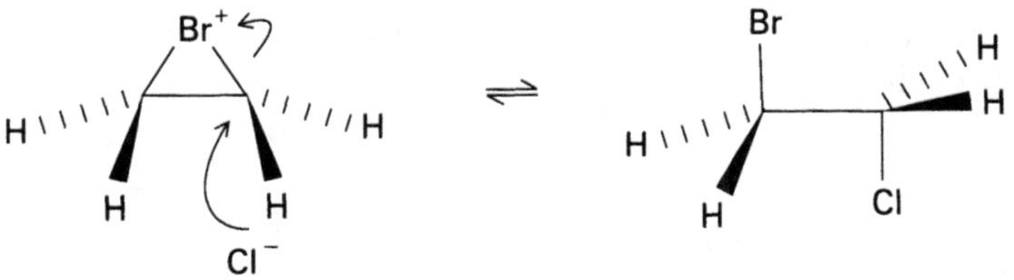

The cyclic bromonium ion could be attacked by a chloride ion instead of a bromide ion, and so the mixed bromochloro product could be formed. This type of side reaction actually occurs, and is indicative that the route suggested above is actually the pathway that is followed in practice.

Now we will consider the bromination of an alkyne. Suggest what will be the product when one molecule of bromine is added to one molecule of ethyne;

write down the equation for such a reaction.

$$H-C \equiv C-H + Br_2 = BrHC = CHBr$$

The 1,2-dibromo product is formed. At first sight, one would expect bromine to react faster with ethyne than with ethene in forming the single addition product. However, this is not the case. Bearing in mind what you already know about the electron distribution in alkynes and also about the nature of the intermediates in the reaction pathway, suggest three reasons to explain why alkenes react faster with bromine than do alkynes.

The first possible reason is that the electrons in the alkynes are held more closely to the nucleus than is the case in alkenes, as evidenced by the difference in acidity of hydrogens that are attached to each type of carbon. The second possible reason is that in the reaction pathway of the alkyne, there is a possible hybridisation of the bromine atom in the cyclic bromonium species that gives rise to an anti-aromatic species and which is hence unstable. This would in turn result in a higher activation energy for this reaction pathway. However, probably the most important reason is that the cyclic bromonium ion that contains a carbon/carbon double bond is far more strained than the saturated version, and hence is more unstable, and so less readily formed.

This analysis of the simple addition of an electrophilic bromine molecule to a symmetrical alkene or alkyne has highlighted many points. First, there is the induction of a temporary dipole of the soft electrophile by the π electrons of the carbon/carbon double bond. Second, there is the heterolytic fission of the bromine molecule, and the subsequent formation of the cyclic bromonium ion. Third, this cyclic intermediate places certain restrictions on the potential line of attack for the second reagent, and so controls the structural and stereochemical consequences for the product.

Finally in this section, we will look at the addition of symmetrical molecules to unsymmetrical substrates and study some of the further stereochemical consequences of *anti*-addition.

Maleic acid (more properly called *cis*-butendioic acid) reacts with bromine to give 2,3-dibromosuccinic acid (2,3-dibromobutandioic acid). Write out the possible permutations for this reaction, and so deduce the resultant stereochemistry of the products.

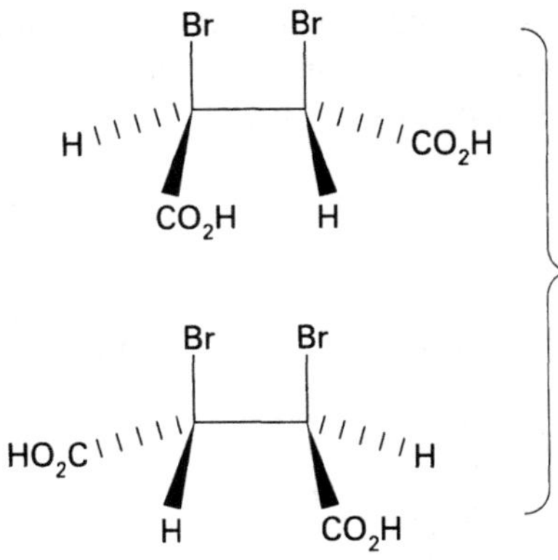

threo *d,l* pair

This reaction proceeds via a cyclic bromonium ion. The *anti*-addition of the bromide anion may take place at either of the two carbon atoms that form part of the ring, and so two products are formed in a 1:1 ratio. These are the *threo d,l* pair of stereoisomers. The resultant mixture is not optically active, because each enantiomer is produced in equal amounts.

Now predict the outcome of the addition of bromine to fumaric acid, which is the *trans* isomer of maleic acid.

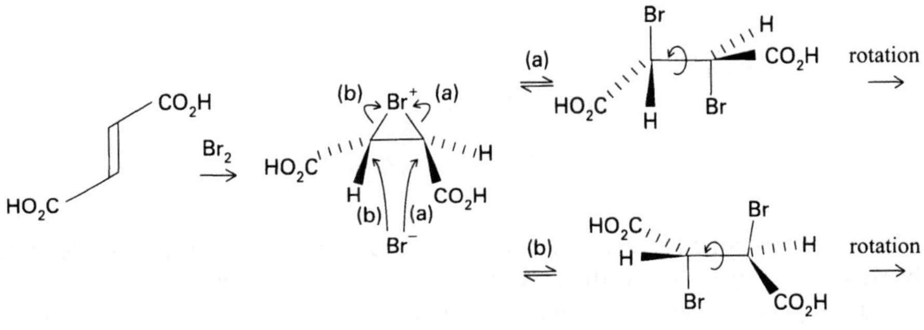

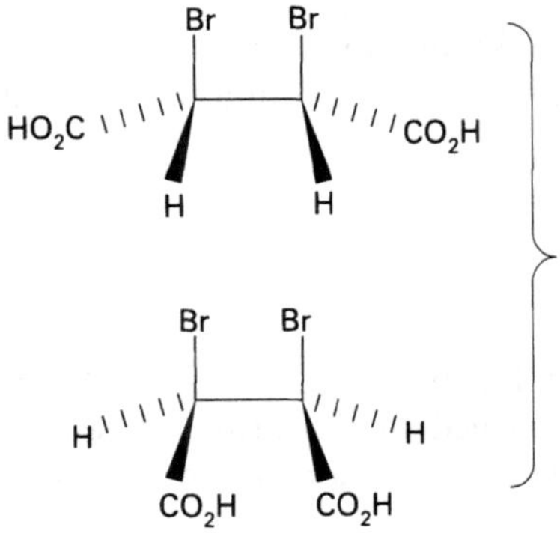

meso compound

227

In this case, the *meso* isomer is formed. For the sake of completeness, we will continue this exercise by adding an unsymmetrical molecule, AB, to fumaric acid. Suggest the stereochemistry of this product.

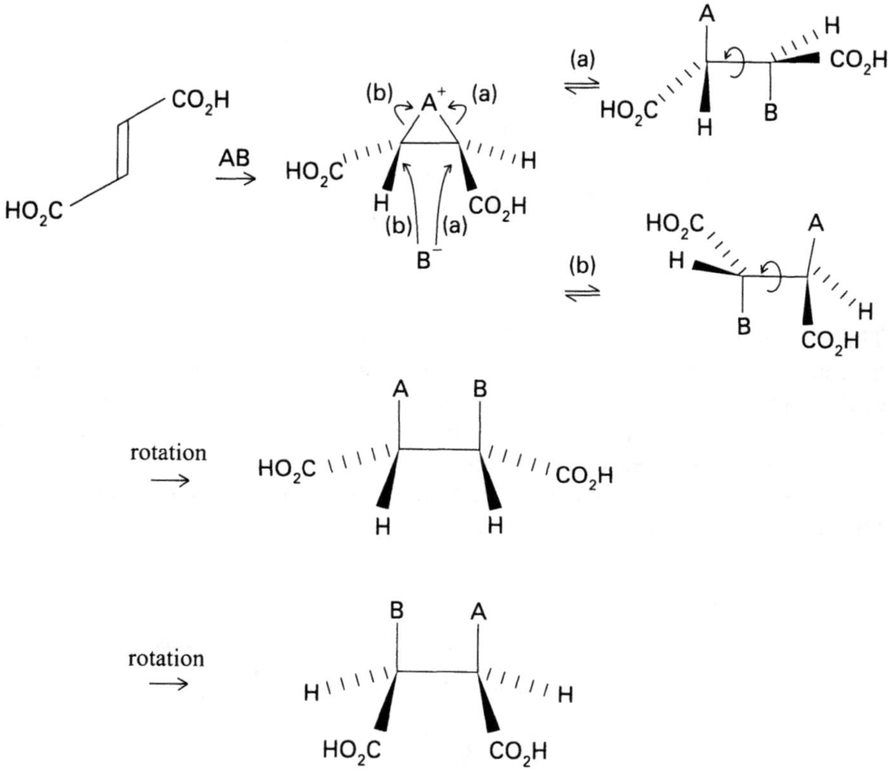

The result this time is the *erythro d,l* pair. Thus, this process is stereoselective, because only one set of products is produced from each reaction; but, further, it is stereospecific, because a given isomer leads to one product pair while the other isomer leads to the opposite product pair.

We will now move on and look more closely at the rate of the addition, and the stereochemical consequences of the addition of unsymmetrical molecules to unsymmetrical substrates.

12.3.2 Rate of addition

In the foregoing reactions, the attacking species has been an electrophile, with a positively charged intermediate. Bearing these two points in mind, suggest what would be the effect of an electron donating substituent on the double bond.

The electron donating groups will have the effect of increasing the electron density in the π orbital, and so increase the attraction towards the incoming electrophile, and also result in greater stabilisation of the resultant cation.

Suggest an order of reactivity for ethene, propene and phenylethene.

phenylethene > propene > ethene

Suggest why the phenyl group enhances the rate so much.

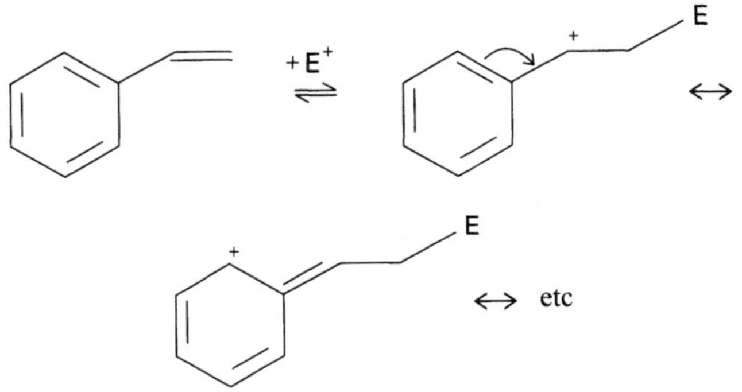

The phenyl group is capable of delocalising the positive charge much more effectively by mesomeric interactions than the alkyl groups may do by inductive interactions alone. It will be readily apparent that the more inductive groups there are increasing the electron density on the double bond, the more likely there is to be an attack by an electrophile and the less likely there is to be an attack by a nucleophile.

Draw out the two intermediate carbonium ions that result from the addition of a proton to ethene and phenylethene. Then suggest which of these two intermediates will have the longer half life.

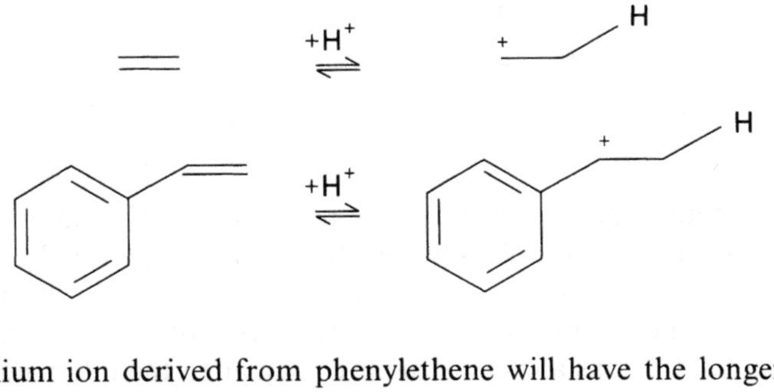

The carbonium ion derived from phenylethene will have the longer half life, because it is the more stable of the two.

Suggest what may be a direct stereochemical consequence of this longer half life.

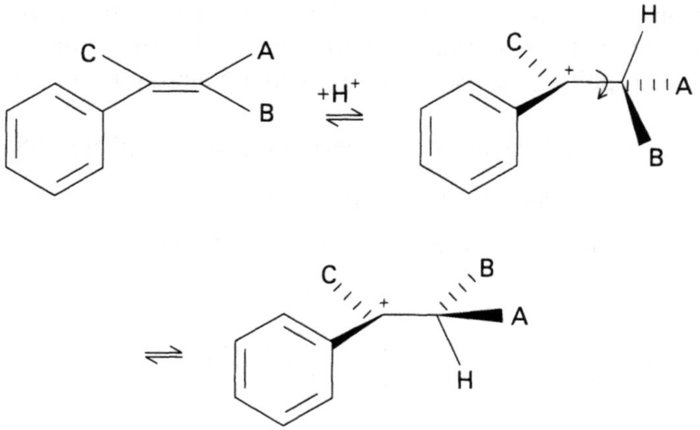

As the carbonium ion has a longer half life, there will be more time for rotation to occur about the carbon/carbon bond. This in turn will lead to a more even mix of the *syn*- and *anti*-addition products.

12.3.3 Orientation of addition

We will use the addition of hydrogen bromide to ethene as an example of the addition of an unsymmetrical molecule to a symmetrical substrate.

Write down the overall equation for this addition reaction.

$$H_2C{=}CH_2 + HBr \rightleftharpoons H_3C{-}CH_2Br$$

Indicate the direction and nature of the dipole in the hydrogen bromide molecule, and so suggest which end of this molecule is attacked by the electron rich π orbital of the ethene. Then draw a reaction mechanism that illustrates the movement of electrons that results in the formation of the charged intermediate.

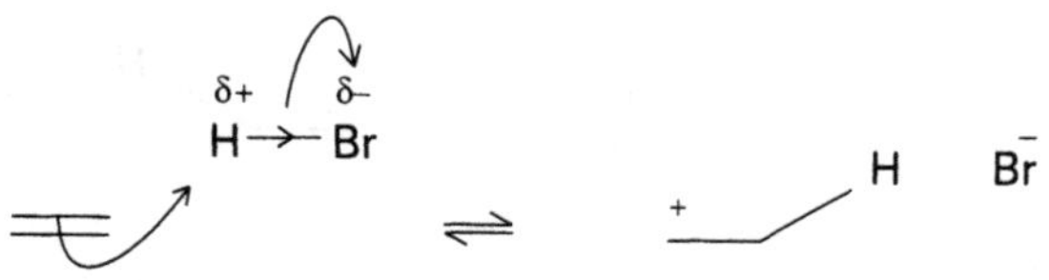

The bromine atom is more electronegative than the hydrogen atom, and so the hydrogen carries a $\delta+$ charge. This means that it is the hydrogen that is attacked by the π electrons to form the carbonium ion. As a result, a bromide ion is formed. Furthermore, the lowest energy form of this carbonium ion intermediate, $C_2H_5{}^+$, has a cyclic structure, similar to the intermediate that is formed in the bromination of ethene. Yet (obviously), on this occasion it is a hydrogen atom that forms the bridge. However, commonly the carbonium ion intermediate is not shown as a cyclic structure.

Suggest how this reaction pathway might continue, and draw the structure of the product.

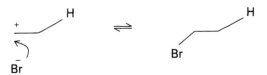

The addition of the bromide ion results in the final product, i.e. bromoethane.

The hydration of an alkene may be catalysed using an acid that has a weakly nucleophilic conjugated base, such as dilute aqueous sulphuric acid. Write down the overall equation for the hydration of ethene, and name the product.

$$H_2C=CH_2 + H_2O \rightleftharpoons H_3C-CH_2OH$$

Ethanol is the product. Suggest a mechanism for this reaction.

In this case the acid donates a proton to the ethene to form the carbonium ion, which is then attacked by a water molecule to give a protonated form of the ethanol molecule. This adduct loses its extra proton to the solvent to result in the desired product being formed. The acid acts merely as a catalyst.

Now let us look at the addition of an unsymmetrical molecule to an unsymmetrical substrate, e.g. the addition of hydrogen bromide to propene. Suggest the two possible isomeric products of this addition reaction.

Hypothetically, the bromine might join to the terminal carbon or to the central carbon. In reality, only one product is formed in this reaction, and to understand why, we need to look more closely at the possible intermediates. Suggest the structure of the two possible carbonium ions that might result from the addition of a proton to this alkene.

One is a primary carbonium ion, while the other is a secondary carbonium ion. Which of these two carbonium ions is the more stable, and why?

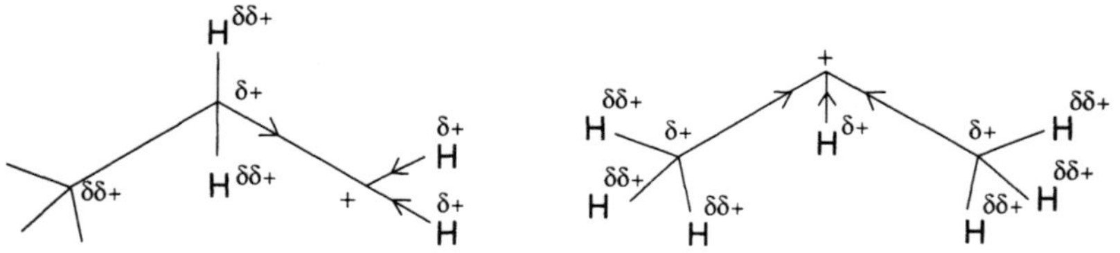

The secondary carbonium ion is the more stable, because it has two electron donating groups inductively feeding electrons towards it. Furthermore, these groups are able to interact favourably with the positive charge by hyperconjugation, and this increases the stabilisation.

Given this difference in stability of the charged intermediate, suggest what will be the final product when hydrogen bromide is reacted with propene.

$$H_3C-CH=CH_2 + HBr \rightleftharpoons H_3C-CHBr-CH_3$$

The 2-bromopropane molecule is the result, and not the 1-bromopropane isomer. We may invoke the Hammond postulate to explain this result. Due to the greater stability of the secondary carbonium ion, there is a lower activation energy barrier that needs to be overcome in order for it to be formed, in contrast to the higher activation barrier that must surmounted in order to form the less stable primary carbonium ion. Hence, the secondary carbonium ion is formed much more quickly, and once formed it reacts to give the product.

This explanation for the formation of the more stable carbonium ion intermediate underlies one of the basic guidelines of electrophilic addition reactions, namely Markovnikov's rule. This rule states that the positive part of the reagent goes to the side of the double or triple bond that has more hydrogens. The positive part of a molecule is the part that is electron deficient. In practice, this is often hydrogen, and so the orientation of the addition process can be remembered by using the phrase 'the rich get richer, while the poor remain poor' in relation to the number of hydrogen atoms that a particular atom bears. This is an example of regioselectivity, i.e. the orientation of the addition product is determined by the structure of the substrate.

A further illustration of this may be seen in the addition of an interhalogen compound, e.g. bromochloride, to propene. Suggest what is the product, and draw out the mechanism for this reaction.

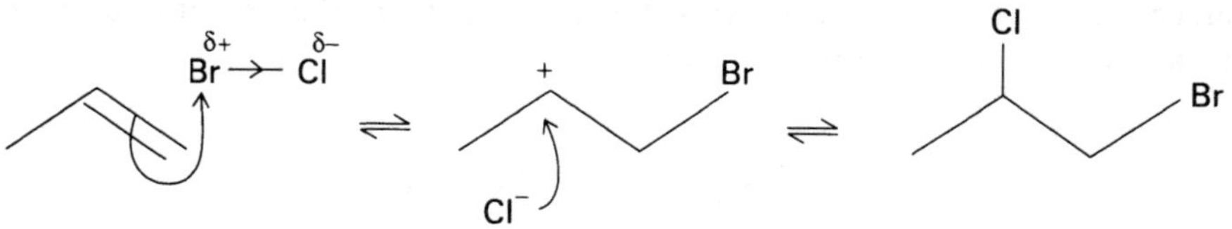

The bromine is more positive than the chlorine, and so becomes attached to the terminal carbon, forming the more stable secondary carbonium ion. This is then attacked by the chloride ion to form the 1-bromo-2-chloropropane.

Hydrogen halides may also add to alkynes. Write down the mechanism for the addition of one molecule of hydrogen bromide to one molecule of ethyne.

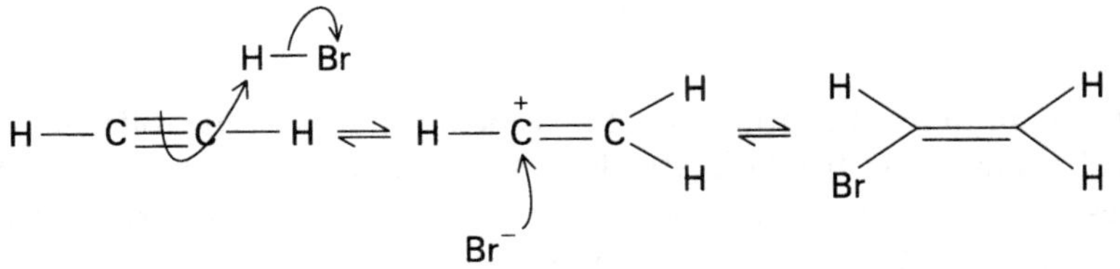

This molecule is still unsaturated, and so it may add another molecule of hydrogen bromide. Write down the mechanism for this step, clearly indicating the orientation of the addition of this second molecule of hydrogen bromide, and also suggest whether this second step is faster or slower than the corresponding addition of hydrogen bromide to ethene.

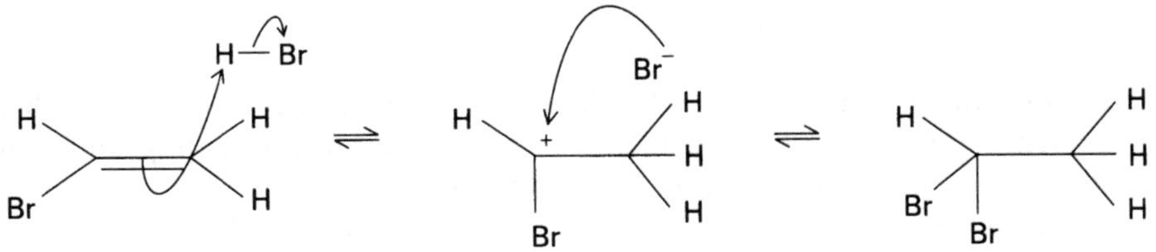

The second addition follows Markovnikov's rule as well. However, this second addition is about 30 times slower than the corresponding addition of hydrogen bromide to ethene, because of the electron withdrawing properties of the halogeno substituent. Notice that the *geminal* product is formed, rather than the *vicinal* product.

12.3.4 Some other electrophilic additions

Another interesting and useful reaction is hydroxylation. This is the addition of a molecule of hydrogen peroxide across a double bond to give a 1,2-diol. Write down the overall equation for the hydroxylation of propene.

$$H_3C{-}CH{=}CH_2 + H_2O_2 \rightleftharpoons H_3C{-}CHOH{-}CH_2OH$$

This reaction may be performed in a number of different ways, and so a degree of control over the resultant stereochemistry may be achieved. One method involves the initial formation of the epoxide, followed by hydration to give the 1,2-diol.

Write a balanced equation for the reaction of propene with a general peroxyacid, RCO_2OH, given that such a reaction results in the formation of the epoxide.

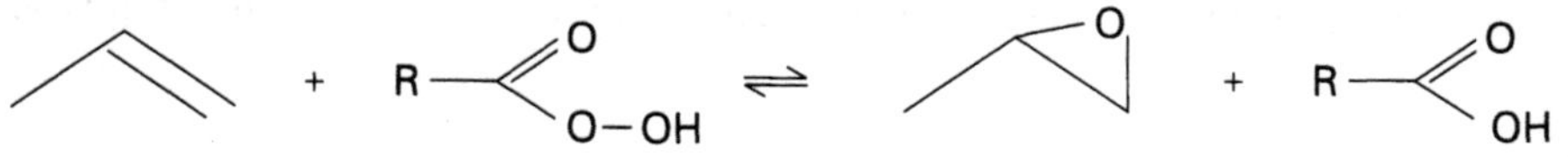

The mechanism for the formation of this epoxide intermediate follows the same general route of electrophilic attack as the addition of a bromine molecule. Suggest a mechanism for the formation of the epoxide, and clearly indicate the movement of electrons in the reaction scheme.

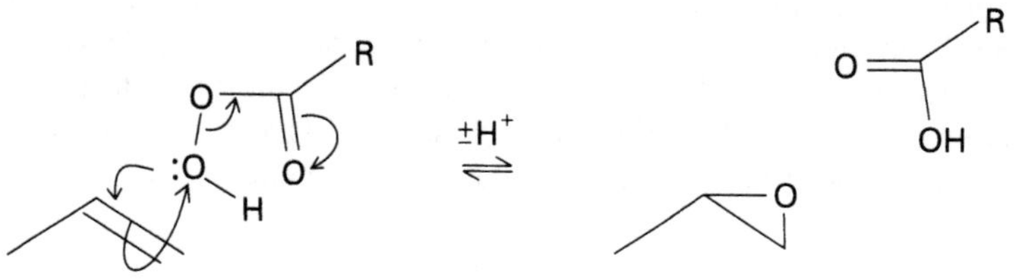

Note that the leaving group in this case is the carboxylate anion, which then reacts further with the protonated epoxide to form the corresponding conjugate acid, i.e. the carboxylic acid, and the neutral epoxide.

This epoxide may undergo nucleophilic attack, under either acid or base catalysed conditions to yield the 1,2-diol.

Write down the reaction mechanism for the formation of the 1,2-diol under these different conditions.

acidic conditions

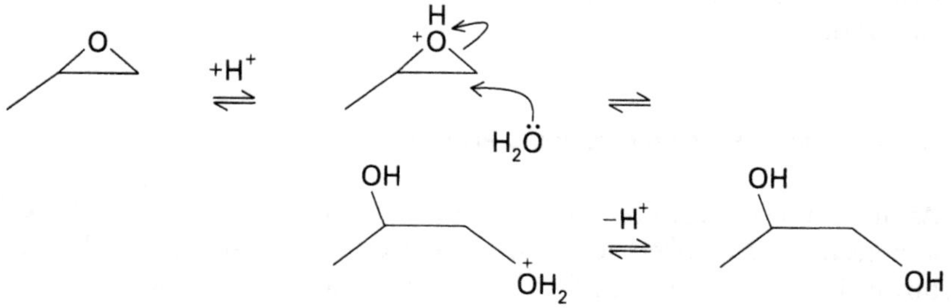

basic conditions

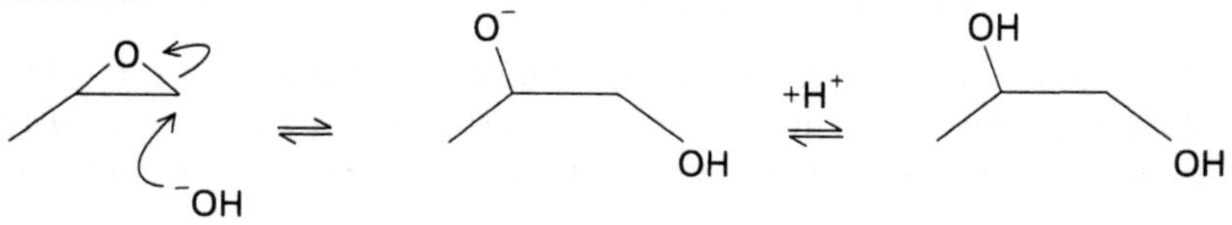

In each case, note that the attack of the nucleophile is from the opposite side to the bridging oxygen atom, and so the resultant addition of the two hydroxyl units is *anti*. Under basic conditions, the nucleophile is a hydroxide anion, and the resultant anion gains a proton from the solvent. Under acidic conditions, the epoxide is activated by protonation of the epoxide oxygen. This in turn makes the epoxide carbons more electrophilic and so more susceptible to attack by a nucleophile, namely in this case a water molecule. After the nucleophilic attack, the resultant adduct loses a proton to the solvent, and so yields the desired product. Overall, there has been an *anti*-addition of the symmetrical molecule H_2O_2.

Earlier we looked at the formation of 2-bromopropane when hydrogen bromide was added to propene, and also the formation of ethanol from ethene reacting with dilute sulphuric acid. Suggest what would be the product of the reaction between propene and dilute sulphuric acid.

$$H_3C-CH=CH_2 + H_2O \rightleftharpoons H_3C-CHOH-CH_3$$

Synthetically, it would be useful to be able to control the regiochemistry of this addition so as to produce propan-1-ol, rather than propan-2-ol. In the next section, we see that there is a way to control the regiochemistry of the addition of hydrogen bromide to give 1-bromopropane. Using this intermediate, it would be possible to form the corresponding alcohol by substituting the bromine functional group for the hydroxyl functional group.

However, there is another way in which the overall conversion from propene to propan-1-ol may be accomplished. This proceeds by the indirect route of hydroboration, using diborane, followed by the oxidation of the resultant trialkylboron with alkaline hydrogen peroxide. For the present purposes, we are only concerned with the reaction between diborane and the alkene. Diborane is generated *in situ*, but reacts as if it were the monomer BH_3, which is a Lewis acid. Write down the mechanism for the addition of BH_3 to propene.

The electrons from the π orbital attack the empty orbital on the boron tri-hydride to form a secondary carbonium ion, i.e. the normal Markovnikov addition; but note that this time the centre to which the electrons are being attracted is a boron atom and not a hydrogen atom. This is because the boron has an empty orbital into which the electrons from the double bond can readily move.

To complete this reaction, there is first a transfer of a hydride ion from the boron to the positive carbon; second, this monoalkylated boron compound undergoes two further alkylations to give the trialkylated product; and third, on the addition of alkaline hydrogen peroxide, the carbon/boron bond is broken and a hydroxyl group is substituted. Write down a structure for each of these intermediates.

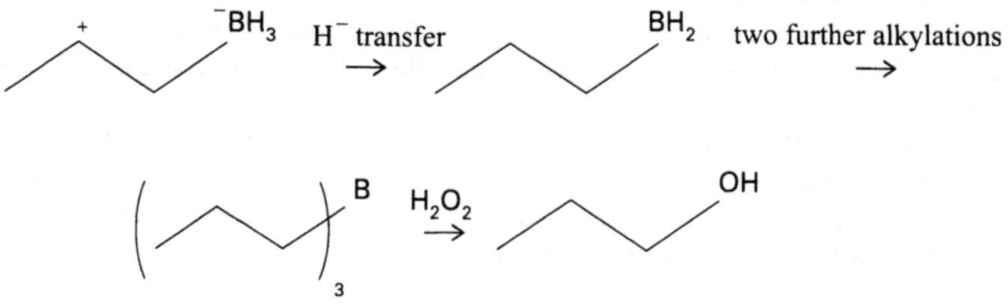

Hence, the overall regiochemistry is the addition of water in an *anti*-Markovnikov manner, but it was brought about by a Markovnikov addition of a boron derivative to the double bond, followed by the replacement of that boron derivative with a hydroxyl group.

This polar mechanism is in contrast to the non-polar radical mechanism for the addition of hydrogen bromide that we will study in the next section. We will see that the radical mechanism gives rise to addition with the *anti*-Markovnikov regiochemistry. There is a further difference between the polar and the non-polar reactions, in that the radical addition is usually *syn* in nature, assuming that the radical has a short half life, while, as we have seen, the polar reaction is usually *anti* in nature.

12.4 Radical addition

So far we have looked at the addition of electrophiles to carbon/carbon double bonds that have followed the Markovnikov regioselectivity. Now we will study reactions that adopt the opposite regioselectivity, i.e. an *anti*-Markovnikov addition. From a synthetic point of view, obviously, it is useful to have the choice.

If we take the addition of hydrogen bromide to propene as an example, this change in regiochemistry may be achieved by the use of a peroxide, e.g. (RO)$_2$, which promotes the formation of radicals. Given that (RO)$_2$ readily undergoes homolytic fission to give two RO· radicals, suggest how one of these might react with HBr to give another reactive radical. Once this new radical is formed, suggest how it might react with propene, paying particular attention to the relative stability of any possible carbon radical intermediates, and so predict the regiochemistry of the final product.

$$(RO)_2 \overset{\Delta}{\rightleftharpoons} 2RO\bullet$$

$$RO\bullet + HBr \rightleftharpoons ROH + Br\bullet$$

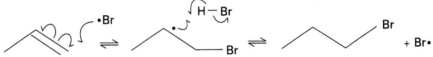

In this case, the formation of, and subsequent attack by, the bromine radical on the alkene leads to the most stable carbon radical species, namely the secondary radical. This in turn picks up a hydrogen atom from another hydrogen bromide molecule and forms another bromine radical that might continue the reaction. This route produces the 1-bromopropane isomer, and not 2-bromopropane, i.e. the *anti*-Markovnikov orientation. This change in regiochemistry, which is caused by the introduction of some radical initiator, is called the peroxide effect, because peroxides are often used as the radical initiator.

12.5 Nucleophilic addition

So far we have looked at reactions where the attacking species on the alkene was either an electrophile, which sought out the electron rich π orbitals, or was a radical. However, it is possible for a nucleophile to attack an alkene.

Suggest what sort of substituents on a carbon/carbon double bond would increase the likelihood of an attack by a nucleophile on such a double bond.

The approaching nucleophile is electron rich and would naturally seek a positive centre, thus anything that reduces the π electron density of the double bond would favour the attack by a nucleophile. Thus, groups such as cyano, nitro, or other $-M$ groups, or even strongly $-I$ groups such as fluorine, would favour nucleophilic attack.

Write down the general equation for the formation of the intermediate after the attack by a nucleophile on an alkene.

$$R_2C{=}CR_2 + X^- \rightleftharpoons [R_2C{-}CXR_2]^-$$

The resultant anion needs to be stabilised, and this may be done, as usual, more effectively by $-M$ groups rather than $-I$ groups.

Suggest the intermediate that is formed by the attack of the cyanide ion, CN^-, on $PhCH{=}C(CN)Ph$, taking particular care to illustrate which end of the double bond the cyanide anion attacks.

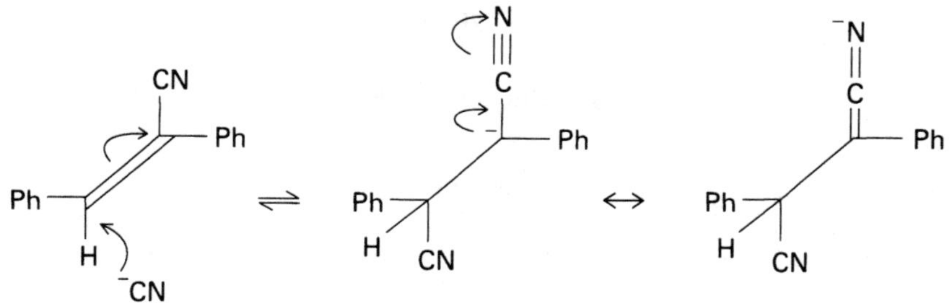

The orientation of the addition of the cyanide ion is determined by the preferential formation of the more stabilised carbanion, which in this case means that the resultant negative charge must be adjacent to the cyano group, which is capable of delocalising it.

One reaction of particular synthetic utility is when a nucleophile is made to attack the cyanoethene molecule (or acrylonitrile). Write down the mechanism for the attack of phenol on this molecule when the reaction is performed under basic conditions.

$$PhOH + B: \rightleftharpoons PhO^- + BH^+$$

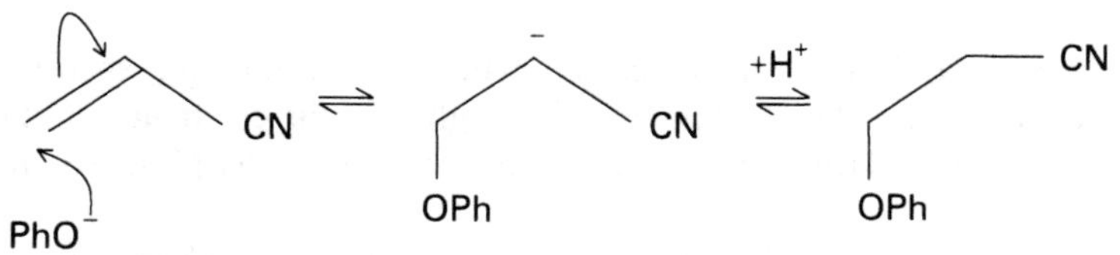

The phenoxide anion is formed by the action of the base on phenol. This resultant anion then attacks the cyanoethene to give an adduct, which (after reprotonation from the solvent) yields a product in which a three-carbon unit has in effect been added to the phenol. This three-carbon unit has a terminal functionality that might then be used in further synthetic steps. This reaction is called cyanoethylation and may be performed with many other nucleophiles such as amines and the related anions, alkoxide anions, and even carbon anions. When it is performed with carbon anions it is called the Michael reaction, which in general is the attack by a carbanion on a substituted alkene. This reaction is particularly useful in forming carbon/carbon bonds. If this reaction is viewed from the perspective of the cyanoethene molecule, instead of the other molecule involved, then it becomes apparent that there has been an addition to a conjugated system, namely to the $CH_2=CH-C\equiv N$ system. We will now study such additions in a little more detail.

12.6 Conjugate addition

We saw in the introduction that where the original substrate contained more than one double bond along a chain, and if there was only partial hydrogenation, a number of different products could result, depending upon exactly where the hydrogen had been added. If the double bonds are not conjugated to each other, then the mechanism of addition across any one of them follows a similar mechanism to the addition across an isolated double bond. However, if two double bonds are conjugated with each other, i.e. separated by only one single carbon/carbon bond, then a new possibility arises.

Write down the two possible isomeric products that might be formed from the addition of one molecule of bromine to one molecule of 1,3-butadiene.

These are the 1,2- and the 1,4-dibromoalkenes, namely $H_2C=CH–CHBr–CH_2Br$ and $BrH_2C–CH=CH–CH_2Br$. The bromine molecule might add across one of the double bonds as originally depicted in the starting molecule: this would result in the 1,2-dibromobut-3-ene. Alternatively, it might add across the terminal carbons, and, with a concomitant reorganisation of the residual double bond, yield the 1,4-dibromobut-2-ene. The former addition is called 1,2-addition; the latter is called 1,4-, or conjugate, addition. It is addition reactions of the latter type with which we are concerned in this section.

Write down the initial intermediate formed by the reaction between a bromine molecule and 1,3-butadiene.

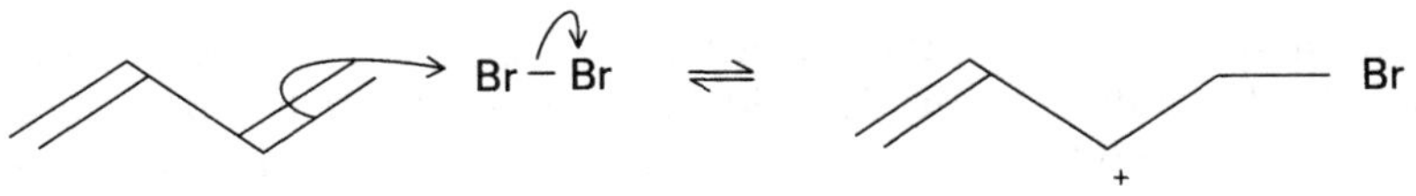

The initial attack always occurs on the carbon at one end of the conjugated system, and so in this case it occurs at a terminal carbon. Consider the structure of the carbonium ion that would result if the bromine cation had attacked one of the non-terminal carbons, and so suggest a reason why the terminally substituted carbonium ion is favoured.

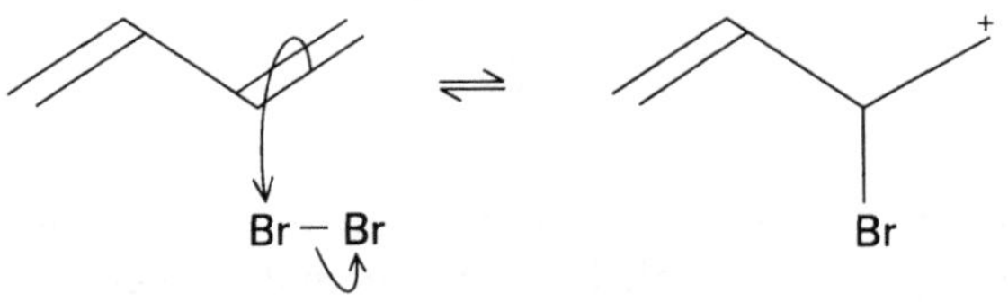

In this latter case, the carbonium ion formed cannot be stabilised by delocalisation with the other double bond. Hence, it may be deduced that the terminally substituted carbonium ion will be formed more quickly.

Write down the delocalised structure for the terminally substituted carbonium ion.

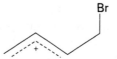

The C2 and C4 carbons both bear a partial positive charge, which together add up to a single unit of positive charge. Suggest what will be the next step in the addition reaction.

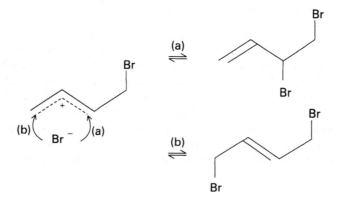

The bromide anion may attack either the C2 or the C4 carbon and so form either the 1,2-adduct or the 1,4-adduct, respectively. It has been discovered experimentally that the 1,2-adduct is the kinetic product, while the 1,4-adduct is the thermodynamic product.

In the simple addition reaction of a bromine molecule to a carbon/carbon double bond, it has been suggested that a cyclic bromonium ion intermediate exists. In the present case, a different cyclic bromonium ion could possibly exist. Suggest a structure for this alternative intermediate.

If this five-membered cyclic bromonium ion did exist, what would be the resultant geometry of the double bond in the final adduct?

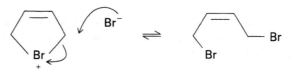

Such a five-membered cyclic intermediate would give rise to a *cis* double bond in the final adduct. However, it is observed experimentally that only the adduct with the *trans* geometry is formed. Thus, even though this five-membered cyclic intermediate might appear attractive on paper, it is not formed in practice. It is a salutary lesson that nature does not always proceed via a structure that looks plausible, and further that one must always base a proposed mechanism on experimental observations.

However, what is the nature of the intermediate in this type of conjugate addition? It is observed that if a solution containing pure 1,4-dibromobut-2-ene or pure 1,2-dibromobut-3-ene is heated, then the same equilibrium mixture of products is obtained in each case. A possible explanation for this is that there is an interconversion which occurs via the same intermediate. Suggest a possible structure for this intermediate and how it might be formed from the two different starting materials.

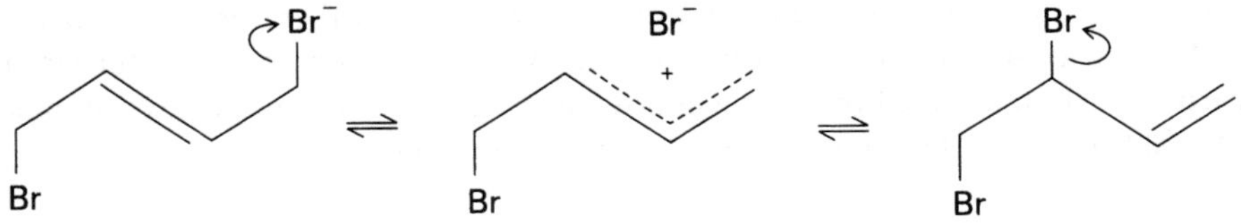

The potential intermediate could be an ion pair, which would be capable of forming either the 1,2-adduct or the 1,4-adduct, as well as providing an intermediate structure through which either of the two final adducts could pass when forming an equilibrium mixture on heating.

Finally, in this section we will consider what happens if an unsymmetrical molecule adds to an unsymmetrical conjugated olefin. In such a scenario, the issue of the orientation of addition arises.

Draw out the possible intermediate structures, and the related canonical structures that would result from the addition of a proton to 1,3-pentadiene.

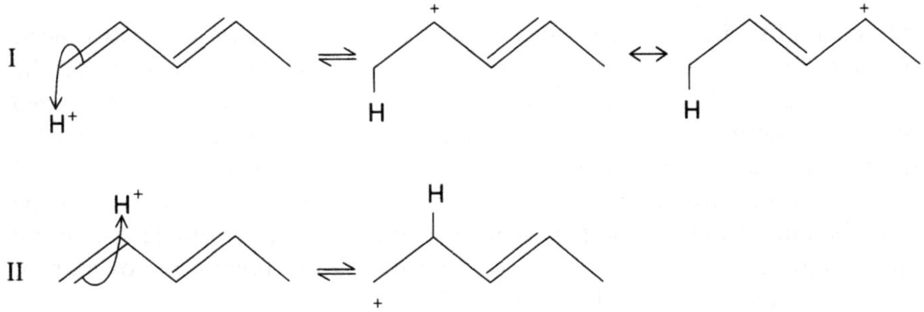

First, it is again apparent that terminal addition, e.g. I and IV, gives rise to intermediates that have the more favourable delocalisation possibilities. Secondly, one may distinguish between the two terminal addition adducts by the fact that I is more stable, because there is greater alkyl substitution on the carbons bearing the positive charge. It is observed that such an intermediate is preferred, and hence the orientation of the addition may therefore be deduced.

Having dealt with the various ways in which addition might occur on a carbon/carbon double bond, we will now turn our attention in the next chapter to addition reactions on a carbon/oxygen double bond.

12.7 Summary of addition reactions to carbon/carbon multiple bonds

Unsaturated molecules undergo addition reactions to give adducts, which if they contain no more multiple bonds are saturated. A polyunsaturated molecule is one that can undergo more than one addition reaction. The addition may be *syn* or *anti*. The attacking species may be an electrophile, nucleophile or a free radical. The orientation of the addition of an unsymmetrical molecule to an unsymmetrical unsaturated molecule may occur in two different ways.

The commonest example of *syn*-addition is the heterogeneously catalysed hydrogenation of an alkene, using either Ni, Pt or Pd, or in the case of the reduction from an alkyne to an alkene, using the Lindlar catalyst. The reduction of an alkene can also be achieved using diimide.

The treatment of an alkene with osmium tetroxide results in the *syn*-addition of hydrogen peroxide.

Molecules like 1,3-butadiene may exist in the *cisoid* or *transoid* conformers, as a result of the small barrier to free rotation that exists due to the fact that the central single bond has some double bond characteristics. Only the *cisoid* conformer undergoes the Diels–Alder reaction. Usually, the rate of the Diels–Alder reaction is increased by electron donating substituents on the diene, and by electron withdrawing groups on the alkene, or dienophile. Thus, the diene acts as the nucleophile, while the dienophile acts as the electrophile. Under all normal conditions, the *endo* adduct is preferred due to favourable overlap of secondary orbitals. This results in *syn*-addition of an R–R unit.

In the ene synthesis, the addition of an R–H molecule occurs.

In pericyclic reactions, there is a concerted movement of electrons. There are three principal types of pericyclic reactions, namely, electrocyclic, cycloaddition and sigmatropic.

The addition of bromine to a double bond occurs in two steps, the first of which forms a bridged bromonium ion intermediate. This intermediate accounts for the fact that *geminal*, rather than *vicinal*, addition occurs, and that the addition is *anti*.

The addition of a symmetrical molecule to an unsymmetrical substrate via an *anti*-addition route, results in a *threo d,l* pair from a *cis* starting material, and a *meso* isomer from a *trans* starting material. If an unsymmetrical molecule were added to a *trans* starting material, then an *erythro d,l* pair results. Thus, this reaction is stereoselective, because preferentially one set of products is in each case produced; but, further, it is stereospecific, because a given isomer leads to one product pair while the other isomer leads to the other product pair.

Electron donating groups increase the rate of addition, as do groups that can delocalise the positive charge. Carbonium ion intermediates that are more stable will have more time in which to undergo rotation, and so will give rise to a more even distribution of *anti*- and *syn*-addition products.

The addition of an unsymmetrical molecule to an unsymmetrical alkene is governed by Markovnikov's rule, which states that the positive portion of the reagent goes to the side of the double or triple bond that has more hydrogens. This is an example of regioselectivity, i.e. the orientation of the addition product is determined by the structure of the substrate.

The addition of hydrogen peroxide, i.e. double hydroxylation, may occur in a number of different ways. For example, an epoxide may be initially formed using a peroxyacid, and this can then be opened up under acid or base catalysed conditions to yield the *anti*-diol.

Hydroboration, followed by oxidation using alkaline hydrogen peroxide, results in overall *anti*-Markovnikov addition of water.

Radical addition is usually *syn* in nature, and *anti*-Markovnikov in regiochemistry. This result is often called the peroxide effect, because peroxides are often used as the radical initiator.

Nucleophilic attack can occur if there are sufficient electron withdrawing groups attached to the double bond so as to reduce its electron density. e.g. $-M$ or very strong $-I$ groups. Cyanoethylation is an example of a nucleophilic addition, in which a cyanoethene molecule (or acrylonitrile) is added to a suitable nucleophile such as a phenoxide, amino, or alkoxide anion. When it is performed with carbon anions it is called the Michael reaction, and is particularly useful in forming carbon/carbon bonds.

In a conjugated system, the possibility of 1,4- rather than 1,2-addition arises. However, this still proceeds via a three-membered intermediate if possible.

13

Addition Reactions to Carbon/Oxygen Double Bonds

13.1 Introduction

In the previous chapter, we looked at addition reactions to carbon/carbon multiple bonds. In this chapter, we will look at addition reactions to another very common multiple bond system, namely the carbon/oxygen double bond. This system differs quite significantly from the carbon/carbon multiple bond system, primarily because there is an inherent dipole across the bond. In the carbon/carbon double bond system such a dipole only occurs if it is caused by the substituents that are attached to it, while in the carbonyl system it arises inherently due to the differences in the electronegativities of the component parts.

13.2 Structure and reactivity

The general formula for a molecule that contains a carbon/oxygen double bond is $R^1R^2C=O$. If either or both of the substituents is substituted by a hydrogen atom, then the molecule becomes an aldehyde; but if both are alkyl groups then the molecule is a ketone. The $C=O$ group itself is called the carbonyl group. In this chapter, we will look at the reactions of a carbonyl group that is joined to hydrogen atoms or alkyl groups, i.e. aldehydes and ketones. Such carbonyl compounds undergo addition reactions. However, if one of the groups attached to the carbonyl carbon is strongly electronegative, such as a chloro substituent or hydroxyl group, then the carbonyl group will perform different types of reactions, which we will study later.

In the previous chapter, we looked at addition reactions to the carbon/carbon double and triple bonds in a variety of species. When the carbon/carbon double bond is not symmetrical, Markovnikov's rule is invoked in order to determine the orientation of the addition of an unsymmetrical reagent. The situation is rather different when we come to consider the carbon/oxygen double bond. This heteroatomic multiple bond has an inherent, and clearly defined, polarity. Write down the structure for this system and indicate the polarity, and so deduce the nature of the more important canonical structure in which charge separation has occurred.

Note that the oxygen is both $-I$ and $-M$. Thus, there is a dipole along the σ bond. Furthermore, the electrons in the π bond may be redistributed so as to give a canonical structure that has a full charge separation. The carbon is thus significantly $\delta+$, with the oxygen correspondingly $\delta-$.

Given this charge distribution, suggest which end of the carbon/oxygen double bond will be attacked by an incoming nucleophile and which by an incoming electrophile.

The oxygen will always be attacked by the electrophile, while the carbon will always be attacked by the nucleophile. A consequence of this is that there is never any doubt about the orientation of attack of an unsymmetrical molecule, and so there is no need for a rule similar to Markovnikov's rule in order to predict the regiochemistry of addition.

In principle, the attack could be initiated either by a nucleophile on the carbonyl carbon or by an electrophile on the carbonyl oxygen. In practice, the only electrophile of any significance is an acid, either a proton or a Lewis acid. Write down the intermediate that results from the addition of, first, a proton and second, a Lewis acid such as $AlCl_3$.

Proton addition

Lewis acid addition

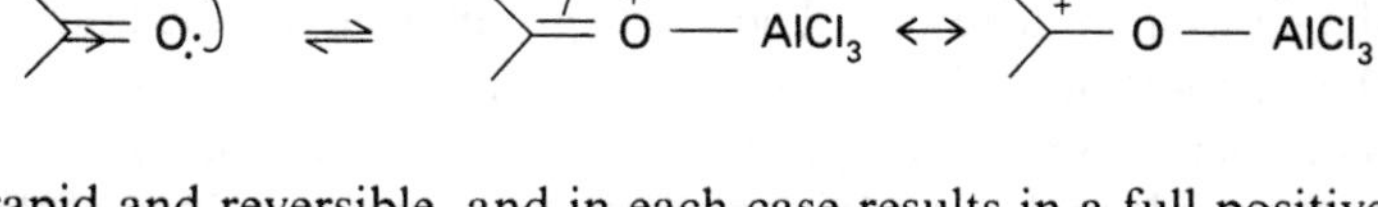

This addition is rapid and reversible, and in each case results in a full positive charge residing on the carbonyl carbon. Accordingly, this atom has been activated towards a subsequent nucleophilic attack.

Not only may an addition reaction to a carbonyl group be catalysed by an acid, but it may also be catalysed by a base. In this case, the base reacts with the initial reagent, HX, to form the anion, X^-, which clearly is a stronger nucleophile than HX, and so more likely to attack the carbonyl carbon. Correspondingly, an acid has the opposite effect and increases the concentration of the weaker nucleophile, HX, by protonation of X^-. However, it would also make the carbonyl carbon more prone to nucleophilic attack as we saw above, because of the increased degree of protonation of the carbonyl oxygen. The consequence of this dependence on the acid/base concentration means that for most carbonyl addition reactions, there is an optimum pH at which they should be performed.

Taking the common case of the carbonyl group, $R_2C=O$, being attacked by a nucleophile, X^-, write down the result of the first step in the addition reaction.

The rate of the initial attack by the nucleophile is governed by Coulombic factors, such as the size of the $\delta+$ charge on the carbonyl carbon. Hence, suggest whether the rate of attack by a nucleophile would be favoured by electron withdrawing groups or electron donating groups attached to the carbonyl carbon, and so suggest whether methanal would react more or less quickly than propanone.

The rate of nucleophilic attack on the carbonyl carbon is increased if electron withdrawing groups are attached to that carbon, because they enhance its electron deficiency and so it exerts a greater attraction towards the incoming nucleophile. Thus, methanal reacts more quickly than propanone, because the methyl groups of the latter exert a $+I$ effect and a hyperconjugative effect, and so reduce the electron deficiency on the carbonyl carbon.

Suggest what would be the effect upon the rate of attack if an aryl group such as phenyl were substituted for an alkyl group, and give your reasons for any change in the observed rate.

The rate would decrease. In the carbonyl compound, it is possible for the positive charge that resides on the carbonyl carbon in one of the canonical structures to be delocalised around the aromatic ring. Thus, the carbonyl carbon is less electrophilic than would otherwise be the case. As such, it is less readily attacked, and so the rate of addition is reduced. The decrease in the rate is also observed for any other substituent that can conjugate with the carbonyl group.

Apart from the electronic consequences of the substituents, there may also be steric ramifications. Taking the same example of a general nucleophile, X^-, attacking a general carbonyl compound, $R_2C=O$, consider the changes that occur in the steric relationships between the groups as the nucleophile approaches and forms the anionic intermediate.

In the original carbonyl compound, there is a standard sp^2 hybridisation with a bond angle of 120° between the groups; while in the anionic intermediate there is sp^3 hybridisation with a bond angle of approximately 109°. Hence, in the intermediate, the groups are closer together, and so there is an increase in the steric crowding of the molecule. As a consequence, if there are bulky substituents on the carbonyl group, then the intermediate will suffer more severely than the original carbonyl compound from steric crowding and so will be less favoured than otherwise would be the case. Accordingly, it would be formed more slowly.

Now consider what will be the line of attack that a nucleophile will follow when it approaches the carbonyl carbon.

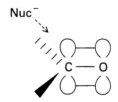

The incoming nucleophile will approach along the line of least steric and electronic resistance. Hence, it will enter from either above or below the plane that contains the carbonyl group, and approach along a line that passes between the substituents attached to the carbonyl carbon.

Now that we have discussed the reaction in general terms, we will examine some individual reactions in more detail.

13.3 Hydration

Many of the addition reactions that a carbonyl group undergoes are reversible, and so the thermodynamic product is formed. However, in addition to examining which is the favoured product, it is interesting to study the kinetics of the reaction. Accordingly, we will look at both aspects.

One of the simplest addition reactions is when one molecule of water adds to a ketone, $R^1R^2C{=}O$, resulting in the formation of a hydrate. Write down the equation for this reaction.

$$R^1R^2C{=}O + H_2O \rightleftharpoons R^1R^2C(OH)_2$$

The rate of this reaction may be increased by either raising or lowering the pH of the solution. If the reaction is performed at high pH, suggest what would be the first step.

At high pH there is a high concentration of hydroxide ions, which are very nucleophilic and so readily attack the positive carbonyl carbon. Suggest what is the second step of this pathway.

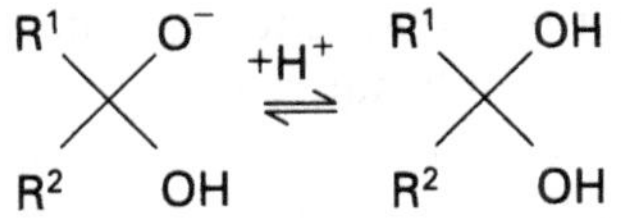

$$\underset{R^2 \quad OH}{\overset{R^1 \quad O^-}{\diagdown\!\!\diagup}} \quad \overset{+H^+}{\rightleftharpoons} \quad \underset{R^2 \quad OH}{\overset{R^1 \quad OH}{\diagdown\!\!\diagup}}$$

The ionic intermediate is then protonated rapidly from the solvent to form the neutral hydrate.

Under acidic conditions, suggest what would be the first step.

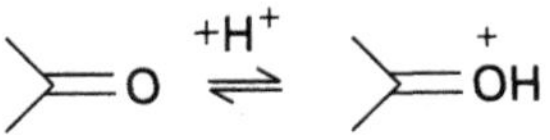

$$\diagup\!\!\!\diagdown\!\!=\!O \quad \overset{+H^+}{\rightleftharpoons} \quad \diagup\!\!\!\diagdown\!\!=\!\overset{+}{O}H$$

In this case, the carbonyl oxygen is attacked first by the electrophilic proton. Draw another canonical structure for this cationic species.

$$\diagup\!\!\!\diagdown\!\!=\!\overset{+}{O}H \quad \leftrightarrow \quad \diagup\!\!\!\overset{+}{\diagdown}\!\!-\!OH$$

It is clear from these structures that by protonating the carbonyl oxygen, the carbonyl carbon has been made more electron deficient, and so more susceptible to attack by a nucleophile. Suggest what will be the second step of this mechanism.

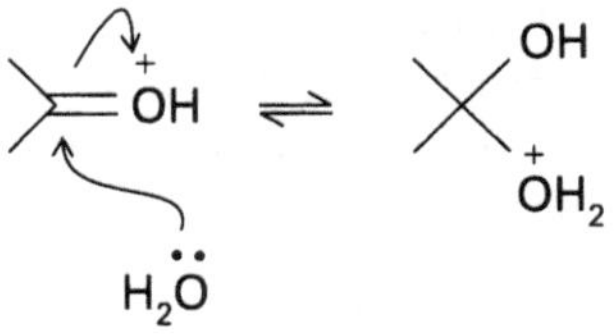

$$\diagup\!\!\!\diagdown\!\!=\!\overset{+}{O}H \quad \rightleftharpoons \quad \underset{\overset{+}{O}H_2}{\overset{OH}{\diagdown\!\!\diagup}}$$

$$\underset{H_2\ddot{O}}{}$$

Now that the carbonyl carbon has been made more susceptible to attack as a result of the protonation of the carbonyl oxygen, one of the solvent molecules, i.e. water, is nucleophilic enough to attack it. Finally, a proton is lost from the water molecule that attacked the carbonyl carbon. The overall result is the addition of a water molecule.

This illustrates that the rate of attack on the carbonyl carbon is dependent on both the nucleophilicity of the incoming reagent and the positive nature of the carbonyl carbon (i.e. its electrophilicity): the more positive it is, the greater the rate of attack. Thus, suggest whether, generally, an aldehyde will react faster or slower than a ketone.

Generally, an aldehyde would react faster, because it has only one electron donating group reducing the positive nature of the carbonyl carbon. Thus, taking this line of reasoning a step further, methanal reacts faster than ethanal. In fact, methanal hydrates quite rapidly even at pH 7, while propanone only hydrates very slowly at that pH, and requires either acid or base catalysis for this reaction to proceed at a significant rate.

Now turning our attention to the effect of various substituents on the carbonyl group, suggest whether or not the hydration of trichloroethanal would occur more or less quickly than the hydration of a carbonyl group to which either simple alkyl or aryl groups were attached. Write down the mechanism for this reaction.

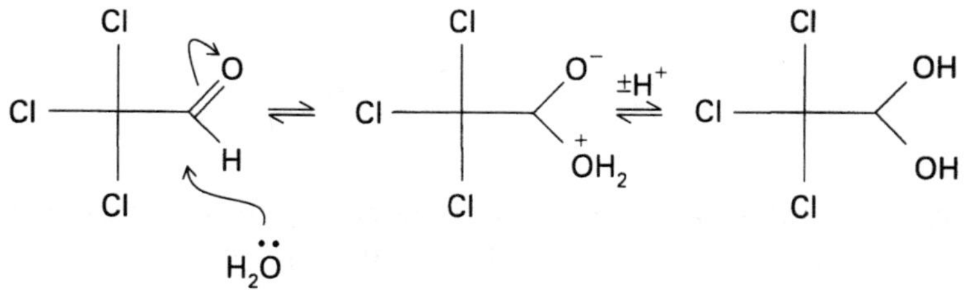

We saw earlier that electron releasing groups, such as alkyl groups, reduce the rate of attack by reducing the positive nature of the carbonyl carbon. Similarly, aryl groups also decrease the rate. However, with respect to aryl groups, it is because they reduce the electrophilicity of the carbonyl carbon by reason of the fact that they delocalise the positive charge that resides on the carbonyl carbon in one of the possible canonical structures. In contrast, electron withdrawing groups increase the rate of attack, because they increase the electrophilicity of the carbonyl carbon by making it more positive in nature. So, in the case of trichloroethanal, the rate of formation of the hydrate is faster than the unsubstituted derivative. Furthermore, the final product is quite stable in comparison to the reagents. In fact, in this particular case, it is possible to isolate the hydrate as a crystalline product.

Another interesting hydrate is that formed from cyclopropanone. Suggest why this hydrate is so stable.

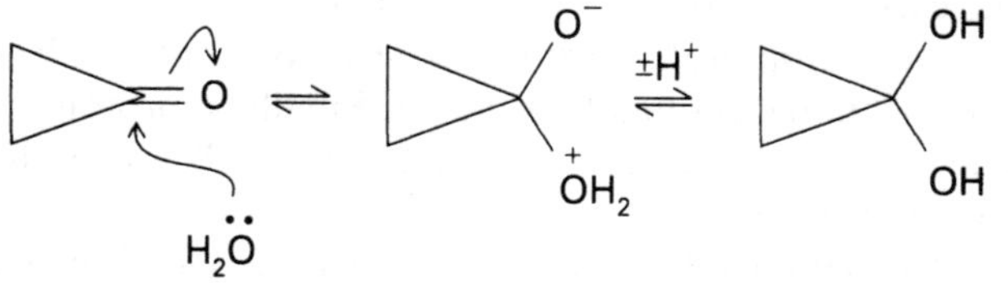

In this case, even though there has been an increase in the number of groups around the former carbonyl carbon, there has been a decrease in the bond strain that was associated with having an sp^2 hybridised carbon in a three-membered ring. This is because there is less strain in having sp^3 hybridised carbons in a three-membered ring. Note that this is in direct contrast to the situation for a normal (open chain) carbonyl compound in which the steric crowding would increase in going from an sp^2 arrangement to an sp^3 arrangement.

One further example illustrates yet another reason why the hydrate may be favoured in certain circumstances. The hydrate of diphenylpropantrione may also be isolated as a crystalline adduct. Suggest a reason for the stability of this hydrate.

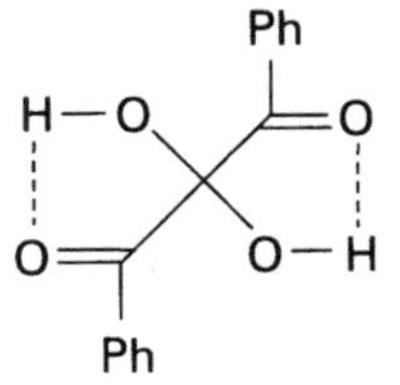

In this case, once the middle carbonyl group has been hydrated, it is possible for both hydroxyl groups to hydrogen bond with the remaining carbonyl groups, which stabilises this hydrate.

So, in summary, for any general carbonyl compound the rate of addition may be affected by acid or base catalysis. This might arise in one of two ways: by deprotonating the original nucleophile so as to form one that is more nucleophilic; or by increasing the electrophilicity of the carbonyl carbon by protonating the carbonyl oxygen. Furthermore, electron donating groups reduce the rate, while electron withdrawing groups increase the rate of addition. This arises by altering the electron density on the carbonyl carbon, and hence its electrophilicity.

In addition to these kinetic effects, there are a number of thermodynamic effects that affect the equilibrium position, for example, the increase in the steric crowding in going from sp^2 to sp^3 hybridisation in an open chain system; or the decrease in strain when going from sp^2 to sp^3 hybridisation in a cyclopropanone system; or the presence of hydrogen bonding in the product.

13.4 Other addition reactions

The addition of an alcohol molecule follows a similar pattern to the addition of water. Draw the mechanism for the addition of ethanol to ethanal under general acid catalysis conditions.

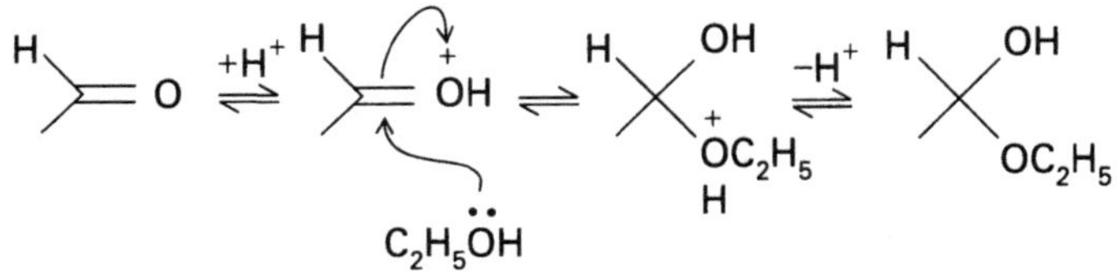

The product is called a hemiacetal, and this is more stable if there are electron withdrawing groups attached to what was the carbonyl carbon. Interestingly, even though the formation of the hemiacetal is general acid catalysed, the subsequent formation of the acetal, in which both hydroxyl groups of the hydrate have been substituted for alkoxy groups, requires a specific acid catalyst. Thus, instead of the rate determining step being the protonation of the carbonyl oxygen, it is now the step following the rapid protonation, i.e. the loss of a water molecule. Write the mechanism for the formation of the acetal from the hemiacetal formed above, indicating the fast and slow steps.

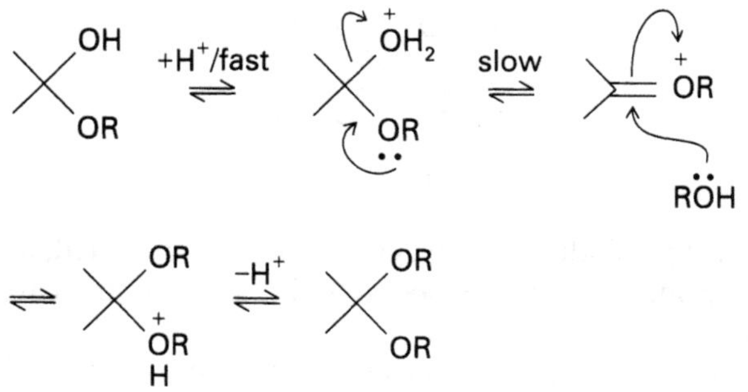

Note that the positively charged oxonium ion intermediate has a canonical structure that is the alkyl equivalent of a protonated carbonyl species. The subsequent attack by the alcohol molecule and the loss of a proton are both fast steps.

When this reaction is conducted on ketones, instead of aldehydes, the equivalent species to the hemiacetal and acetal are now called hemiketal and ketal respectively. However, when simple ketones are reacted with normal monofunctional alcohols, the reaction rarely goes to completion, unlike the situation with aldehydes where the reaction proceeds smoothly. Yet, when a 1,2-diol is reacted with a ketone, then the reaction goes to completion much more readily. Write the complete reaction sequence for the formation of the cyclic ketal starting from propanone and 1,2-ethandiol, under acidic conditions. Also, suggest why the product between one molecule of each reagent is favoured over the alternative product that would result from the reaction between two molecules of the diol and one of the ketone.

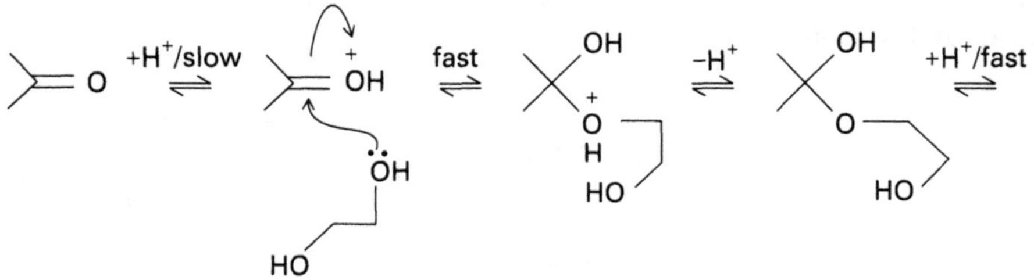

The formation of the hemiketal is general acid catalysed, while the formation of the ketal is specific acid catalysed as indicated above. The formation of the cyclic ketal is favoured over the formation of ketal derived from one ketone and two molecules of the alcohol for three reasons: first, the formation of a five-membered ring is sterically easy; secondly, the approach of a third molecule would be hindered by the presence of the crowded intermediate; and thirdly, ring closure is entropically favoured compared to the addition of a further alcohol molecule. The formation of a cyclic ketal in this way is a very useful reaction, because, once formed, the ketal is resistant to attack by a base and so it acts as a protecting group for the carbonyl group. Yet this protecting group can be removed easily at a later stage by treatment with dilute aqueous acid so as to regenerate the carbonyl group once again.

Another very common addition reaction to the carbonyl group is that of the bisulphite anion. Historically, these derivatives are important, because as they readily yield a crystalline product, they are easily purified. Suggest the mechanism for this addition and hence the nature of the product.

The attacking species is the sulphur atom. This is because, as mentioned before, the attack on a carbonyl carbon is determined by Coulombic factors. Thus, as the sulphur has the highest electron density due to the three oxygen atoms mesomerically feeding electrons onto it, it is the sulphur centre that attacks the carbonyl carbon. Furthermore, it appears that the attacking species is the sulphite dianion, rather than the bisulphite anion, since even though the latter is present in much larger concentrations, the former is far more nucleophilic. The final product is a derivative of the sulphonic acid and not the alkyl ester of sulphite, which would have resulted from an attack by one of the oxygen atoms.

We will now turn our attention to the addition of carbon nucleophiles to the carbonyl group. This type of reaction represents a very important group of reactions that have wide synthetic utility, because they allow for the extension of the carbon backbone of the starting material.

13.5 Addition of carbon nucleophiles

13.5.1 Cyanide, alkynyl and alkyl anions

The multitude of reactions that occur in organic chemistry may be divided into many different categories along many different lines; however, of all the divisions along synthetic lines, there are two principal groups. The first concerns the conversion of one functional group into another, the so-called functional group interchange or interconversion, FGI. The second division concerns the formation of new carbon/carbon bonds, and is of particular interest, because it is by this means that small simple molecules may be built into larger and more complex molecules of greater value.

Much of synthetic organic chemistry is concerned with the methods of executing and controlling carbon/carbon bond formation. One of the simplest of such reactions involves the addition of a cyanide ion to a carbonyl group. Write the first step in the addition reaction between a cyanide ion and a general ketone, $R_2C=O$.

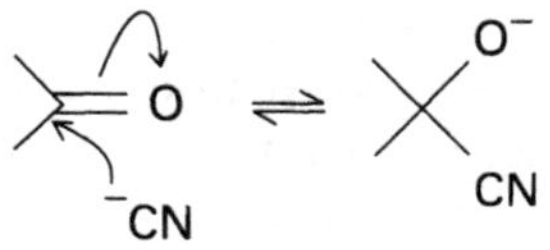

The addition of the cyanide anion is slow, but it is followed by the fast addition of a proton, from either HCN or a protic solvent, to form the cyanohydrin.

It is observed that this reaction requires base catalysis; suggest a reason why this may be so.

The base is needed to form the cyanide anion, because its conjugate acid, HCN, is not a strong enough nucleophile to attack the carbonyl carbon.

This simple reaction has introduced an extra carbon atom into the original molecule. Furthermore, this extra carbon is contained within a functional group that may be subsequently developed, for instance, by hydrolysis to reveal a carboxylic acid functionality, or by reduction to an amine. This synthetic route is the basis of the Kiliani method for extending the carbon chain of a sugar.

Another simple, yet important, reaction is the addition of an acetylide anion, C_2H^-, to a carbonyl group. The anion may be formed by the reaction between a strong base, such as $NaNH_2$, and an alkyne. Write down the complete reaction sequence from the starting materials to the final product.

$$RC_2H + \bar{N}H_2 \longrightarrow RC_2^- + NH_3 \uparrow$$

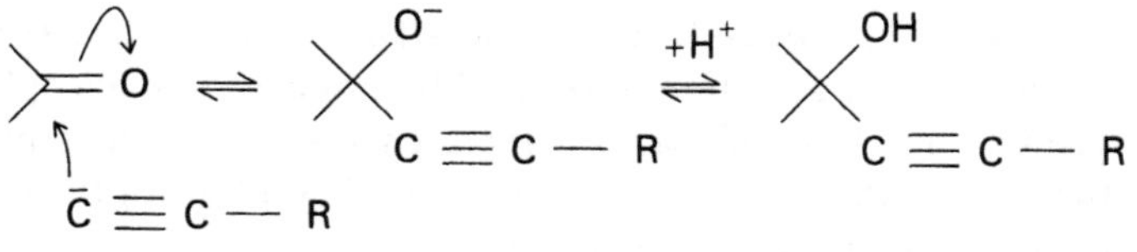

The alkynyl moiety of the adduct may subsequently undergo further reactions; for example, by the partial reduction to a double bond by using hydrogen with a Lindlar catalyst, which results in the overall formation of an allylic alcohol from a carbonyl compound.

In both of the previous examples, the carbanion has not required stabilisation by conjugation. Furthermore, the group that has been introduced has been capable of further synthetic elaboration. The cyanide and alkynyl moieties are said to be masked functionalities, because they are capable of revealing other functional groups by further simple reactions.

The last group of reactions that we will consider in this subsection involves alkyl anions, e.g. $C_2H_5^-$. Unlike the two previous anions that have been considered, it is not very easy to form these anions directly from the conjugate acid, e.g. C_2H_6. This is because there is no stabilisation of the negative charge. However, a synthetic equivalent may be formed easily. The reaction between metallic magnesium and an alkyl halide gives rise to an organo-metallic compound that may be represented by the formula RMgX. These compounds are called Grignard reagents. Their precise structure is open to debate but they react as a source of negatively polarised carbon moieties, e.g. $R^{\delta-}Mg^{\delta+}X$.

It appears that two molecules of the Grignard reagent react with one molecule of the carbonyl moiety when adding an alkyl group. Suggest a possible mechanism for this addition.

If you are having difficulty writing this mechanism, try arranging the three molecules in a cyclic six-membered transition state.

In this cyclic transition state, one Grignard molecule donates its alkyl group to the carbonyl carbon, while the other Grignard molecule activates this carbon by acting as a Lewis acid in complexing to the carbonyl oxygen. The resultant magnesium alkoxide is hydrolysed in the aqueous acidic work-up to yield the alcohol.

The involvement of this cyclic transition state receives support from the observation that Grignard reagents that have β-hydrogens tend to reduce the carbonyl compound to the corresponding alcohol without adding the alkyl group. Suggest a mechanism for this alternative reaction.

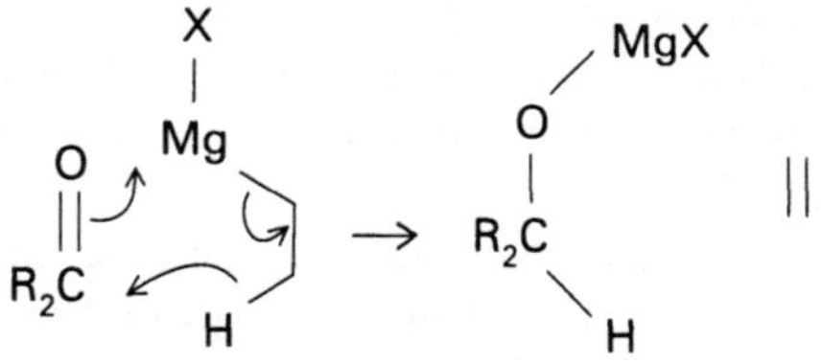

Note that in this case the Grignard reagent is converted to the corresponding alkene.

A different side reaction may occur when a Grignard reagent reacts with a sterically hindered carbonyl compound that possesses an α-hydrogen. Suggest what may be the transition state for this reaction, and so suggest what would be the products.

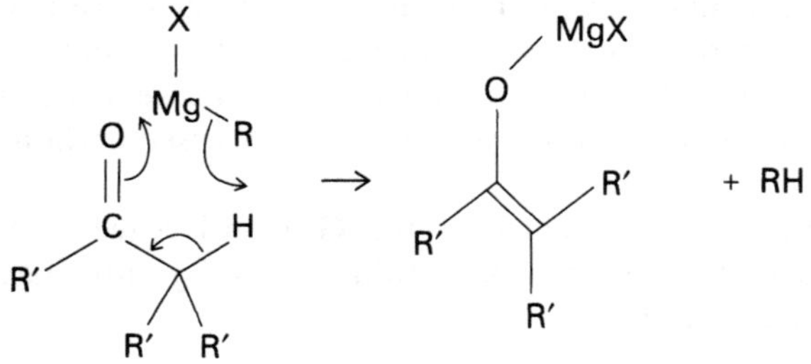

In this case, the Grignard reagent complexes with the carbonyl oxygen as usual, but, because of the steric crowding due to the bulky side group on the carbonyl carbon, there is not sufficient room for a second Grignard reagent to approach. Instead, the α-hydrogen from the carbonyl compound is transferred to the Grignard reagent, resulting in the loss of an alkane and the formation of a magnesium enolate derivative.

The addition of the alkyl anion is usually irreversible and proceeds in high yield. The use of organometallic reagents in synthetic organic chemistry is common, and other examples would include those derived from zinc or lithium. The latter in particular are useful, because they tend to undergo fewer side reactions, preferring instead to perform the simple addition across the carbonyl double bond.

13.5.2 *Carbonyl stabilised carbon anions*

In the previous sub-section, the carbon anion was either inherently stable, e.g. the cyanide or ethynyl ions, or arose from the heterolytic cleavage of an organometallic bond, e.g. the Grignard reagent. In this sub-section, we will look at reactions that involve carbanions in which additional stabilisation is provided by a carbonyl group. This is one of the commonest methods of providing such stabilisation, and as such there are many examples, often identified by the name of the discoverer, or else by the trivial name of the compound that is formed. This tendency to give particular reactions the names of individuals is common in organic chemistry. Once the names are learnt, it becomes a very quick way of communicating information, even though initially it can be somewhat confusing and mystifying.

The first example we will consider is the aldol reaction, in which two molecules of ethanal, on treatment with a base such as NaOH, react together to form aldol, 3-hydroxybutanal. Write down the mechanism for the removal of a proton from ethanal to give the carbanion, and also indicate the stabilisation that is available for this ion.

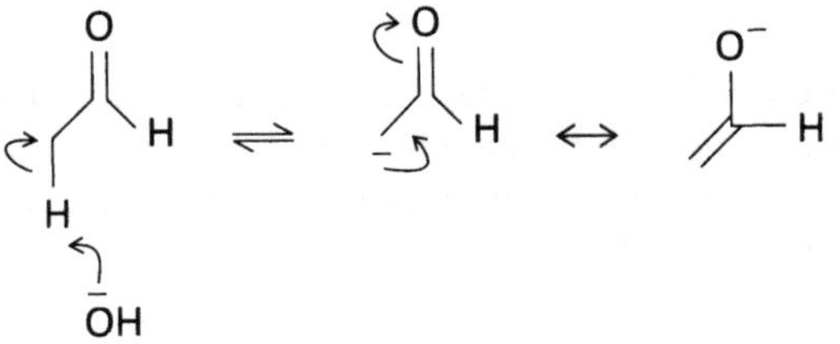

First, note that this is a simple Brönsted–Lowry acid/base reaction, which gives rise to the enolate anion. Secondly, the proton that is removed is the one that is adjacent to the carbonyl group. This is an absolute requirement: no α-hydrogen, no formation of an enolate. Thirdly, the initial carbanion that is formed is stabilised by the $-M$ effect of the carbonyl group, which leads to the negative charge residing on the carbonyl oxygen. It is possible to protonate this canonical structure, and the resultant compound is called an enol, which is a tautomer of the related keto form. Write down both tautomers of ethanal, and label the enol and keto form respectively.

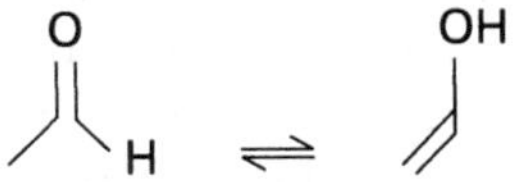

keto   enol

The anion is potentially capable of reacting as a nucleophile either through the oxygen or the carbon, i.e. it is an ambident nucleophile. When this anion is used to attack another carbonyl compound, it invariably reacts via the carbon. Write the mechanism for the attack of the anion of ethanal on another molecule of ethanal, and the subsequent protonation to form the neutral adduct.

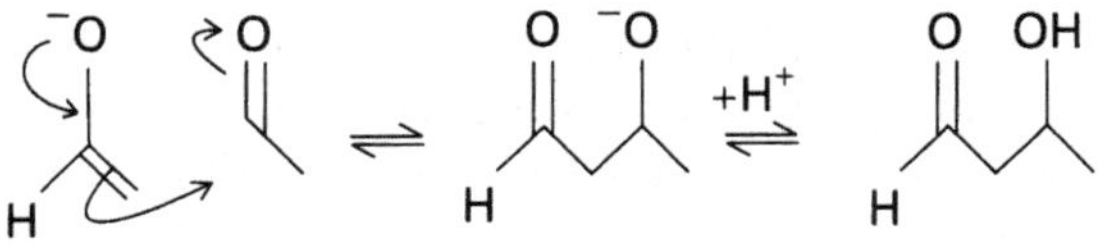

In this case, a 3-hydroxyl carbonyl compound has been formed. This product has another α-hydrogen, which may be subsequently removed by the base. Then one of two things may happen: either this new anion may react with another molecule of ethanal and so form a trimer; or a hydroxide anion may be eliminated to form an α,β-unsaturated carbonyl compound. The detailed mechanism for the elimination step is covered in the next chapter; so, for the present purposes, simply indicate the product that would result from the elimination of the hydroxide anion.

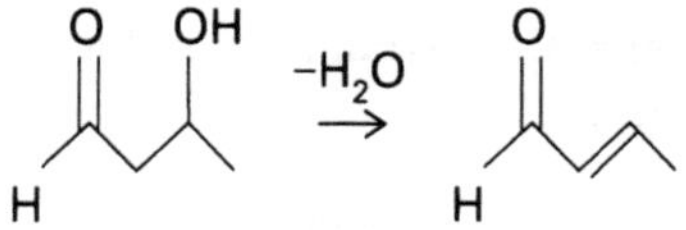

This α,β-unsaturated carbonyl compound is called crotonaldehyde, and so, if an aldol reaction is followed by elimination, it is called the crotonaldehyde reaction. In summary, the formation of crotonaldehyde and aldol from ethanal under basic conditions is as follows:

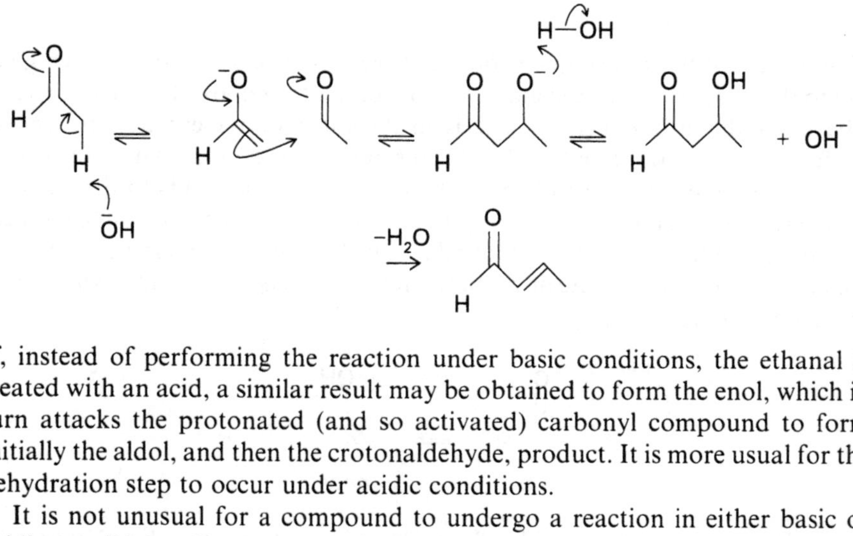

If, instead of performing the reaction under basic conditions, the ethanal is treated with an acid, a similar result may be obtained to form the enol, which in turn attacks the protonated (and so activated) carbonyl compound to form initially the aldol, and then the crotonaldehyde, product. It is more usual for the dehydration step to occur under acidic conditions.

It is not unusual for a compound to undergo a reaction in either basic or acidic conditions. However, particular care must be taken to ensure that the intermediates invoked are appropriate to the conditions used. In this case, the end result is the same, but this is not necessarily so, and it is possible for different products to be formed depending upon the pH of the reaction mixture.

If there are two carbonyl compounds present, then a cross condensation might occur, but usually it is of little synthetic value as there are four possible products. However, if one of the carbonyl compounds lacks an α-hydrogen, then the reaction might prove useful, because this compound cannot form the enol intermediate. An example is the Claisen–Schmidt condensation involving an aromatic aldehyde, e.g. benzaldehyde, with a simple aldehyde or ketone, e.g. ethanal, in the presence of 10% aqueous KOH. Write the mechanism for this reaction, and also suggest whether dehydration would occur and why.

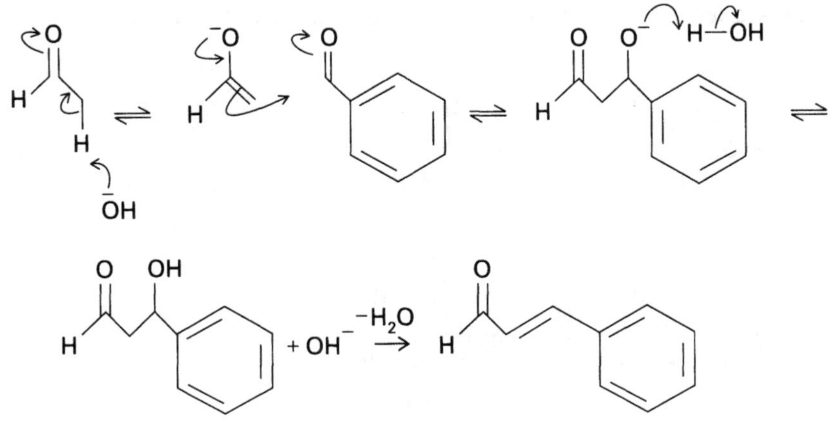

Dehydration always occurs in this situation, because of the advantageous possibility of forming a conjugated system, which connects the aromatic ring to the carbonyl group.

As was mentioned earlier in this sub-section, this area of synthetic chemistry is replete with reactions that are named after their discoverers. A few more examples will now be considered.

In the Claisen condensation, the anion is formed directly from an ester by the action of a base. This anion then reacts with another ester. Write down the reaction sequence, and suggest what will be the product. Also, suggest what would be a suitable base.

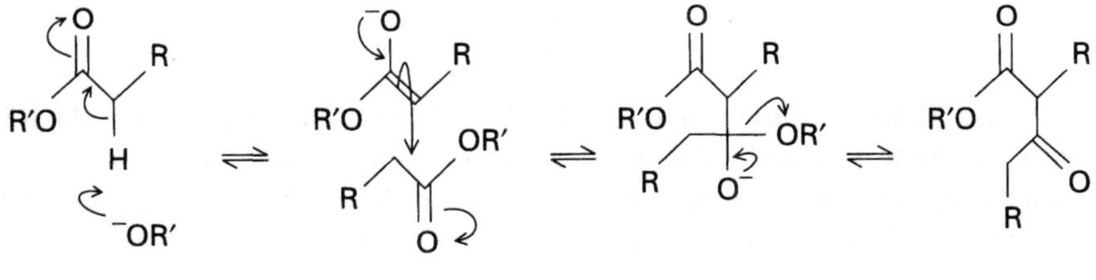

The product this time is a 3-ketoester. A suitable base would be the alkoxide that is derived from the parent alcohol of the ester, because it does not matter if the ester undergoes transesterification; that is, where one alkoxy moiety is exchanged for another. Note that, instead of the anionic adduct just picking up a proton, a alkoxide anion is eliminated on the formation of the carbonyl group. If both ester groups are in the same molecule, then an internal condensation reaction is possible. This is called the Dieckmann cyclisation, and works best when the ring formed contains five, six or seven members.

If the ester has a halogen at the α position, then there are two common variations that are synthetically useful. The first is the Reformatsky reaction in which an α-halogenoester is treated with zinc to form a masked carbanion, which in turn can react with an aldehyde or ketone. Write down the general reaction sequence for this reaction and suggest what will be the product.

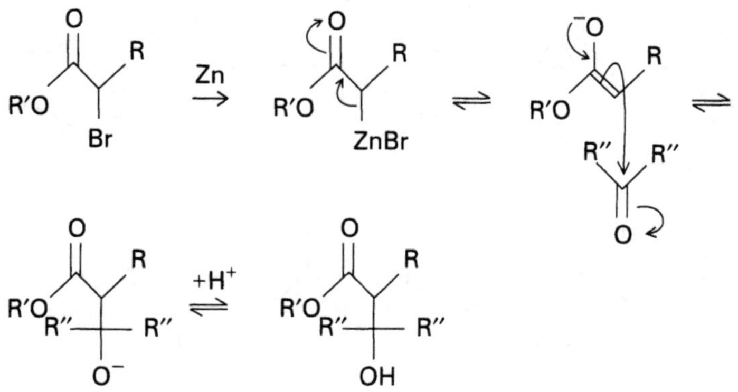

The product is a 3-hydroxyester.

In the second reaction sequence, the anion is formed from the α-halogenoester and not by way of an organozinc derivative, but directly by the action of a base. Write down the reaction sequence for the addition of this anion to a general ketone, $R_2C=O$.

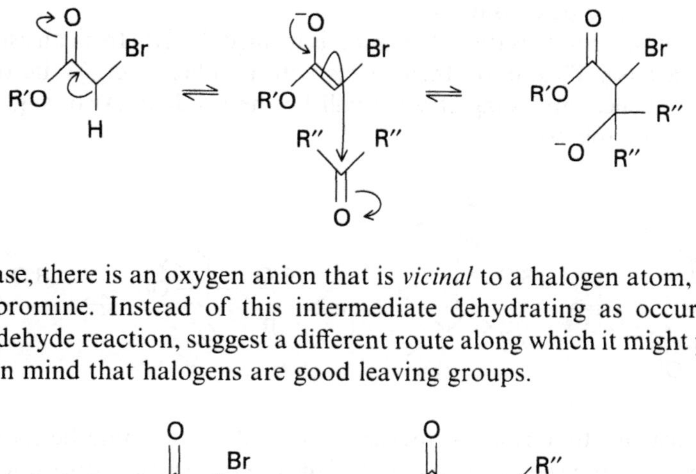

In this case, there is an oxygen anion that is *vicinal* to a halogen atom, which is usually bromine. Instead of this intermediate dehydrating as occurs in the crotonaldehyde reaction, suggest a different route along which it might proceed, bearing in mind that halogens are good leaving groups.

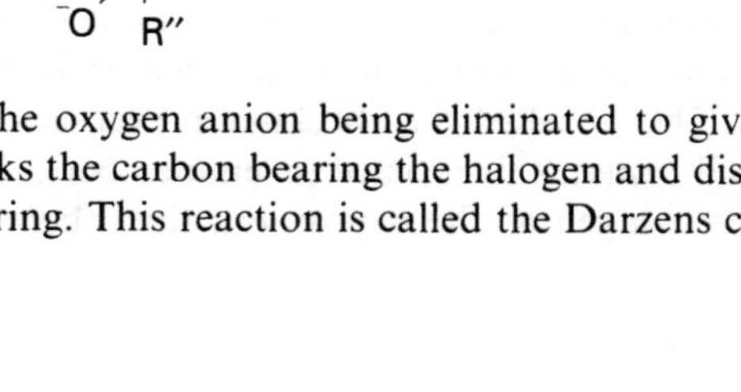

Here, instead of the oxygen anion being eliminated to give rise to an alkene derivative, it attacks the carbon bearing the halogen and displaces that so as to form the epoxide ring. This reaction is called the Darzens condensation.

In the Knoevenagel condensation reaction, the anion is derived from a compound of the general formula CH_2X_2, where X is a $-M$ group, such as an ester. The resultant anion then reacts with a different carbonyl compound. If the $-M$ group is an ester, then the base employed is usually derived from the parent alcohol of the ester. Normally, dehydration occurs so that the α,β-unsaturated diester product is ultimately formed. Write the general reaction sequence for this condensation.

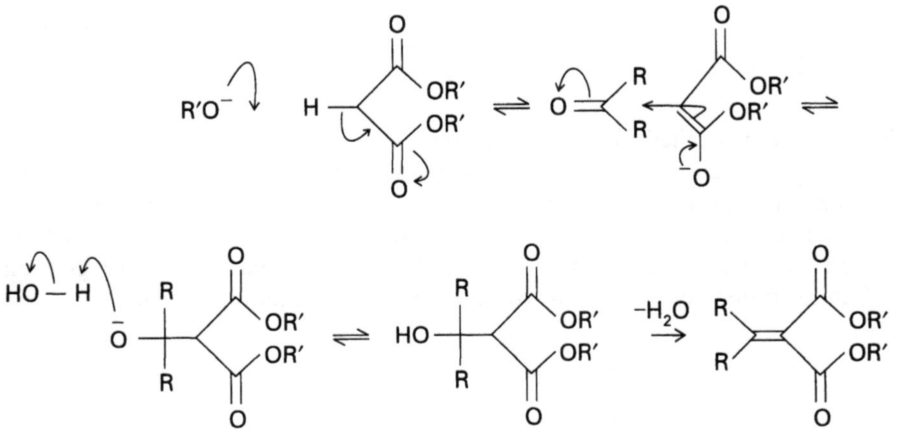

A variation on this reaction is when the anion is formed from the diester derivative of 4-butandioic acid (succinic acid). In this case, the reaction proceeds via an intermediate that is a cyclic ester (i.e. a lactone). Write down the mechanism for the reaction between a general ketone, $R_2C=O$, and the anion of succinic acid, suggesting how the lactone intermediate might be formed.

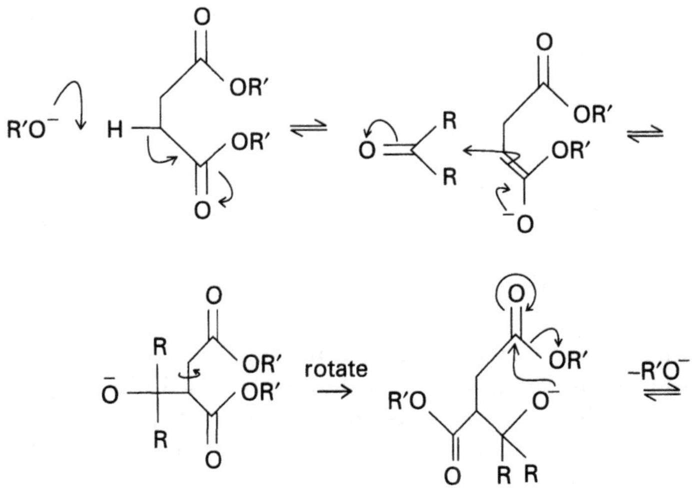

The remaining α-hydrogen on the side chain ester can then be removed, and the ring opened up to reveal an α,β-unsaturated ester with a carboxylate group on the methylene side chain. Write the mechanism for these remaining steps.

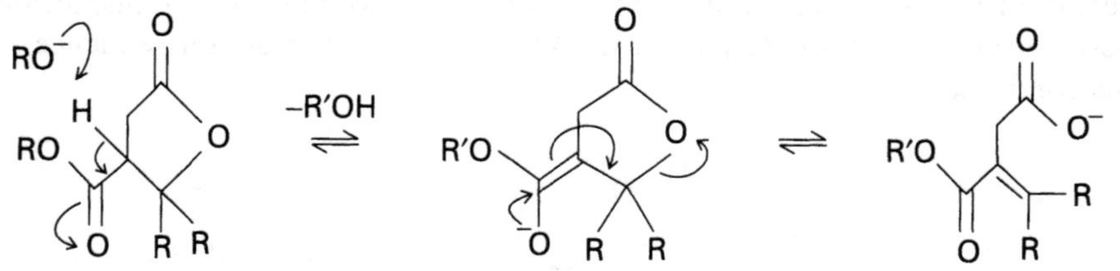

This reaction is called the Stobbe synthesis.

An anion may be formed from a symmetrical acid anhydride by using the carboxylate anion of the corresponding acid as the base. This anion may then be reacted with an aldehyde, such as benzaldehyde, to yield a mixed anhydride as the initial product. Dehydration and hydrolysis may follow to give an α,β-unsaturated acid. Write down the steps of this reaction sequence.

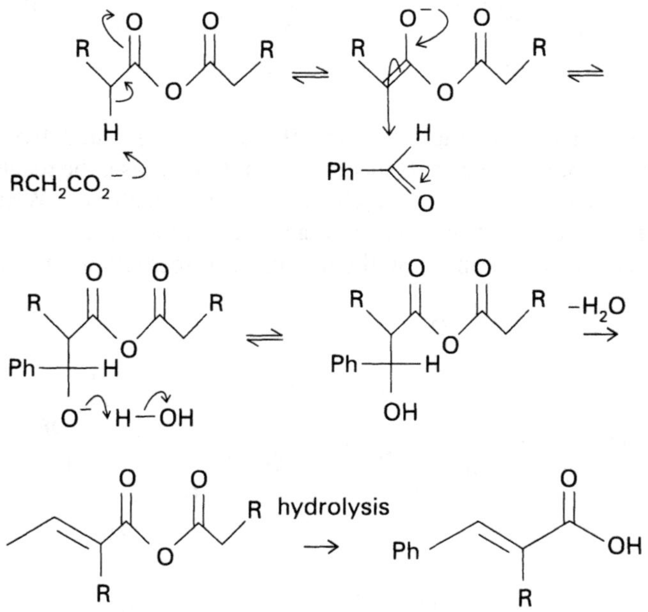

This synthetic route is called the Perkin reaction.

The last reaction we will study, like the first, is named after the product that is formed, rather than its discoverer. Benzaldehyde may undergo attack by a cyanide ion to yield an anion. Write down the mechanism for the formation of this anionic intermediate.

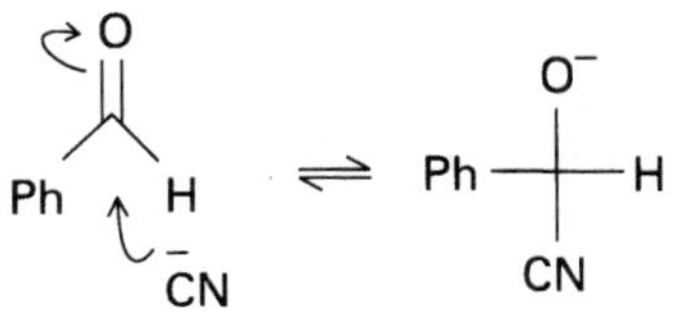

So far, this reaction is proceeding along the same lines as the cyanohydrin formation that we looked at in the previous sub-section. In that case, there was then a simple protonation by a solvent molecule of the anionic oxygen. However, in this case, the hydrogen on what was the carbonyl carbon is removed, and is used to protonate that oxygen. This is possible because that hydrogen is fairly acidic. The resultant carbanion may then attack another benzaldehyde molecule. Write down these reaction steps.

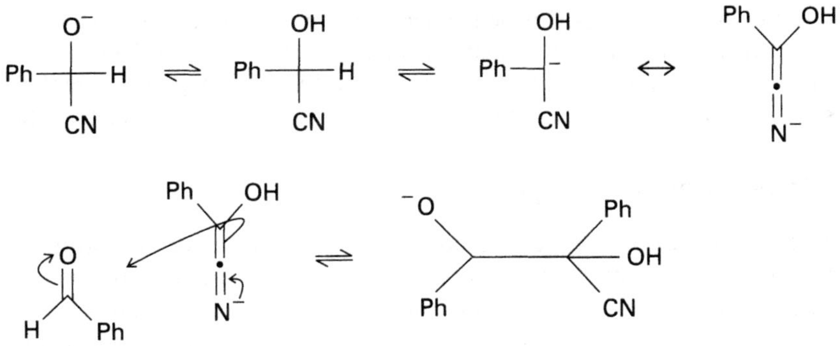

There is now another rearrangement of the acidic hydrogens so as to place the anionic charge on the oxygen *geminal* to the cyanide group. This then reforms the initial carbonyl group, and in so doing eliminates the cyanide anion. Write these final steps and identify the product.

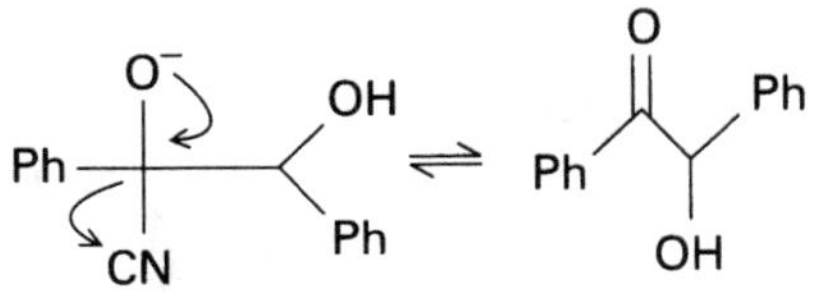

The product is a 2-hydroxyketone, and when starting with benzaldehyde is called benzoin, which gives this reaction its name, namely the benzoin condensation. Notice that the cyanide ion is acting as a catalyst in this reaction, and so is not consumed.

In this sub-section we have seen how a carbonyl group has facilitated the formation of a carbanion. This function may also be performed by other groups, such as a nitro group. Write down the tautomeric forms of nitroethane.

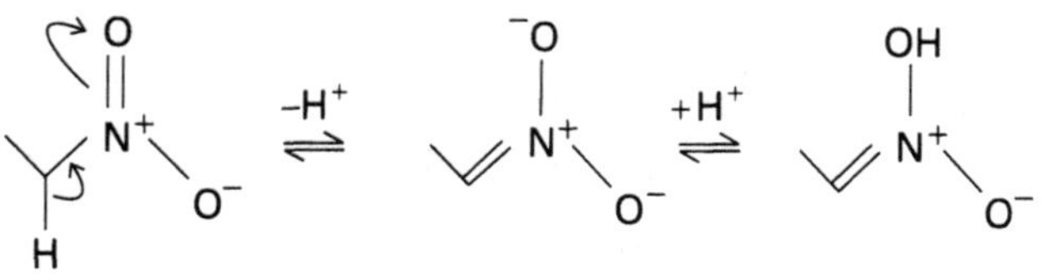

nitro   aci

Anions stabilised in this way may undergo similar reactions to the ones that have been outlined above in which the anions were stabilised by one or more α-carbonyl groups.

Finally, it should be noted that nearly all of the above reactions, with the notable exception of the Grignard additions, are reversible under certain conditions, and so under the appropriate conditions α,β-unsaturated carbonyl compounds can be cleaved. However, even though these reactions are reversible in theory, they are normally performed under basic conditions that lead to the kinetic product being formed so as to maximise their synthetic utility. When performed under acidic conditions, the thermodynamic product is formed, because then all the steps are reversible, which is in contrast to the situation under basic conditions.

13.6 Stereochemical consequences

At the beginning of this chapter we saw that the orientation of the addition of a nucleophile to a carbonyl group is readily determined, because of the clearly defined polarity of the carbonyl group. Thus, unlike additions to carbon/carbon bonds, the addition to a carbon/oxygen bond invariably only occurs with one regiochemistry. Another simplification is that the stereochemistry of the addition is of less importance, because it is impossible to tell if the addition has been *syn* or *anti*. Show that this is true by depicting the addition of HY to a general ketone $R^1R^2C=O$ in both the *syn* and *anti* manners.

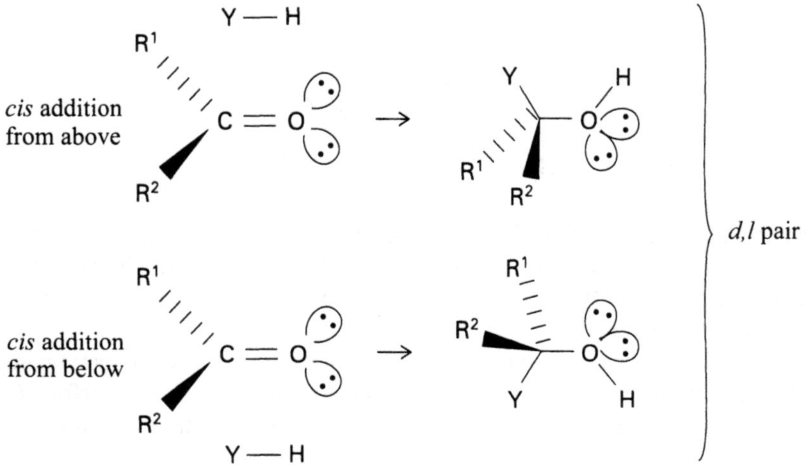

cis addition from above

cis addition from below

d,l pair

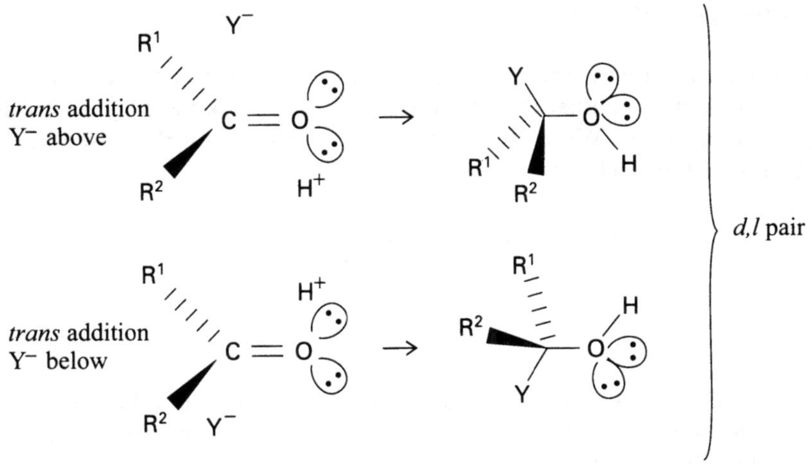

The product is the racemic mixture in each case.

However, if either of the groups that are attached to the carbonyl carbon is itself chiral, then there is a possibility of differentiating between the attack from one side or the other of the carbonyl group. It is observed that the carbonyl oxygen orientates itself so to be diametrically opposed to the largest substituent. Draw the Newman projection along the bond between the carbonyl carbon and the chiral α-carbon. Then assign to the groups that are attached to that carbon the letters S, M and L to designate the small, medium and large groups, respectively. Having done this, suggest from which direction the incoming nucleophile will prefer to approach.

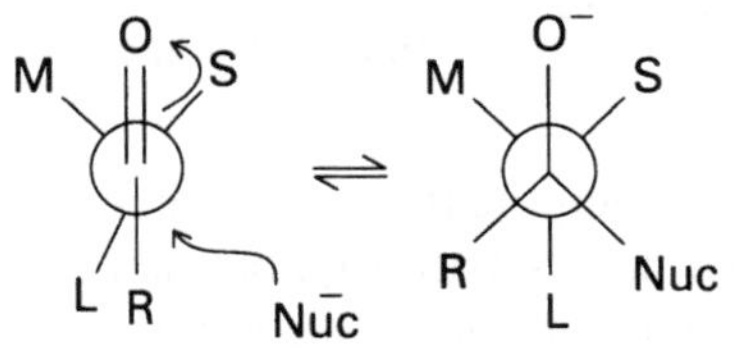

The nucleophile will prefer to approach along the line that offers least steric hindrance, and this is between the large and small groups. This empirical observation is called Cram's rule, and has been widely investigated for Grignard reactions. The greater the degree of steric interference, the greater the degree of selection is found.

Other empirical guidelines have been formulated that attempt to predict and rationalise the addition of HY to such a system. However, for all of these it must be remembered that only the relative direction of addition of the anionic component with respect to the rest of the substrate is considered. The orientation of the addition of the proton is ignored in all cases.

265

13.7 Conjugate addition

In α,β-unsaturated carbonyl compounds, the question arises, as it did with conjugated dienes, whether the addition will be 1,2 or 1,4, i.e. whether it will be simple or conjugated. Write down the reaction sequence for the 1,4-addition of RH to $H_2C=CH-CH=O$.

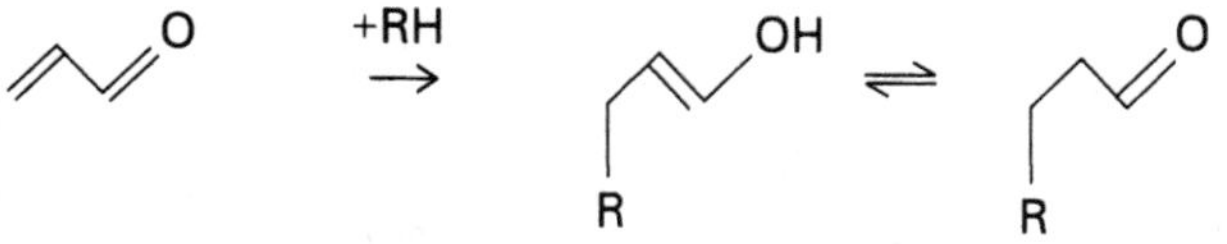

Whether 1,2 or 1,4-addition occurs depends upon a number of factors, such as the precise reagent used, and any steric factors present on the substrate. Thus, for example, lithium dialkylcopper reagents usually undergo 1,4-addition. However, Grignard reagents usually add in a 1,2-manner to aldehydes; but with conjugated ketones, steric factors are important, as illustrated below.

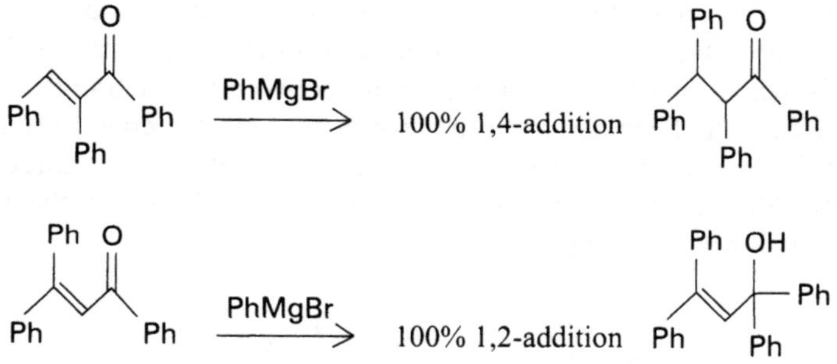

If the reaction is reversible, then the thermodynamic product will predominate, which usually results from the 1,4-pathway, because the carbon/oxygen double bond is stronger than carbon/carbon double bond. Consequently, it is better to break the weaker carbon/carbon double bond than the stronger carbon/oxygen double bond in the overall reaction. If, however, the reaction is irreversible then the kinetic product will dominate, and this may be influenced by the steric demands at the respective sites.

When an α,β-unsaturated carbonyl compound is treated with a carbanion, particularly one that is stabilised by a carbonyl group as well, the resultant 1,4-addition with the formation of a new carbon/carbon bond is called the Michael reaction. Like the Claisen condensation, it is possible for this reaction to occur in an intramolecular manner with a suitably substituted compound. This reaction leads to the formation of a new ring and is called the Robinson annulation reaction. Write down the complete reaction sequence for the reaction between cyclohexanone and but-1-en-3-one.

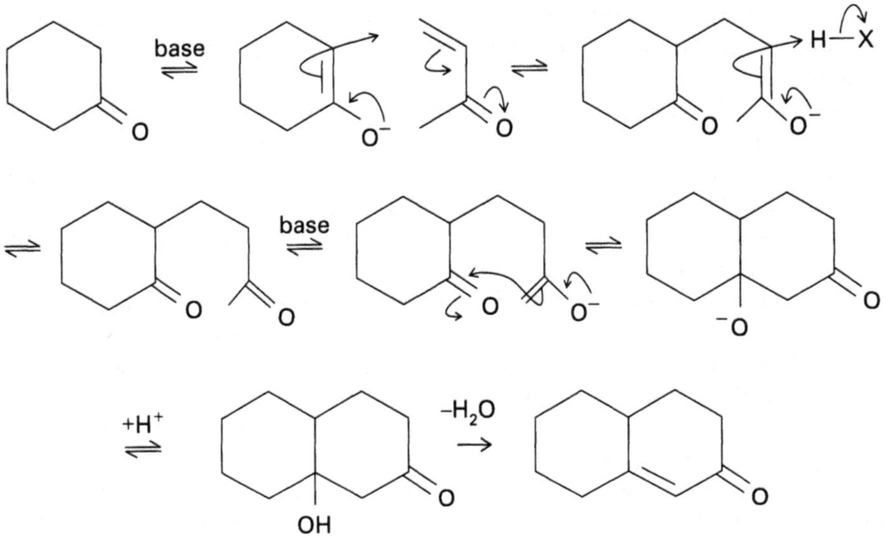

The loss of the water molecule in the final step is similar to the dehydration step that occurred in the crotonaldehyde reaction that we studied earlier. Furthermore, in the context of this sub-section, it should be noted that it is the reverse of a 1,4-addition of a hydroxide anion to an enone system. This reaction concludes not only this chapter on addition reactions to the carbon/oxygen system, but also additions reactions in general. We will now turn our attention to those reactions that perform the opposite synthetic function, i.e. elimination of a part of the molecule.

13.8 Summary of addition reactions to carbon/oxygen double bonds

The general formula for a molecule that contains a carbon/oxygen double bond is $R^1R^2C=O$. If either or both of the substituents is a hydrogen atom, the molecule is an aldehyde; if both substituents are alkyl groups, the molecule is a ketone. The $C=O$ group itself is called a carbonyl group.

The oxygen of a carbonyl group is always attacked by the electrophilic part of the reagent, which is normally either a proton or a Lewis acid; the carbon is always attacked by the nucleophilic part of the reagent.

Most carbonyl addition reactions are sensitive to acid and base catalysis, and so they usually have an optimum pH at which substitution can be performed. Furthermore, electron withdrawing groups attached to the carbonyl carbon enhance the rate of attack by a nucleophile. Large groups attached to the carbonyl group hinder the formation of the anionic tetrahedral intermediate, and so reduce the rate of attack. Any group that can conjugate with the carbonyl group, such as a phenyl group, reduces the rate of attack, because they reduce the electrophilicity of the carbonyl carbon.

Nucleophiles will approach from either above or below the plane that contains the carbonyl group, and along a line that is between the groups attached to the carbonyl carbon.

Many of the addition reactions that a carbonyl group undergoes are reversible, and so the thermodynamic product is formed.

The rate of addition may be affected by acid or base catalysis. This might arise in one of two ways: by deprotonating the original nucleophile so as to form one that is more nucleophilic; or by increasing the electrophilicity of the carbonyl carbon by protonating the carbonyl oxygen.

The addition of water results in a hydrate. The rate of formation of the resultant hydrate can be affected by various factors, such as the presence of electron withdrawing groups, e.g. trichloroethanal. The stability of the resultant hydrate can also be affected by a number of factors such as the change in the steric crowding in going from sp^2 to sp^3 hybridisation or vice versa; or the presence of hydrogen bonding in the product, e.g. diphenylpropantrione. The addition of an alcohol to an aldehyde gives a hemiacetal initially, and then the acetal if a second alcohol unit is added. The latter step is easy if performed with a 1,2-diol, which is capable of forming a cyclic acetal. The equivalent species when a ketone is used instead of an aldehyde are called hemiketal and ketal, respectively.

The addition of carbon nucleophiles is of great synthetic utility. For example, the addition of a cyanide ion is the basis of the Kiliani method for extending the carbon chain of a sugar. The addition of an acetylide anion, followed by partial reduction of the triple bond, results in the formation of an allylic alcohol. Synthetic equivalents for alkyl anions may be formed by using Grignard reagents. They react as a source of negatively polarised carbon moieties, e.g. $R^{\delta-}Mg^{\delta+}X$. Grignard reagents that have β-hydrogens sometimes reduce the carbonyl compound to the corresponding alcohol rather than adding an alkyl group. A further side reaction occurs when a Grignard reagent reacts with a sterically hindered carbonyl compound that possesses an α-hydrogen. In this case, the α-hydrogen from the carbonyl compound is transferred to the Grignard reagent, resulting in the loss of an alkane and the formation of a magnesium enolate derivative.

A common, and synthetically very useful, source of carbanions that are capable of attacking a carbonyl carbon are those that have been stabilised by a $-M$ group, such as a carbonyl functionality. Stabilised carbanions may also be formed from nitro compounds.

In the aldol reaction, two molecules of ethanal, on treatment with a base such as NaOH, react together to form aldol, 3-hydroxybutanal. In the general case, a 3-hydroxycarbonyl compound is formed. There is an absolute requirement for there to be a hydrogen α to the carbonyl carbon in the carbonyl compound from which the prospective carbanion is to be derived. It is possible for the aldol product to undergo dehydration, in which case the reaction is called a crotonaldehyde reaction, and gives rise to an α,β-unsaturated carbonyl compound.

In the Claisen–Schmidt condensation, one of the carbonyl compounds lacks an α-hydrogen, and so cannot form an enolate anion. This reaction involves an aromatic aldehyde, e.g. benzaldehyde, reacting with a simple aldehyde or ketone, e.g. ethanal, in the presence of 10% aqueous KOH. Dehydration always occurs in this situation, because of the possibility of forming an extended conjugated system.

In the Claisen condensation, the anion is formed from an ester directly by the action of a base, e.g. the corresponding alkoxide of the ester, and this stabilised carbanion then reacts with another ester to form a 3-ketoester. If both ester groups are in the same molecule, then an intramolecular condensation reaction is possible, which is called the Dieckmann cyclisation; this works best when the ring formed contains five, six or seven members.

In the Reformatsky reaction, an ester with an α-halogeno substituent is treated with zinc, and so forms a masked carbanion, which in turn can react with a carbonyl compound to form a 3-hydroxyester. In the Darzens condensation, an α-halogenoester, on treatment with a base, reacts with a carbonyl compound to form an α,β-epoxyester.

In the Knoevenagel condensation, the anion is derived from a compound of the general formula CH_2X_2, where X is a −M group, such as an ester. The base employed is usually the alkoxide moiety of the ester. Normally, dehydration occurs so as to result in the formation of the α,β-unsaturated diester. If this reaction is performed using a diester derivative of 1,4-butandioic acid (succinic acid), then it usually proceeds via a lactone to result in an α,β-unsaturated ester with a carboxylate group on the methylene side chain. This is the Stobbe synthesis.

An anion may be formed from a symmetrical acid anhydride by using the carboxylate anion of the corresponding acid as the base. This anion may then be reacted with an aldehyde, such as benzaldehyde, to yield as the initial product a mixed anhydride. Dehydration and hydrolysis often follow to result in an α,β-unsaturated acid. This is the Perkin reaction.

In the benzoin condensation, benzaldehyde is treated with cyanide ion to form an anion, which then attacks another benzaldehyde molecule to form, after the elimination of the original cyanide ion, a 2-hydroxyketone.

Both *syn-* and *anti-*addition to a carbonyl group give rise to the same product, namely the racemic mixture in each case. However, if either of the groups attached to the carbonyl group is itself chiral, then it becomes possible to distinguish between the attack from one side or the other of the carbonyl group by invoking Cram's rule. The carbonyl group positions itself opposite the largest substituent, and the incoming group attacks along a line between the smallest and largest groups attached to the chiral α-carbonyl carbon.

In an α,β-unsaturated carbonyl system, the addition could be either 1,2 or 1,4, i.e. simple or conjugate. With Grignard reagents, the nucleophilic addition can be either 1,2 or 1,4, and often depends upon any steric factors that are present in the substrate. When an α,β-unsaturated carbonyl compound is treated with a carbanion, particularly one that is stabilised by a carbonyl group, the resultant 1,4-addition with the formation of a new carbon/carbon bond is called the Michael reaction. If the reaction occurs intramolecularly, it is called the Robinson annulation reaction.

14

Elimination Reactions

14.1 Introduction

So far, we have looked at substitution reactions and addition reactions. We will now turn our attention to elimination reactions. The overall result of an elimination reaction is, not surprisingly, the opposite of an addition reaction. However, when we come to study the details of the mechanisms that are involved in elimination reactions, we find that they are not always closely related to the reverse of addition reactions. In contrast, elimination reactions are often more closely related mechanistically to substitution reactions than to addition reactions. In fact, elimination and substitution reactions often compete with each other in the reaction mixture.

In a general elimination reaction, usually two new molecules (often of unequal size) are formed from the original molecule. Furthermore, one of the resultant molecules usually has a greater degree of unsaturation than was present initially in the original molecule. Often this is achieved by losing two groups, A and B, that were originally *vicinal* to each other in the starting material. In this case, a double bond is formed in the remaining part of the molecule, and another molecule, A–B is formed from the parts that were eliminated. Write the general equation for the loss of the A–B molecule from the starting material, $R_2AC–CBR_2$.

$$R_2AC–CBR_2 \rightleftharpoons R_2C{=}CR_2 + A–B$$

This type of elimination is called β-elimination, or 1,2-elimination. If there were already a double bond between the α and β atoms, then a triple bond would be formed in the elimination product.

It is possible for both groups, A and B, to be lost from the same atom, i.e.

they were originally *geminal* to each other. Write down the general equation for the loss of the molecule A–B from the starting material, R_3C–CABR, and so identify the initial product.

$$R_3C\text{–}CABR \rightleftharpoons R_3C\text{–}CR + A\text{–}B$$

In this case, the initial product is a carbene. If the groups A and B had been lost from the nitrogen species, R_3C–NAB, then the initial product would have been a nitrene, R_3C–N:. This type of elimination is called an α-elimination, or 1,1-elimination. The carbene or nitrene that is formed can undergo a number of further reactions. If there were a hydrogen on the β-carbon, then a new double bond could be formed by a hydride shift. Write down this step of the reaction.

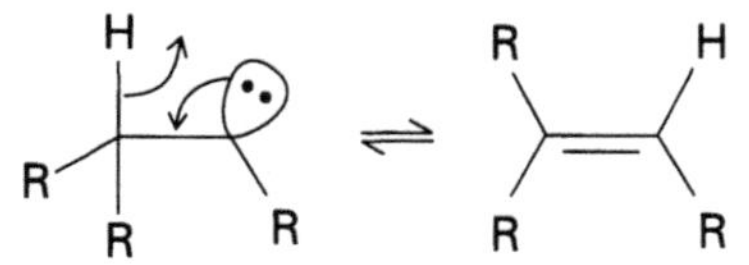

There is a third type of elimination, which is called γ-elimination, or 1,3-elimination, in which a three-membered ring is formed. Write down the general equation for this reaction.

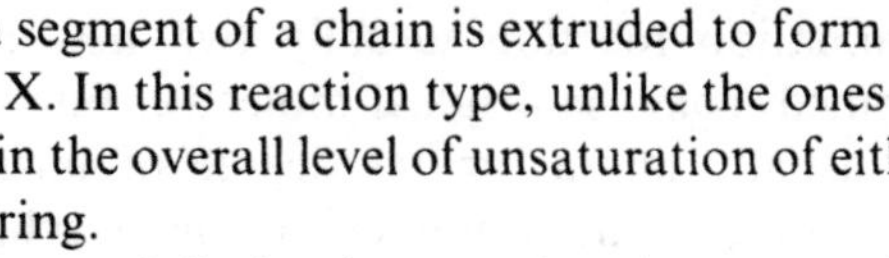

There is a final type of elimination mechanism, which is called an extrusion reaction, in which a segment of a chain is extruded to form a smaller chain, e.g. A–X–B = A–B + X. In this reaction type, unlike the ones we have seen so far, there is no increase in the overall level of unsaturation of either part, nor is there the formation of a ring.

The commonest type of elimination reaction that occurs in solution is the 1,2-elimination. However, it is also possible for 1,1-elimination to occur under certain conditions for reactions that occur in solution.

In the gas phase, when the reaction is normally initiated by heat, and so is called pyrolysis, there are two common pathways. The first is via a cyclic transition state and the second involves a free radical pathway. These two pyrolytic eliminations are rather different in nature from those that occur in solution, and so we will discuss them separately.

We will start by looking in some detail at the 1,2-elimination reaction that occurs in solution, as this forms the principal mechanistic route for elimination reactions.

14.2 Bimolecular 1,2-elimination reactions

14.2.1 Anti-*elimination reactions*

There is a tendency in some elementary organic textbooks to indicate elimination reactions by encircling the two groups that are closest to each other, and then showing the final eliminated product. This type of lasso chemistry predicts that both groups are eliminated from the same side of the starting material, i.e. that there has been *syn*-elimination. In practice, this is found to be rather rare. Instead, what is observed is that the two groups that are eliminated come from opposite sides of the molecule, i.e. *anti*-elimination. We will now examine why this is the case.

The first elimination mechanism that we will study is called the bimolecular elimination reaction, E2. In this mechanism, the segments that are eliminated are *vicinal* to each other. This reaction is very similar to the S$_N$2 reaction that we studied in the chapter on nucleophilic substitution, and like that pathway it is a bimolecular one-step reaction. In substitution reactions, the incoming group arrives at the same time as the leaving group departs (which results in overall substitution); while in elimination reactions, the removal of the first group, A, occurs at the same time as the second group, B, leaves. However, there is an important difference between the S$_N$2 reaction and the E2 reaction: in the substitution reaction, the incoming group attacks the same centre as that from which the departing group is leaving, i.e. only one carbon centre is affected. However, in the elimination reaction, the two departing groups depart from adjacent carbons, i.e. two carbon centres are affected.

Write the equation for the E2 reaction between a general base, B, and the substrate, $R_2CH-CXR_2$.

$$R_2CH-CXR_2 + B \rightleftharpoons R_2C=CR_2 + BH^+ + X^-$$

This reaction has resulted in HX being eliminated. Compare this elimination reaction with the S$_N$2 reaction by writing the equation for the substitution of an X substituent by a nucleophile, Nuc$^-$, on the substrate, $R_2CH-CXR_2$.

$$R_2CH-CXR_2 + Nuc^- \rightleftharpoons R_2CH-CR_2-Nuc + X^-$$

The species X$^-$ is called the nucleofuge. In each case, the attacking species, namely the nucleophile and the Brönsted–Lowry base, are fundamentally very similar. With respect to the electronic characteristics of the attacking species and the species that is being attacked, suggest what are the similarities that are shared by, and the differences that distinguish, these two reactions.

In the case of the substitution reaction, the attacking species is acting like a nucleophile, which means that it is acting as a soft base, and so it attacks the softer centre, which in this case is the $\delta+$ carbon. However, in the case of the elimination reaction, the attacking species is acting as a hard base, and thus attacks the harder centre, which is the α-hydrogen. The σ bond electrons in the carbon/hydrogen bond then act in a similar manner to an internal substitution reaction to eliminate the substituent that is to form the leaving group on the *vicinal* carbon. Thus, the fundamental difference between the substitution reaction and the elimination reaction is that in the former instance the attacking species is soft in nature, while in the latter it is hard. Hence, elimination via the E2 mechanism will be favoured over substitution when a hard basic reagent is employed, and vice versa when a soft nucleophilic reagent is employed.

Now write the mechanism for an E2 elimination reaction, that combines all the points that have been discussed so far.

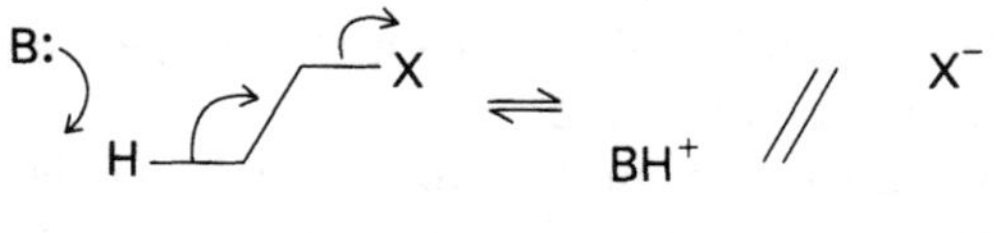

Of course, no attacking species acts purely as a hard or soft base, but rather each has a character that is somewhere along a continuum between one extreme and the other. Accordingly, in any given reaction mixture, substitution and elimination reactions usually compete with each other.

The stereochemistry of this reaction can be investigated by studying the elimination of HBr from 1,2-dibromo-1,2-diphenylethane. It is observed that the *meso* form of this compound produces only the *cis* isomer of the resultant alkene. Suggest which conformation of the transition state would explain this result.

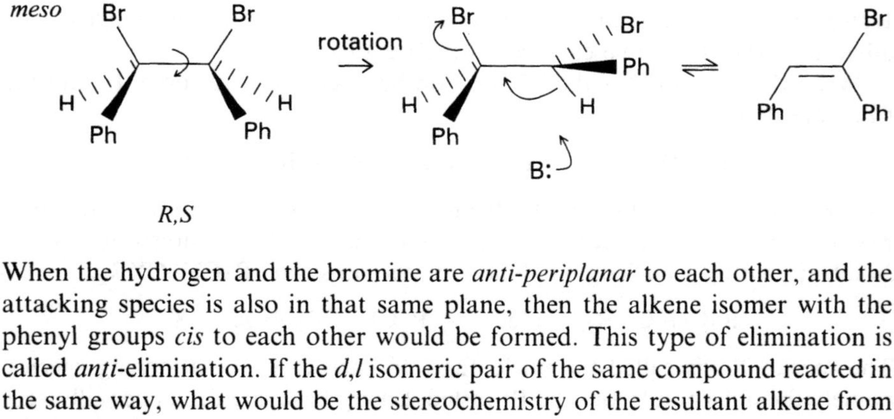

When the hydrogen and the bromine are *anti-periplanar* to each other, and the attacking species is also in that same plane, then the alkene isomer with the phenyl groups *cis* to each other would be formed. This type of elimination is called *anti*-elimination. If the *d,l* isomeric pair of the same compound reacted in the same way, what would be the stereochemistry of the resultant alkene from each optical isomer?

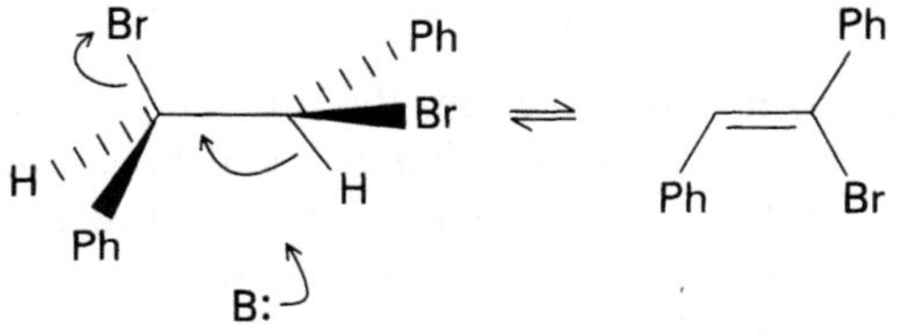

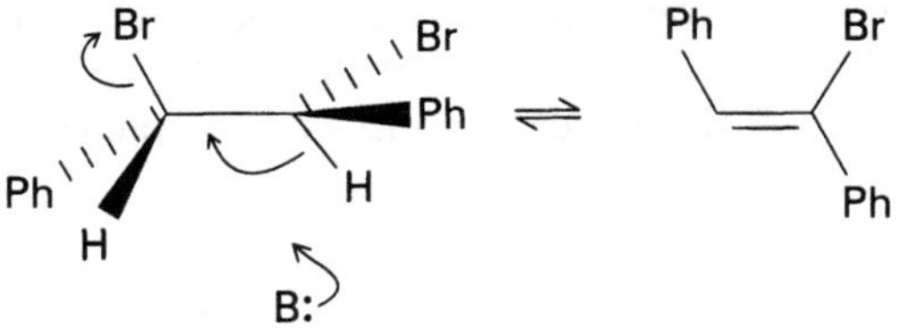

In each case, the phenyl groups would be *trans* to each other. Furthermore, this is the result that is found to occur in practice.

If instead of eliminating HBr from this compound, a bromine molecule is eliminated by the action of an iodide ion, what would be the geometry of the resultant alkene from the *meso* isomer and from the *d,l* isomeric pair?

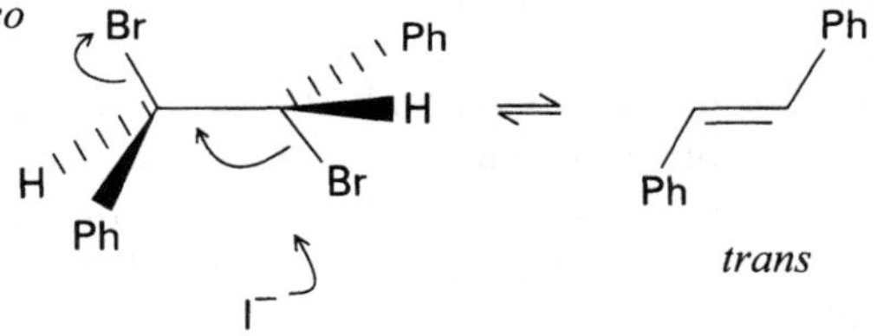

In this case, the *meso* compound gives the *trans* isomer as a result of *anti*-elimination, while the *d,l* pair gives the *cis* isomer.

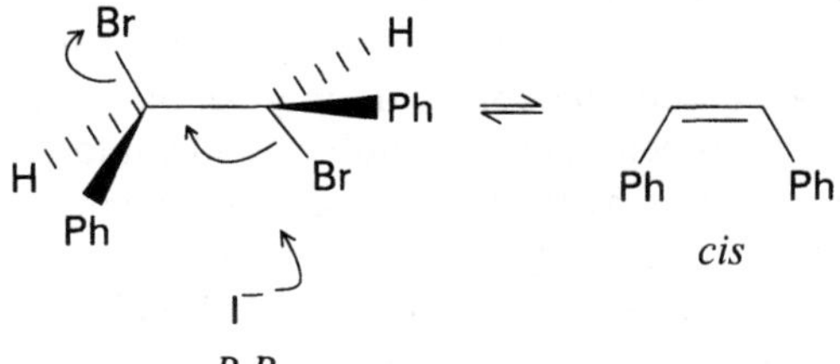

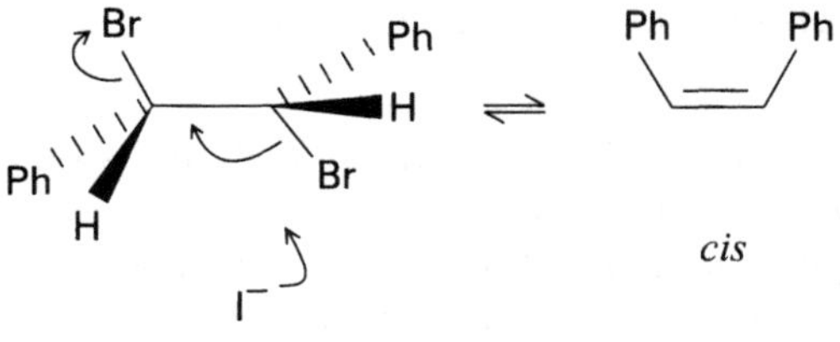

In general, *anti*-elimination will result in an *erythro d,l* pair giving the *cis* isomer, and the *threo d,l* pair results in the *trans* isomer. The *threo* isomer, however, tends to react significantly faster than the *erythro* isomer. Taking the elimination of $HNMe_3^+$ from the (1,2-diphenylpropyl)trimethylammonium ion, $PhMeCH-CHPhNMe_3^+$, by the ethoxide ion as an example, suggest a reason why this is the case. Illustrate your answer using the sawhorse projection.

one *threo* isomer   one *erythro* isomer

In this case, the *threo* isomer can achieve the required *anti-periplanar* geometry with the phenyl groups staggered with respect to each other. In the case of the *erythro* isomer, the phenyl groups are *gauche*. Thus, for elimination to occur in the *erythro* isomer, initially it must adopt an unfavourable conformation, while in the case of the *threo* isomer, elimination may occur from the most favourable conformation. In a solution in which all the conformations are in thermal equilibrium with each other, the most stable will be present in the largest concentration. Accordingly, other things being equal, the reaction that utilises the most stable conformation will proceed the fastest. This is called the eclipsing effect, and has wide applicability.

So far, we have only considered the stereochemistry of the transition state and how that relates to the stereochemistry of the product. This is most readily investigated when there is only one β-hydrogen, because this means that the resultant carbon/carbon double bond can only be formed in one place. If there is more than one β-hydrogen, then the double bond could be formed in different places depending on which hydrogen has been removed.

The first guideline in deciding where the double bond will be placed in the product is Bredt's rule. This states that a double bond will not be formed at a bridgehead, unless the rings are large enough. In order to determine whether the rings are large enough, we use the S number; that is the sum of the number of the atoms in the bridges that form the bicyclic system. What is the S number for the norbornyl system?

The norbornyl system is a [2.2.1] bicyclo system, and so the S number is five. Bridgehead double bonds only form if the S number is seven or greater. There is an additional requirement, in that the double bond may exist only in a ring that contains eight or more atoms, i.e. a derivative of *trans*-cyclooctene or larger.

The next principle may be illustrated by investigating which isomer is formed on the elimination of 4-bromopent-2-one. Suggest what are the two possible alkene products, and then suggest which one will be formed in practice.

Elimination could occur to yield the unconjugated alkene. However, that is not observed in practice; instead elimination nearly always occurs so as to give the conjugated product. The side chain in which the elimination did not occur is distinguished by using a prime symbol, so in this case the C5-methyl group has the β′-hydrogens, while the C3-methylene group has the β-hydrogens, one of which is actually eliminated to give the conjugated product.

In *anti*-E2 reactions that occur on an aliphatic chain, where the nucleofuge is neutral, or if the elimination occurs from a six-membered ring, it is observed that the most highly substituted alkene is preferentially formed. This is called Zaitsev elimination, and is generally true when the nucleofuge can form a good leaving group, i.e. the C–X bond is weak and the resultant anion, X^-, is very stable. When the substituent is positively charged, then the opposite orientation is found, i.e. the double bond that is formed goes towards the least highly substituted carbons. This is the Hofmann elimination, and is generally favoured when the substituent would form a poor leaving group and the C–X bond is strong.

An example of a reaction that shows this latter regiochemistry is the Hofmann exhaustive methylation, which is sometimes called, rather ambiguously, the Hofmann degradation (this is a term best avoided, because it is sometimes also used to describe the Hofmann rearrangement). In this reaction, an amine is methylated with methyl iodide until the quaternary ammonium iodide is formed. This is then treated with moist silver oxide to convert the iodide to the hydroxide. Write out the final elimination step that occurs on heating.

The hydroxide ion removes the β-hydrogen in a normal *anti*-E2 mechanism.

There are many explanations that attempt to rationalise the difference between these two rules for the regiochemistry of the elimination. However, none of them is completely satisfactory. Some suggest that for positive nucleofuges, the E2 reactions become paenecarbanionic in nature, i.e. more like the E1cB process (which is discussed in more detail below), and as such the cleavage of the C–H bond is more important. Hence, those factors that stabilise the incipient carbanion, such as primary carbons, will favour the Hofmann route, and so give rise to the less highly substituted product. When the nucleofuge is neutral, the reaction is more paenecarboniumic in character, and so the cleavage of the C–X bond becomes more important. This means that factors that result in a more stable alkene will favour this route, e.g. the Zaitsev product with the more highly substituted carbon.

Another explanation focuses on the steric interactions, rather than the electronic ones, and proposes a reason similar to the eclipsing effect, which was mentioned earlier. It has also been pointed out earlier that very sterically demanding bases will not always remove the most acidic proton, but are sometimes limited to those that are more accessible, i.e. on the terminal methyl groups. The removal of such a proton would normally result in Hofmann regiochemistry. Hofmann regiochemistry is favoured as the size of the substituent that is to form the leaving group increases. However, Zaitsev elimination is favoured as the number of β-alkyl substituents increases, but if these β-substituents are capable of stabilising a negative charge, then Hofmann elimination once again becomes the favoured option. It is, thus, readily apparent that there is no simple answer to which regiochemistry will be favoured in any particular case.

For a substrate that is undergoing a bimolecular reaction, as we have noted earlier, there is competition between the substitution and elimination pathways. As the amount of branching increases in the carbon chain, would this favour or disfavour the elimination pathway over the corresponding substitution pathway?

For bimolecular reactions, increased carbon branching favours the elimination pathway, to the extent that in highly branched substrates there may be very little substitution.

This may be illustrated by re-examining a reaction that we have already studied. Earlier, we saw that when a 1,2-dibromide was treated with iodine, the reaction followed a normal *anti*-E2 pathway to eliminate a bromine molecule, which meant that the *threo* isomer yielded the *trans* isomer. If this elimination reaction is performed when either, or both, of the bromine atoms is, or are, attached to a primary carbon (instead of a secondary or tertiary carbon as was the case when we first looked at this reaction), then a different stereochemistry results. When this experiment is performed with deuteriated derivatives, it becomes apparent that the *threo* isomer gives the *cis* isomer. Suggest an explanation for this result.

Contrary to first appearances, the elimination mechanism has not changed; instead there is now an initial substitution reaction by the iodide anion for a bromide anion, with a concomitant inversion of configuration at that centre. This substitution is then followed by an elimination reaction with the normal *anti* stereochemistry. This illustrates the point that substitution reactions occur more easily at a primary carbon than at a secondary or tertiary carbon, at which locations elimination reactions are preferred. Suggest why this may be the case.

There are two reasons. First, for a tertiary carbon centre there is a large release of crowding strain on elimination, while for a primary carbon centre there is little such effect. Second, the stability of the resultant alkene increases as the number of substituents increases on the carbon/carbon double bond.

This example also illustrates one of the pitfalls of attempting to elucidate a mechanistic pathway. The use of alkyl groups, such as methyl and ethyl, to label certain carbon atoms changes their character from, say, primary to secondary, and this may have an effect on the mechanism that is being studied. The use of deuterium labels may be preferred, but requires the initial stereoselective synthesis of such compounds, and this might be difficult. Furthermore, it must be possible to distinguish between the various isotopic products, otherwise no useful information will be gained.

We have already briefly mentioned that elimination is favoured by a reagent that is hard in nature, while substitution is favoured by one that is soft. The concentration of the base also has an effect on the relative ratio of substitution to elimination reaction that might occur. For the bimolecular pathway, suggest what would be the effect of increasing the concentration of the external base.

Increasing the concentration of the base favours the elimination reaction over the substitution pathway. The size of the attacking reagent will also affect the distribution of reaction types: the larger the reagent, the more difficult it will be for it to approach the carbon centre to perform a substitution reaction. However, it will still normally be able to remove the β-hydrogen and so be able to perform the E2 elimination.

Furthermore, it is found that decreasing the polarity of the solvent increases the proportion of elimination to substitution. Thus, alcoholic KOH is used for elimination reactions, while aqueous KOH is used to effect the substitution of a functional group.

An example of this concerns the elimination reaction in which a *geminal* dihalide, $RH_2C-CHCl_2$, is treated with alcoholic KOH. Suggest what might be the product from this reaction.

$$RH_2C-CHCl_2 + KOH \rightleftharpoons RCH=CHCl + H_2O + KCl$$
$$RCH=CHCl + KOH \rightleftharpoons RC\equiv CH + H_2O + KCl$$

After the first elimination of HCl, the reaction may be repeated on the alkene to give the alkyne. If, instead of a *geminal* dihalide, a *vicinal* dihalide is used, e.g. $RCH_2CHCl-CH_2Cl$, then the same alkyne is usually produced.

$$RCH_2CHCl-CH_2Cl + KOH \rightleftharpoons RCH_2CH=CHCl + H_2O + KCl$$
$$RCH_2CH=CHCl + KOH \rightleftharpoons RCH_2C\equiv CH + H_2O + KCl$$

Potentially however, there is a different elimination reaction that might compete with the normal second elimination step. Suggest what the alternative product might be that results from the elimination of two HCl molecules from this dihalide, $RCH_2CHCl-CH_2Cl$.

$$RCH_2CHCl-CH_2Cl + KOH \rightleftharpoons RCH_2CCl=CH_2 + H_2O + KCl$$
$$RCH_2CCl=CH_2 + KOH \rightleftharpoons RCH=C=CH_2 + H_2O + KCl$$

The second elimination could occur between the second and third carbons in the chain to give an allene.

A change of solvent may have a great effect on the strength of the base. In particular, if the solvent is changed from one that is polar and hydroxylic to another that is bipolar and aprotic such as DMF or DMSO, then the strength of the base increases. Suggest why this is the case.

In the latter type of solvent there will be no hydrogen bonds to anions, and so there is no envelope of solvent molecules hindering the attack of the base on the proton. This also has the effect of increasing the effective charge density on the base, as it is not dissipated by a solvent shell. This in turn increases the effective hardness of the base, and so favours elimination over substitution. Furthermore, in a solvent that has a low polarity, the stability of any charged intermediate would be reduced, and so bimolecular reactions would be favoured over unimolecular alternatives, and this also favours elimination.

What would be the effect upon the relative ratios of substitution to elimination of raising the temperature at which the reaction is performed?

In a substitution reaction, the number and nature of the products are the same as the starting materials, and so there is little change in entropy, and thus a change in temperature has little effect. For elimination reactions, the entropy is increased as there are more products than there are starting materials; raising the temperature favours their production, and so favours the elimination pathway.

We will now look at the special factors that are involved in the bimolecular *syn*-elimination reaction.

14.2.2 Syn-elimination reactions

In the previous sub-section, we saw that *anti*-elimination requires the two groups that are to be eliminated to be *anti-periplanar* to each other. If there is a deviation from this angle then a different mechanism is observed.

If *exo*-2-bromonorbornane, which is stereospecifically deuteriated in the *exo*-β-position to the halogeno substituent, is treated with a base, then an elimination reaction occurs. However, in the carbon product there is no deuterium, i.e. deuterium bromide has been eliminated. Account for this observation.

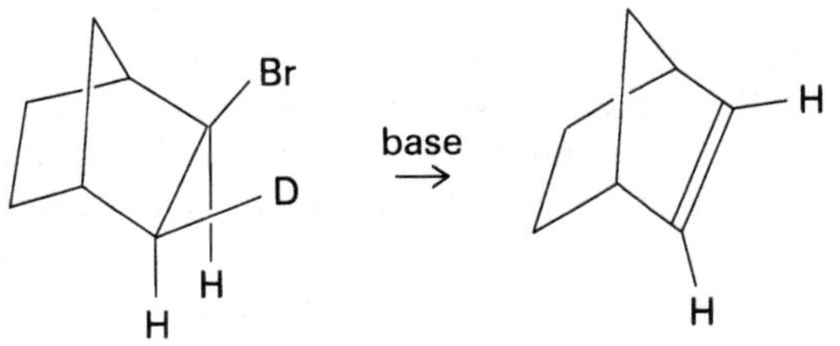

In this case, *syn*-elimination must have occurred in order for the deuterium to be lost from the norbornyl skeleton. The hydrogen atom, which was in the *endo*-β-position, describes a dihedral angle of 120° with the halogeno substituent. As this hydrogen atom was not eliminated with the halogeno substituent in a normal *anti*-elimination mechanism, it appears that for the E2 *anti*-elimination mechanism to operate the dihedral angle between the eliminated groups must be almost exactly 180°. It is generally found that *syn*-elimination is slower than *anti*-elimination, and so it is usually only observed in significant amounts when the *anti*-elimination is so disfavoured, e.g. by the impossibility of achieving a dihedral angle of 180°, that the *syn*-elimination reaction can compete successfully.

There are some exceptions to this general observation. For example, *syn*-elimination reactions have been found to be promoted by weakly ionising solvents, but only when the eliminated substituent is neutral.

For the elimination reaction of HX from the general substrate $R_2CH-CHXR$, where X is neutral, e.g. a bromo substituent, using potassium *t*-butoxide as the base, suggest a reason why *syn*-elimination would be favoured in a solvent of low polarity, e.g. benzene.

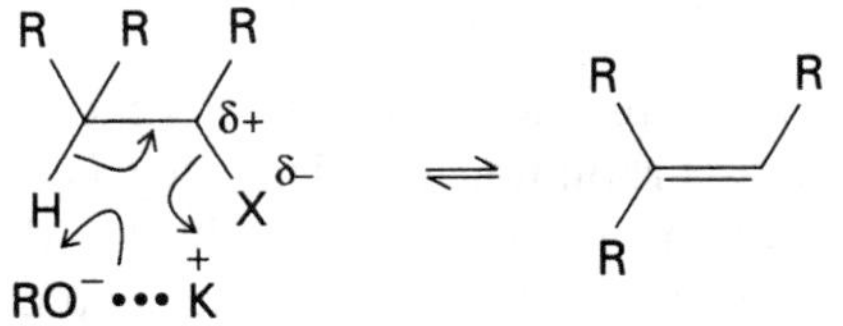

When a solvent of low polarity is used, an increased amount of ion pairing occurs, and as such a six-membered cyclic transition state may easily be formed; this results in *syn*-elimination.

If the substituent is positive, however, the opposite is found to be true, i.e. solvents of low polarity now favour *anti*-elimination. Suggest why this is the case.

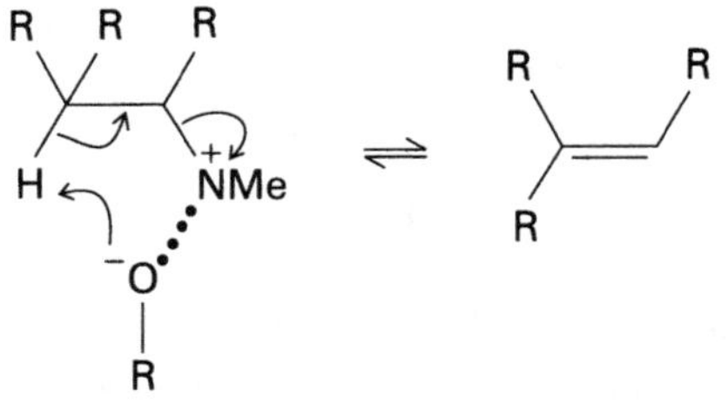

The explanation is the same: a solvent of low polarity favours ion pairing. Thus, in solvents of high polarity, the base forms a cyclic transition state with the substrate directly, without the involvement of its corresponding cation, i.e. the transition state is now a five- and not a six-membered ring, and so the base is now in an optimal position to remove the *syn*-hydrogen. Conversely, in solvents of low polarity with large amounts of ion pairing, the presence of the cation interferes with the abstraction of the proton and so *anti*-elimination once more predominates.

If it is possible for the regiochemistry of the formation of the double bond to be in doubt, then in a *syn*-E2 reaction Hofmann elimination is nearly always followed. This is in contrast to the situation of the *anti*-E2 reaction in which either the Zaitsev or the Hofmann elimination could occur, depending on the circumstances.

Now that we have studied bimolecular elimination reaction in some depth, we will turn our attention to the unimolecular pathway.

14.3 Unimolecular 1,2-elimination reactions

In the previous section, we looked at the bimolecular 1,2-elimination reaction. We will now look at the unimolecular 1,2-elimination reaction, which is given the label E1. This reaction bears many similarities to the SN1 reaction that we studied in an earlier chapter.

Remind yourself of the mechanism for the SN1 substitution reaction.

$$R_3CX \rightleftharpoons R_3C^+ + X^-$$
$$R_3C^+ + Y^- \rightleftharpoons R_3CY$$

This reaction occurs in two steps, the first of which is the rate limiting loss of the leaving group to give the carbonium ion. This step is unimolecular, and so the concentration of the other reagent is not involved in the rate equation. The second step involves the attack by the incoming nucleophile to result in overall substitution.

If, instead of the addition of an incoming nucleophile in the second step, the β-hydrogen is lost, then the overall result would be a unimolecular 1,2-elimination reaction. Write the first ionisation step, and then involve the elimination step in the second stage.

This is the mechanism for the E1 elimination reaction. The similarities with the SN1 reaction are plain. Like the E2 and SN2 reactions that often compete with each other, the E1 and SN1 reactions also often compete with each other.

The first step is exactly the same in the SN1 and E1 mechanisms, and thus those considerations that applied to the first step of the SN1 mechanism apply equally to the first step of the E1 mechanism. These factors include the stability of the carbonium ion intermediate, the loss of stereochemistry at that centre if no other factors intervene, and the possible rearrangement of the carbonium ion intermediate.

In the normal E1 mechanism, the heteroatomic anion leaves first to form the carbonium ion intermediate. However, it is possible for a proton (i.e. a hydrogen cation) to leave first. This would result in a carbanion being formed. Write down this reaction step, and then deduce what the second step must be in order to result in an overall elimination.

This reaction type is called the E1cB mechanism, which stands for unimolecular elimination conjugate base reaction, because the conjugate base of the starting material is being formed as the reactive intermediate. It is sometimes called the carbanion mechanism. As this mechanism results from the removal of a proton, it is not surprising that it is favoured by those substrates that possess an acidic hydrogen atom. Thus, would you expect the E1cB mechanism to be more prevalent in reactions that result in a carbon/carbon double bond or in reactions that result in a carbon/carbon triple bond?

To form a carbon/carbon triple bond, the proton must be removed from an sp^2 hybridised carbon, rather than an sp^3 hybridised carbon, and as the former is more acidic than the latter, one would expect the E1cB mechanism to be preferred in the former case.

The intermediate of the E1cB mechanism is a carbanion, and thus any factors that stabilise such an ion should favour this mechanism. We have already noted above that on the face of it, elimination reactions are the reverse of addition reactions. However, we also noted that the actual mechanistic pathways involved in elimination reactions were more similar to substitution reactions than addition reactions. This is because normally elimination reactions proceed via a carbonium ion or in a single step that has certain similarities to an SN2 substitution reaction. However, there are also addition reactions that proceed via a carbanion intermediate, for example the Michael-type reaction, in which a carbanion adds to an α, β-unsaturated carbonyl compound. Indicate the Michael-type addition between the anion formed from the diester of propandioic acid (or malonic acid) and 2-butenal.

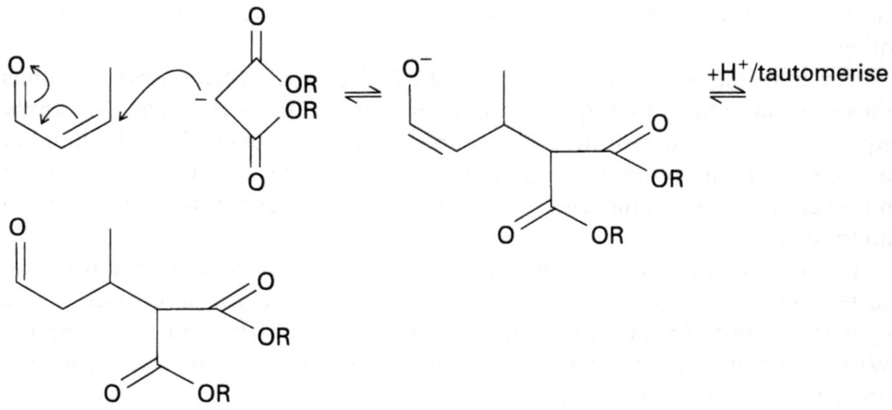

Now write the reverse of this mechanism.

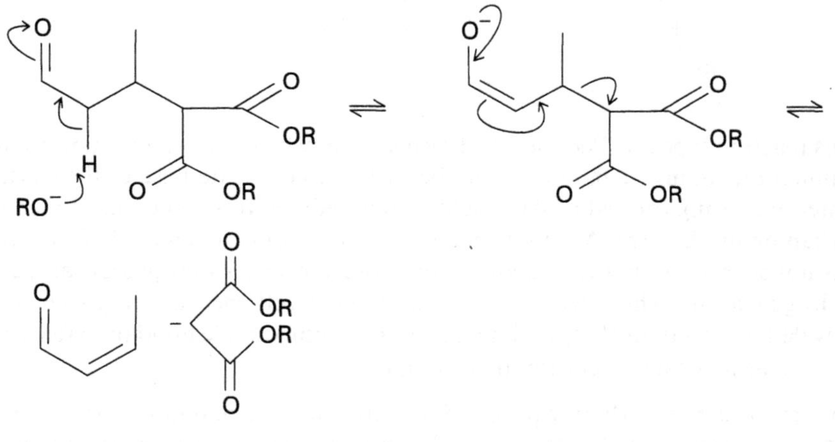

This pathway follows the E1cB route. Furthermore, it is an example of the microreversibility of a reaction, i.e. that a reaction may proceed in either direction, and along exactly the same pathway at every step of the way.

Another factor that favours the E1cB mechanism is a poor nucleofuge. If the nucleofuge left easily then, either a normal E1 mechanism would occur, or it would leave at the same time as the β-hydrogen was being removed by the base.

The E1cB mechanism is rare in practice when the elimination reaction would result in a carbon/carbon double bond. When a carbon/oxygen double bond is to be formed then it is far more common. For example, the E1cB mechanism is found in the reverse of the cyanohydrin formation reaction. You will recall that the forward reaction involves the addition of a cyanide anion to a carbonyl group. Write down the pathway for the reverse reaction, i.e. the elimination reaction.

The starting molecule has an acidic hydrogen atom, and a nucleofuge that is strongly bonded to the carbon atom. In this case, the carbanion that is formed is the cyanide ion.

It is clear from this discussion that the E1, E2 and E1cB mechanisms are all closely related. In all of these mechanisms, one group leaves with its electron pair (the nucleofuge), while the other group leaves without its electron pair, and this second group is often a proton. The difference arises in the exact timing of the departure of these two groups. In the first mechanism, i.e. E1, the nucleofuge leaves first; while in the last, i.e. E1cB, the proton is removed first, and thereafter the nucleofuge. In the E2 mechanism, the two groups depart simultaneously, i.e. it falls half way between the two other extremes. As we have already seen in the case of nucleophilic substitution reactions, it is very rare that a reaction has only the characteristics of the idealised extreme. Instead, it is far more common for any given reaction to exhibit intermediate characteristics. This is particularly true for the E2 reaction, which rarely falls exactly equally between the two extreme cases. Rather, depending on the exact circumstances, it may tend towards either the E1 reaction or the E1cB reaction. In the former case, it is said to be paenecarboniumic, while in the later paenecarbanionic. It is possible to distinguish between these various possibilities by studying the effect on the rate of the elimination reaction of isotopic substitution. Either the hydrogen may be replaced by a deuterium atom, or else the atom within the nucleofuge that is bonded to the carbon may be substituted for a different isotope from its normal one.

An example of an E2 mechanism that is paenecarbanionic is the base elimination of water from a 3-hydroxycarbonyl derivative. Normally, water is eliminated under acidic conditions in which the hydroxyl group is protonated to form a substituent, namely $-OH_2^+$, that in turn would give rise to a more favourable leaving group, i.e. H_2O. When the first stage of the aldol reaction is performed between two molecules of ethanal, the initial product is aldol itself, which contains a 3-hydroxycarbonyl skeleton. Under the normal reaction conditions, elimination of water occurs to yield the conjugated unsaturated carbonyl compound, crotonaldehyde. Suggest the mechanism for this reaction.

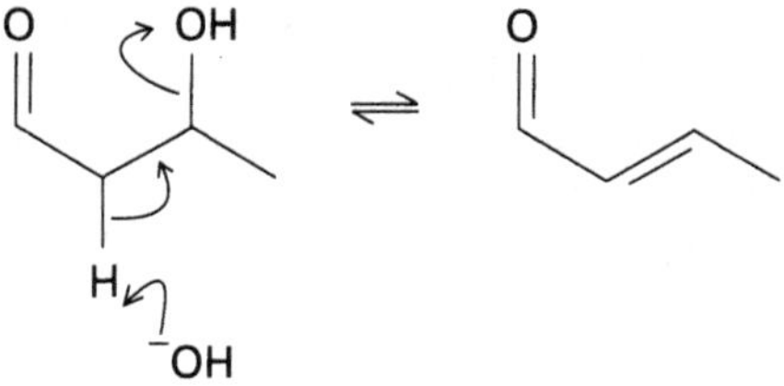

In this case, the aldehyde group, which is −M, makes the hydrogen more acidic. So generally, all −M groups, such as carbonyl, when adjacent to the hydrogen atom, or even −I groups, such as halogen atoms, favour the E1cB route; while +M groups, or even +I groups like alkyl groups, disfavour this pathway.

In E1cB reactions, there is usually little ambiguity as to the regiochemistry of the resultant compound, because usually there is only one site from which an acidic hydrogen atom may be removed.

In the case of the E1 mechanism, where there is doubt over the regio-chemistry of the resultant double bond, the orientation of the double bond in the final product may be deduced from the relative stabilities of the final unsaturated product. From heat of combustion data, it is found that the more highly substituted the double bond, then the greater is the stability of that alkene. Thus, in the case of 2-bromobutane, suggest what will be the resultant alkene after the elimination of HBr.

But-2-ene will be formed in preference to but-1-ene, as the secondary carbonium ion is significantly more stable than the primary carbonium ion, i.e. the former is the thermodynamic product. Thus, in E1 reactions, Zaitsev elimination is preferred. However, it does not apply if the predicted product would suffer steric strain.

What would be the effect of an alkyl or an aryl group at the α-carbon on the E1 pathway?

In the case of the E1 pathway, a carbonium ion intermediate is being formed, and this may be stabilised by either an aryl or an alkyl group; both of these substituents will therefore favour the E1 pathway when attached to the α-carbon.

What would be the effect on the relative rates of elimination to substitution of increasing the amount of branching in the carbon chain of the substrate that is reacting in a unimolecular manner?

Generally, the more branching in the carbon chain, the more elimination occurs, because of the increased importance of releasing steric strain.

In unimolecular reactions, the base that removes the proton in the second stage of the reaction is usually the solvent, and not a separate component of the reaction mixture. Using this information, suggest what would be the effect of adding an external base to the reaction mixture.

The effect would be that the bimolecular pathway would be favoured as the concentration and strength of the base increased. At a low concentration of a weak base, the unimolecular pathway is favoured, so long as the solvent is ionising. Under these latter conditions, substitution is favoured over elimination.

The nature of the substituent that is to give rise to the leaving group has quite a large effect on the particular pathway that is followed. Suggest whether a substituent that would give rise to a very good leaving group would favour the E1, E2 or E1cB pathways.

A substituent that gives rise to a very good leaving group leads to the E1 pathway being favoured. Synthetically, the most important leaving group is the H_2O molecule. If the nucleofuge is Cl^-, Br^-, I^- or NR_3, then the reaction tends to proceed via the E2 route. The E1cB pathway is favoured by substituents that would give rise to poor leaving groups.

The E1, E1cB and E2 elimination pathways are all 1,2-elimination reactions. These account for the vast majority of elimination reactions that occur in solution. We will now briefly examine some other pathways that occur under particular conditions.

14.4   1,1-Elimination reactions

If 1-chloro-1,1-dideuteriobutane is treated with a very strong base such as phenyl sodium, then, even though some of the resultant alkene contains two deuterium atoms on the terminal carbon atom, a higher percentage of the 1-alkene contains only one deuterium atom. This deuterium atom is located on the terminal carbon atom. The expected product, with two deuterium atoms, which is formed in smaller amounts, results from the normal E2 mechanism. Suggest a mechanism that accounts for the formation of the major product.

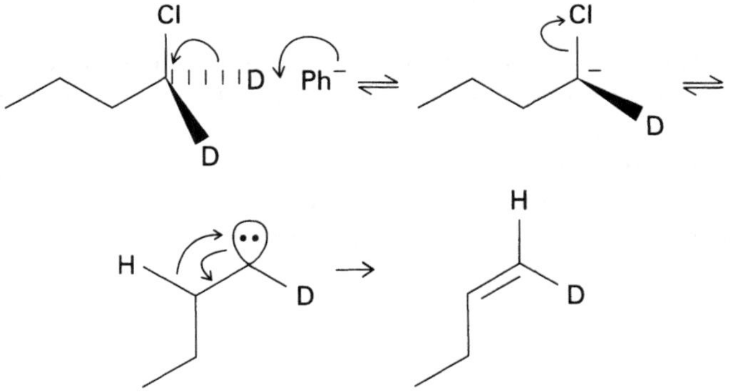

Under strongly basic conditions, we have seen that the E1cB mechanism may operate. In that case, the base removes the β-hydrogen atom. However, in this case, the phenyl sodium has removed the deuterium that is *geminal* to the chlorine. This deuterium is more acidic than any of the other hydrogen (protium) atoms. The resultant anion may then lose a chloride ion to give a carbene, which then inserts its lone pair into the β-carbon C–H bond. This yields the mono-deuteriated alkene. This type of reaction is called 1,1- or α-elimination, and is much less common than 1,2-elimination.

So far, we have mainly studied elimination mechanisms that occur in solution. They are characterised by the involvement of ionic intermediates, such as the E1 and E1cB mechanisms or the 1,1-elimination just mentioned, or at least require an ionic reagent like the E2 mechanism. There is another class of elimination reactions that occur when compounds are heated. These are called pyrolytic elimination reactions, and usually proceed via a cyclic transition state or by a pathway that involves radicals.

14.5 Pyrolytic elimination reactions

This type of elimination reaction often occurs in the gas phase, without the addition of another reagent. This point distinguishes this type of reaction from those reactions that we have already studied, because all of the other types required a base, be it an additional external base or the solvent.

If carboxylate esters are heated, they readily form an alkene and a carboxylic acid. Suggest a possible mechanism for this reaction, illustrating your answer with the general ester, RCH_2-CH_2-OCOR'.

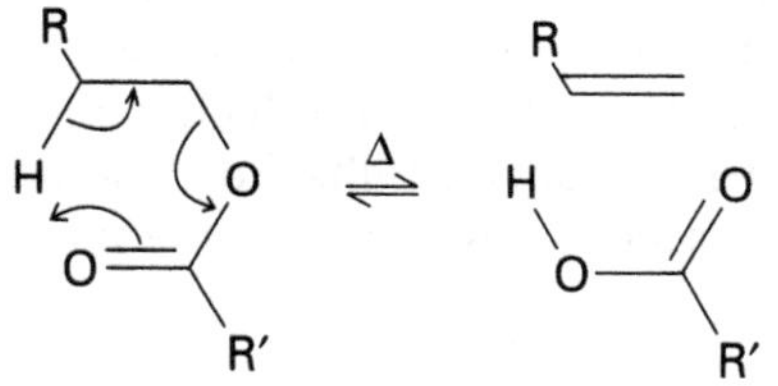

In this case, there is a six-membered cyclic transition state, in which an acid and alkene are formed. There is a requirement for a *cis*-β-hydrogen in order for this reaction pathway to proceed.

We saw earlier that for E2 *anti*-elimination, an *erythro* isomer would result in the *cis* alkene isomer. For *erythro*-1-acetoxy-2-deuterio-1,2-diphenylethane, suggest which isomer of the alkene will be formed on pyrolysis.

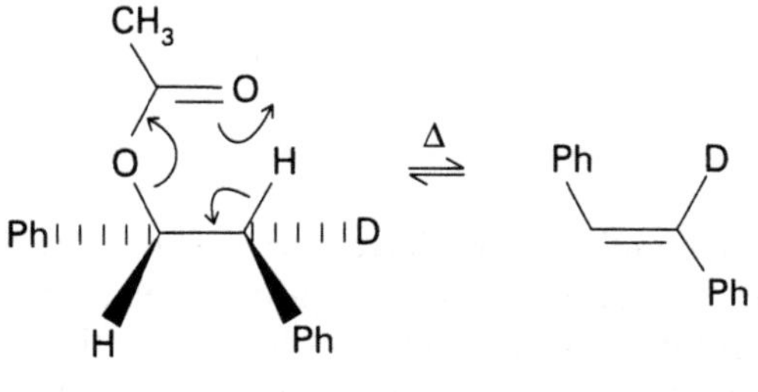

The *trans* isomer of the alkene stilbene is formed, as indicated by the presence of the deuterium atom in the product. In this case, elimination is via a cyclic transition state, and so the opposite stereochemistry is obtained than was achieved in the E2 mechanism. This cyclic pyrolytic elimination is labelled the Ei mechanism, which stands for intramolecular, or internal, elimination.

In those versions of the Ei mechanism that involve a four- or five-membered cyclic transition state, there is a requirement that all the atoms are co-planar or very nearly so. However, this restriction does not apply for the six-membered cyclic transition state.

Methyl xanthates, $R_2CH–O–CSSCH_3$, may be easily synthesized from the corresponding alcohol by treatment with NaOH and CS_2, followed by CH_3I. These compounds readily undergo pyrolytic elimination to give an alkene, together with $CH_3SC(=O)SH$, which subsequently fragments to CH_3SH and COS. This is called the Chugaev reaction. The xanthate group is quite large, and so when it is present in a six-membered ring, it adopts an equatorial, rather than an axial, position. Bearing this in mind, suggest what will be the products from the *cis*-β-methyl derivative of the methyl xanthate cyclohexane.

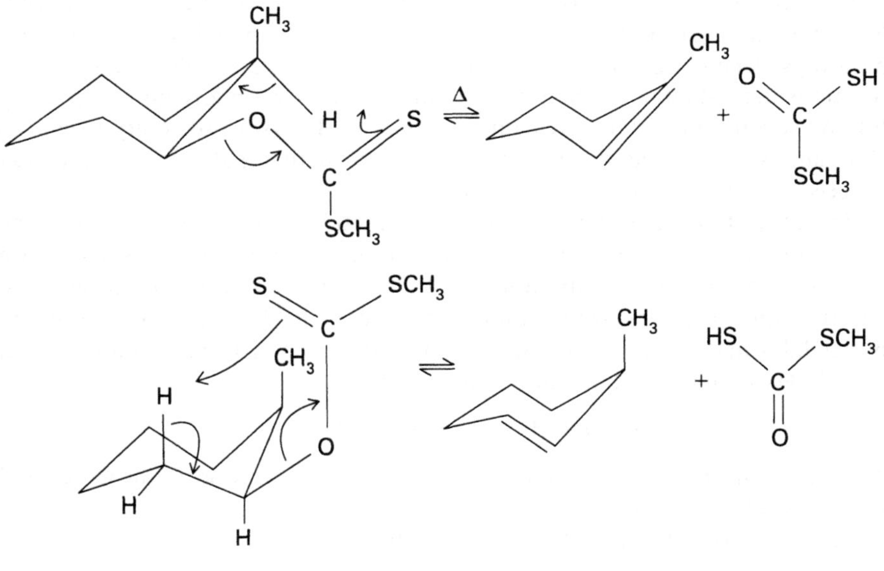

On the unsubstituted side of the cyclohexane ring, there is both a *cis* and *trans* hydrogen with which the xanthate might form the cyclic transition state for the Ei mechanism. However, on the methyl substituted side, there is only a *trans* hydrogen, and yet the xanthate is still able to form a cyclic transition state. This latter transition state resembles the *trans* decalin molecule. So, six-membered cyclic transition states may be formed with both *cis* and *trans* β-hydrogens, because of the capability of a six-membered ring to form a puckered ring of either configuration.

When we looked at *anti*-E2 bimolecular eliminations, we saw that the Hofmann exhaustive methylation followed that pathway. There are occasions, however, when it proceeds via different route. Normally, the hydroxide anion removes the β-hydrogen, which then initiates the E2 mechanism. If, however, the quaternary ammonium derivative is so highly hindered that this is not possible, then the hydroxide anion removes a proton from one of the methyl groups that forms part of the ammonium cation. Suggest how this reaction proceeds, and then write the whole reaction pathway.

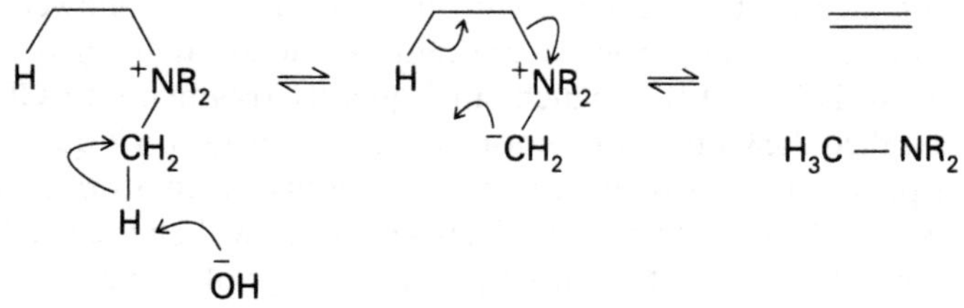

This is an example of a five-membered ring version of the Ei mechanism. Moreover, it is an example of a synthetic reaction that proceeds via two different pathways depending upon the exact conditions. These pyrolytic elimination reactions are valuable synthetically, because there is normally no opportunity for the starting material to rearrange, which is always a possibility when the elimination occurs via the E1 route, because in that pathway the reaction proceeds via a carbonium ion intermediate.

The other type of pyrolytic elimination reaction involves free radicals. The steps are similar to those that we studied in the free radical substitution reactions, i.e. there is an initiation step, followed by several propagation steps, and then there are some termination steps. Free radical elimination is found in polyhalides and primary monohalides. For the general primary monohalide, R_2CHCH_2X, suggest what will be the first step.

$$R_2CHCH_2X \rightleftharpoons R_2CHCH_2\cdot + X\cdot$$

The initiation step is the thermally induced homolytic cleavage of the C–X bond to give two free radicals.

Suggest a couple of possible propagation steps.

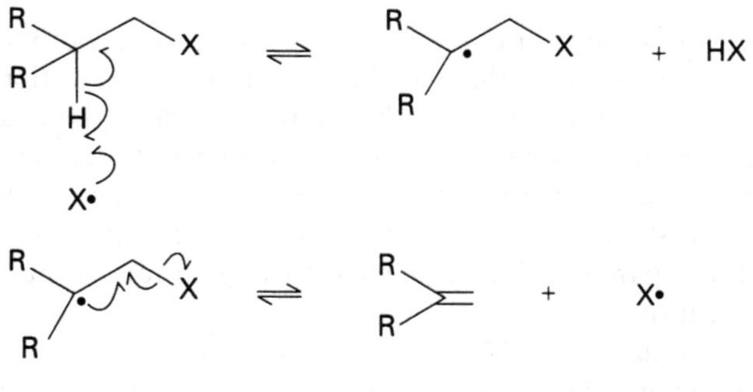

The two that have been chosen here represent, first, a typical hydrogen atom abstraction that gives rise to the alkyl radical, and secondly, a fragmentation reaction of that radical to regenerate more halide radicals and the desired final product.

Suggest a termination reaction that also produces the desired product.

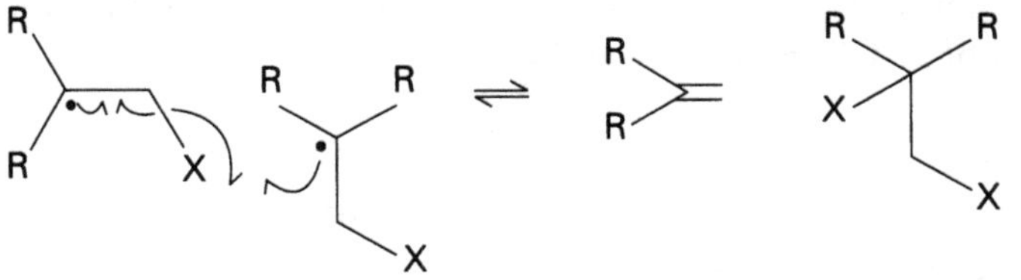

This disproportionation reaction produces some *vicinal* dihalide, and also some of the final product.

Free radical elimination reactions in solution are rather rare; instead they tend to be confined to the pyrolysis of polyhalides and primary monohalides as illustrated above.

14.6 γ-Elimination and extrusion reactions

We will now briefly examine γ-elimination and extrusion reactions. An example of the former is the formation of a cyclopropane from 1-pyrazoline initiated either by heat or light. Suggest a mechanism for this reaction.

The cleavage of both of the weak C–N bonds and the formation of the nitrogen/nitrogen triple bond, coupled with the great increase in entropy, help drive this reaction. The initial biradical, which is formed by the loss of the nitrogen molecule, then closes the ring to give the cyclic product.

The great driving force, both enthalpically and entropically, that is provided by the loss of a nitrogen molecule may be used to produce some strained compounds that would otherwise be difficult to synthesise. For example, 3H-pyrazole on photolysis yields cyclopropene. This reaction is thought to proceed via a diazo intermediate, which may then form a carbene. Suggest what is the mechanistic pathway for this reaction.

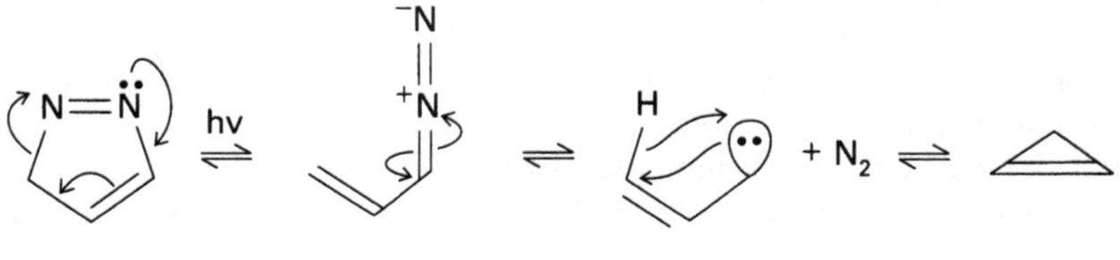

In this case, the 3H-pyrazole is stable to heat, and the reaction only proceeds on irradiation with light. Light may be used to initiate an extrusion reaction in which the carbonyl group is lost from a chain. Suggest the route for the conversion of R–CO–R to R–R and C=O.

In this case, the C–C bonds on each side of the carbonyl group cleave, in what is called a Norrish Type I reaction, and the resultant radicals then combine and so effect the extrusion of the CO molecule.

That ends this chapter on elimination reactions. It is apparent from the foregoing discussion that elimination reactions are far less well defined than either the substitution or addition reactions that we had studied previously. However, it was still possible to rationalise the production of the experimentally observed products. We will now look at sequential reactions, in which an addition reaction is followed by an elimination, or vice versa, and so leads to an overall substitution reaction.

14.7 Summary of elimination reactions

Elimination reactions are closely related mechanistically to substitution reactions, and are often in competition with them in the reaction mixture.

In a general elimination reaction, a small part of the molecule is lost so that two new molecules result. Usually, one part has a greater degree of unsaturation than was present initially in the original molecule. The principal types of elimination reactions are α-elimination (1,1-elimination), in which the eliminated parts are *geminal*; β-elimination (1,2-elimination), in which they are *vicinal*; γ-elimination (1,3-elimination), in which a three membered-ring is formed; and an extrusion reaction, in which there is no overall change in the degree of unsaturation, nor is there the formation of a ring.

The commonest mechanism for an elimination reaction is the E2 reaction, which stands for bimolecular elimination reaction. In this reaction, the removal of one group occurs at the same time as the other group is leaving, which results in an overall elimination. The leaving group is called a nucleofuge. In an elimination reaction, the attacking species is acting as a Brönsted–Lowry base, i.e. a hard base, and so attacks the harder centre, namely the α-hydrogen. This is in contrast to a substitution reaction, in which the attacking species acts as a nucleophile, i.e. a soft base, and so attacks the softer δ+ carbon rather than the harder adjoining hydrogen atom.

In the E2 mechanism, the eliminated groups are *anti-periplanar* to each other, and this gives rise to *anti*-elimination. In general, *anti*-elimination will result in an *erythro d,l* pair giving the *cis* isomer, while the *threo d,l* pair gives the *trans* isomer. The *threo* isomer, however, tends to react significantly faster than the *erythro* isomer, and this is called the eclipsing effect.

When there is a choice as to which β-hydrogen to eliminate, there are a number of guidelines. Bredt's rule states that a double bond will not be formed at a bridgehead, unless the S number is seven or greater and the double bond is in a ring that contains eight or more atoms. Furthermore, elimination usually occurs so as to form the conjugated derivative. In *anti*-E2 reactions that occur on an aliphatic chain, where the nucleofuge is anionic, or if elimination occurs from a six-membered ring, it is observed that the most highly substituted alkene is preferentially formed. This is called Zaitsev elimination and is generally true when the nucleofuge is a good leaving group, i.e. the C–X bond is weak and the resultant anion, X^-, is very stable. When the leaving group is neutral, then the opposite orientation is found, i.e. the double bond that is formed goes towards the least highly substituted carbons. This is called Hofmann elimination, and is generally favoured when the leaving group is poor and the C–X bond is strong. For example, in the Hofmann exhaustive methylation an amine is methylated with methyl iodide until a quaternary ammonium iodide is formed, and this is then treated with moist silver oxide to convert the iodide to the hydroxide, which, after elimination, results in a double bond forming between the least highly substituted carbons.

As the level of substitution is increased, so the degree of elimination increases. This is also the case if the concentration of the base is increased. The larger the base, the more elimination will occur. Decreasing the polarity of the solvent increases the proportion of elimination to substitution. If the solvent is changed from one which is polar and hydroxylic to another that is bipolar and aprotic such as DMF or DMSO, then the strength of the base increases. This results in more elimination occurring over substitution. Furthermore, in a solvent that has a low polarity, the stability of any charged intermediate would be reduced, and so bimolecular over unimolecular reactions would be favoured, and this in turn favours elimination. Raising the temperature favours the elimination pathway.

If there is a deviation from the *anti-periplanar* orientation of the eliminated groups then *syn*-elimination occurs. It is generally found that *syn*-elimination is slower than *anti*-elimination, and so it is usually only observed in significant amounts when the *anti*-elimination is significantly disfavoured. There are some exceptions to this general observation. For example, *syn*-elimination reactions have been found to be promoted by weakly ionising solvents, but only when the substituent that will form the leaving group is neutral. This is a result of ion pairing.

If it is possible for the regiochemistry of the formation of the double bond to be in doubt, then in a *syn*-E2 reaction, Hofmann elimination is nearly always followed.

The unimolecular 1,2-elimination reaction is given the label E1. This reaction bears many similarities to the SN1 reaction. The E1 and SN1 reactions often compete with each other. In the normal E1 mechanism, the heteroatom leaves first to form the carbonium ion intermediate. However, if the hydrogen leaves first, then this would result in a carbanion being formed. This reaction type is called the E1cB mechanism, which stands for unimolecular elimination conjugate base reaction. An example would be the reverse of a Michael addition reaction. Furthermore, a poor nucleofuge favours an E1cB reaction pathway, e.g. in the reverse of the cyanohydrin formation reaction.

The E1, E2 and E1cB mechanisms are all closely related. In all of these mechanisms, one group leaves with its electron pair (the nucleofuge), while the other group, which is often a proton, leaves without its electron pair. In the E1 mechanism, the nucleofuge leaves first; while in the E1cB, the proton leaves first, and thereafter the nucleofuge. In the E2 mechanism, the two groups depart simultaneously. All −M groups, such as carbonyl, when adjacent to the hydrogen atom, or even −I groups, such as the halogens, favour the E1cB route; while +M groups, or even +I groups like alkyl groups, disfavour this pathway.

In E1cB reactions, there is usually little ambiguity as to the regiochemistry of the resultant compound, because normally there is only one site from which an acidic hydrogen may be removed as a proton. Thus, in E1 reactions, Zaitsev elimination is preferred. Both alkyl or aryl groups at the α-carbon favour the E1 pathway. Further, generally, the more branching in the carbon chain, the more elimination occurs, because of the increased importance of alleviating steric strain.

The addition of an external base alters the reaction pathway to an E2 mechanism. At a low concentration of a weak base, the unimolecular pathway is favoured, so long as the solvent is ionising. Under these conditions substitution is favoured over elimination. The better the leaving group, the more the E1 pathway is favoured. Synthetically, the most important leaving group is the H_2O molecule. If the nucleofuge is Cl^-, Br^-, I^- or NR_3, then the reaction tends to proceed via the E2 route.

Under strongly basic conditions, the hydrogen that is *geminal* to the substituent that is to form the leaving group can be removed, so long as it is the most acidic. The resultant anion may then lose the nucleofuge to give a carbene, which then inserts its lone pair into the β-carbon C–H bond. This type of reaction is called the 1,1- or α-elimination, and is much less common than the 1,2-elimination.

Pyrolytic reactions often occur in the gas phase, without the addition of another reagent. For example, heating carboxylate esters readily produces an alkene and a carboxylic acid. There is a requirement for a *cis*-β-hydrogen in order for this reaction pathway to proceed, because the reaction proceeds via a six-membered cyclic transition state. This cyclic pyrolytic elimination is labelled the Ei mechanism, which stands for intramolecular, or internal, elimination. In those versions of the Ei mechanism that involve a four- or five-membered cyclic transition state, there is a requirement that all the atoms are co-planar or virtually so.

The Chugaev reaction involves the pyrolysis of a methyl xanthate, $R_2CH-O-CSSCH_3$ (which may be easily synthesised from the corresponding alcohol by treatment with NaOH and CS_2) to give an alkene, and $CH_3SC(=O)SH$, which subsequently fragments to CH_3SH and COS. This reaction illustrates the point that six-membered cyclic transition states may be formed with both *cis* and *trans* β-hydrogens, because of the capability of a six-membered ring to adopt a relatively strain-free puckered conformation of either type.

The Hofmann exhaustive methylation usually follows the *anti*-E2 bimolecular elimination pathway. Normally, the hydroxide anion removes the β-hydrogen and so initiates the E2 mechanism. If, however, the quaternary ammonium species is so highly hindered that this is not possible, then the hydroxide anion removes a proton on one of the methyl groups that forms part of the ammonium cation and reacts via a five-membered ring version of the Ei mechanism.

The other type of pyrolytic elimination reaction involves free radicals, for example in polyhalides and primary monohalides. These reactions involve the normal sequence for radical reactions, namely: initiation, propagation and termination steps.

An example of a γ-elimination reaction is the formation of a cyclopropane from 1-pyrazoline initiated either by heat or light. The initial biradical, which is formed by the loss of the nitrogen molecule, then closes the ring to give the cyclic product. The conversion of R–CO–R to R–R and C=O is called a Norrish Type I reaction. In this case, the C–C bonds on each side of the carbonyl group cleave, and the resultant radicals then combine and so effect the extrusion of the CO molecule.

15

Sequential Addition/Elimination Reactions

15.1 Introduction

We have now dealt with the three principal reaction mechanisms that occur in organic chemistry, i.e. substitution, addition and elimination reactions. The remaining chapters in this part of the book deal with some of the other types of reactions that may be encountered. In this chapter, we will look at the sequential addition/elimination reactions that result in an overall substitution occurring on the original substrate. The division between those simple substitution reactions that we have already studied and the sequential variety is sometimes rather fine. For example, in the S$_N$1 unimolecular substitution reaction, there is an initial loss of the leaving group to form the carbonium ion intermediate. This intermediate is then attacked by the incoming group to result in the final substituted product. This is normally considered under simple substitution reactions, because the intermediate is charged and only exists for a short time. Many of the reactions in this chapter have intermediates that are both neutral and possess a full complement of electrons, i.e. they are neither radical nor carbene intermediates. Some of these intermediates may be isolated; however, many are very reactive and so may only be produced *in situ*.

It is apparent that an addition followed by an elimination reaction will result in an overall substitution occurring. However, the order of the reactions may be reversed, i.e. an elimination reaction followed by an addition reaction will also have the same result. Examples of both types of sequence exist, and will be considered in turn.

297

15.2 Addition/elimination reactions

15.2.1 *Carbon/carbon double bonds*

In an earlier chapter, we dealt with SN2 reactions in aliphatic compounds. In that chapter it was stated that the normal nucleophilic pathway could not operate at carbon centres that were unsaturated, because of the impossibility of inversion of the configuration of the carbon involved. This does not mean that nucleophilic substitution at a vinyl carbon is impossible; if it is to occur, however, then there must be a different pathway by which it must proceed.

We have already seen in the chapter on nucleophilic substitution that carbon/carbon double bonds, e.g. vinyl compounds, do not readily undergo the simple addition of a nucleophile so as to form a tetrahedral intermediate. This is because, in such a case, the negative charge would have to reside on a carbon atom. However, if there are sufficient electron withdrawing groups attached to this carbon, then it is possible to form this intermediate. Once formed, it could then eliminate another group so as to re-form the carbon/carbon double bond, resulting in an overall substitution reaction. However, in practice this pathway is rarely observed. Instead, it is found that an alternative pathway is followed, in which the anionic intermediate that is formed initially picks up a proton, which results in a formal addition reaction as the first step.

Thus, for example, when 1,1-dichloroethene is treated with ArS$^-$, in the presence of a catalytic amount of ethoxide anion, the product is the 1,2-dithiophenoxy derivative. Suggest a route for this reaction.

The alkane and alkene intermediates have been isolated, which supports the contention that an addition/elimination sequence has been followed. After the first addition, which yields the saturated compound, two elimination reactions occur to give the alkyne derivative. This then undergoes a second addition to result in the final, rearranged, substitution product.

We will now turn our attention to SN2 reactions in aromatic compounds. It is again immediately apparent that any suggested mechanistic pathway must be different from that observed in saturated aliphatic compounds, because of the impossibility of inversion occurring at the unsaturated carbon. Furthermore, the approach of the nucleophile is inhibited by the delocalised electrons that are present above and below the aromatic ring. It is for this latter reason that electrophilic attack on an aromatic ring is usually the favoured pathway. However, nucleophilic attack on an aromatic ring does occur under some conditions. Suggest what may be the intermediate when the methoxide nucleophile, CH_3O^-, attacks the ethyl ether derivative of 2,4,6-trinitrophenol.

The incoming nucleophile can only approach from the side, i.e. above or below the plane of the ring. The extra negative charge is delocalised into the ring with the help of the strongly $-M$ nitro groups. This reaction pathway is labelled the SN2 (aromatic) mechanism. The intermediate that is formed is called a Meisenheimer complex, which may be isolated as a red crystalline solid with the counterion of the alkoxide. On the addition of some protic acid, the complex may fragment with the elimination of either the ethoxide or methoxide anion. If the methoxide anion is eliminated, then the starting material is regenerated; if the ethoxide anion is eliminated, then this would result in an overall substitution of an ethoxy group for the methoxy group. In this latter case, an overall substitution at the *ipso* position (i.e. the 1 position) in the aromatic ring has been achieved.

A further example is the attack of an amine on a 2,4-dinitrohalogenobenzene. For this reaction, suggest whether the fluoro derivative would react more or less quickly than the bromo derivative and why.

The fluoro derivative reacts about 1000 times faster than the bromo derivative, which is the opposite to that which would occur in a normal SN2 reaction. This is because the highly electronegative fluorine significantly increases the positive character of the carbon to which it is attached, and this increases the Coulombic attraction for the incoming nucleophile. Furthermore, the fluorine helps in stabilising the anionic intermediate. Indeed, 2,4-dinitrofluorobenzene is used to label the amino terminal of polypeptides, because this reaction occurs so readily. Furthermore, once it has reacted, the product is very resistant to subsequent hydrolysis.

15.2.2 *Carbon/oxygen double bonds*

As well as vinyl compounds undergoing sequential addition/elimination reactions, compounds containing a carbon/oxygen double bond, i.e. carbonyl compounds, often perform this type of reaction as well. There are, however, many significant differences between the vinyl and carbonyl pathways. Not least, is that in the latter case, there is often the possibility of substituting the carbonyl oxygen atom itself, rather than just substituting one of the groups joined to the carbon/oxygen double bond system.

Carbonyl compounds react via the tetrahedral mechanism. In this case, on the addition of the nucleophile, the negative charge is borne by the oxygen atom, which is an inherently more stable intermediate than the carbanion version. The tetrahedral intermediate may be isolated when a suitable substrate is used. An example is the formation of the cyanohydrin on the addition of a cyanide ion, which we studied in the chapter on addition reactions to a carbon/oxygen system.

If, instead of the addition product being stable, a subsequent elimination occurs, then there would be an overall substitution reaction. We have already looked at some of these reactions in the chapter on nucleophilic substitution. For example, in ester hydrolysis, the alkoxy group, –OR, is replaced by the hydroxyl group, –OH, so as to form the carboxylic acid. Some of the mechanisms for ester hydrolysis are simple substitution reactions. However, others proceed via a different route, an example of which also occurs in the hydrolysis of an acid chloride, $R(C=O)Cl$, when it is treated with water. Suggest what the product will be in this case, and the mechanism whereby this product might be reached.

In the first step, the water adds to the carbonyl group, and as the chloride anion is the best leaving group, it is the one that subsequently departs. This results in the re-formation of the carbon/oxygen double bond. The overall result is the substitution of one of the groups that was originally attached to the carbonyl carbon. This type of substitution reaction is common so long as the fourth bond to the carbonyl carbon is not from another carbon or hydrogen, i.e. the compound is not a ketone or aldehyde. This is the general mechanism whereby the various carboxylic acid derivatives, such as acid chlorides, acid anhydrides and amides, are interconverted.

So far, we have looked at the substitution of one of the groups that was attached to the carbonyl carbon. In addition to this type of reaction, there are many in which the carbonyl oxygen itself is substituted. For example, suggest what is the initial stable product that results from the addition of hydroxylamine, NH_2OH, to a general carbonyl compound.

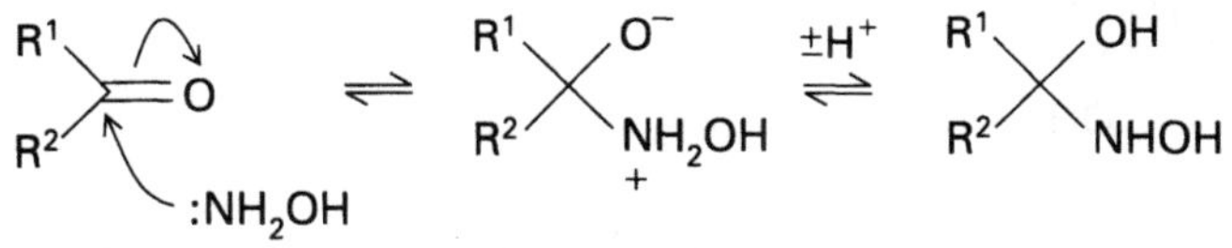

This intermediate is similar in nature to the cyanohydrin product that is formed by the addition of a cyanide anion to a carbonyl compound. However, in the latter case, no further reaction may occur, yet in this case there is the potential for further reaction. Suggest what are the next steps, and what is the final product.

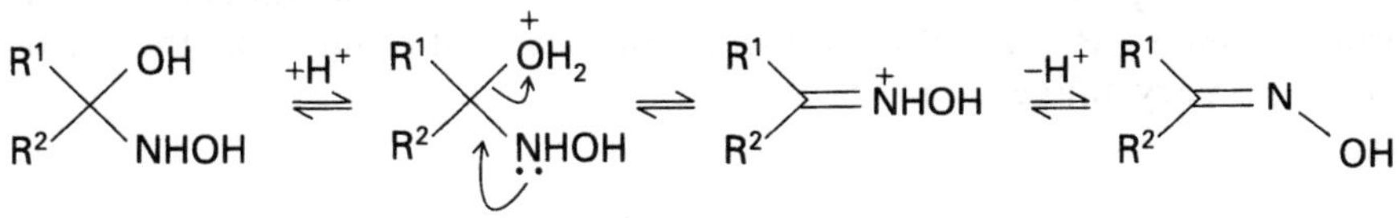

After the elimination of water, the carbonyl oxygen has been substituted for the =NOH group in the final product. This compound is called an oxime. This reaction can also be performed with base catalysis. This type of tetrahedral substitution reaction may be performed with many nitrogen derivatives, such as hydrazine, NH_2NH_2. Write down the product that is formed from the reaction of a carbonyl compound with hydrazine.

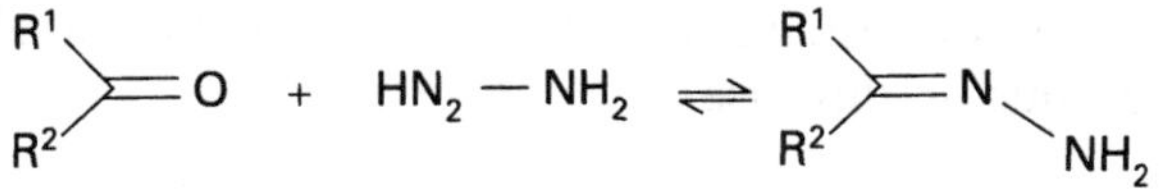

This type of compound is called a hydrazone, and is the first intermediate in a very important reaction called the Wolff–Kishner reduction. This reaction, along with the Clemmensen reduction (which we will discuss in more detail in the chapter on redox reactions), is one of the principal methods for reducing a carbonyl compound to the corresponding alkane, i.e. $R_2C{=}O$ to R_2CH_2. In the Wolff–Kishner method, the carbonyl compound is heated with hydrazine hydrate and a base, usually either sodium or potassium hydroxide. After the initial formation of the hydrazone, suggest what are the subsequent steps in this reduction.

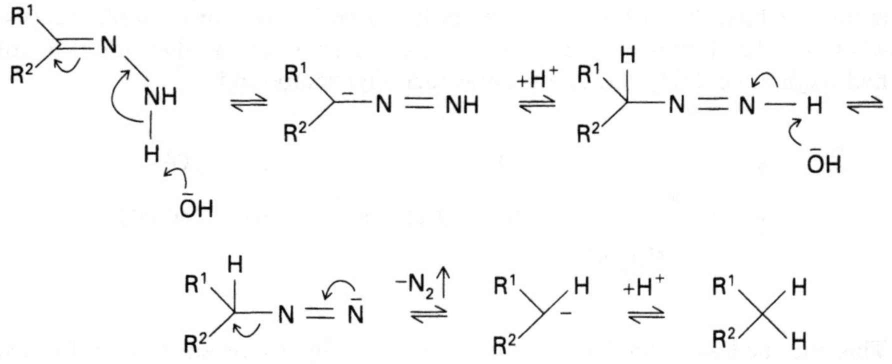

The loss of a nitrogen molecule helps drive this reaction. The reaction is usually performed in refluxing diethylene glycol, and is then called the Huang–Minlon modification.

Another important class of derivatives is formed from ammonia, NH_3, and primary amines, R^3NH_2. Write down the product that results from the reaction with a primary amine.

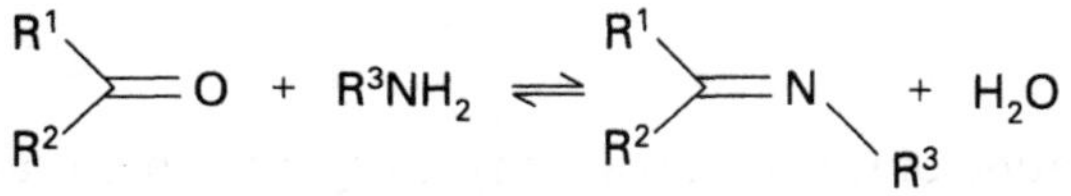

This class of compound is called the imines, and usually they are not very stable. However, if one of the substituents on the carbon is an aryl group then they become quite stable. Imines are also known as Schiff bases. For all of these derivatives, the original nitrogen compound possessed at least two hydrogen atoms, so that the nitrogen could end up doubly bonded to what was the carbonyl carbon. In contrast, if a secondary amine, NHR_2, is used, suggest what would be the product in this case.

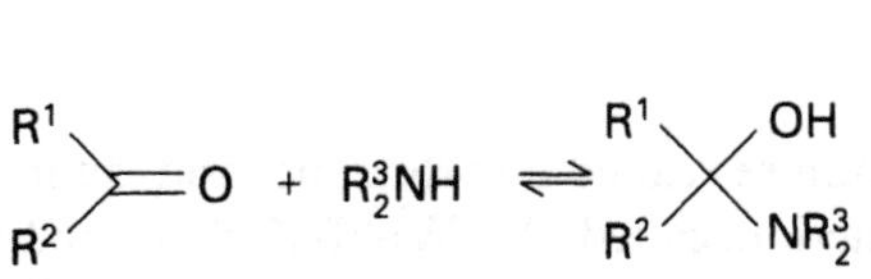

This intermediate is also capable of losing a water molecule. However, in this case, the nitrogen is not involved in a double bond in the final product. Suggest the mechanism for this dehydration, and so suggest what would be the final product.

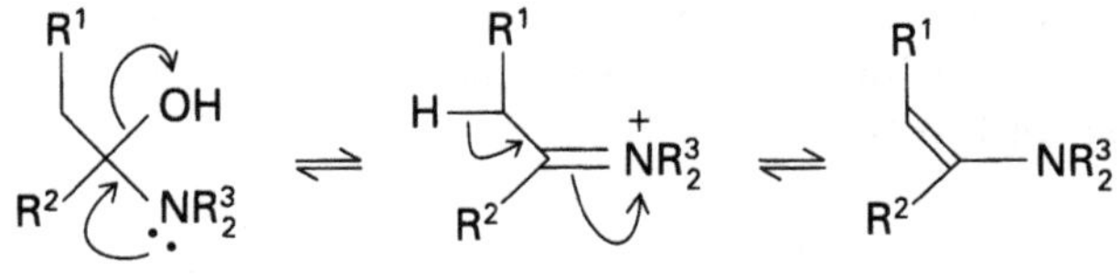

This compound has some similarities to the enol form of a carbonyl compound, and it is called an enamine. As this reaction is performed under reversible conditions, the thermodynamic isomer is always formed, i.e. the double bond is formed on the most highly substituted side. These compounds are of great synthetic importance. Furthermore, their use complements the use of the anions formed from carbonyl compounds. This is because enolate anions are usually produced under strongly basic, i.e. irreversible, conditions, and so the proton that is normally removed is the kinetically favoured one. Hence, the hydrogen atom that is removed in the formation of the enolate anion may be different from the one that is eliminated under thermodynamic conditions prevalent in the formation of the carbon/carbon double bond of the enamine.

Carbonyl compounds are capable of undergoing another completely different sort of reaction, in which the carbonyl oxygen is replaced, not by a heteroatom as we saw above, but by an sp^2 carbon atom. When triphenylphosphine is treated with a secondary aliphatic halide, a normal substitution reaction occurs, resulting in the replacement of the halide by the phosphorus compound. This phosphonium salt adduct may then be treated with a very strong base, such as phenyllithium, to produce an ylid. An ylid is a compound in which there is a positive charge adjacent to a negative charge. Write down this reaction sequence.

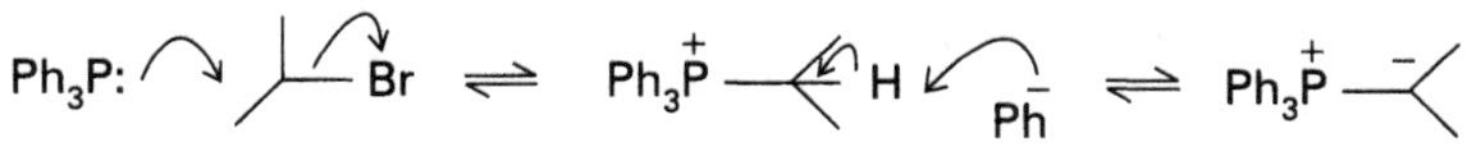

The ylid may then be used to attack a carbonyl compound. Suggest how this occurs, and the nature of the intermediate.

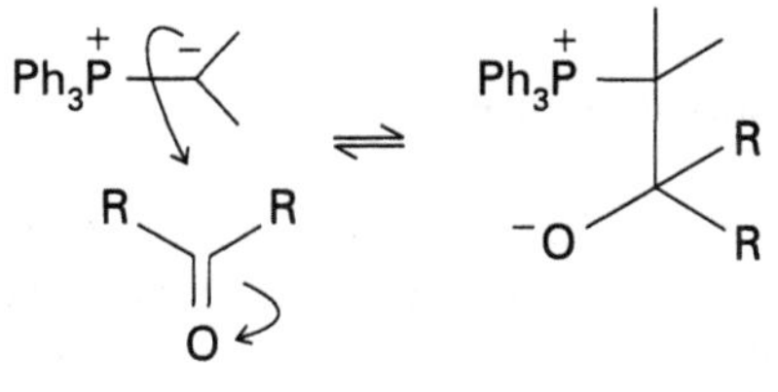

The zwitterionic intermediate, which is called a betaine, then forms a four-membered cyclic intermediate. This is possible because the phosphorus is a third row element, and thus has accessible d orbitals, and so is not limited by the

Octet rule as is the carbon atom. This intermediate, which is called an oxaphosphetane, then fragments to form the desired compound. Write down the mechanism for this step.

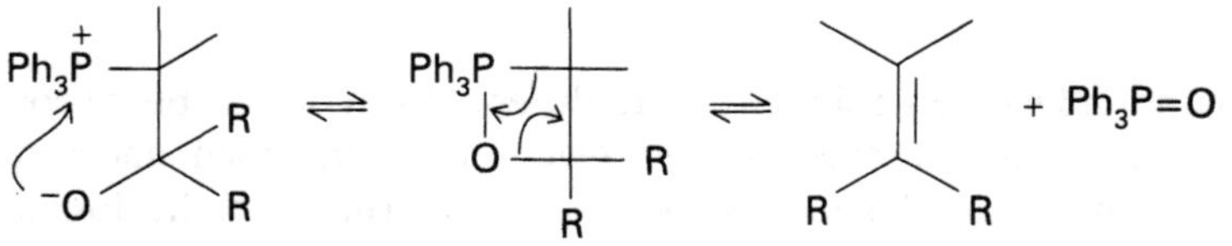

The complete synthetic pathway is called the Wittig reaction, and it is a very useful synthetic tool, because the position of the carbon/carbon double bond is known. This method may also be used to produce a carbon/carbon double bond in a position that would be very difficult to achieve by any other means, e.g. *exo*-cyclic bonds, or 1,4-dienes, i.e. unconjugated.

The attack by the incoming reagent does not always occur directly at the carbonyl carbon atom. It is possible for the attack to occur at the conjugated position. You will recall that this was also possible for nucleophilic substitution. For example, suggest what is the intermediate when tetrachloroquinone reacts with a hydroxide anion.

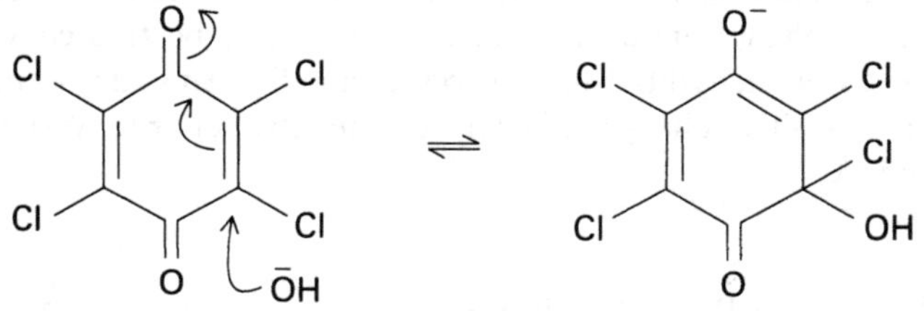

Now that we have studied sequential addition/elimination reactions, we will look at a few reaction sequences that occur in the opposite order, i.e. elimination followed by addition.

15.3 Elimination/addition reactions

The first reaction we will study in this section is similar to one that we looked at in the previous section. We saw earlier a reaction that involved the substitution of 1,1-dichloroethene by the ArS^- anion, using a catalytic amount of ethoxide anion. If the concentration of the ethoxide anion is increased, and the substrate altered to *cis*-1,2-dichloroethene, then the first reaction is an elimination step, instead of an addition step. Suggest what the subsequent steps might be, and the geometry of the final product.

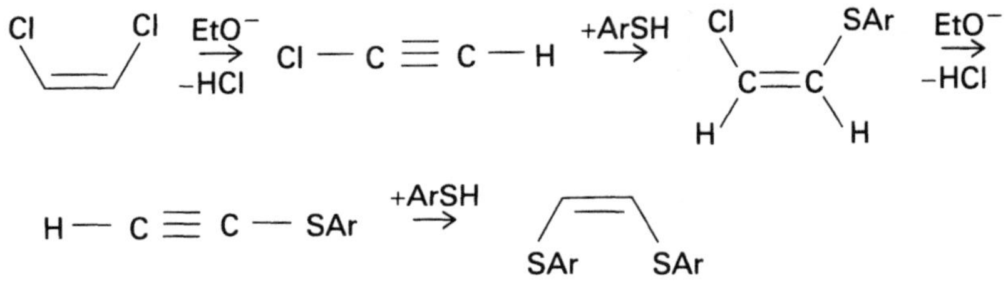

Each elimination is followed by an addition reaction to result in the overall production of the *cis*-1,2-dithiophenyloxy derivative. There is overall retention of the configuration, because both the elimination and the addition reactions are *anti* in nature. This reaction pathway is called the elimination/addition mechanism.

In aromatic systems, a rather similar sequence is found under certain extreme conditions. Thus, chlorobenzene may be converted to aniline when treated with sodium amide in liquid ammonia. If this reaction is performed upon *p*-chloromethylbenzene, then two compounds are produced. Both the *meta* and the expected *para* amino derivatives may be isolated. The substitution in the unexpected position is called *cine* substitution. Suggest a mechanism whereby this result may be explained.

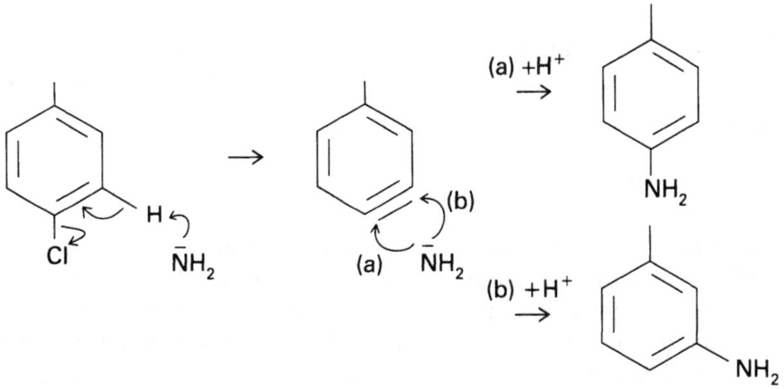

The NH_2^- anion is a very strong base and removes a proton from the aromatic ring. This anion then eliminates a chloride anion, to leave an aromatic species that also contains a triple bond. This is called an aryne intermediate. This may be attacked by another NH_2^- anion (or possibly, by NH_3) in either the *meta* or *para* positions to give, after protonation, the two products. This pathway is called the benzyne mechanism.

Benzyne intermediates may be formed in a large number of different ways. If *o*-aminobenzoic acid is treated with nitrous acid, the resultant diazonium salt may decompose to form benzyne itself. Write out this reaction sequence.

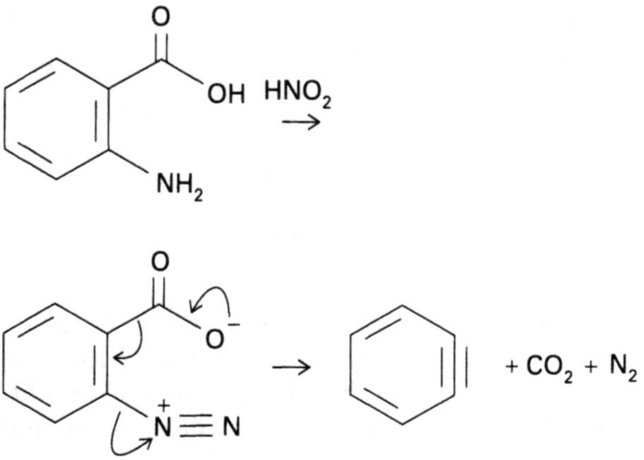

Here, two gaseous products are formed, which help drive the reaction.

Another method of producing benzyne is from the decomposition of 1-aminobenzotriazole, after it has been oxidised with lead tetraacetate. The oxidation converts the amino group to the nitrene intermediate, and this fragments. Even though there is some doubt over whether nitrenes can exist as true intermediates in these circumstances, suggest a mechanism that utilises their involvement in order to achieve the desired result.

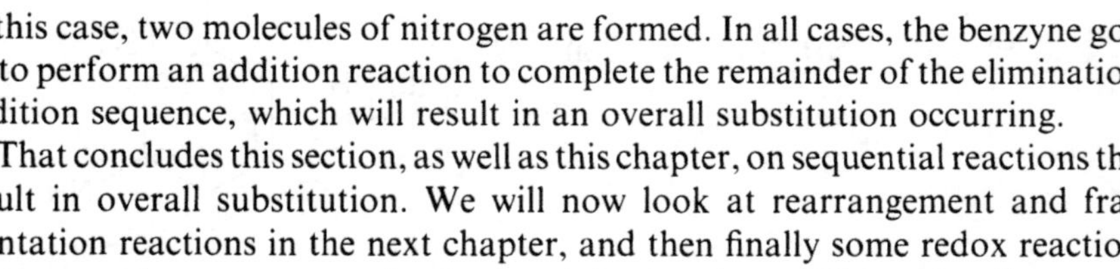

In this case, two molecules of nitrogen are formed. In all cases, the benzyne goes on to perform an addition reaction to complete the remainder of the elimination/addition sequence, which will result in an overall substitution occurring.

That concludes this section, as well as this chapter, on sequential reactions that result in overall substitution. We will now look at rearrangement and fragmentation reactions in the next chapter, and then finally some redox reactions that cannot be more conveniently categorised under a different heading.

15.4 Summary of sequential addition/elimination reactions

Nucleophilic substitution at an sp^2 carbon usually proceeds via an addition/elimination sequential pathway. Thus, the reaction of ArS^- to, say, 1,1-dichloroethene, in the presence of a catalytic amount ethoxide anion, proceeds via an addition followed by a double elimination, with finally a further addition to yield the 1,2-dithiophenoxy derivative.

Nucleophilic attack on an aromatic ring proceeds via the S$_N$2 (aromatic) mechanism. For example, the addition of the nucleophile CH_3O^- to 2,4,6-trinitrophenetole yields initially the Meisenheimer complex. On the addition of a protic acid, the complex may fragment with the elimination of either the ethoxide or methoxide anions. If the ethoxide anion is eliminated this results in an overall substitution of an ethoxy group for methoxy group at the *ipso* position of the aromatic ring. Another example would involve the use of 2,4-dinitrofluorobenzene, which is used to form the derivative of the amino terminal of polypeptides.

Nucleophilic substitution of compounds containing carbon/oxygen double bonds, i.e. carbonyl compounds, proceed via the tetrahedral mechanism. The tetrahedral intermediate may be isolated when a suitable substrate is used, e.g. the addition of HCN to a carbonyl compound yields a cyanohydrin. If one of the groups attached to the carbonyl carbon is eliminated, then an overall substitution occurs, e.g. in the hydrolysis of an acid chloride.

In contrast, the carbonyl oxygen can itself be substituted. For example, when hydroxylamine, NH_2OH, is added to a general carbonyl compound, the tetrahedral intermediate may react further to eliminate a water molecule, yielding the oxime, i.e., in effect substituting the $=NOH$ group for the $=O$ substituent. A similar reaction is exploited in the Wolff–Kishner reduction of carbonyl compounds to the corresponding alkane, i.e. $R_2C=O$ to R_2CH_2. In this method, the carbonyl compound is heated with hydrazine hydrate and a base, usually sodium or potassium hydroxide, to form initially the hydrazone. If the reaction is performed in refluxing diethylene glycol, it is called the Huang–Minlon modification.

Imines, or Schiff bases, are formed by the reaction between ammonia or an amine and a carbonyl compound. If the nitrogen has an aryl substituent, then the product is quite stable. If a secondary amine is used, then water cannot be eliminated as before to yield the carbon/nitrogen double bond. However, water can still be eliminated, but this time in the side chain to yield the enamine. The thermodynamic isomer is always formed (i.e. usually the imine over the enamine).

Further, the carbonyl oxygen can be replaced, instead of by a heteroatom, by a carbon atom, for example, in the Wittig reaction.

Conjugate attack is also possible.

If *cis*-1,2-dichloroethene is treated with the ArS^- anion, and at a high concentration of ethoxide ion, then the first reaction is an elimination step, instead of an addition step. There is overall retention of the configuration, because both the elimination and the addition reactions are *anti* in nature. This reaction pathway is called the elimination/addition mechanism.

In aromatic systems, substitution can occur under extremely basic conditions. Thus, *p*-chloromethylbenzene is converted to a mixture of the *meta* and the expected *para* amino derivatives when treated with sodium amide in liquid ammonia. The substitution at the former, unexpected, position is called *cine* substitution. The mechanism proceeds via an aryne intermediate, and is called the benzyne mechanism.

Rearrangement and Fragmentation Reactions

16.1 Introduction

Sometimes a reaction occurs in which the final product is totally different in structure from the starting materials. This may have resulted from a rearrangement of the carbon skeleton of the original molecule. Often such reactions appear at first sight to be very complicated. This, however, is a false impression. Rearrangement reactions follow the same basic principles that have been already illustrated in the preceding chapters. Thus, rearrangement reactions require a source of excess electrons and a sink to receive them; there is still the overall 'push-me–pull-you' picture that characterises curly arrow mechanistic chemistry.

We will also consider fragmentation reactions in this chapter, i.e. those reactions in which the molecule falls apart to yield smaller units. Again, these reactions often appear difficult at first sight because the skeleton of the molecule has been disrupted. However, yet again, this type of reaction follows the same basic principles that have been already covered.

For the purpose of this chapter, the rearrangement reactions are divided by the nature of the principal bond that is being broken in the first part of the reaction, e.g. carbon/carbon, carbon/nitrogen or carbon/oxygen. This chapter gives a large number of examples, because this is the easiest way to become familiar with this type of reaction, and so to become confident in drawing the associated mechanisms.

16.2 Carbon/carbon rearrangements

16.2.1 *Allylic rearrangements*

An important and common rearrangement reaction involves the relocation of a double bond in an alkene derivative. This rearrangement does not actually involve the carbon skeleton isomerising; rather, the location of the unsaturation within the existing carbon skeleton merely changes. What has 'rearranged' is the position of the carbon/substituent bond between the starting material and the product. Write down the carbonium that initially results from the loss of a bromide ion from 1-bromobut-2-ene.

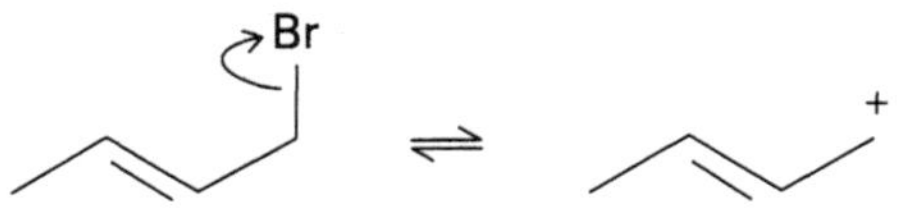

This carbonium ion is stabilised by the conjugated double bond, and so it is a delocalised carbonium ion. The delocalised cation may now be attacked in one of two possible positions by an incoming nucleophile. Suggest what are these two positions, and so identify the two substitution products.

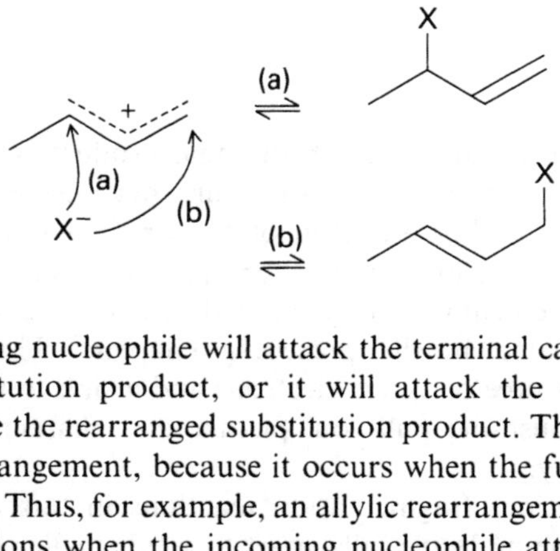

Either the incoming nucleophile will attack the terminal carbon to give the un-rearranged substitution product, or it will attack the C-3 of the original compound, to give the rearranged substitution product. This type of reaction is called allylic rearrangement, because it occurs when the functional group is in an allylic position. Thus, for example, an allylic rearrangement will take place in substitution reactions when the incoming nucleophile attacks the sp^2 carbon atom rather than the sp^3 carbon to which the leaving group is originally attached. This variation is indicated by a prime sign ($'$), i.e. S_N1' or S_N2'. In the case of the S_N1' reaction, it usually follows that the longer the half life of the intermediate carbonium ion, the greater the amount of conjugated attack that will occur. This will result in a smaller product spread, i.e. the un-rearranged to rearranged products will be produced in more equal amounts.

When $CH_3CH=CHCD(OH)CH_3$ is treated with thionyl chloride, $SOCl_2$, the monochloro product has all of the deuterium in the vinylic position. Suggest a mechanistic pathway that accounts for this observation.

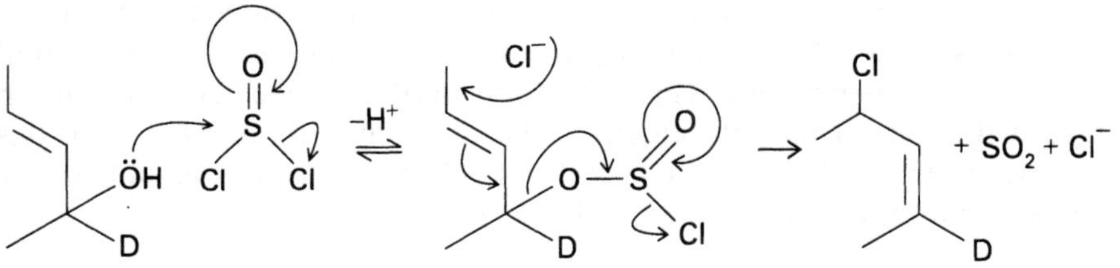

You will recall that thionyl chloride usually reacts via an SNi mechanism. In this case, the intermediate is attacked by the displaced chloride ion at the 3-position, i.e. in an SN2′ manner, before the sulphur complex has had a chance to fragment in order to allow internal attack by the second chloride ion.

There is an alternative mechanism that would also account for the observed placement of the deuterium atom. Suggest what this might be.

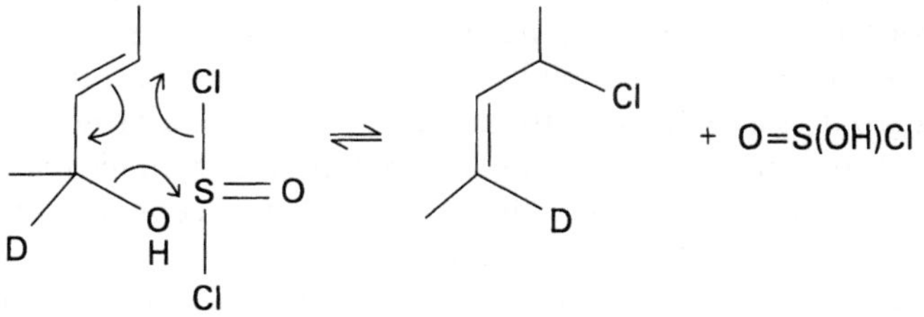

In this case the thionyl chloride reacts in a concerted manner in a pericyclic reaction to give the desired product directly along with O=S(OH)Cl, which decomposes to form HCl and SO_2. To distinguish between these two possible routes would require more information. However, it is salutary that a plausible alternative route can be readily postulated. One should never be too dogmatic that one's pet mechanism actually occurs in practice.

If this chloro product is refluxed with ethanoic acid, two products are formed in approximately equal proportions. In one, the deuterium is still in the vinylic position; while in the other, the deuterium label is on the carbon attached to the oxygen. Suggest a pathway that accounts for this observation.

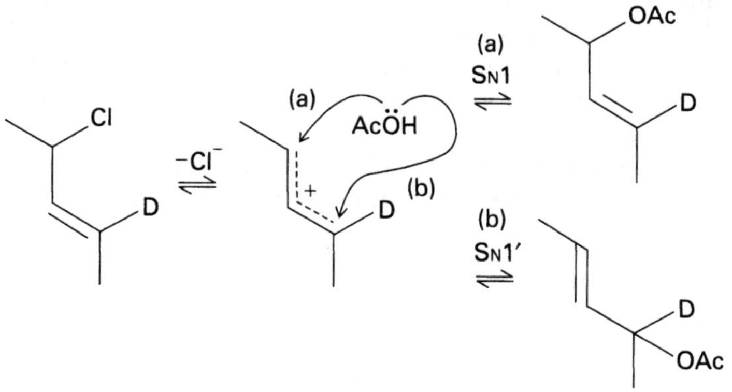

In this case, the ethanoic acid is a solvent with a high dielectric constant, but it is also a weak nucleophile, and so it provides ideal conditions for a long lived carbonium ion intermediate. This is then statistically attacked at either carbon atom at the ends of the delocalised 1,3-dimethyl allyl cation. As a result this gives rise to the 1 : 1 ratio of substituted products, i.e. equal amounts of the products that result from the SN1 and SN1' reactions.

16.2.2 *Whitmore 1,2-shifts*

Rearrangement reactions that involve the carbon skeleton isomerising usually involve a 1,2-shift, i.e. the migrating group, M, goes from the migration origin, A, to the adjacent atom, which is called the migration terminus, B. The vast majority of rearrangements are nucleophilic in character. This means that the migrating group moves with its electron pair. Such rearrangements may also be called anionotropic. If the migrating group moves without its electron pair, then it is described as an electrophilic or cationotropic rearrangement; if the migrating group is a hydrogen cation, then it is labelled a prototropic reaction. Radical rearrangements are also possible, but like electrophilic rearrangements, they are rare. In this chapter, we will consider only nucleophilic rearrangements, because they are by far the most common.

In the actual migration step, it is possible for the migrating group to become detached from the original molecule and join another molecule before it moves to the migration terminus. Such a reaction is called an intermolecular rearrangement. If the migrating group does not become detached, then an intramolecular rearrangement occurs. These two types may be distinguished by cross-over experiments, where two similar sets of reagents are allowed to react in the same vessel; the occurrence of intermolecular rearrangements can be deduced from the presence of a product that contains a group from the other starting material.

When there is a choice of which group may migrate, it has been observed that aryl groups tend to migrate more easily than alkyl groups.

Nucleophilic rearrangements formally occur in three steps. These steps may be illustrated by the rearrangement of a carbonium ion. A typical example is provided by the neopentyl system. It will be recalled that the SN2 mechanism at such a centre is slow, because of the steric hindrance caused by the *t*-butyl group in the β-position. Furthermore, normal elimination is impossible, because there is no β-hydrogen.

Write down the first formed carbonium ion that is derived from the protonation, and subsequent dehydration, of neopentyl alcohol.

This is the first step, i.e. formation of the initial carbonium ion. These ions may be formed in many different ways: protonation of an alcohol; loss of nitrogen from a diazonium ion; addition of a proton to an alkene; or the interaction of a Lewis acid with a halide.

The resultant carbonium has only six electrons, i.e. it is an open sextet. This provides the electron deficient sink that will accommodate the excess electrons that are at present contained by a nucleophilic migratory group. Suggest what will be the next step.

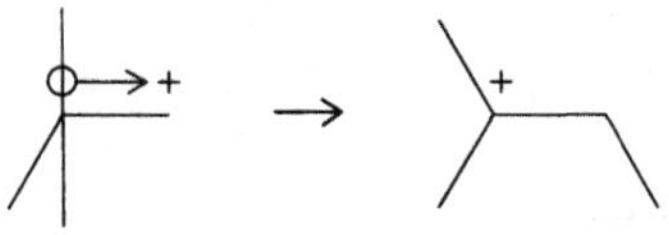

The second step is the actual migration. The migration origin becomes electron deficient, while the migration terminus regains its octet. In simple molecules, it is usually clear which bond is migrating, but in more complicated molecules it may not be immediately obvious. In order to avoid ambiguity when drawing the reaction mechanism, a double headed arrow with a small ring at its tail is used. The ring encloses the bond that is migrating.

Suggest what will occur on the final step. There are two common endings.

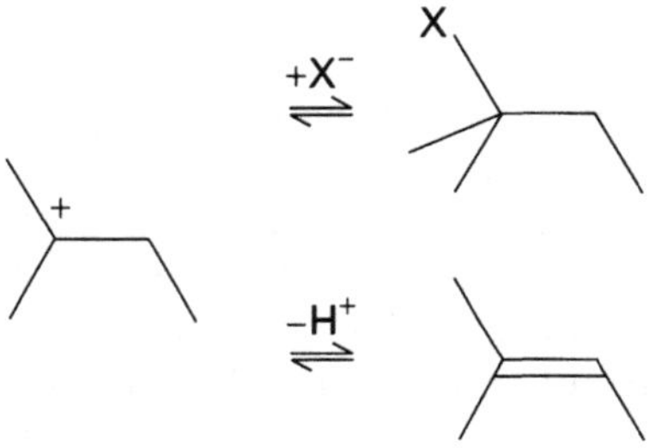

Either the migration origin will complete its octet by combining with a incoming nucleophile, i.e. to result in overall nucleophilic substitution; or there will be the loss of a proton to result in overall elimination.

This rearrangement of a carbonium ion is called the Wagner–Meerwein rearrangement, and proceeds via a Whitmore 1,2-shift. In this case, there is no obvious electron source that might provide the push; however, the open sextet of the migration terminus does provide some pull. The stability of the product, which is what drives this rearrangement, is due to the possibility of hyperconjugation now occurring with the electrons in the β-bonds.

Wagner–Meerwein and allylic rearrangements are very common whenever a carbonium ion is formed. For example, in the simple Friedel–Crafts alkylation, the alkyl group always tends to rearrange to give the most stable carbonium ion, which then adds to the aromatic ring. In order to prevent this rearrangement, the Friedel–Crafts acylation process is used instead. Here, the acyl cation does not undergo the Wagner–Meerwein rearrangement, and so the resulting substituent on the aromatic ring has the same carbon skeleton as the starting material. Another advantage is that there is greater control of the resultant product, because the acyl derivative de-activates the aromatic ring, and so usually only one group is added to the ring. In the case of normal alkylation, the ring becomes more activated with the addition of each alkyl group, and so poly-alkylation is favoured.

There are other possible systems in which there is no obvious excess of electrons to provide the impetus for migration. Instead, the driving force comes from the stability of the final product. Thus, for example, a cyclohexadienone system that has two alkyl groups in the 4-position undergoes a rearrangement on treatment with acid. Suggest what will be the product, and also what is the driving force behind this reaction.

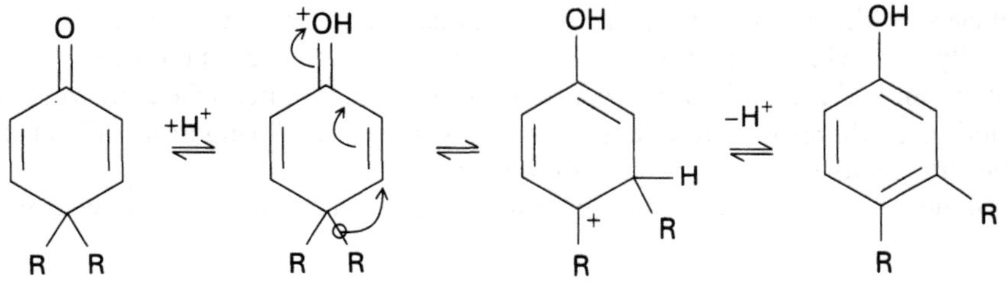

This reaction is called the dienone–phenol rearrangement. The driving force in this case is the formation of the aromatic ring.

In both the Wagner–Meerwein and dienone–phenol rearrangement reactions, there was no clear source of excess electrons to provide the impetus for migration. However, in the next example, the lone pair of electrons located on an oxygen atom are clearly capable of acting as the driving force behind the migration. *Vicinal*-diols, e.g. pinacols in symmetrical cases, rearrange in the presence of acid to give the related carbonyl compound. Thus, write down the mechanistic pathway for the rearrangement of pinacol, $(CH_3)_2C(OH)C(OH)(CH_3)_2$, under acidic conditions.

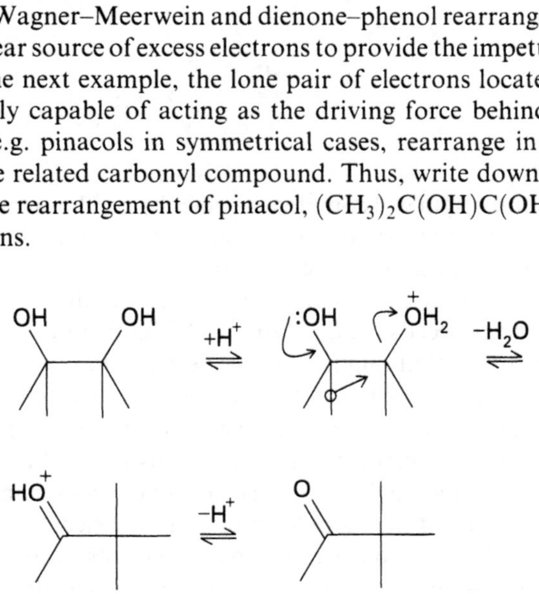

The product is called pinacolone, and so this reaction is called the pinacol–pinacolone rearrangement. There are many variations of this reaction. They all depend upon placing a positive charge on the carbon that is adjacent to the carbon bearing the hydroxyl group. In the original example this was achieved by the departure of the protonated hydroxyl group that was adjacent to the hydroxyl group providing the excess of electrons. So, for example, a variation would be to treat β-amino alcohols with nitrous acid. This version is called the semipinacol rearrangement. Write down the mechanistic pathway for this reaction when $Ph_2C(OH)CPh(CH_3)NH_2$ is so treated. What are the stereochemical consequences at the migration terminus?

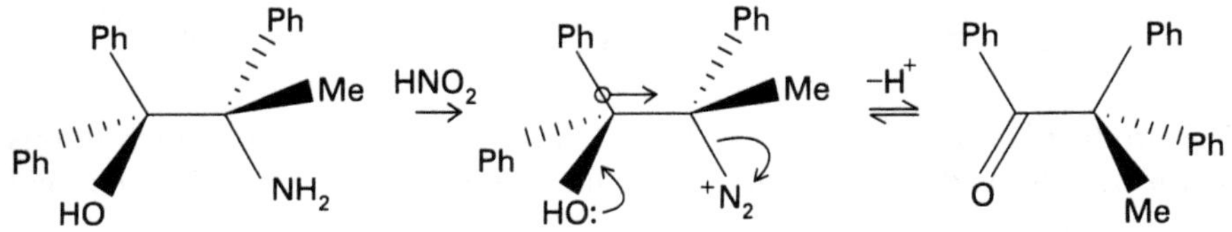

There is inversion of the configuration at the migration terminus in this reaction. What does this indicate about the first two steps of the migration process, i.e. the formal formation of the carbonium ion and the subsequent migration?

In this case, as no racemisation occurred at the migration terminus, then there cannot have existed an independent carbonium ion. Instead, the departure of the leaving group is assisted by the migrating group moving across simultaneously, in the same way that a neighbouring group would interact. However, unlike a neighbouring group interaction, the migrating group cleaves from the migration origin and bonds to the migration terminus. This results in an inversion of the configuration at the migration terminus similar to an internal S_N2 type mechanism. This anchimeric assistance often results in an increased rate of reaction.

The semipinacol reaction may be modified so that a ring expansion takes place. For example, draw the mechanism for the ring expansion that occurs when 1-aminomethylcyclopentanol is treated with nitrous acid.

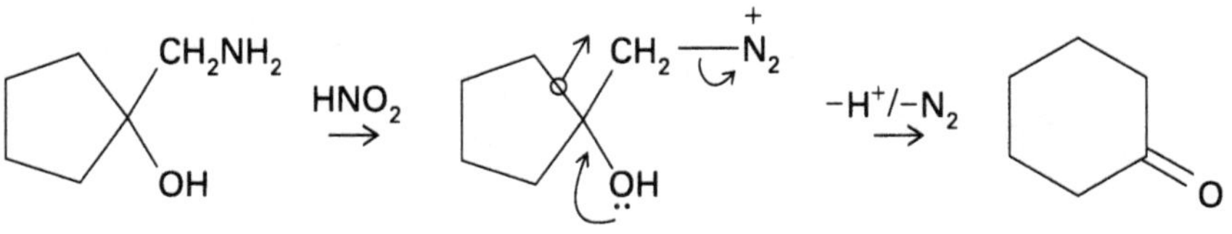

This version is called the Tiffeneu–Demyanov ring expansion reaction.

Earlier, we saw that *vicinal* diols perform a rearrangement when treated with acid. If an α-diketone is treated with a base, it will undergo a rearrangement.

Suggest what will be the reaction pathway when PhCOCOPh, benzil, is treated with sodium hydroxide solution.

The product is the sodium salt of the α-hydroxycarboxylic acid, which in this case is benzilic acid. Accordingly, this reaction is called the benzil–benzilic acid rearrangement.

The closely related α-halogenoketone, when treated with an alkoxide anion, undergoes a rearrangement. Suggest what is the pathway for this mechanism, and so deduce what will be the product.

This reaction is called the Favorskii rearrangement. Again, there are many variations; for example, instead of an alkoxide anion, a hydroxide anion or even an amine may be used, in which case the salt of the carboxylic acid or the amide will be formed, respectively. This reaction may also be used so as to result in a ring contraction. Write down the mechanism for the reaction between an alkoxide anion and 2-chlorocyclohexanone.

The product is the expected ester.

The following example is rather unusual in that superficially it appears as if a ring expansion has occurred, namely a five-membered ring expanding to a six-membered ring. However, on a closer examination, there has been a disruption of two small rings to give one larger ring. Take the *geminal* dibromide compound, 1,1-dibromobicyclohexane [3.1.0], and predict what will occur when it is treated with aqueous silver nitrate.

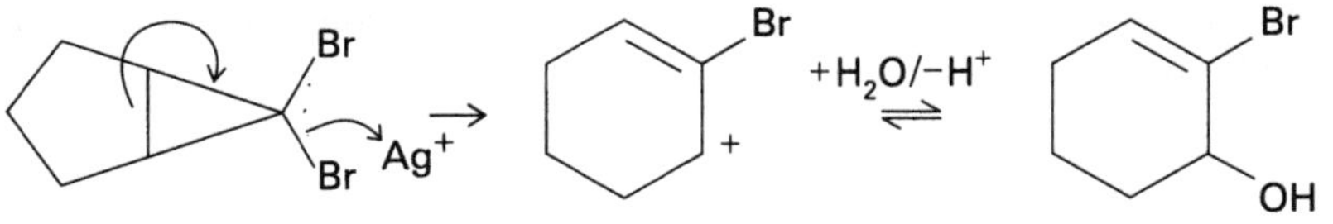

It is possible to view this reaction as involving a ring contraction, because the three-membered ring has formed an allylic cation, i.e. the double bond can be considered as a two-membered ring system, which is then attacked by a solvent molecule. This is an example of a cyclopropyl/allylic cation rearrangement, which we first encountered in the chapter on nucleophilic substitution reactions.

In an earlier chapter, we saw how to reduce the carbon chain by one methylene unit using the Hunsdiecker reaction. The opposite of this, namely the extension of the carbon chain by one unit, may be achieved with the Arndt–Eistert synthesis. In the first step, an acyl halide is treated with diazomethane to form the α-diazo ketone. This is then treated with water and silver oxide. The resultant product is the free acid. If an alcohol is used instead of water, then the related ester is formed. Suggest what is the pathway for this reaction.

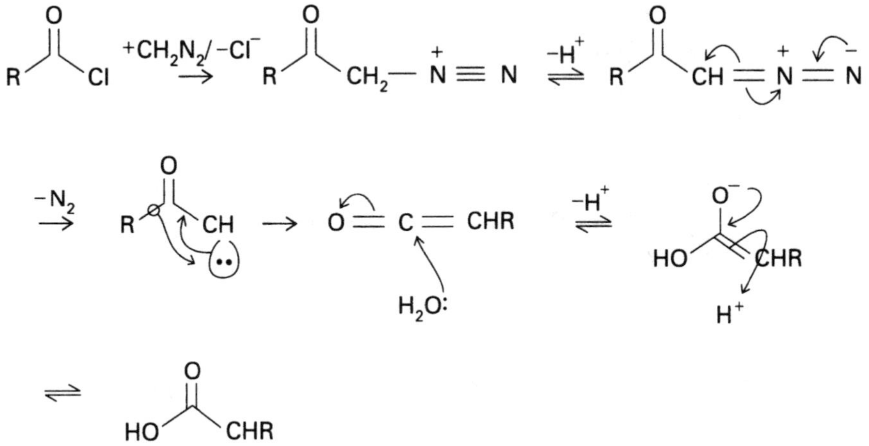

Note that in this reaction, the migration step occurs at the same time as the completion of the octet of the atom at the migration origin. The electron deficient carbon species is a carbene, which after rearranging forms the ketene, $RCH=C=O$. This is the best way of extending a chain by one unit if the carboxylic acid is available. The process of extension in this manner is called homologation. Again, like the Favorskii rearrangement, it may be used to effect a ring contraction if one starts with the appropriate α-diazo ketone.

317

The actual rearrangement part of the Arndt–Eistert synthesis, which involves insertion of the carbene to give the ketene, is called the Wolff rearrangement. The ketene then undergoes addition of water to yield the final desired product after suitable tautomerisation has occurred.

We will now look at a few reactions in which there is migration from a carbon atom to a nitrogen atom.

16.3 Carbon/nitrogen rearrangements

In the Hofmann rearrangement, when an unsubstituted amide is treated with sodium hydroxide and bromine, the corresponding amine with one fewer carbon atom is produced. The first intermediate is the N-bromo amide, which then undergoes the three steps of the rearrangement reaction simultaneously. Suggest the pathway for this reaction, and how the final product is achieved.

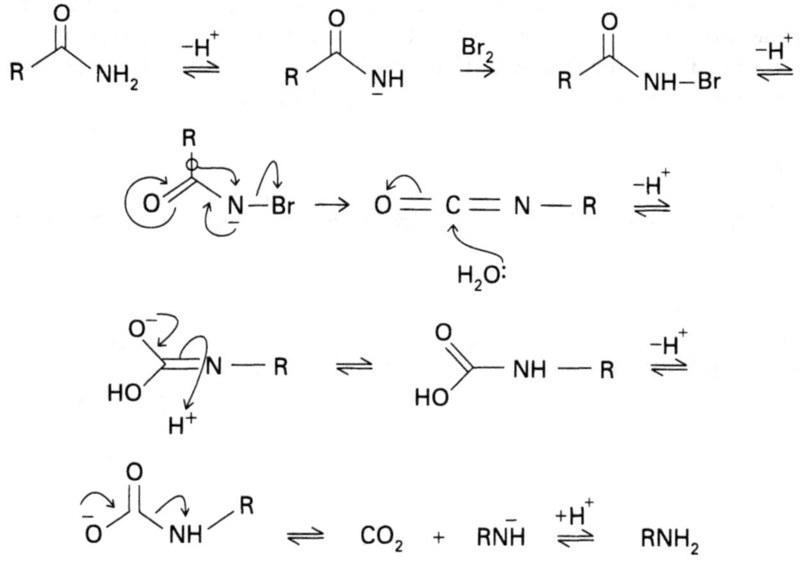

The product of the rearrangement reaction is the isocyanate, which is similar to the ketene intermediate that was obtained in the Wolff rearrangement. In this case, the isocyanate is hydrolysed under the reaction conditions to yield eventually an amine. If $NaOCH_3$ is used instead of NaOH then an carbamate, $RNHCOOCH_3$, is formed instead of an amine. This reaction is sometimes called the Hofmann degradation, but this causes confusion with the Hofmann exhaustive methylation, in which a quaternary ammonium hydroxide is cleaved thermally to yield an alkene, because this latter reaction is also sometimes called the Hofmann degradation. It is better to avoid the term Hofmann degradation altogether.

The Hofmann rearrangement proceeds quite smoothly, even when the amide carbon is joined to a bridgehead carbon, as in 1-norbornyl carboxyamide, which may be easily converted to 1-aminonorbornane. The ease of this reaction indicates that the migrating group moves with retention of configuration, because in this case there is no possibility of inversion or racemisation. If inversion were required, then the reaction would be impossible to perform on such a substrate.

There are a multitude of variations on the basic Hofmann rearrangement. The simplest is the pyrolysis of acyl azides to isocyanates, which only involves the central rearrangement step. Write down the pathway for this reaction.

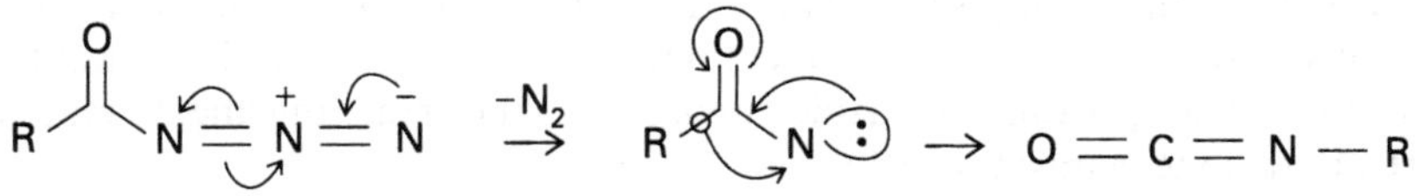

This is called the Curtius rearrangement. There is no evidence for the existence of the free nitrene in this reaction, unlike in the Wolff rearrangement where there is some evidence for the existence of the free carbene, so probably the steps are concerted in this case. The driving force for this reaction is the production of molecular nitrogen. This provides so much energy, that it is possible to harness it to expand an aromatic ring. Indicate the pathway that is followed when an aryl azide is heated with phenylamine.

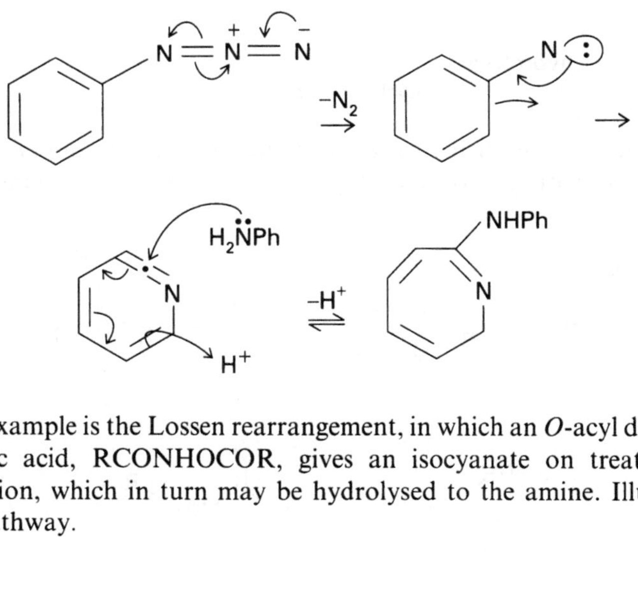

A further example is the Lossen rearrangement, in which an *O*-acyl derivative of hydroxamic acid, RCONHOCOR, gives an isocyanate on treatment with hydroxide ion, which in turn may be hydrolysed to the amine. Illustrate this reaction pathway.

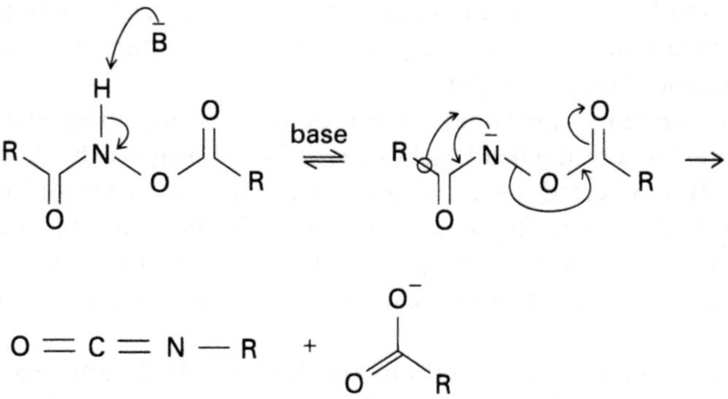

In the Schmidt rearrangement, the treatment of a carboxylic acid with hydrogen azide (hydrazoic acid) also gives the amine, via the isocyanate, when catalysed by an acid, such as sulphuric acid. The first step is the same as the AAc1 mechanism to form the acylium cation, and so is favoured by hindered substrates. The protonated azide undergoes the rearrangement reaction. Illustrate this mechanism as well.

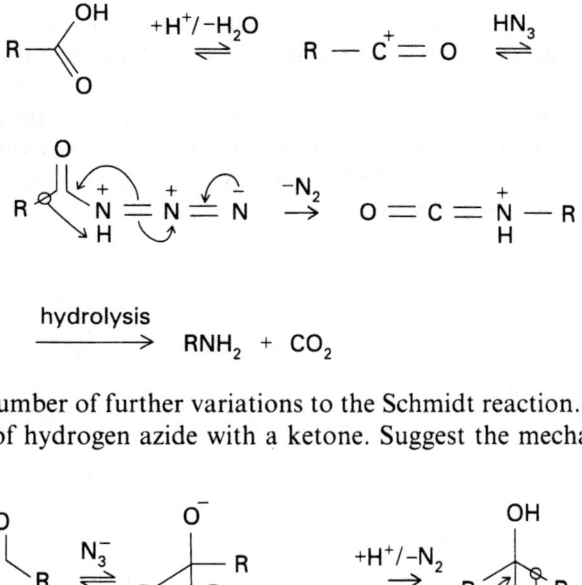

There are a number of further variations to the Schmidt reaction. For example, the reaction of hydrogen azide with a ketone. Suggest the mechanism for this reaction.

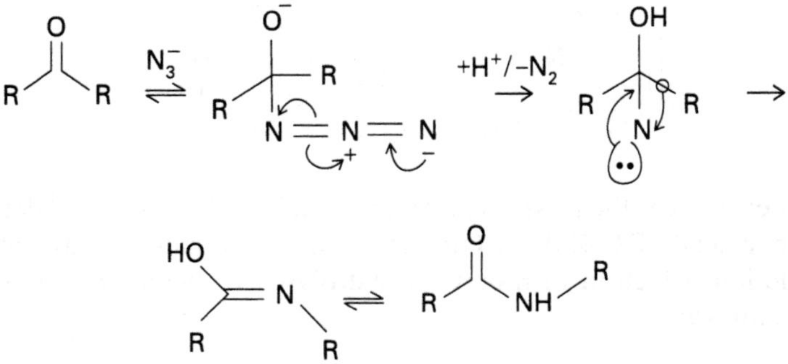

In this case, there is an elimination reaction to form the nitrene, which then rearranges to a product that tautomerises to give the desired amide. In essence, this reaction has resulted in the insertion of a NH unit in the carbon chain. If the ketone were cyclic, then this reaction might be used to expand the ring and so form a lactam.

The last carbon/nitrogen reaction that we will study is slightly different in its mechanism, but it still yields an amide as the product. In this case, an oxime is treated with PCl_5 in concentrated sulphuric acid. Suggest what is the pathway in this case.

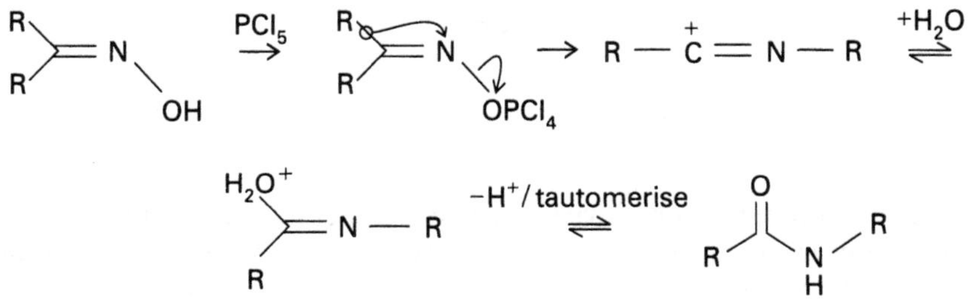

This is called the Beckmann rearrangement. It is also possible to perform this reaction by using sulphuric acid alone, even though classically PCl_5 was used. Suggest whether the *syn* or the *anti* alkyl group migrates to the hydroxyl group.

The *anti* group is the one that migrates in this reaction. If the reaction is performed on an oxime of a cyclic ketone, then ring enlargement occurs and gives the corresponding lactam.

We will now look at some reactions in which the migration terminus is an oxygen atom instead of a nitrogen atom.

16.4 Carbon/oxygen rearrangements

When a ketone is treated with a peracid (or peroxyacid), RCO_3H, in the presence of an acid catalyst, the related ester of the ketone is produced by the equivalent of an insertion of an oxygen atom. This is the Baeyer–Villiger rearrangement. If a cyclic ketone is treated in this manner, a lactone with a ring size one larger than the original ketone results. The mechanism is similar to the pathway that occurs when either diazomethane (Arndt–Eistert synthesis) or hydrogen azide (Schmidt reaction) reacts with a carbonyl species. So suggest the mechanistic route for the formation of the ester.

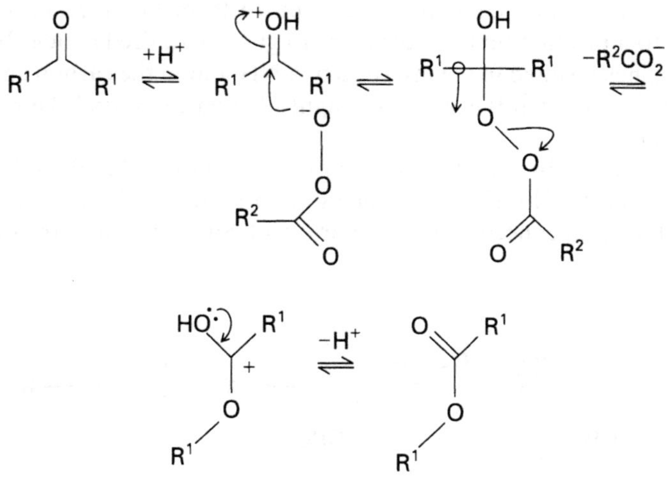

This reaction proceeds in high yield, and so is a useful synthetic tool.

Another example of carbon/oxygen rearrangement is the Claisen rearrangement in which an allyl aryl ether on heating gives an *o*-allylphenol. Suggest a mechanism for this reaction.

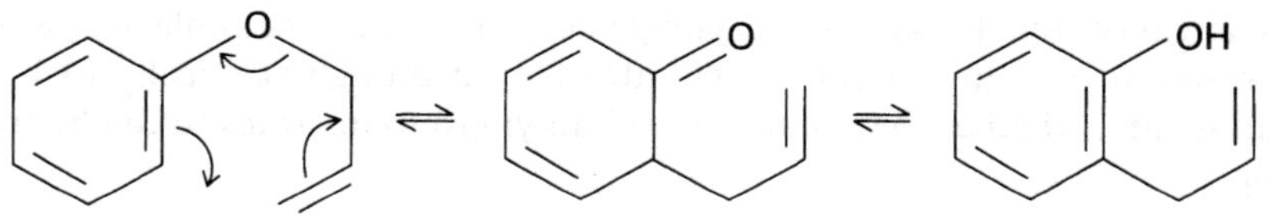

This reaction does not proceed via a Whitmore 1,2-shift, but instead is a concerted pericyclic [3,3] sigmatropic rearrangement that results in a cyclic ketone, which after tautomerisation, gives the phenol. Furthermore, in this reaction a C–O bond is broken, rather than made, and a C–C bond is formed; thus, it could have been considered under carbon/carbon rearrangements.

Now that we have covered the major rearrangement reactions, we will now look at a few fragmentation reactions.

16.5 Fragmentation reactions

Fragmentation of a molecule may occur in many different ways. The following examples will give you a feel for some of the different mechanisms that can be invoked.

The fragmentation reaction that is most often first encountered is the iodoform reaction. This is because it forms the basis of a very easy qualitative test for the presence of a methyl ketone, or alternatively the $CH_3CH(OH)$– grouping. If a methyl ketone is treated with iodine and sodium hydroxide, then a yellow precipitate of iodoform, CHI_3, is rapidly produced. Write down the mechanistic steps for this reaction.

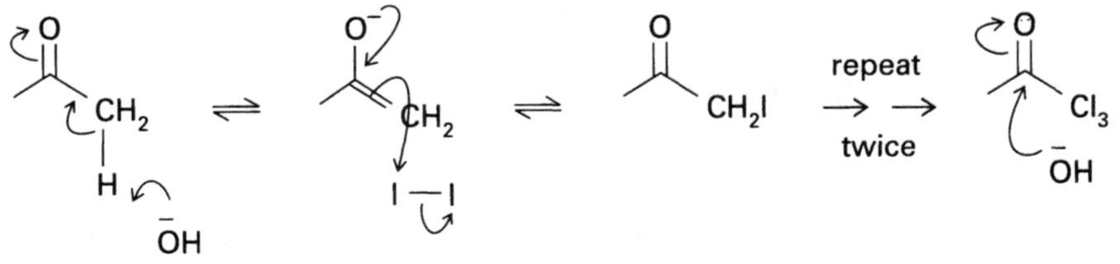

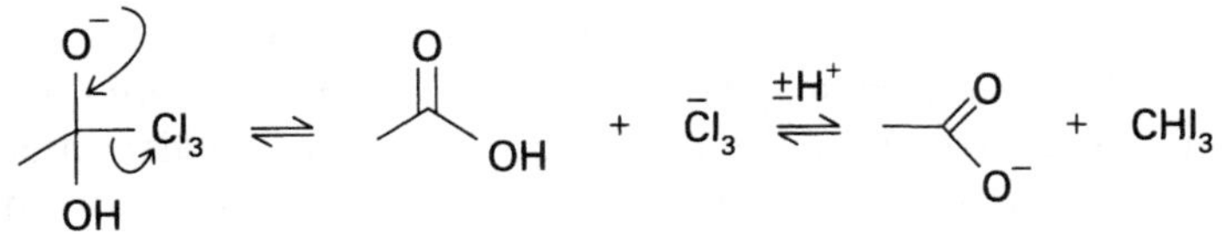

In each iodination step, the terminal methyl group has a hydrogen substituted for another iodine atom. After each such step, any remaining hydrogen atom(s) on the iodinated carbon atom become more acidic due to the $-I$ effect of the halogen. Eventually the tri-iodinated methyl group is capable of departing as an anion, and so may act as a leaving group. This method may be used to cleave methyl ketones to yield the derived carboxylic acid. This type of reaction is called anionic cleavage.

   Another example of this type of reaction is the decarboxylation of aliphatic acids. For the decarboxylation to proceed easily there must be an electron withdrawing group capable of stabilising the negative charge. An example would be an α-ketoacid, e.g. RCOCOOH. Suggest a mechanism for this reaction.

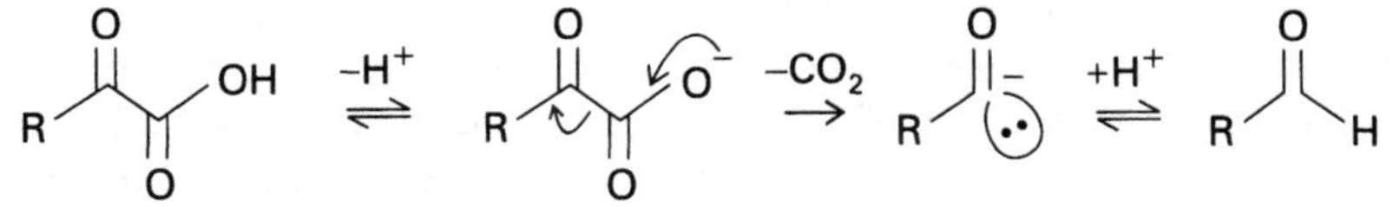

The mechanism is essentially either an SE1 or an SE2 process. There are many variations on this theme. For example, the Darzens condensation produces an α,β-epoxy carboxylic acid, which when heated often undergoes decarboxylation. Suggest a mechanism for this reaction.

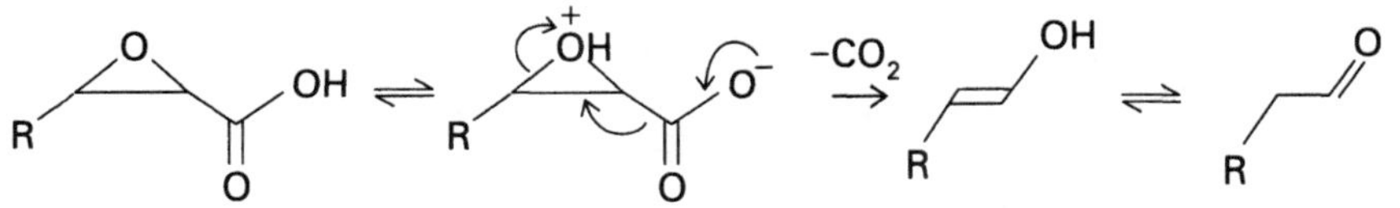

The initial product of the decarboxylation tautomerises to give the carbonyl compound. This is called the glycidic acid rearrangement.

If the electron withdrawing group is a β-carbonyl group, then a different mechanism operates. Suggest what this might be.

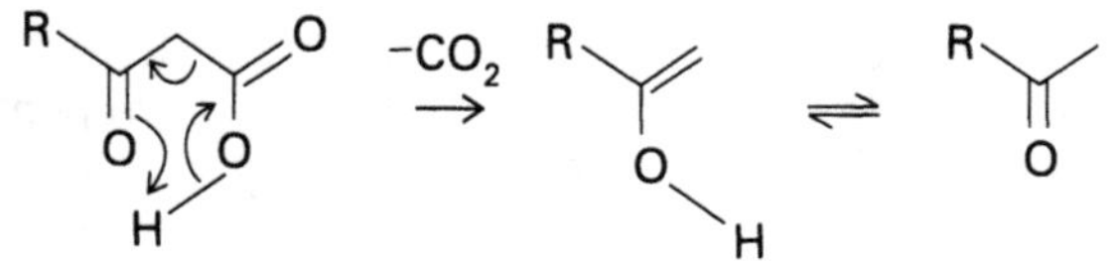

In this case, it is possible for there to be a cyclic transition state that results in the loss of carbon dioxide directly. The compound that is produced directly from the decarboxylation then tautomerises to yield the final product.

1,3-Diketones and β-keto esters may be cleaved under basic conditions, in a reaction pathway that is essentially the reverse of the Claisen condensation. Write the pathway for the ring opening of 1,3-cyclohexadione by hydroxide anion.

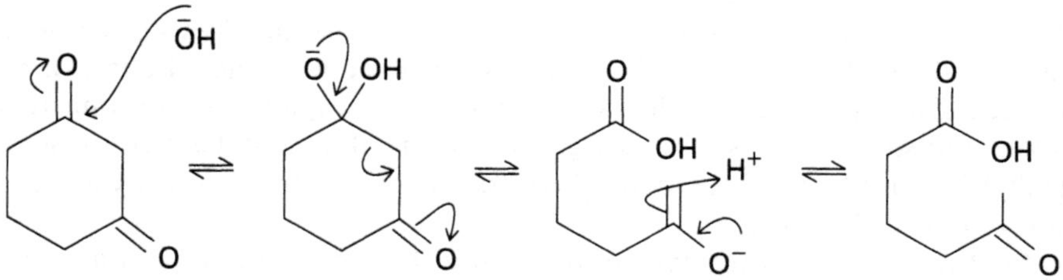

This cleavage reaction can occur for any compound that has a 1,3 arrangement of carbonyl groups. Notice that even though a mechanism can be readily proposed for this reverse reaction, whether or not the reaction actually proceeds in this direction or the other will depend upon a number of factors.

Primary and secondary aliphatic nitro compounds may tautomerise under basic conditions to the salt of the *aci* isomer. These isomers may in turn be hydrolysed by sulphuric acid, and so result in the fragmentation of the original molecule. Suggest how this reaction might proceed, and what is the final product.

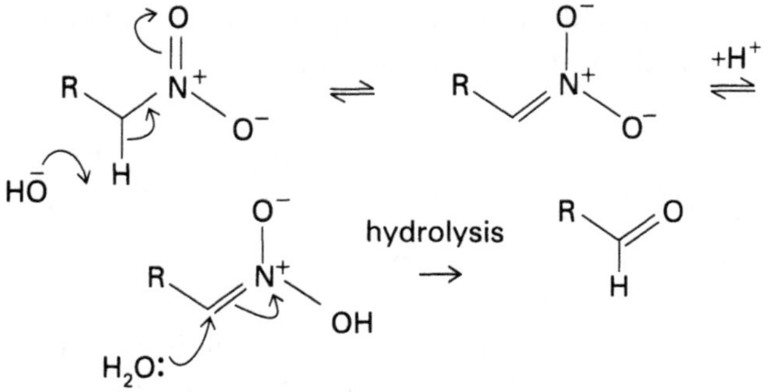

First, the base produces the salt of the nitro compound. Then, on addition of the sulphuric acid, the *aci* form of the original nitro compound is formed, which in turn is hydrolysed to the carbonyl compound. This process is called the Nef reaction.

The last example is a rather unusual fragmentation reaction that involves the rearrangement of a hydroperoxide. In this case, a hydroperoxide, $R_3C–O–O–H$, under acidic conditions, gives a ketone and an alcohol. Suggest a mechanism for this reaction.

$$R_3C–O–OH \;\overset{+H^+}{\rightleftharpoons}\; R_3C–O–\overset{+}{O}H_2 \;\overset{-H_2O}{\rightleftharpoons}\;$$

$$R_2\overset{+}{C}–O–R \;\overset{+H_2O}{\rightleftharpoons}\; R_3C–OR\;(\overset{+}{O}H_2) \;\overset{\pm H^+/-H^+}{\rightleftharpoons}\; R_2C{=}O \;+\; ROH$$

The migration terminus is an oxygen atom. The unstable hemiketal intermediate readily hydrolyses to give the final products. This, not surprisingly, is called the hydroperoxide rearrangement.

That concludes this chapter on rearrangement and fragmentation reactions. The final chapter is on redox reactions and gathers together those reactions that do not fit more conveniently elsewhere.

16.6 Summary of rearrangement and fragmentation reactions

Rearrangement reactions usually involve a 1,2-shift, i.e. the migrating group, M, goes from the migration origin, A, to the adjacent atom, which is called the migration terminus, B. The vast majority of rearrangements are nucleophilic or anionic in character. This means that the migrating group moves with its electron pair. Such rearrangements may also be called anionotropic. If the migrating group moves without its electron pair, then it is described as an electrophilic or cationotropic rearrangement; if the migrating group is a hydrogen cation, then it is called a prototropic rearrangement.

Nucleophilic rearrangements formally occur in three steps: the formation of the carbonium ion; the actual migration; and then either the migration origin will complete its octet by undergoing the second part of a nucleophilic substitution, or there will be the loss of a proton resulting in overall elimination. There is usually inversion of the configuration at the migration terminus. The departure of the leaving group is usually assisted by the migrating group moving across simultaneously, in the same way that a neighbouring group would interact. This anchimeric assistance often results in an increased rate of reaction.

The migration can be either intramolecular or intermolecular, and these can be distinguished by cross-over experiments. Aryl groups tend to migrate more easily than alkyl groups.

Allylic rearrangements are very common, and result in substitution at the 3-position. The rearrangement of the carbon skeleton of a carbonium ion is called the Wagner–Meerwein rearrangement, and proceeds via a Whitmore 1,2-shift. To avoid the problems associated with these rearrangement reactions, the Friedel–Crafts acylation method is used in preference to the normal Friedel–Crafts alkylation process. The driving force behind the Wagner–Meerwein rearrangement is the possibility of hyperconjugation in the rearranged carbonium ion, while in the dienone–phenol rearrangement it is the formation of an aromatic ring.

In the pinacol–pinacolone rearrangement a 1,2-diol rearranges to give a carbonyl group that is adjacent to a quaternary carbon. There are many variations of this reaction; for example, the semipinacol rearrangement, in which a β-amino alcohol is treated with nitrous acid. This reaction may be modified so that a ring expansion takes place, in which case it is called the Tiffeneu–Demyanov ring expansion.

In the benzil–benzilic acid rearrangement, an α-diketone is treated with a base to give the sodium salt of an α-hydroxy carboxylic acid. In the Favorskii rearrangement, an α-halogenoketone is treated with an alkoxide anion to give the α-alkyl ester. This reaction may also be used to effect a ring contraction.

A carbon chain may be extended by one unit by using the Arndt–Eistert synthesis. In the first step, an acyl halide is treated with diazomethane to form the α-diazo ketone. This is then treated with water and silver oxide. The resultant product is the free acid. If an alcohol is used instead of water, then the related ester is formed. This is the best way of extending a chain by one unit if the carboxylic acid is available. The process of extension in this manner is called homologation.

In the Hofmann rearrangement, a primary amide upon treatment with sodium hydroxide and bromine yields the corresponding amine with one fewer carbon atoms. There are many variations on this theme, for example, the pyrolysis of acyl azides to isocyanates. In the Lossen rearrangement, an *O*-acyl derivative of hydroxamic acid, RCONHOCOR, gives on treatment with hydroxide ions an isocyanate, which in turn may be hydrolysed to the amine. In the Schmidt rearrangement, the treatment of a carboxylic acid with hydrogen azide (hydrazoic acid) also gives the amine, via the isocyanate, when catalysed by an acid, such as sulphuric acid. All these reactions proceed via a nitrene intermediate, or related species, that reacts in a similar manner to the carbene intermediate found in the Wolff rearrangement that forms the central part of the Arndt–Eistert synthesis.

In the Beckmann rearrangement, an oxime is classically treated with PCl_5 in concentrated sulphuric acid to yield an amide. The *anti* group is the one that migrates in this reaction. If the reaction is performed on an oxime of a cyclic ketone, then ring enlargement occurs and gives the corresponding lactam.

In the Baeyer–Villiger rearrangement, a ketone is treated with a peracid (or peroxyacid), in the presence of an acid catalyst, and the related ester of the ketone is produced by the equivalent of an insertion of an oxygen atom. If a cyclic ketone is treated in this manner, a lactone with a ring size one larger than the original ketone results.

In the Claisen rearrangement, an allyl aryl ether gives an *o*-allylphenol on heating. The mechanism here is not a Whitmore 1,2-shift, but is a concerted pericyclic [3,3] sigmatropic rearrangement, that results in a cyclohexadienone, which, after tautomerisation, gives the phenol.

If a methyl ketone is treated with iodine and sodium hydroxide, then a yellow precipitate of iodoform, CHI_3, is rapidly produced. This is the iodoform reaction. This method may be used to cleave methyl ketones to yield the carboxylic acid. This type of reaction is called an anionic cleavage.

Another example of this type of reaction is the decarboxylation of aliphatic acids. For the decarboxylation to proceed easily there must be an electron withdrawing group capable of stabilising the negative charge. An example would be an α-carbonyl group, e.g. RCOCOOH. The Darzens condensation produces an α,β-epoxy carboxylic acid, which when heated often undergoes decarboxylation. The initial product formed by the decarboxylation tautomerises to give the carbonyl compound. This is called the glycidic acid rearrangement.

If the electron withdrawing group is a β-carbonyl group, then it is possible for a cyclic transition state to be formed that results in the loss of carbon dioxide directly. The resultant intermediate tautomerises to yield the final product, namely the carbonyl compound.

1,3-Diketones and β-keto esters may be cleaved under basic conditions, in a reaction pathway that is essentially the reverse of the Claisen condensation. This cleavage reaction can occur for any compound that has a 1,3 arrangement of carbonyl groups.

Primary and secondary aliphatic nitro compounds may tautomerise under basic conditions to give the salt of the aci isomer. Then, on addition of sulphuric acid, the aci form of the original nitro compound is formed, which is hydrolysed to the carbonyl compound. This reaction is called the Nef reaction.

In the hydroperoxide rearrangement, a hydroperoxide, $R_3C–O–O–H$, fragments under acidic conditions to give a ketone and an alcohol.

Redox Reactions

17.1 Introduction

This is the final chapter in this part of the book. We have already come across some reactions that formally involve oxidation or reduction of the carbon atoms in the molecule, e.g. the formation of a *vicinal* diol from a double bond is an oxidation reaction, while the addition of a hydrogen molecule to an alkene is a reduction reaction. However, such reactions proceed by clearly defined mechanisms that could be conveniently placed into other general mechanistic divisions, e.g. addition or elimination reactions, and so they were considered under those headings. There remain, however, a number of reactions that do not comfortably fit into any of the previous reaction types that we have studied so far.

This chapter is sub-divided along the lines of whether the overall process involves an oxidation or reduction of the organic component in which we are interested. Of course there must always be a corresponding reduction or oxidation of another reagent in order to ensure that there is an overall balancing of reduction and oxidation. In some organic reactions the organic component of interest undergoes both oxidation and reduction, i.e. there is a disproportionation reaction. Such reactions will be considered at the end of this chapter.

It is not intended to give an exhaustive coverage of all possible reduction and oxidation reactions that are of synthetic utility in organic chemistry. Instead, the aim is to give a selection of reactions that will illustrate the major mechanistic pathways. In the case of redox reactions for organic molecules, there is a large number of cases for which the mechanism has not been studied in any detail, or if it has, no consensus has arisen as to the true pathway. This is even true for such an important synthetic reaction as the Clemmensen reduction of a carbonyl group to a methylene unit by a zinc–mercury amalgam with concentrated HCl, i.e. a typical metal/acid couple. In such a case, there is little point in examining such a reaction in a text of this type. However, there are still many reactions that may be studied from a mechanistic viewpoint, and it is these upon which we will concentrate.

17.2 Reduction reactions

17.2.1 *Electron donors*

Many redox reactions in organic chemistry involve single electron transfers, i.e. are radical reactions. We have already looked at a number of radical reactions, for example those that resulted in overall substitution. We will now study a few more radical reactions that do not conveniently fit under that heading. The addition of a single electron to an atom amounts to the reduction of that atom's oxidation state by one unit.

In the previous chapter, one of the first reactions that we studied was the pinacol–pinacolone rearrangement, in which a symmetrical diol gave a carbonyl compound. There are many ways to make a *vicinal* diol, e.g. the addition of H_2O_2 to a carbon/carbon double bond. Such reactions were considered in the chapter on addition reactions, even though they are clearly oxidation reactions. Here, we consider a fundamentally different approach. In this case, in order to illustrate this mechanism, we will start by working backwards from the desired product, i.e. we will adopt a retrosynthetic approach. This is a very useful technique in designing the synthesis of a novel compound. Thus, indicate the products that would result from a homolytic cleavage of the central C–C bond in a symmetrical 1,2-diol, $R_2C(OH)C(OH)R_2$.

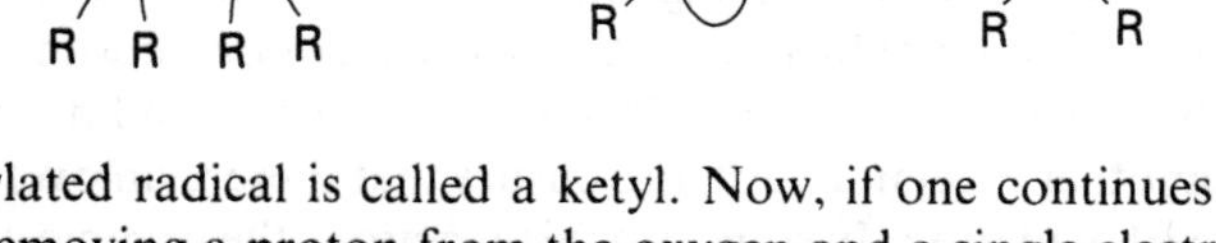

The hydroxylated radical is called a ketyl. Now, if one continues this reverse analysis by removing a proton from the oxygen and a single electron from the resultant radical anion, suggest what might be the starting material.

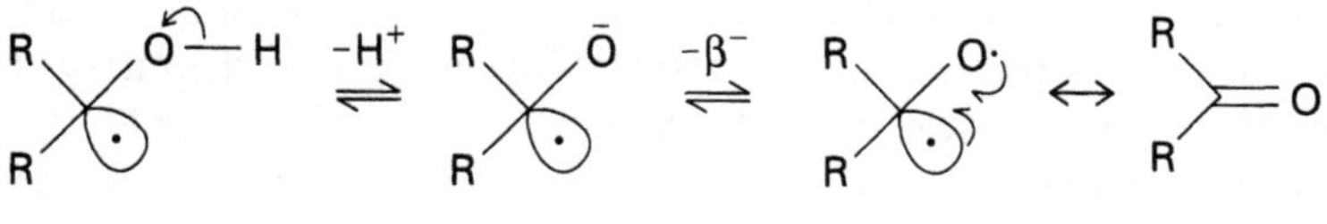

After the departure of a proton and a single electron from the hydroxylated radical, a biradical is produced, which is a canonical form of a carbonyl compound, $R_2C=O$. If this analysis is reversed, then the desired *vicinal* diol would be synthesised. This might be achieved by reacting a ketone with magnesium amalgam in dry ether, under anaerobic conditions, i.e. without oxygen present. Accordingly, write down the complete reaction pathway for the synthesis of pinacol itself, $(CH_3)_2C(OH)C(OH)(CH_3)_2$, starting with Mg and propanone. However, it should be remembered that magnesium is capable of forming a divalent cation. As such it is capable of reacting with two ketone molecules, and so form a cyclic intermediate. This can then be opened up by the hydrolysis of the organometallic intermediate with aqueous acid.

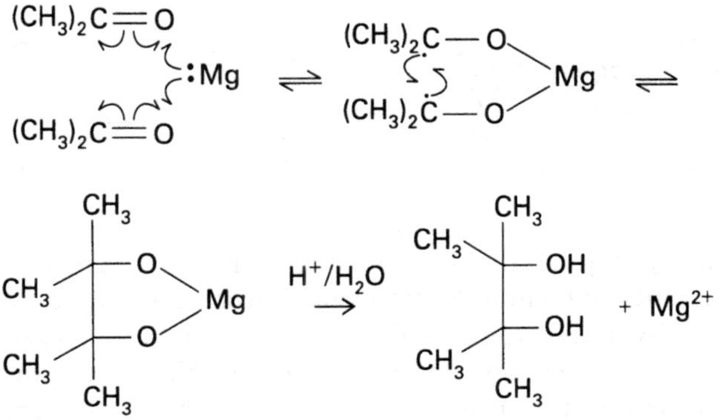

This reaction is called the pinacol synthesis, and is of general applicability. It works best for symmetrical aromatic ketones, because of the high stability of the aromatic ketyl intermediate, which incidentally is blue in colour.

Simple ketones are not the only carbonyl compounds that will accept single electrons. Thus, for example, esters may also pick up an electron when sodium metal is dissolved in an inert solvent such as xylene. Write down the pathway for this reaction, up to the stage where a dimer has been formed.

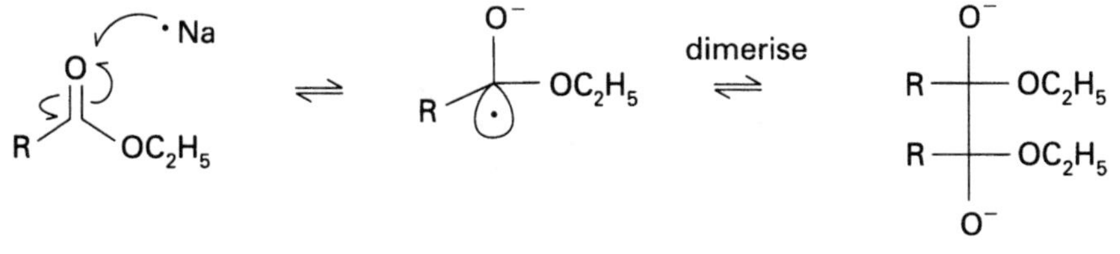

So far, the reaction has proceeded along similar lines to the pinacol reaction. However, in the present case, the dimer contains two good leaving groups. Write down this next double elimination step.

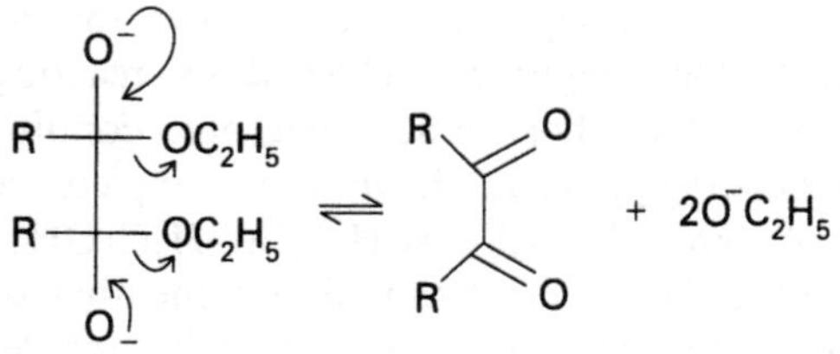

A diketone has now been formed, but such a compound may itself react with single electrons. Suggest what may be the remaining steps for this compound, and so suggest what will be the final product.

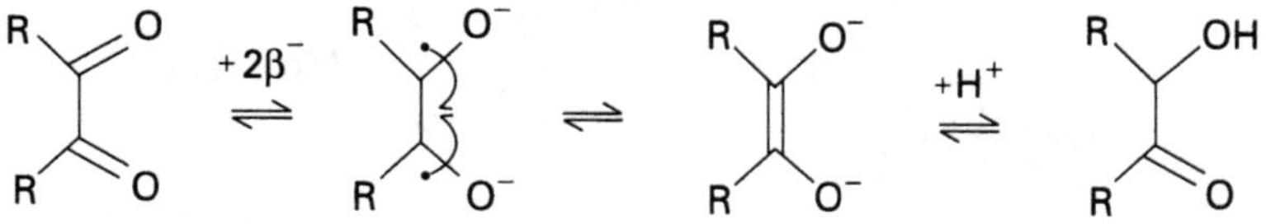

Each carbonyl group of the 1,2-diketone picks up a single electron, thus undergoing further reduction. Subsequent protonation and tautomerism then yields the 1,2-hydroxyketone. This coupling reaction between two ester molecules is called the acyloin condensation. Note that in the Claisen condensation the resultant product has a 1,3 distribution of carbonyl/oxygen functional groups, while in this case the product has a 1,2 distribution of carbonyl/oxygen functional groups. This reaction may be used to effect the cyclisation of long chain diesters in order to form very large rings.

Sodium metal is a good source of free electrons. It is so electropositive, it readily gives up one electron to form the cation. At room temperature, sodium reacts with ammonia to form sodamide, $NaNH_2$, and hydrogen in a simple inorganic redox reaction. However, if the temperature of the reaction mixture is reduced so as to produce liquid ammonia, then the sodium metal reacts to form the solvated cation and free electron. This solvated electron may be put to various uses. For example, it is possible to effect the reduction of a benzene ring. Suggest what may be the first step of such a reduction.

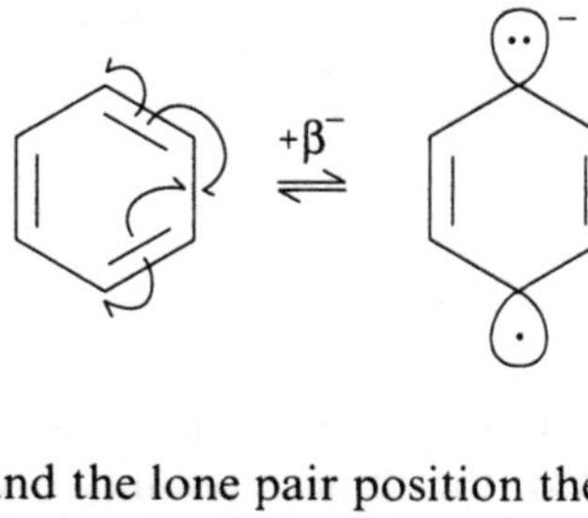

The unpaired electron and the lone pair position themselves as far apart on the ring as possible, i.e. *para* to each other. This reaction is usually performed in the presence of an alcohol, which acts as a mild proton donor. Suggest what would be the remaining steps of this reaction, and so suggest what will be the product and what is unusual about it.

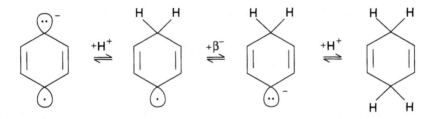

The alcohol acts as a source of protons, which react with the lone pairs. This reaction is called the Birch reduction. The final product is the cyclohexadiene in which the double bonds are unconjugated. If the alcohol is not present, then dimers may be obtained as the radical anions react with each other. In some substituted aromatic systems, both electrons are added before the ionic intermediate picks up any protons so that a dianion is formed initially.

The presence of a substituent on the aromatic ring will influence the final position of the double bonds, and also the overall rate of reduction. Suggest what will be the product when anisole is reduced by sodium in liquid ammonia, and whether the rate will be slower or faster than that shown by benzene.

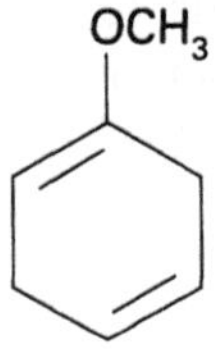

OCH$_3$

The methoxy group is electron donating, and so it disfavours the addition of a further electron to the aromatic ring, reducing the rate of reduction. Furthermore, it disfavours the positioning of the lone pair on the carbon to which it is attached. Thus, the methoxy group is found on the non-reduced carbon in the final product.

Similarly, suggest what will be the effect of an amido group on the rate and regiospecificity of the Birch reaction.

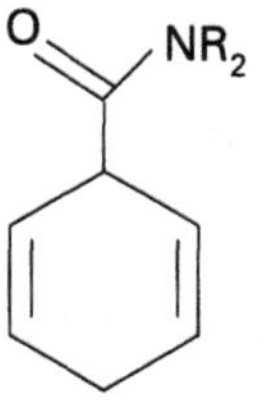

O   NR$_2$

333

The rate is increased, and the ring carbon to which the amido group is joined is reduced to an sp^3 carbon.

The Birch reaction may also be used to reduce carbon/carbon triple bonds that are contained within a carbon chain. Suggest what will be the geometry of the resultant double bond after it has been protonated.

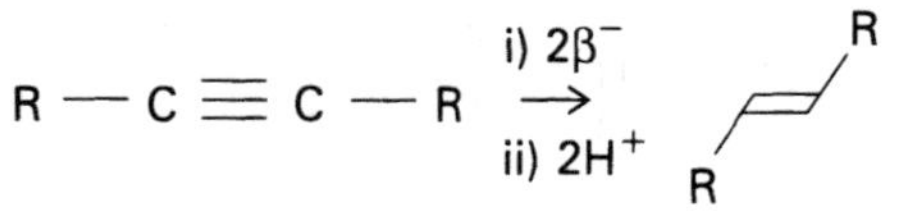

$$R - C \equiv C - R \quad \begin{array}{c} \text{i) } 2\beta^- \\ \longrightarrow \\ \text{ii) } 2H^+ \end{array}$$

The resultant double bond is *trans*, because in the intermediate the lone pairs repel each other, and so are *trans* to each other when they are protonated. This is a way of effecting *trans* hydrogenation. Terminal carbon/carbon triple bonds cannot be reduced by simply using sodium and ammonia, because the acetylide anion is formed under these conditions. Instead, ammonium sulphate must be added to the reaction mixture in order to achieve the reduction of the triple bond.

17.2.2 *Hydride anion donors*

We will now look at some reactions that are performed by hydride ion donors. A very important synthetic reaction involves the reduction of the carbonyl group by a complex metal hydride to give an alcohol moiety. One of the commonest examples is the reduction of a ketone by lithium aluminium hydride, LiAlH$_4$, to give a secondary alcohol. Suggest what is the first step of this reaction.

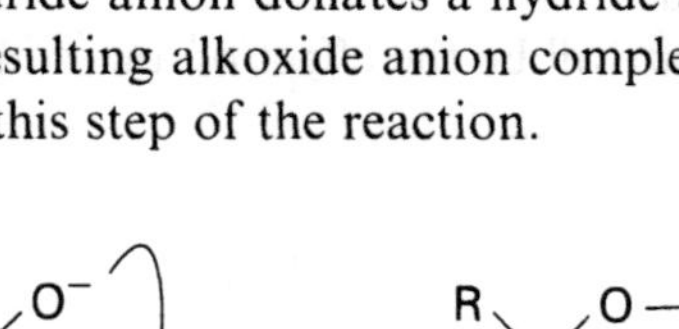

The aluminium tetrahydride anion donates a hydride anion irreversibly to the carbonyl carbon. The resulting alkoxide anion complexes with the aluminium trihydride. Write down this step of the reaction.

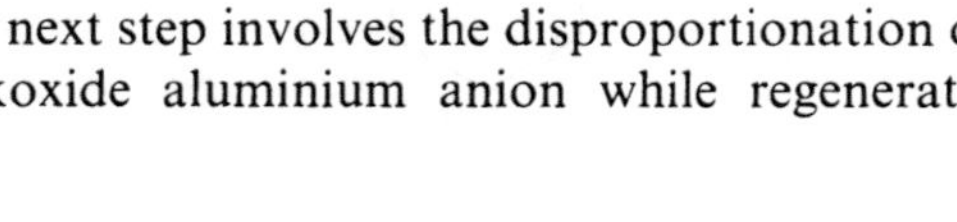

It appears that the next step involves the disproportionation of this complex to yield the tetra-alkoxide aluminium anion while regenerating more of the aluminium tetrahydride anion. The tetrahydride seems to be the only species that donates hydride ions to the carbonyl carbon. In the work-up of the reaction mixture, the desired alcohol is produced by protonation of the aluminium tetra-alkoxide anion by a protic solvent.

Another common variation is the reduction of an ester to a primary alcohol. Suggest what is the mechanistic pathway for this reaction.

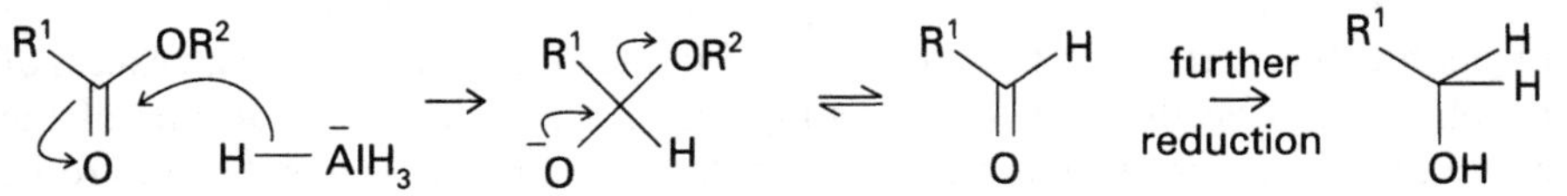

The reaction is facilitated by the fact that the alkoxide ion is a reasonable leaving group under these conditions. The intermediate carbonyl compound is then attacked again so as to effect the addition of the second hydride ion. This now forms an alkoxide anion that is capable of complexation with the aluminium, which is destroyed upon the acidic work-up, to liberate the desired product. If lithium aluminium deuteriide is used, then this method may be used to produce isotopically labelled compounds.

The formation of the intermediate complex may cause problems on occasions. For example, suggest what is the structure of the intermediate complex that is formed during the synthesis of ethanolamine from the ester of glycine.

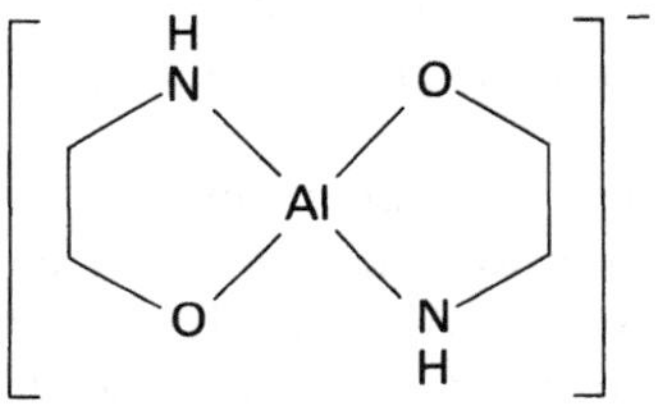

This cyclic complex is fairly stable, but it can be hydrolysed by the addition of aqueous sodium hydroxide or hydrochloric acid. If the latter reagent is used, then the hydrochloride salt of ethanolamine is formed.

Lithium aluminium hydride is a very reactive hydride ion donor, and may be used to effect a large number of reductions. For example, it may be used to reduce nitriles, $R-C \equiv N$, and primary amides, $RCONH_2$, to the related amines, RCH_2NH_2. It is thought that the reduction of the primary amide proceeds via the nitrile.

One consequence of $LiAlH_4$ being very reactive, is that it is rather unselective. In order to improve its selectivity, various derivatives have been made that are less reactive. These derivatives usually have three of the hydride ions replaced by alkoxide ions, i.e. $AlH(OR)_3^-$. We saw earlier that some mixed alkoxide/hydride aluminium anions disproportionate to form the tetra-alkoxide anion and tetrahydride anion. Whether or not this disproportionation is favoured depends upon the exact nature of the alkoxide under consideration, and hence the reducing power of the aluminium complex may be controlled by judicious selection of the alkoxy group.

Instead of using a complex metal hydride as the source of the hydride anion, the hydride anion may be donated from a carbon atom. In such a case, the reaction is called the Meerwein–Ponndorf–Verley reduction, or MPV reduction. In contrast to the reduction by a complex metal hydride, the MPV reduction is reversible. It is performed by heating aluminium isoproproxide with an excess of propan-2-ol. Two different routes compete: one involves one molecule of the aluminium isoproproxide for each molecule containing a carbonyl group to be reduced, while the other route uses two. Suggest the mechanism that involves only one molecule of the aluminium compound.

In this case, propanone is the constituent of the reaction mixture that has the lowest boiling point. Hence, by continuously distilling it out of the system, the equilibrium may be effectively displaced to completion. The excess of propan-2-ol exchanges with the mixed aluminium alkoxide to liberate the reduction product.

As this reaction involves steps that are reversible, it is not surprising that the reaction may be forced to proceed in the reverse direction, i.e. to achieve the oxidation of an alcohol by using $Al(OC(CH_3)_3)_3$ as a catalyst with an excess of propanone. In this case, the reaction is called the Oppenauer oxidation.

In the version that involves two molecules of aluminium alkoxide, one attacks the carbonyl carbon, while the other attacks the carbonyl oxygen. Otherwise, this pathway is similar to the one illustrated previously.

We will now look at a few oxidation reactions.

17.3 Oxidation reactions

Oxidation reactions include those reactions that involve the removal of hydrogen from a molecule. In the chapter on addition reactions to a carbon/carbon double bond, we saw that hydrogen molecules may be added across a carbon/carbon double bond in the presence of a metal catalyst such as platinum or palladium. The reverse reaction is also possible, e.g. heating cyclohexane with platinum at 300°C gives benzene. It is thought that the mechanism of dehydrogenation is simply the reverse of the addition of hydrogen by the hydrogenation process. The reaction is easier if the ring already contains some unsaturation. Dehydrogenation may also be performed by quinone, which is reduced to hydroquinone in the process. Suggest the reaction pathway for the oxidation of cyclohexadi-1,3-ene by quinone.

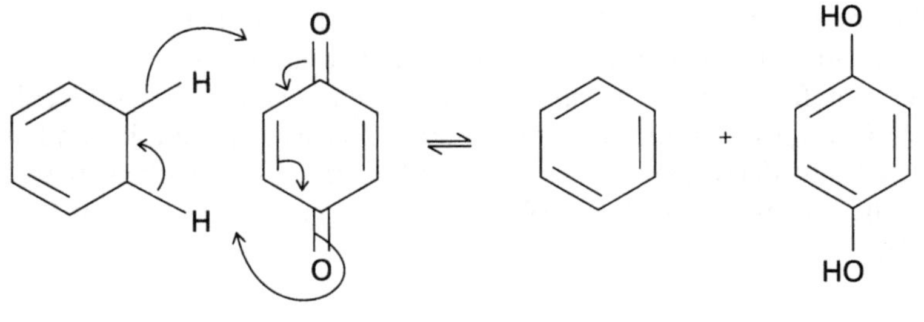

Usually the tetrachloro or dichloro-dicyano derivatives of quinone are used in order to effect this reduction; they are referred to as chloranil and DDQ respectively.

Oxidation reactions also include those reactions that involve the formation of new carbon/oxygen bonds. We have already seen an example: the addition of hydrogen peroxide to a carbon/carbon double bond to give the 1,2-diol. Such compounds may undergo further oxidation by treatment with periodic acid, HIO_4, or lead acetate, $Pb(OAc)_4$. These two reagents are complementary, as the former is best used in an aqueous medium, while the latter is usually used in organic solvents. The mechanistic pathway for both reagents is similar, and involves a cyclic intermediate. Suggest the nature of the intermediate that is formed when $Pb(OAc)_4$ is used.

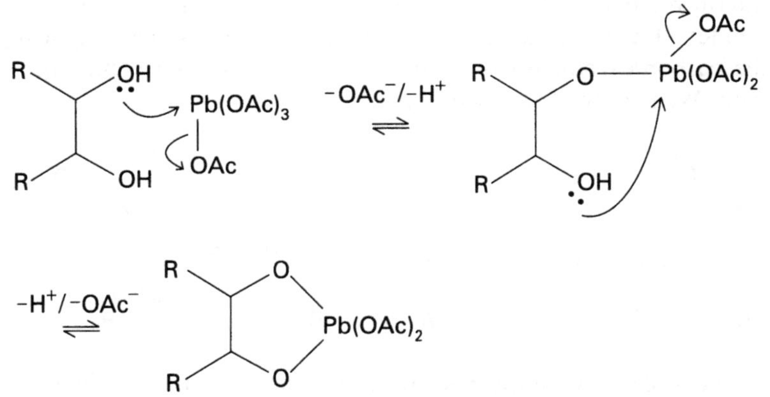

Once formed, this cyclic intermediate is capable of fragmentation to produce two carbonyl compounds. Suggest how this may occur.

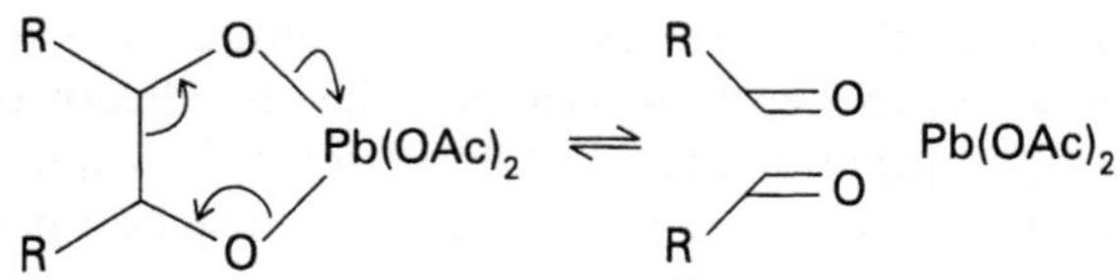

The pathway from a double bond to the *vicinal* diol to the ketone compounds is very efficient, and may be used preparatively. The cleavage of *vicinal* diols is also used to elucidate the structure of carbohydrate compounds.

Instead of forming the *vicinal* diol, and then cleaving that molecule with either Pb(OAc)$_4$ or periodic acid, it is possible to cleave the double bond directly to give carbonyl derivatives. This is achieved by reacting the alkene with ozone, O$_3$, and the process is called ozonolysis. First, suggest a structure for the ozone molecule.

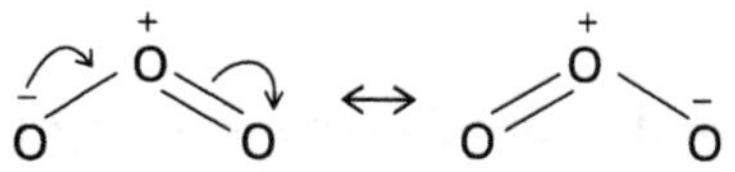

Ozone is a dipolar molecule, and is capable of performing a 1,3-dipolar addition to a carbon/carbon double bond. Write down this step.

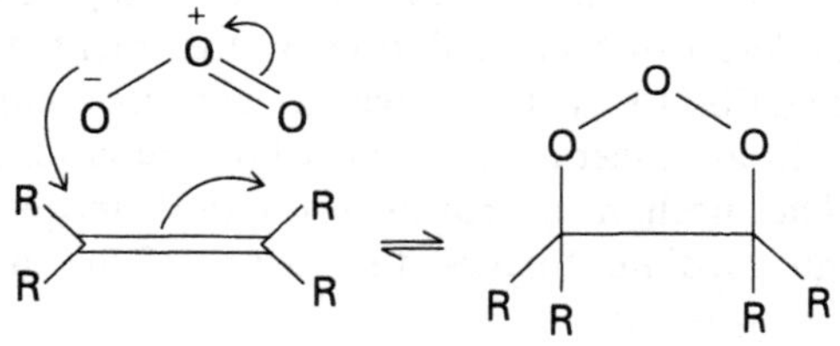

This intermediate is called an initial or primary ozonide (or sometimes a molozonide). One of the oxygen/oxygen bonds now breaks heterolytically, and then the carbon/carbon bond breaks to yield a carbonyl compound and a zwitterion. Write down these two steps.

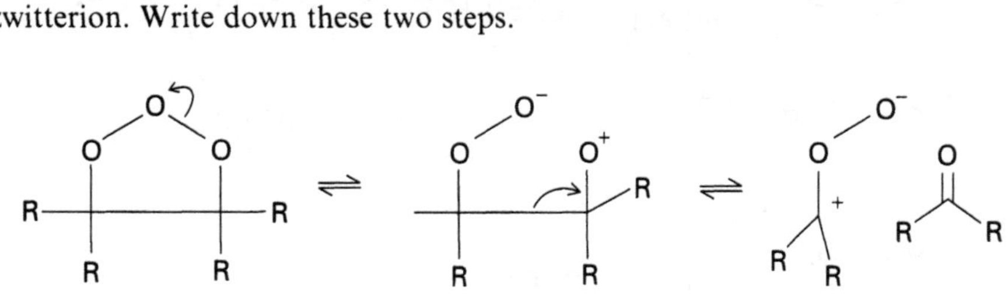

It is also possible to write the mechanism for the formation of these products in a single step, i.e. a cycloreversion. However formed, these intermediates may then recombine in a different 1,3-dipolar addition to give the normal ozonide.

338

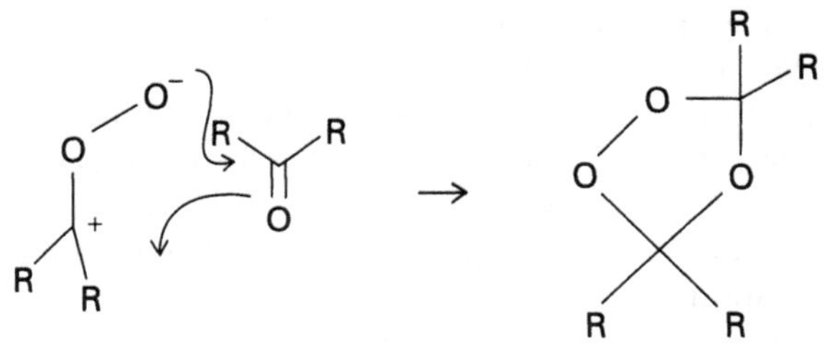

This route is called the Criegee mechanism. Various refinements have been suggested in order to account for further experimental results that do not fit the scheme given above. However, once formed, this new ozonide may be decomposed by zinc and ethanoic acid to give two carbonyl compounds, either ketones or aldehydes depending on the substituents on the original carbon/carbon double bond. The ozonide may also be reduced by LiAlH$_4$ to give two hydroxyl compounds, or oxidised by H$_2$O$_2$ to give ketones or carboxylic acids. Thus, by controlling the work-up conditions, the oxidation state of the final products can be manipulated.

In a totally different reaction, hydrogen peroxide may also be used to effect a rather unusual outcome in which an aromatic aldehyde, e.g. a benzaldehyde derivative, is oxidised to the corresponding phenol. The reaction occurs under alkaline conditions. Suggest what the first step in this pathway might be.

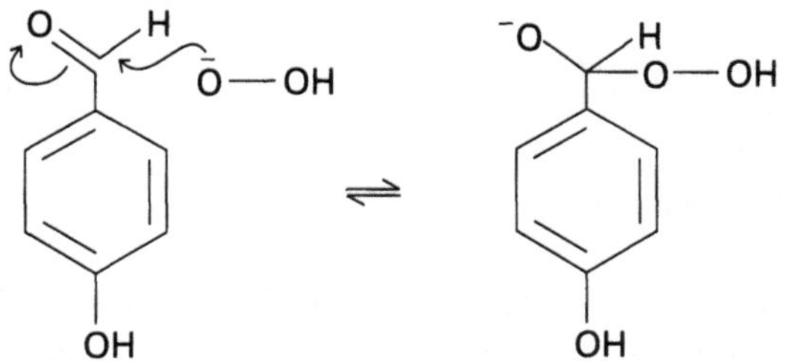

What occurs next is a rearrangement that is similar to the Baeyer–Villiger rearrangement, and results in the formation of an ester. This intermediate is hydrolysed to yield the phenol derivative. Write down these steps of the mechanism.

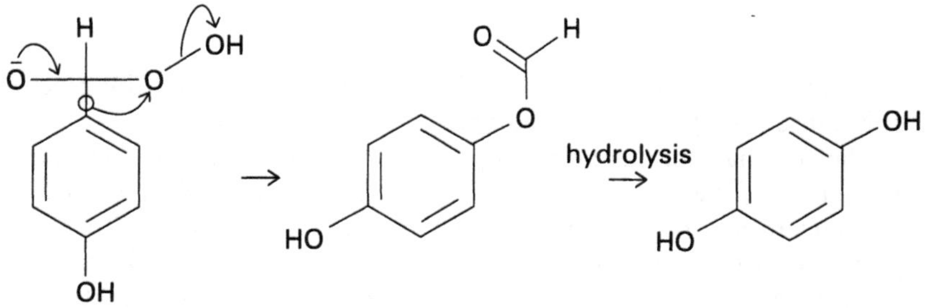

This reaction is called the Dakin reaction, and requires a hydroxyl or amino group in the *ortho* or *para* position of the aromatic aldehyde. One possible reason why such substituents might help this reaction is the hypothesis that a cyclic intermediate is formed that comprises the former aldehyde carbon, the carbon in the aromatic ring to which it is attached and the oxygen atom that is the migration terminus. A +M group in either the *ortho* or *para* position would facilitate the formation of such a ring.

It is possible to oxidise a methylene group that is adjacent to a carbonyl group to yield another carbonyl group. This transformation is achieved using selenium dioxide. The presence of a base is necessary. Suggest how a selenium intermediate may be formed.

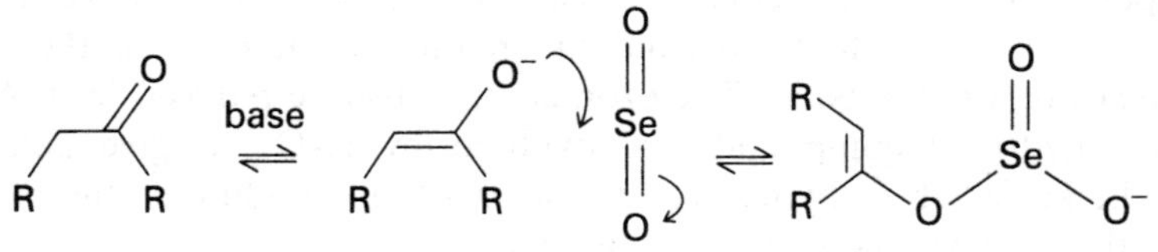

The enolate may attack the selenium atom of the selenium dioxide to form a selenate ester of the enol. This intermediate may rearrange to re-form the carbonyl group, and transfer an oxygen atom to the α-carbon. Suggest how this may occur.

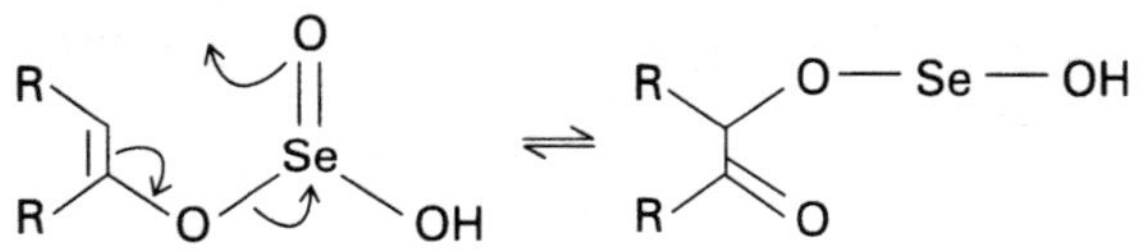

The last remaining α-hydrogen atom is removed by the action of the base, which initiates a fragmentation reaction to yield the α-keto carbonyl compound. Suggest what is the mechanism for this step, and so suggest what are the products for this reaction.

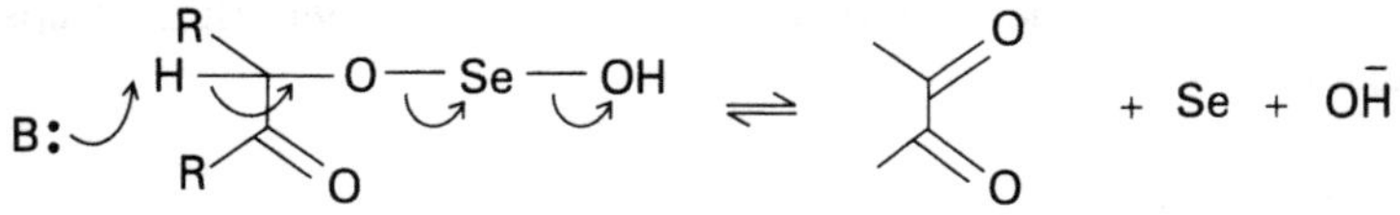

The 1,2-diketone product and elemental selenium are produced. This reaction is useful because it may be performed under very mild conditions.

Another mechanism has been proposed for this reaction. In this case, it is suggested that a β-ketoseleninic acid, $-C(=O)CHSe(=O)OH$ is involved as the intermediate, rather than the selenate ester. This just highlights the fact that very few proposed mechanisms are actually proved in reality. Often, all that may be said for any particular mechanistic pathway is that it accounts for all the kinetic data, including isotopic substitution experiments, and also for all the observed products (in particular, unusual side products), including structural and stereo isomers. Intermediates can rarely be isolated and characterised, but it is sometimes possible to detect and characterise them spectroscopically.

We will now look at reactions in which both oxidation and reduction occur in the organic components of interest.

17.4 Disproportionation reactions

In any reaction that involves the oxidation or reduction of an organic component, there must also be a corresponding reduction or oxidation of the other reagent, be it organic or inorganic. Normally, we are not interested in the other component of the reaction mixture, so long as the organic substrate in which we are interested is modified in the manner that we desire. Occasionally, however, the organic component undergoes both oxidation and reduction, sometimes even within the same molecule at different functional groups. When there are two identical functional groups, and one is oxidised while the other is reduced, then this redox reaction is called disproportionation. This obviously requires a functional group that can be either oxidised or reduced depending upon the conditions, e.g. an aldehyde group, which is capable of being oxidised to a carboxylic acid or reduced to an alcohol. We have already seen a variation of this theme in the chapter on rearrangement reactions. In the benzil–benzilic acid rearrangement, a molecule that initially contained two ketone groups rearranged to give a molecule with an α-hydroxycarboxylic acid arrangement of functional groups. This was an example of an intramolecular disproportionation reaction.

Earlier in this chapter, we saw how a carbonyl group may be reduced by a hydride ion donor. In the examples given, the hydride anion came from either a complex metal hydride such as LiAlH$_4$, or from a carbon reagent, such as aluminium isoproproxide with an excess of propan-2-ol. Under certain circumstances, the hydride ion that is being used to effect this reduction may come from an aldehyde, especially one that lacks an α-hydrogen atom, e.g. methanal or benzaldehyde. The receiving molecule may be a second molecule of the same aldehyde or a different one. In the former case disproportionation occurs, and the reaction is called the Cannizzaro reaction; in the latter case it is called a crossed Cannizzaro reaction. The reaction requires a strong base, and the rate law is found to depend on the concentration of the base and the square of the concentration of the aldehyde. In the initial reaction between a hydroxide anion and benzaldehyde, suggest what might be the first step.

The anion that results from the addition of the hydroxide ion then donates a hydride ion to another aldehyde; write down the mechanism of this step.

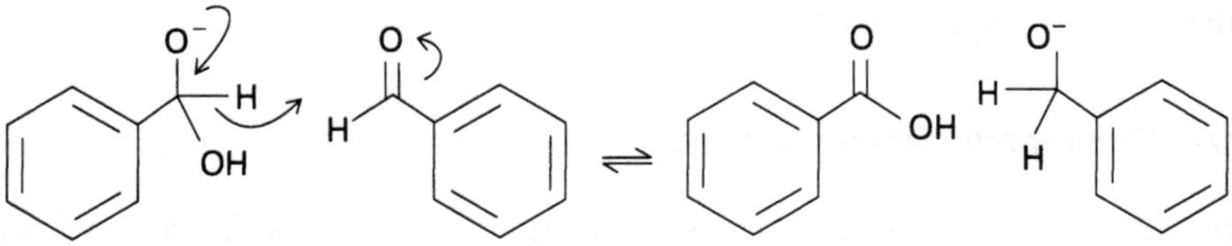

Overall a carboxylate anion and an alcohol have been formed from two molecules of aldehydes, and so a disproportionation reaction has been effected. There is an alternative mechanism in which the rate is found to depend upon the square of the concentration of the base and the square of the concentration of the aldehyde, i.e. it is a fourth order reaction. In this pathway, the base removes both the hydroxyl protons to form the dianion, which then donates a hydride ion to the other aldehyde.

The reaction may occur intramolecularly, e.g. an α-ketoaldehyde will give an α-hydroxycarboxylic acid on treatment with hydroxide ion.

A variation of this reaction is called the Tollens reaction. In this case, a ketone or aldehyde that contains an α-hydrogen is treated with formaldehyde in the presence of $Ca(OH)_2$. If the initial product of the cross aldol reaction contains another α-hydrogen, then another cross aldol reaction occurs. This continues until there are no more α-hydrogens left. Write down the product that results from the exhaustive condensation of ethanal with methanal.

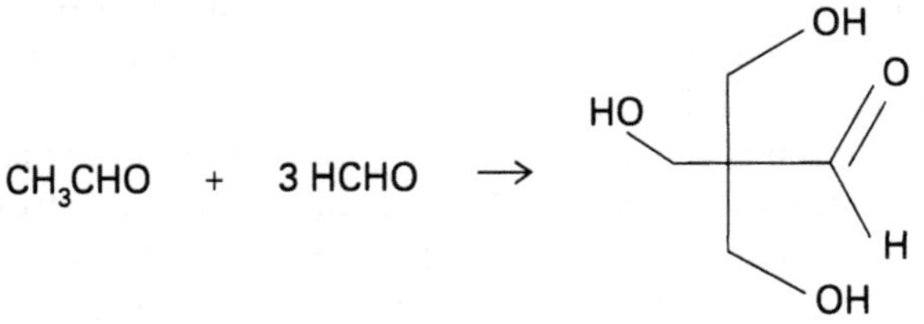

This intermediate has three hydroxyl groups and one carbonyl group, but has no α-hydrogens left. Now a cross Cannizzaro reaction occurs that reduces the aldehyde group to give a fourth hydroxyl group. Write down the equation for this step.

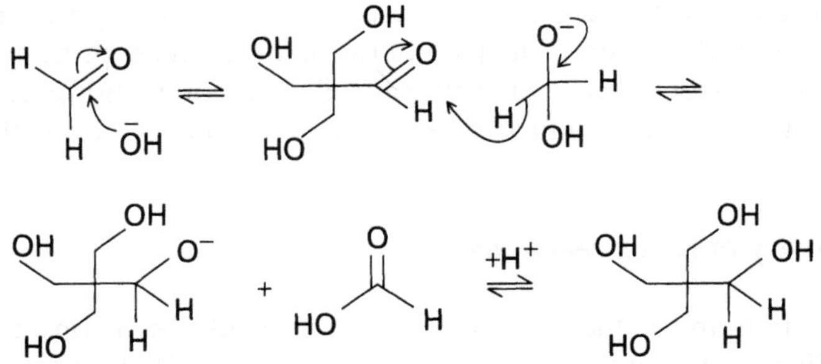

This is an example of a disproportionation reaction between two different molecules, but each contained the same functional group initially.

Aldehydes provide a ready example of organic molecules capable of undergoing disproportionation reactions, because they occupy the midway position in the oxidation sequence from primary alcohols to carboxylic acids. However, they are not the only examples of organic species that are capable of undergoing disproportionation. All that is necessary is to find a species that occupies a similar midway position in some redox sequence; e.g. suggest what is the disproportionation reaction which occurs between two alkyl radicals.

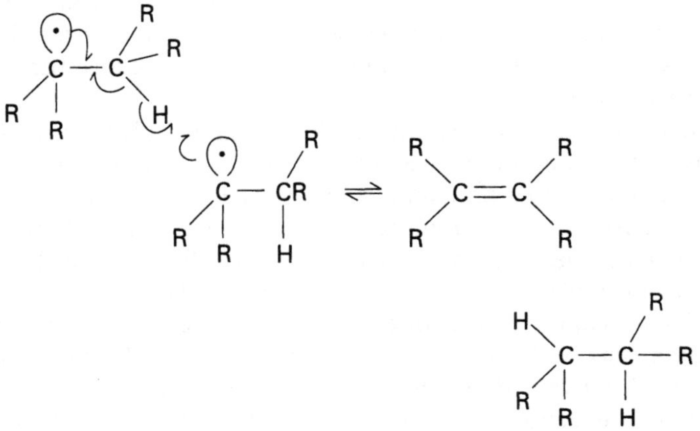

In this case, they may react together to form an alkane and an alkene. This reaction occurs in competition with dimerisation and other termination steps in free radical substitution reactions.

That concludes this part of the book in which we have studied the basic mechanisms that are thought to account for the vast array of organic reactions that have been observed. In all of the reaction pathways examined, the basic principles that were outlined originally have been applied. Now that these fundamental principles have been learnt and put into the context of the basic mechanism, you should be able to write a sensible mechanistic sequence for most of the reactions that you will encounter. There are, of course, more advanced principles that help explain results which would appear anomalous when using just the basic principles outlined here. Such principles include frontier orbital theory and symmetry controlled reactions; however, these are beyond the aim of this introductory text, and so can be left for further study.

17.5 Summary of redox reactions

Many oxidation and reduction reactions have already been considered under other headings, e.g. addition or elimination reactions. What remains are those reactions that do not fit conveniently into the previous chapters.

A 1,2-diol may be formed by the action of magnesium amalgam on a ketone. This reaction is called the pinacol synthesis. Another single electron reduction occurs with the action of sodium metal on an ester to yield a 2-hydroxyketone. This coupling reaction between two ester molecules is called the acyloin condensation. This reaction may be used to effect the cyclisation of long chain diesters in order to form very large rings. In the Birch reduction, sodium metal dissolved in liquid ammonia may be used to reduce benzene derivatives to give the unconjugated cyclohexadienes. Groups that are $-M$ reduce the rate of reduction, and direct the incoming electrons so that such groups are on one of the sp^2 carbon atoms of the final product. Conversely, $+M$ groups activate the ring to reduction and the ring carbon to which they are attached is reduced to an sp^3 hybridisation. The Birch procedure can be used to reduce triple bonds to yield the *trans* double bond.

The reduction of a ketone by lithium aluminium hydride to give a secondary alcohol is a very important synthetic route. A common variation is the reduction of an ester to a primary alcohol. $LiAlH_4$ is very reactive, and hence rather unselective. In order to improve its selectivity, various derivatives have been made that are less reactive; usually, these modified reagents have three of the hydride ions replaced by alkoxide ions, e.g. $AlH(OR)_3^-$.

Instead of a complex metal hydride being the source of the hydride anion, the hydride anion may be donated from a carbon atom. In such a case, the reaction is called the Meerwein–Ponndorf–Verley reduction, or MPV reduction. In contrast to the reduction by a complex metal hydride, the MPV reduction is reversible. The MPV reduction is performed by heating aluminium isopro-proxide with an excess of propan-2-ol. The reaction may be forced to proceed in the reverse direction, i.e. to achieve the oxidation of an alcohol by using $Al(OC(CH_3)_3)_3$ as a catalyst with an excess of propanone. In this case, the reaction is called the Oppenauer oxidation.

Dehydrogenation of cyclohexane can be achieved by heating with a metal catalyst such as platinum. Dehydrogenation may also be performed by quinone, which is reduced to hydroquinone in the process.

The addition of hydrogen peroxide to a carbon/carbon double bond gives the 1,2-diol. Such compounds may undergo further oxidation by treatment with periodic acid, HIO_4, or lead acetate, $Pb(OAc)_4$. These two reagents are comple-mentary, as the former is best used in an aqueous medium, while the latter is used in organic solvents.

Instead of going via the *vicinal* diol, and then cleaving that molecule with either $Pb(OAc)_4$ or periodic acid, it is possible to cleave the double bond directly to give the carbonyl derivatives. This is achieved by reacting the alkene with ozone, O_3. The Criegee mechanism has been proposed to account for the formation of the ozonide. This intermediate may be decomposed by zinc and ethanoic acid to give two carbonyl compounds, either ketones or aldehydes depending on the substituents on the original carbon/carbon double bond. The intermediate may also be reduced by $LiAlH_4$ to give two hydroxyl compounds; or oxidised by H_2O_2 to give ketones or carboxylic acids.

In the Dakin reaction, an aromatic aldehyde, which has a hydroxyl or amino group in the *ortho* or *para* position of the aromatic aldehyde, can be oxidised by hydrogen peroxide to yield the corresponding phenol.

It is possible to oxidise a methylene group that is adjacent to a carbonyl group to yield another carbonyl group. This transformation is achieved using selenium dioxide. The presence of a base is necessary; however, the reaction conditions are very mild.

When there are two identical functional groups, and one is oxidised while the other is reduced, then this redox reaction is called disproportionation. This obviously requires a functional group that can be either oxidised or reduced depending upon the conditions, e.g. an aldehyde group, which is capable of being oxidised to a carboxylic acid or reduced to an alcohol.

In the Cannizzaro reaction, the hydride ion that is being used to effect this reduction may come from an aldehyde that lacks an α-hydrogen atom, e.g. methanal or benzaldehyde. The receiving molecule may be a second molecule of the same aldehyde or a different one. The reaction requires a strong base, and the rate law is found to depend on the square of the concentration of the aldehyde and either the concentration or the square of the concentration of the base used. Overall a carboxylate anion and an alcohol are formed from two molecules of the aldehyde(s). The reaction may occur intramolecularly, i.e. α-ketoaldehydes give the α-hydroxycarboxylic acids on treatment with hydroxide ions. A variation of this process is called the Tollens reaction. In this case, a ketone or aldehyde that contains an α-hydrogen is treated with formaldehyde in the presence of $Ca(OH)_2$. The eventual product is the cross Cannizzaro derivative, with all the carbonyl groups reduced to alcohols.

Disproportionation reactions can occur with alkyl derivatives, e.g. two alkyl radicals can react together to form an alkane and an alkene. This reaction occurs in competition with dimerisation and other termination steps in free radical substitution reactions.

PART III

Appendices

Appendices

18

Glossary

Absolute configuration: Defines the real position in space of all the molecular co-ordinates. Previously referred to the fact that the configuration of the molecule had been related to D or L glyceraldehyde.

Absorption: The process whereby the chemical in question enters the molecular lattice of the compound doing the absorption. In contrast to adsorption, where the chemical in question remains only on the surface.

Acetal: The functional group $RCH(OR)_2$. Sometimes also used for the functional group $R'_2C(OR)_2$ which should more correctly be called a ketal.

Acetylide: Contains the $RC\equiv C^-$ anion.

Achiral: A molecule which is not chiral.

Aci tautomer: A tautomeric form, $R_2C=N^+(-OH)O^-$, of a nitro compound, R_2CH-NO_2, hence the nitro/aci tautomerism.

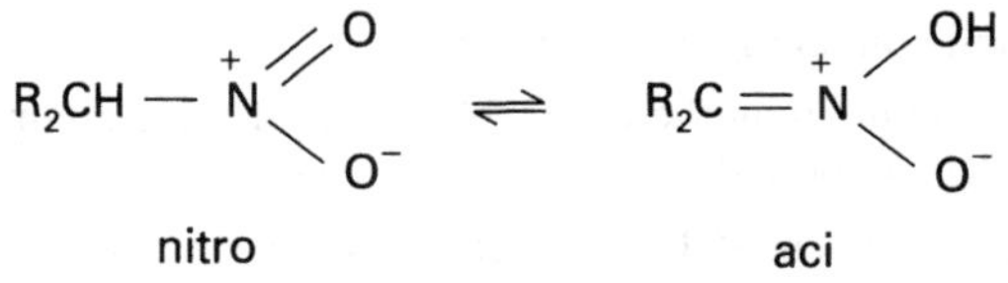

$$R_2CH - \overset{+}{N}\underset{O^-}{\overset{O}{\diagup\diagdown}} \rightleftharpoons R_2C = \overset{+}{N}\underset{O^-}{\overset{OH}{\diagup\diagdown}}$$

nitro   aci

Acid: Usually either a proton donor (Brönsted–Lowry definition) or an electron pair acceptor (Lewis definition), but there are other definitions, such as Arrhenius and Cady–Elsey.

Acid anhydride: Compound of the general type $R^1CO_2COR^2$. May be symmetrical or mixed.

Acid halide: Compound of the general type RCOX.

Acrylonitrile: Cyanoethene, $CH_2=CH-C\equiv N$.

Glossary

Acyl: The functional group $-C(=O)R$.

Acyloin condensation: The reaction between two esters in the presence of sodium dissolved in xylene to give a 2-hydroxyketone. Useful for forming large rings. The acyloin group is $R_2C(OH)-C(=O)R$.

Addend: The product of an addition reaction.

Adsorption: The process whereby the molecule in question only reacts with the surface of the compound which is doing the adsorption. In contrast to absorption, which is a bulk phenomenon.

Alcohol: Compound of the general type ROH. Sub-divided into primary, secondary and tertiary depending on the number of carbons attached to the carbon that is joined to the hydroxyl, OH, group.

Alcoholysis: The cleavage of a large molecule by an alcohol to give two other molecules with the overall addition of the alcohol to the original molecule.

Aldehyde: Compound of the general type $RCH=O$.

Aldimine: Compound of the general type $RCH=NR$. Also called an azomethine.

Aldoketene: Compound of the general type $RCH=C=O$.

Aldol condensation: The reaction between an aldehyde, ketone or ester with another aldehyde or ketone in the presence of a base to give a 3-hydroxyaldehyde or ketone. Dehydration may follow to give the α, β-unsaturated aldehyde or ketone, then called the crotonaldehyde reaction.

Aldoxime: Compound of the general type $RCH=NOH$.

Alicyclic: A carbon ring compound, which may be saturated or unsaturated, but is not aromatic in nature.

Aliphatic: An open-chained carbon compound, i.e. not aromatic or alicyclic. It may be saturated or unsaturated.

Alkali: A compound that gives rise to the hydroxide anion in an aqueous solution.

Alkene: A compound containing the $C=C$ group.

Alkoxide: The oxygen anion of an alcohol, i.e. RO^-.

Alkyl: The generic term for an aliphatic carbon substituent, often abbreviated to $R-$.

Alkylation: The addition of an alkyl group, to a carbon (C-alkylation), nitrogen (N-alkylation), oxygen (O-alkylation), or sulphur (S-alkylation) atom.

Alkyne: A compound containing the $C\equiv C$ group.

Allene: A compound containing the $C=C=C$ group. If there are more than two such double bonds then the compound is called a cumulene.

Allylic position: The methylene (CH_2) or methine (CH) group next to a $C=C$ bond.

Ambident nucleophile: A nucleophile that is capable of attacking in two or more different ways to give different products, e.g. NCO^- may attack via the oxygen to give cyanates, ROCN, or via the nitrogen to give isocyanates, RNCO.

Amide: The $R-CONR^1R^2$ group, which is the condensation product of a carboxylic acid and an amine. It may be primary, secondary or tertiary depending on whether there are zero, one or two alkyl groups attached to the nitrogen.

Amidine: The $NH_2-CR=NH$ group.

Amidoxime: The $RC(=NOH)-NH_2$ group.

Amine: An alkylated or arylated derivative of ammonia. It may be primary, secondary, tertiary or quaternary, e.g. RNH_2, R_2NH, R_3N, or R_4N^+.

Amino acid: A molecule that has both an amino group and a carboxylic acid group in it. If these two groups are attached to the same carbon it is called an α-amino acid, which on polymerisation forms proteins.

Aminonitrene: The $R_2N–N$: molecule.

Ammonia: The NH_3 molecule.

Amphoteric: A species that may act as both an acid or a base, e.g. water.

Anchimeric assistance: The assistance rendered by a neighbouring group that results in the rate of substitution being faster than if the attacking species were attacking directly.

Angle strain: The strain imposed on a bond when an sp^3 hybridised carbon is in either a large or a small ring so that the angle between the bonds in the ring deviates from 109°. Also applies to the strain on sp^2 or sp hybridised carbons in a ring, when the bond angle deviates from 120° or 180°, respectively.

Anhydride: A condensation product that has resulted from the loss of a water molecule.

Aniline: The molecule $C_6H_5NH_2$.

Anion: A negatively charged ion.

Anisole: The molecule $C_6H_5OCH_3$.

Annulation: Used to describe a reaction that results in the formation of a ring.

Anti: A mechanistic description of the addition or elimination of a molecule, AB, to or from a double bond, where A joins or leaves from the opposite orientation to B. Opposite of *syn*.

Anti-aromatic: *See* aromaticity.

Anti-bonding orbital: An orbital that if occupied by electrons, leads to the molecule being less stable than if it remained unoccupied.

Anticlinal: Used to describe the conformer of a 1,2-disubstituted ethane derivative in which the dihedral angle between the groups under consideration is 120°. The other conformations are called *antiperiplanar, gauche* (or *synclinal*), and *synperiplanar*, to represent the dihedral angles of 180°, 60° and 0°, respectively.

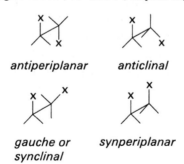

antiperiplanar   *anticlinal*

gauche or   *synperiplanar*
synclinal

Anti-Markovnikov: The addition to a double bond of a molecule in which the negative part joins to the carbon that initially had the most hydrogens, usually the result of a radical mechanism.

Antiperiplanar: *See anticlinal.*

Glossary

Apofacial: The attacking species enters from the opposite side as the leaving groups departs, as opposed to synfacial.

Aprotic: Used to describe a solvent that does not readily donate protons, e.g. THF. Usually such a solvent has no hydrogens attached to heteroatoms.

Arbuzov reaction: The formation of a phosphonate from a phosphite, $P(OR)_3$, and an alkyl halide.

Arene: A compound based on an aromatic ring.

Arenium ion: The addition of an electrophile to an aromatic ring gives rise to a cation of the general formula $C_6H_6E^+$, which is called a Wheland intermediate, or a σ complex or an arenium ion.

Arndt–Eistert reaction: The conversion of an acid chloride to the carboxylic acid that is its homologue, i.e. a compound with an extra carbon in the chain. Performed using diazomethane to give the diazo ketone, followed by heating with silver oxide and then acid hydrolysis. The second step involves the Wolff rearrangement.

Aromaticity: The property of having $(4n + 2)$ π electrons in a closed circuit of molecular orbitals. The commonest example is benzene, where n equals one, around a six-carbon ring. A more precise definition is the ability to sustain an induced diamagnetic ring current, i.e. one that is diatropic.

Arrhenius: A definition of an acid/base system, in which an acid is a hydrogen cation donor, while a base is a hydroxide anion donor. Really only of use in aqueous conditions.

Aryl: The generic term for a substituted or unsubstituted aromatic substituent.

Aryne intermediate: An intermediate formed in the substitution reaction of a chlorobenzene derivative when treated with a very strong base such as sodamide. The intermediate has a triple bond in the ring.

Ate complex: When a Lewis acid combines with a base to give a negative ion in which the central atom has a higher than normal valence, e.g. $(CH_3)_4B^-Li^+$. *See also* onium salt.

Atomic number: The number of protons in the nucleus of an element.

Atomic orbital: An electron orbital that is centred on only one nucleus.

Atropisomers: Stereoisomers that may be separated, because they do not readily interconvert by reason of restricted rotation about a single bond.

Axial: Used to describe a bond that projects above or below the plane of a cyclohexane ring. In contrast to equatorial.

Azide: A molecule of the general formula RN_3.

Azide anion: The N_3^- anion.

Aziridine: A saturated three-membered ring comprising two carbons and one nitrogen.

Azirine: The 1-azirine has a three-membered ring comprising two carbons and doubly bonded nitrogen. The 2-azirine, in which the double bond is between the carbon atoms, does not exist, possibly because it is predicted to be antiaromatic.

1–azirine   2–azirine

Azlactone: A five-membered ring compound, which is a useful synthetic intermediate, also called 5-oxazolone.

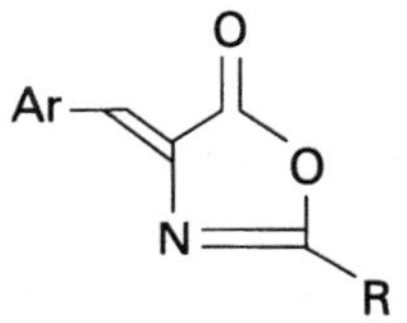

Azo compound: A compound of the general formula $RN=NR$.

Azomethine: *See* aldimine.

B strain: Stands for back strain, and is the compression strain resulting from the steric crowding of non-bonding groups.

Baeyer–Villiger rearrangement: The rearrangement of a ketone with a peroxyacid, e.g. *m*-CPBA, to give an ester.

Banana bond: *See* small angle strain.

Base: Usually either a proton acceptor (Brönsted–Lowry definition) or an electron pair donor (Lewis definition), but there are other definitions.

Bathochromic shift: When a given chromophore absorbs at a certain wavelength, and the substitution of one group for another causes the absorption to be at a higher wavelength. The opposite to a hypsochromic shift.

Beckmann rearrangement: The rearrangement reaction that occurs when a ketone is treated with hydroxylamine, NH_2OH, followed by, for example, PCl_5. The reaction proceeds via an oxime to result in a secondary amide.

Bent bonds: *See* small angle strain.

Benzidine rearrangement: The rearrangement of hydrazobenzene under acidic conditions to give 4,4'-diaminobiphenyl, benzidine. General for *N*,*N*'-diarylhydrazines.

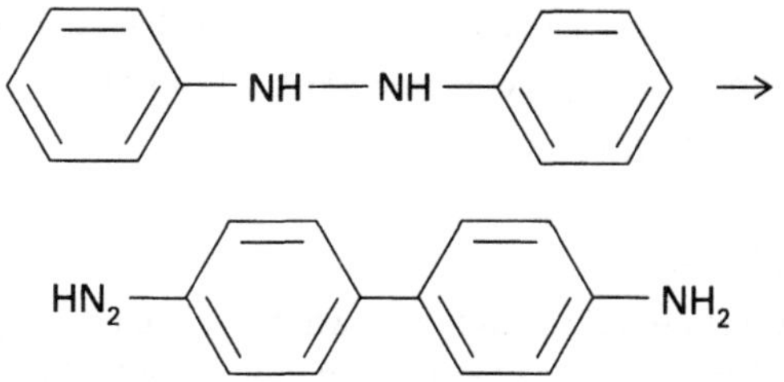

Benzil–benzilic acid rearrangement: The rearrangement of a 1,2-diketone aromatic species in the presence of hydroxide anion to give the α-hydroxycarboxylic acid.

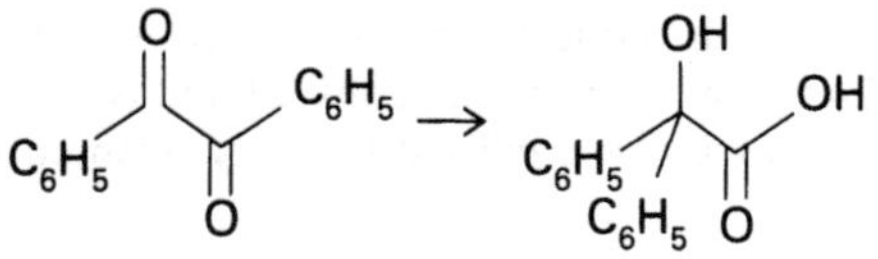

Benzoic acid: The molecule $C_6H_5CO_2H$.

Benzoin condensation: The reaction between two aromatic aldehydes with no α-hydrogens in the presence of NaCN, to give the 2-hydroxyketone adduct.

Benzophenone: The molecule $C_6H_5COC_6H_5$.

Benzoyl: The C_6H_5CO- group.

Benzyl: The $C_6H_5CH_2-$ group.

Birch reduction: The reduction of an aromatic ring using sodium in liquid ammonia, to result in a non-conjugated system of two double bonds.

Bisulphite: The product that results from the addition of an HSO_3^- anion to a carbonyl group.

Boat conformer: A conformer of cyclohexane.

Bond energy: The average energy needed to break all the bonds of a given type in a certain molecule. In water, it is the average of the two O–H dissociation energies, while in methane it is the average of four C–H dissociation energies.

Bredt's rule: In small bridged bicyclic compounds, double bonds are impossible at the bridgehead. Small means that the S number must be less than or equal to six. The S number is the sum of the number of atoms in the bridges of a bicyclic system. If the S number is equal to seven, then if the double bond is in a seven-membered ring it cannot be formed; it can be formed, however, if it is in an eight-membered ring.

Bromonium ion: The cyclic version of the β-bromo carbonium ion, $R_2C^+-CBrR_2$. The bromine atom forms a bridged bromonium ion with the adjacent carbon atoms.

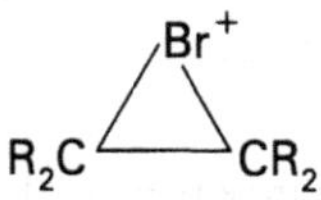

Brönsted–Lowry: A theory that defines an acid as a proton donor, and a base as a proton acceptor. Every acid and base is related to its conjugate base and acid respectively.

Brosylates: Compounds of the general type $p\text{-}BrC_6H_4SO_2OR$.

Bunte salts: Anions of the general type $RSSO_3^-$.

C-alkylation: *See* alkylation.

Cady–Elsey: A rare and esoteric definition of what constitutes an acid or base. An acid is defined as a species that raises the concentration of the cation of the solvent, while a base raises the concentration of the anion of the solvent. This definition is of use when dealing with autolytic solvents such as liquid ammonia.

Cahn–Ingold–Prelog: This system has essentially replaced the DL system for representing stereoisomers. The four groups around an asymmetric carbon are ranked according to a set of sequence rules, according to the atomic number of the constituent atoms. Used to distinguish asymmetric centres.

Cannizzaro reaction: A disproportionation reaction between two molecules of aldehydes that usually have no α-hydrogens in the presence of base to give an alcohol and an acid. Called a crossed Cannizzaro reaction if the reaction occurs between different aldehydes.

Canonical form: A hypothetical representation of a molecule, ion or radical, in which the electrons are portrayed as being present in discrete localised orbitals, i.e. a Lewis structure. The real molecule, ion or radical actually exists as the weighted average of all the possible canonical forms, which is called the resonance hybrid.

canonical forms   resonance hybrid

Carbamate: A molecule of the general formula R^1R^2NCOOR.

Carbamic acid: The molecule H_2NCOOH.

Carbanion: A negatively charged carbon species.

Carbene: The bivalent neutral carbon species that has only six electrons around it, e.g. dichlorocarbene, $:CCl_2$.

Carbenium ion: Strictly, the name for a trivalent positive carbon species, CR_3^+. However, in practice, this species is usually referred to as a carbonium ion.

Carbides: Containing either the C^{4-} or C_2^{2-} unit: the former should, more strictly, be called methylides, while the latter should be called acetylides.

Carbocation: Any positively charged carbon species.

Carbodiimide: A molecule with the general formula $RN=C=NR$.

Carbonate: An ester of carbonic acid, $O=C(OR)_2$.

Carbonic acid: The molecule H_2CO_3.

Carbonium ion: Strictly, the name for tetra- and penta-coordinated positive ions of carbon. However, in general use as the name for any positive carbon species.

Carbonyl: A carbon doubly bonded to an oxygen, i.e. $R^1R^2C=O$. The R groups are usually hydrogen or carbon chains. If the R groups are otherwise, then the entity is given a new name, e.g. carboxylic acid, ester, acid halide, acid anhydride, amide, etc.

Carbonylation: The formation of a carbonyl functionality, e.g. adjacent to an existing carbonyl group using SeO_2.

Carboxylate: The oxygen anion of a carboxylic acid, i.e. RCO_2^-.

Carboxylation: The addition of CO_2.

Carboxylic acid: A compound of the general type R–COOH.

Carboxylic ester: A compound of the general type R^1COOR^2, which is the condensation product between a carboxylic acid and an alcohol.

Glossary

Catalysis: The process of altering the rate of a reaction by means of a catalyst.

Catalyst: A substance that alters only the rate at which the equilibrium of a reaction is reached without altering the position of the resulting equilibrium. The catalyst is not consumed during the reaction. It may be either heterogeneous or homogeneous, depending on whether or not the catalyst is in a different or the same physical state as the reaction mixture it is catalysing.

Catenanes: Compounds made up of two or more rings interlaced as links in a chain.

Cation: A positively charged ion.

Chain reaction: A reaction that is self-perpetuating, e.g. the photochlorination of methane. The product quantum yield is greater than one, which means that more molecules of product are formed than quanta of light are absorbed by the reagent. There is typically an initiation step, followed by a large number of propagation steps, ending with a series of termination steps.

Chair conformer: The most stable conformer of cyclohexane.

Cheletropic reaction: A reaction in which two σ bonds that terminate at a single atom are made or broken in concert.

Chiral: Used to describe a molecule that does not possess a centre of symmetry, a plane of symmetry, or an alternating axis of symmetry. Such a molecule will rotate plane polarised light.

Chloral hydrate: The water addition product of chloral, i.e. $Cl_3CCH(OH)_2$.

Chloroform: The trivial name for trichloromethane.

Chloromethylation: The addition of a $-CH_2Cl$ group, e.g. by the action of methanal and HCl on an arene.

Chromophore: A group that causes the absorption of light by a molecule.

Chugaev reaction: The thermal pyrolysis of a methyl xanthate to an alkene.

Cine: When substitution occurs at a different position on the aromatic ring than at the position of the original group.

Cinnamic acid: The molecule 3-phenylpropenoic acid.

CIP system: *See* Cahn–Ingold–Prelog.

Cis: A term used to describe the geometric isomer that has the two substituents in question on the same side of the double bond. In the CIP system the symbol would be *Z*. The opposite of *trans*.

Cisoid: Used to describe a conformer of a diene. *See transoid*.

Claisen condensation: The reaction of two ester molecules in the presence of the appropriate alkoxide base, followed by mild acid hydrolysis to yield the 3-ketoester. The cyclic version is called the Dieckmann reaction.

356

Claisen rearrangement: The thermal rearrangement of allylarylethers to *ortho*-allylphenols. If both *ortho* positions are occupied, then the allyl group migrates to the *para* position.

Claisen–Schmidt reaction: The cross condensation reaction of an aromatic aldehyde with a simple aliphatic aldehyde or ketone with 10% KOH to yield the dehydrated product, i.e. an α, β-unsaturated arenone.

Clathrate: A host compound that forms a crystal lattice in which a guest molecule fits, and in which the only interaction is van der Waals forces. The spaces formed by the host compound completely enclose the guest molecule, unlike inclusion compounds.

Clemmensen reduction: The reduction of a carbonyl group using a zinc/mercury amalgam and concentrated HCl to give the methylene unit.

Condensation: A reaction in which two approximately equally sized molecules react to form one larger molecule with the elimination of a much smaller molecule, often water.

Configuration: Two isomers that can only be interconverted by the breaking and reforming of covalent bonds.

Conformation: Two isomers that may be interconverted merely by free rotation about bonds.

Conjugate acid/base: Two molecules that are related by the addition of a single proton to the Brönsted–Lowry base to give rise to the Brönsted–Lowry acid or vice versa.

Conjugated system: A system containing alternating single and multiple bonds.

Cope reaction: The thermal cleavage of an amine oxide to produce an alkene and a hydroxylamine. The amine oxide is usually formed *in situ* from the amine and an oxidising agent. Stereoselectively *syn*, via a five-membered Ei mechanism.

Cope rearrangement: The [3,3] sigmatropic rearrangement of 1,5-dienes. If the reagent is symmetrical about the 3,4 bond then the product is indistinguishable from the starting material.

Correlation diagram: *See* Woodward–Hoffmann rules.

Covalent bond: A bond that comprises electrons that are shared more or less equally, usually between two nuclei. The electrons occupy molecular orbitals.

Cram's rule: The less hindered side of a carbonyl group will be preferentially attacked by an incoming nucleophile.

L, M, S, are large, medium and small groups

Cresol: The compound with the formula $CH_3C_6H_4OH$, of which there are three isomers, namely with the methyl and the hydroxyl groups *ortho*, *meta* and *para* to each other.

Criegee mechanism: A proposed route for the formation of an ozonide when ozone reacts with an alkene.

Cross-conjugation: A system in which there are at least three unsaturated groups, two of which are not conjugated with each other.

Crown ether: A large ring ether compound, which contains several oxygen atoms, and which possesses the property of forming complexes with positive ions, often metallic ions.

Cumulene: A compound containing more than two carbon/carbon double bonds joined directly to one another, i.e. without an intervening carbon/carbon single bond, e.g. $C=C=C=C$. If there are just two such double bonds, then the compound is an allene.

Curly arrow mechanism: The movement of electrons, usually in pairs, is indicated by curly arrows to indicate which bonds are being formed and which are being broken within the reacting molecules.

Curtius rearrangement: The pyrolytic rearrangement of an acyl azide to an isocyanate.

Cyanate: A compound of the general type $R–O–C≡N$.

Cyanic acid: The molecule HNCO.

Cyanide: A compound of the general type $R–C≡N$; also called a nitrile.

Cyanide anion: The univalent $[C≡N]^-$ anion.

Cyanoethylation: The addition of a $–CH_2CH_2C≡N$ group.

Cyanogen: The molecule $N≡C–C≡N$.

Cyanohydrin: Compound of the general type $R_2C(OH)CN$; also called a hydroxynitrile.

Cycloaddition: A type of pericyclic reaction in which there is a concerted addition of two molecules, i.e. addition takes place in only one step.

Cycloalkane: A cyclic version of an open chain alkane.

Cyclopentadienyl anion: The aromatic cyclic anion $C_5H_5^-$. Its salts are called cyclopentadienide salts.

Cyclopropenyl cation: The aromatic cyclic cation $C_3H_3^+$. The salts are called cyclopropenium salts.

Dakin rearrangement: The oxidation of an aromatic aldehyde with hydrogen peroxide to yield the corresponding phenol under alkaline conditions. The reaction proceeds via a rearrangement that is akin to the Baeyer–Villiger rearrangement. There is a requirement for a hydroxy or amino group to be in the *ortho* or *para* position to the aldehyde group.

Darzens condensation: The coupling of an α-bromocarboxylate ester with an aldehyde or ketone under basic conditions to give a glycidic ester, i.e. an ester with an α, β-epoxide group.

Decarboxylation: The loss of a molecule of carbon dioxide from a molecule.

Dehydration: The loss of a molecule of water from a molecule.

Delocalisation: The situation in which a molecular orbital is centred on more than two atomic nuclei.

Diastereomer: A stereoisomer that is not an enantiomer.

Diastereotopic: The situation when two groups cannot be brought within chemically identical positions by rotation or any other symmetry operation, i.e. the opposite of enantiotopic.

Diatropic: *See* aromaticity.

Diaziridine: A saturated three-membered ring comprising one carbon atom and two nitrogen atoms.

Diazoalkane: A compound of the general type $R_2C=N^+=N^-$.

Diazonium species: A molecule containing the $-N_2^+$ group.

Dieckmann: The cyclic version of the Claisen condensation.

Diels–Alder reaction: The 1,4-addition of a double bond to a conjugated diene (a 4 + 2 cycloaddition) to produce a six-membered ring.

Dieneophile: A species that likes to attack dienes.

Dienone–phenol rearrangement: A rearrangement in which a cyclohexadienone that has two substituents at the 4 position, yields on treatment with an acid a phenol substituted at the 3 and 4 positions.

Dihedral angle: The angle described between two *geminal* substituents when viewed along the connecting bond.

Diimide: The HN=NH molecule.

Dimerisation: The formation of a dimer, i.e. a molecule that has the same empirical formula, but double the RMM.

Di-π-methane: The photochemical rearrangement of 1,4-dienes to vinylcyclopropanes.

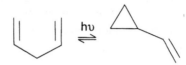

Diol: A compound that contains two hydroxyl groups, usually on different carbons, e.g. a *vicinal* diol such as $HOCH_2$–CH_2OH. If the diol is *geminal*, then it is called a hydrate.

Dismutation: *See* metathesis.

Dissociation energy: The energy necessary to cleave a particular bond to give the constituent radicals, e.g. HO–H to give HO· and H· requires 495 kJ mol^{-1}; while H–O to give ·O· and H· requires 420 kJ mol^{-1}. The average of (approximately) 460 kJ mol^{-1} is taken as the O–H bond association energy in water.

Eclipsed: The dihedral angle is zero between atoms or groups that are bonded to adjacent carbon atoms located in a saturated backbone. When the dihedral angle is 60°, then the conformation is called staggered.

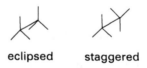

eclipsed   staggered

Eglinton reaction: The symmetrical coupling of terminal alkynes by heating with stoichiometric amounts of cupric salts in pyridine, or a similar base.

***E* isomer:** A stereochemical term used to indicate that the two groups with the highest ranking under the CIP system are diametrically positioned across a double bond. If they are on the same side, then the system is designated *Z*. *E* and *Z* are abbreviations of *entgegen* and *zusammen*, meaning opposite and together respectively.

Electrocyclic: A type of pericyclic reaction in which there is a concerted cyclisation of a polyene, or the reverse, which results in ring opening.

Electrofuge: A leaving group that departs without its electron pair, i.e. the leaving group equivalent of an attacking electrophile.

Electronegativity: The property of an element that represents the degree of attraction that that element has for electrons. The greater the electronegativity, the greater the attraction.

Electrophile: A species that likes to attack areas of high electron density, i.e. negative sites. The electrophile is usually electron deficient or positively charged or both.

Elimination: A reaction where a small part of the original molecule is eliminated to leave behind the bulk of the molecule, which usually now has some higher degree of unsaturation.

Elimination/addition: Overall a reaction that results in substitution of a group, but which proceeds via an initial elimination followed by an addition.

Empirical formula: The formula that indicates the relative proportions of the constituent elements in one molecule of the compound.

Enamine: A compound that contains the $R_2C=C(R)-NR_2$ configuration. The nitrogen atom is usually alkylated, or else the imine is more stable.

Enantiomer: A stereoisomer that is not superimposable on its mirror image.

Enantiomorph: An alternative name for an enantiomer.

Enantiotropic: Used to describe two atoms or groups that upon replacement with a third group give enantiomers.

Endo **addition:** In the Diels–Alder reaction, if part of the dienophile is under the diene such that the hydrogens on the dienophile adopt the equatorial position in the product, then the addition is *endo*. This allows for maximum overlap of secondary orbitals, and is thus favoured; hence this is the kinetic product. The opposite of *exo* addition.

Endo **cyclic:** Used to describe a double bond that is between two atoms within a ring, as opposed to one that is external to a ring, which is called *exo*.

Ene reaction: The addition of a reactive dienophile to an allylic compound that has a hydrogen in the 3 position.

Enol: The tautomeric isomer of a ketone or aldehyde, e.g. the enol of propanone is $CH_2=C(OH)CH_3$.

Enol ether: The *O*-alkyl derivative of an enol, e.g. $R_2C=CHOR$, which is also called a vinyl ether.

Enolate: The anion of an enol, in which the negative charge formally resides on the oxygen.

Enthalpy: The energy of a system.

Entropy: The disorder of a system.

Envelope conformer: One of the puckered conformations of cyclopentane, the other being the half chair conformation.

Glossary

Epimer: A diastereomer having the opposite configuration at only one chiral centre.

Episulphide: The sulphur equivalent of an epoxide, which is also called a thiirane.

Episulphone: A three-membered ring that contains two carbons and an SO_2 group.

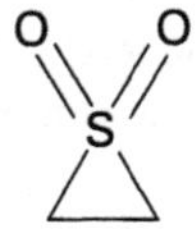

Epoxide: A three-membered ring compound in which two corners are occupied by carbon atoms and the third by an oxygen. Also called an oxirane.

Equatorial bond: In a six-membered ring system, those bonds that project approximately along the plane of the six atoms. The other bonds are axial.

Erlenmeyer reaction: The condensation reaction between an aromatic aldehyde and an *N*-acyl derivative of glycine, in the presence of acetic anhydride and sodium acetate, to give an azlactone.

***Erythro* isomer:** The stereoisomer of $C_{abx}C_{aby}$ that is represented by the following diagram using the Fischer projection. *See threo* isomer.

<pre>
      x   x
      | |
a ----+---- b   b ----+---- a
      | |
a ----+---- b   b ----+---- a
      | |
      y   y
</pre>

Eschweiler–Clarke reaction: The reductive methylation reaction of an amine with methanal and methanoic acid to give the *N*-methyl derivative.

Ester: The condensation product of a carboxylic acid and an alcohol, i.e. $R^1CO_2R^2$. Strictly, a carboxylic ester, as it is possible for alcohols to form esters with inorganic acids, e.g. with carbonic acid to form carbonate esters, $O=C(OR)_2$. However, even this particular example is open to debate as carbonic acid is sometimes considered to be an organic compound.

Ether: A molecule of the general formula R^1OR^2.

Exhaustive alkylation: Used in reference to the alkylation of, for example, an amine, when the nitrogen becomes fully alkylated.

***Exo* addition:** The opposite of *endo* addition, in that the smaller side of the dieneophile is under the cyclic diene and hence there is no opportunity for secondary orbital overlap. This leads to the thermodynamic product being formed.

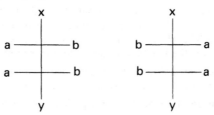

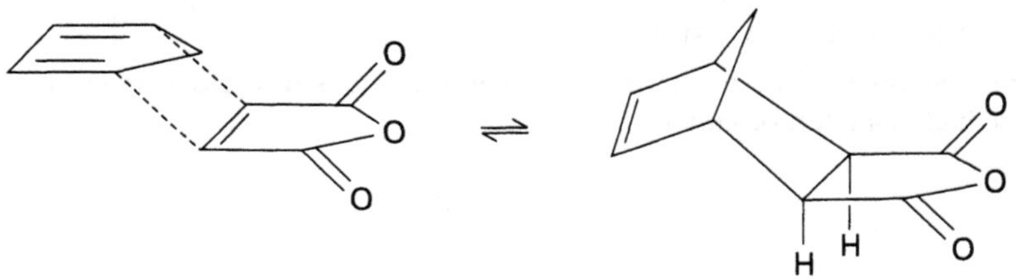

Exo **cyclic:** Used to describe a double bond that joins an atom in a ring to an atom that is not within the ring. The opposite of an *endo* cyclic double bond.

Favorskii rearrangement: The rearrangement that occurs when an α-halogenoketone, under basic conditions, reacts to form an ester.

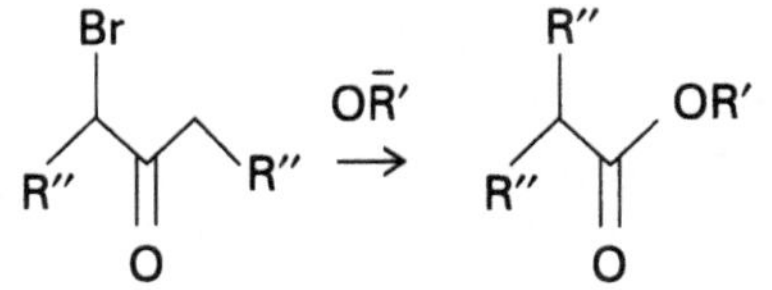

Fehling's solution: Made by dissolving Rochelle salt (sodium potassium 2,3-dihydroxy-butanedioate tetrahydrate) and NaOH in water and adding the mixture to a solution of $CuSO_4$. The Rochelle salt prevents the precipitation of the $Cu(OH)_2$, and the mixture formed is violet due to complexed Cu^{2+} ions. It is reduced by aldehydes to give a red precipitate of Cu_2O. Usually used as a qualitative test for an aldehyde (Fehling's test).

Fenton's reagent: A mixture of H_2O_2 and $FeSO_4$. Used as an oxidising reagent, e.g. to hydroxylate aromatic rings, but the yields are often poor. *See also* Udenfriend reagent.

Ferrocene: A well known metallocene that comprises the sandwich compound between an Fe^{2+} ion and two cyclopentadienyl anions.

Field effect: The direct through space polarisation of a bond by another charged centre.

Finkelstein reaction: The exchange of one halide for another in aliphatic compounds, resulting in the equilibrium mixture being formed.

Fischer projection: A notation used to portray the three-dimensional arrangement of the constituent parts of a chiral molecule. The horizontal lines project forward from the paper, while the vertical lines project backwards.

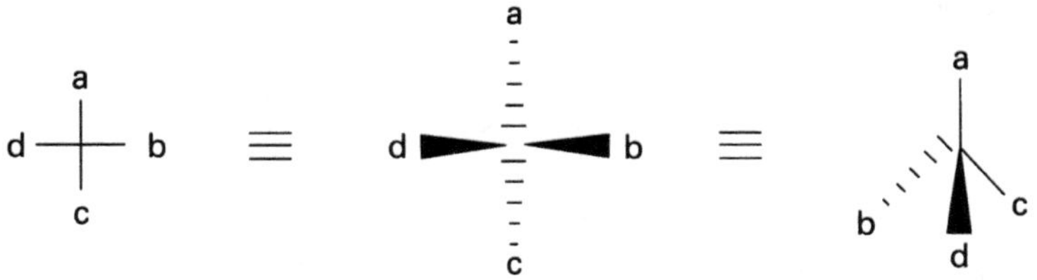

Formamide: The molecule H_2NCHO.

Formamidine: A compound of the general type $RN=CHNR_2$.

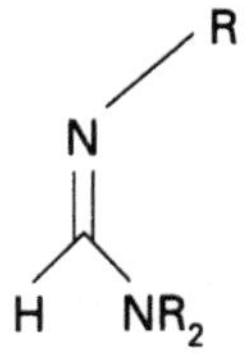

Formylation: The replacement of a hydrogen atom with a CHO group.

Fragmentation: A reaction in which the carbon moiety is the positive leaving group (i.e. the electrofuge) in an elimination reaction.

Free radical: A species that has an unpaired electron.

Friedel–Crafts acylation: The reaction of an acyl halide, using stoichiometric amounts of a Lewis acid such as aluminium trihalide, with an aromatic compound to form the acylated derivative. This variation avoids the polysubstitution that commonly occurs in the Friedel–Crafts alkylation reaction.

Friedel–Crafts alkylation: The reaction of an alkyl halide, using stoichiometric amounts of a Lewis acid such as aluminium trihalide, with an aromatic compound to form the alkylated derivative.

Fries rearrangement: The rearrangement of a phenolic ester by heating with a Lewis acid catalyst to produce *ortho-* and *para*-acylphenols.

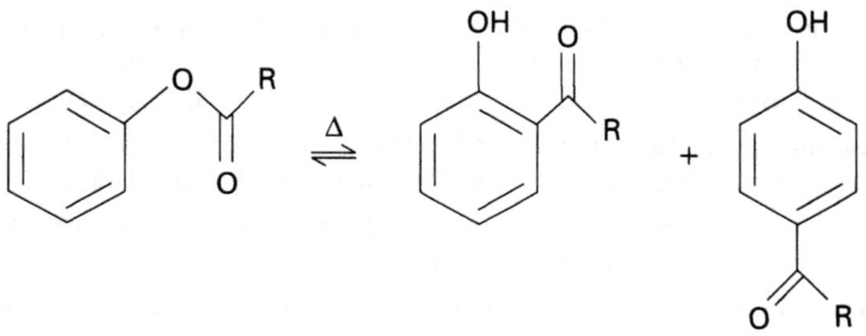

Frontier orbital: The orbital that is either the highest energy orbital that is occupied (HOMO), or the lowest in energy that is empty (LUMO). *See also* Woodward–Hoffmann rules.

F-strain: Face strain is the steric strain present when a covalent bond is formed between two atoms each of which has three large groups.

Fulvene: A cyclopentadiene with an additional *exo*-cyclic double bond.

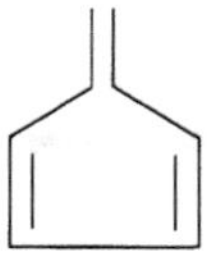

Fumaric acid: *Trans*-butenedioic acid. The *cis* isomer is called maleic acid.

Functional group: The atom, or group of atoms, that will react as a single moiety. Nearly always refers to a heteroatomic atom or group such as a hydroxyl, nitro or keto group.

Functional group interchange: Also known as the functional group interconversion, refers to the conversion of one functional group to another.

Furan: A five-membered heterocyclic ring containing two carbon/carbon double bonds and an oxygen.

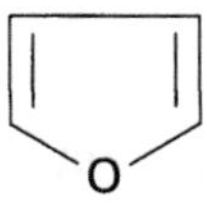

Gatterman amide synthesis: The synthesis of an amide by the reaction between an aromatic molecule and carbamoyl chloride, H_2NCOCl, in the presence of $AlCl_3$.

Gatterman–Koch reaction: The reaction of an aromatic molecule with CO and HCl to give an aromatic aldehyde, in the presence of $AlCl_3$ and CuCl.

Gatterman reaction: The conversion of a diazonium salt to the aryl chloride or bromide using copper metal and HCl or HBr.

Gauche: *See anticlinal.* Also called *synclinal.*

Geminal: The two groups under consideration are attached to the same carbon atom, e.g. a *geminal* dichloride, $-CCl_2-$. *See also vicinal.*

General acid catalysis: In the two step reaction, where the first step is the protonation of A by the conjugate acid of the solvent to form AH^+, and the second step is the reaction of AH^+ to form the products, the reaction is described as being general acid catalysed if the second step is fast and the first step is rate determining. This reflects the fact that an increase in the concentration of any acid will increase the rate of the reaction. General acid catalysis is in contrast to specific acid catalysis.

General base catalysis: The same as general acid catalysis, *mutatis mutandis*, but with the first step being the deprotonation of AH by a general base.

Gibbs free energy: The change in the total energy of the system as calculated from the equation $\Delta G = \Delta H - T\Delta S$, where ΔH is the change in the enthalpy, and ΔS is the change in the entropy of the system.

Glyceraldehyde: The *dextro* or (+) isomer of glyceraldehyde was arbitrarily assigned the configuration shown below by Rosanoff and given the label D. The *laevo* or (−) isomer was given the label L. In 1951, using phase lagged X-ray crystallography, Bijvoet was able to determine that Rosanoff had made the correct choice. When a molecule has been assigned to either the D or L series it is said that its absolute configuration is known. The DL system is little used now having been replaced by the Cahn–Ingold–Prelog system which uses the symbols R and S.

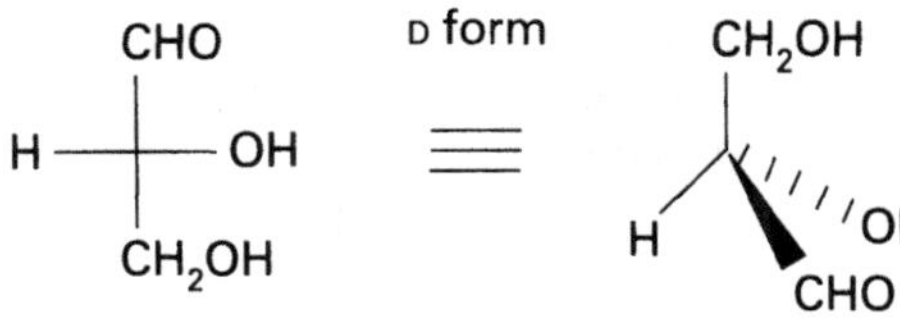

Glycidic ester: The reaction in which a glycidic ester (an α,β-epoxycarboxylic ester) rearranges under acidic conditions to give, after heating, an aldehyde.

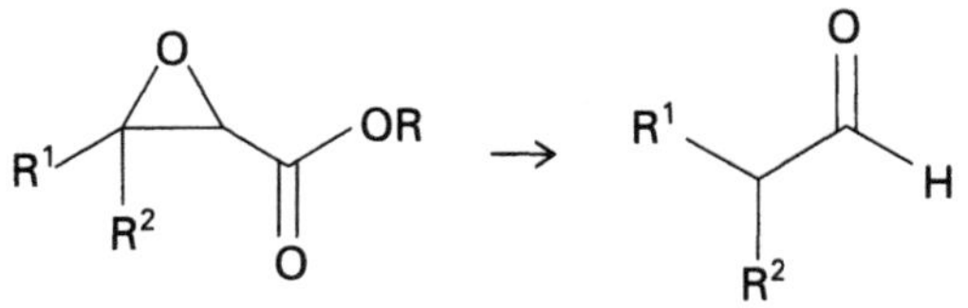

Gomberg–Bachmann reaction: When an acidic solution of a diazonium salt is made alkaline, the aryl portion of the diazonium salt may couple with another aromatic ring, i.e. ArH reacting with ArN_2^+ would give Ar–Ar.

Grignard reagent: An organohalogenomagnesium compound with the generic formula of RMgX; it may more accurately be represented by the Schlenk equilibrium that exists between the following species: RMgX; R_2Mg and MgX_2, and '$R_2Mg \cdot MgX_2$'.

Guanidine: The compound with the formula $(H_2N)_2C=NH$. A very strong organic base.

Half chair conformer: One of the puckered conformations of cyclopentane, the other being the envelope.

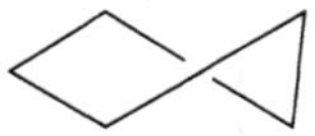

Haloform: A molecule of the general formula CHX_3, where X is a halogen. Used also as the general name for the reaction that is illustrated by the iodoform reaction.

Halogen: The general term for a group VII element, e.g. F, Cl, Br or I. The adjective is halogeno.

Halogenation: The addition of a halogen atom to a molecule.

Halogen dance: The rearrangement of polyhalogenobenzenes catalysed by very strong bases, which proceeds via an aryl carbanion intermediate (i.e. an S_E1 mechanism).

Halohydrin: A molecule of the general formula $R_2C(OH)X$.

Hammond postulate: This states that for any single reaction step, the geometry of the transition state for that step resembles the compound to which it is closer in Gibbs free energy.

Hard electrophile/nucleophile: A species whose behaviour as an electrophile or nucleophile is mainly governed by Coulombic (i.e. electric charge) interactions. Tends to be difficult to polarise, and so usually acts as a Brönsted–Lowry acid or base.

Haworth notation: A type of notation used to describe the three-dimensional configuration of a compound.

Haworth reaction: A reaction in which an aryl compound is treated with a cyclic anhydride, such as succinic anhydride, and the intermediate Friedel–Crafts product is reduced and then cyclised via an internal Friedel–Crafts reaction to give a 1,2-disubstituted aromatic compound with a carbonyl group in the new ring. The whole sequence is called the Haworth reaction.

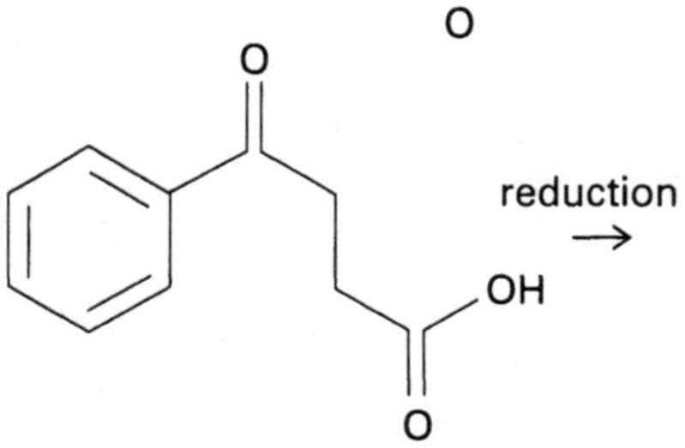

Helicene: A molecule which is chiral due to the whole molecule having a helical shape, e.g. hexahelicene, in which one end of the molecule must lie above the other, because of steric crowding.

Hell–Volhard–Zelinskii reaction: The reaction in which a carboxylic acid is treated with Br_2 and PBr_3 (often formed *in situ* from red phosphorus and bromine) to give the α-bromocarboxylic acid bromide.

Hemiacetal: A molecule of the general type RCH(OR)OH. Also used to describe a molecule of the general type $R_2C(OR)OH$, which is more correctly called a hemiketal.

Hemihydrate: A species formed by the condensation of two hydrates of carbonyl compounds to give a compound of general formula RCH(OH)–O–CHR(OH).

Heteroatom: Either any atom that is different from that under consideration; or any atom that is not a carbon or a hydrogen, but is usually restricted to non-metals, e.g. N, O, P, S or Si.

Heterocyclic: A cyclic compound that contains a heteroatom within the ring. If the ring is aromatic, the molecule is heteroaromatic.

Heterogeneous: An adjective that is usually applied to a catalyst to mean that the catalyst is in a different physical state from that of the reaction mixture it is catalysing, e.g. the catalytic hydrogenation of gaseous alkenes by platinum. In contrast, if the catalyst is homogeneous, then it is in the same physical state as the reaction mixture, e.g. the acid catalysis of an acetal in solution by an acid.

Heterolytic fission: The unequal breaking of a covalent bond so that both electrons are taken by one part, and hence the products are charged.

Hofmann degradation: *See* Hofmann exhaustive methylation and Hofmann rearrangement. A term best avoided, because it is ambiguous.

Hofmann elimination: The less highly substituted alkene derivative is formed.

Hofmann exhaustive methylation: The treatment of an amine with excess methyl iodide to form the quaternary ammonium iodide, which is then converted into the hydroxide by treatment with moist silver oxide; on heating in an aqueous or alcoholic solution, it decomposes to give an alkene. Also called the Hofmann degradation.

Hofmann rearrangement: The reaction of an amide with bromine and sodium methoxide in methanol to give the corresponding amine, with the loss of the carbonyl carbon atom as carbon dioxide. Also called the Hofmann degradation.

Homoallylic: Used to describe the situation where there is one carbon atom, usually part of a methylene group, between the group in question and the carbon/carbon double bond.

$$\begin{array}{c} H_v \quad H_h \; H_h \\ \diagdown \diagup \diagup \\ H_v \diagup \quad \diagdown \\ \diagup \diagdown \\ H_v \quad H_a \; H_a \end{array}$$

v vinylic
a allylic
h homoallylic

Homoaromatic: A compound that contains one or more sp^3 hybridised atoms, usually carbon, in an otherwise conjugated cyclic system, e.g. the homotropylium cation, $C_8H_9^+$, in which there is one sp^3 carbon, and an aromatic sextet of electrons spread over seven carbon atoms.

Homogeneous: *See* heterogeneous.

Homologisation: The addition/removal of a methylene unit to form the homologue.

Homologue: A compound of the same class, but with either one more, or one less, methylene unit in the carbon chain.

Homolytic fission: The equal breaking of a covalent bond so that both parts take one electron each to form two radical products.

Homotopic: *See* stereoheterotopic.

Huang–Minlon reduction: The modification of the Wolff–Kishner reaction in which the reaction is carried out in refluxing diethylene glycol. This version has completely replaced the original method.

Hückel's rule: Electron rings that have $(4n + 2)$ π electrons are aromatic, while those that have $4n$ π electrons are anti-aromatic.

Hund's rule: When a number of degenerate orbitals are available and there are not enough electrons to fill them all, then all the orbitals will be half-filled before any of them are fully filled.

Hunsdiecker: The reaction of the silver salt of a carboxylic acid with Br$_2$ to give the bromoalkane with one carbon less than the original compound.

Hybridisation: The hypothetical mixing of atomic orbitals to form hybrid orbitals.

Hydrate: A molecule of the general formula $R_2C(OH)_2$, i.e. the product of the addition of water to a carbonyl compound.

Hydration: The addition of water to a molecule.

Hydrazide: A molecule of general formula RCONHNH$_2$.

Hydrazide anion: The H_2NNH^- anion.

Hydrazine: The $H_2N–NH_2$ molecule, which forms hydrazo compounds when one hydrogen on each nitrogen is substituted, RHN–NHR, and hydrazines, when the substitution is only on one nitrogen, RHN–NH$_2$. Similarly, R_2NNH_2 is often referred to as a *syn*-dialkylhydrazine.

Hydrazoic acid: The molecule HN$_3$.

Hydrazone: A compound with the general formula $R^1R^2C=N–NR^3_2$.

Hydride: The H$^-$ anion.

Hydroacylation: The addition of an RCHO group across a double bond, resulting in the addition of hydrogen to one end and an acyl, $R(C=O)-$, group to the other.

Hydroboration: The addition of an HBR_2 group across a double bond.

Hydrocarbon: A compound composed of just carbon and hydrogen atoms; may be alicyclic, cyclic, aromatic, saturated or unsaturated.

Hydrocarboxylation: The addition of HCO_2H across a double bond.

Hydrocyanic acid: The molecule HCN, also known as hydrogen cyanide.

Hydroformylation: The addition of HCHO across a double bond, achieved by the reaction of CO and H_2 with a cobalt catalyst.

Hydrogenation: The addition of a hydrogen molecule across an unsaturated bond, usually a carbon/carbon bond.

Hydrogen bonding: The weak bond between a functional group, X–H, and an atom or group of atoms, Y, in the same or different molecules, i.e. *intra-* or *inter-*molecular respectively. X and Y are usually oxygen, nitrogen, chlorine or fluorine.

Hydrolysis: The splitting of a large molecule, usually into two fragments by the overall addition of water, e.g. ester plus water gives alcohol plus carboxylic acid.

Hydroperoxide: A molecule with the general structure ROOH.

Hydroxamic acid: A compound of the general type RCONHOH. The enol form is called hydroximic acid.

keto   enol

Hydroxide anion: The HO^- anion.

Hydroxonium cation: The H_3O^+ cation.

Hydroxyl: The –OH group.

Hydroxylamine: The molecule NH_2OH.

Hydroxylation: The replacement of a hydrogen atom with the OH group. If a molecule of H_2O_2 is added to give a 1,2-diol, then that is called dihydroxylation.

Hydroxynitrile: *See* cyanohydrin.

Hyperconjugation: The delocalisation of electrons using the σ molecular bond electrons from a hydrogen/carbon bond, also called a no-bond resonance or σ-conjugation. Muller and Mulliken call the hyperconjugation found in neutral molecule sacrificial hyperconjugation, and this involves no-bond resonance and charge separation not found in the ground state. In the isovalent hyperconjugation found in free radicals and carbonium ions, the canonical forms display no more charge separation than is found in the main form.

Hypohalous acid: An acid of the general formula HOX, where X is a halogen.

Hypsochromic shift: The opposite to a bathochromic shift.

Imidate: *See* imidic ester.

Glossary

Imidazole: An unsaturated heterocycle, also called iminazole or glyoxaline.

Imidazolide: The *N*-acyl derivative of an imidazole.

Imide: A compound of the general type RC(=O)–NHC(=O)R.

Imidic ester: A compound of the general type RC(OR)=NR, also called imidate, imino ester or imino ether.

Imidine: A compound of the general type RC(=NH)–NH–C(=NH)R.

Imido sulphur: A compound of the general type R–N=S=N–R.

Imidoyl chloride: A compound of the general type R–N=C(R)–Cl, also called imine chloride.

Imine: A compound of the general structure $R^1R^2C=NR^3$, also called a ketimine or a Schiff base.

Iminium cation: A cation of the general formula $R_2C=NR_2{}^+$.

Imino chloride: *See* imidoyl chloride.

Imino ether: *See* imidic ester.

Iminonitrile: A compound of the general type $RC(C\equiv N)=NH$.

Inclusion compound: A host compound that forms a crystal lattice that has spaces large enough for guest molecules. The only bonding between the host and the guest compounds is van der Waals forces. The spaces in the lattice are in the form of long tunnels or channels, unlike clathrate compounds in which the spaces are completely enclosed.

Indicator: Usually a water soluble acid/base conjugate pair in which one or both of the species is highly coloured. The pK_a is around the pH that is of interest to the reaction being studied. Used to indicate whether the solution under investigation is more or less acidic than the pH that corresponds to the pK_a value of the indicator.

Inductive effect: The polarisation of a bond caused by the electronegativity of an adjacent atom or group, which acts through the σ bond system, and not through either the π bond system or directly through space.

Initiation: The first step of a free radical reaction in which the number of free radicals increases.

Insertion: A reaction in which another atom is introduced into a chain. Often the species that is introduced is a carbene, nitrene or such like.

***In situ*:**   Used to describe the generation of a reagent in the reaction vessel, often because the reagent is unstable and so cannot be stored.

Intermediate: A molecular species that actually exists, if only briefly, along a reaction pathway from the starting material to the final product. Such a species exists in a well in a free-energy profile, i.e. there is an activation barrier before and after the intermediate, and so, in principle, it could be isolated. The deeper the well the more stable the intermediate.

370

Inversion: Used in respect to the configuration of a chiral centre, when in the course of a reaction the stereochemistry of the chiral atom is inverted, i.e. changes from R to S, or vice versa.

Iodoform reaction: The reaction of a secondary alcohol that is adjacent to a terminal methyl group, i.e. $RCH(OH)CH_3$, which after oxidation *in situ* to the related ketone, i.e. $RCOCH_3$, by an iodine/hydroxide mixture, reacts further to form the carboxylic acid and CHI_3, and the latter precipitates as a yellow solid. It is a test for the general structure: $RCHOHCH_3$ or $RC(=O)CH_3$.

Iodonium ion: The iodo equivalent of a bromonium ion.

Ionic bond: A Coulombic, i.e. electrostatic, attraction between two or more atomic nuclei that are charged as a result of having lost or gained electrons in their valence shells, i.e. are ions.

Isochronous: Hydrogens that are indistinguishable in the NMR spectrum.

Isocyanate: The univalent $-N=C=O$ group.

Isocyanide: *See* isonitrile.

Isoinversion: *See* isoracemisation.

Isomer: A configuration of the molecule that corresponds to the molecular formula. There may be several isomers that each correspond to the molecular formula, but each is different from the other in the exact type of bonds, or the distribution of those bonds.

Isonitrile: The univalent $-N=C$ group. Note that the carbon atom is only divalent in this moiety.

Isoracemisation: The situation in which racemisation occurs faster than isotope exchange at the chiral centre. The first step must be isoinversion.

Isothiocyanate: The univalent $-N=C=S$ group.

Isotope: A variant of an element that has the same atomic number, but a different atomic mass, i.e. a different number of neutrons within the nucleus. For example, hydrogen and deuterium are isotopes, each with one proton, but with zero and one neutron, respectively.

Isotope effect: The primary isotope effect is where the substitution of, for example, a deuterium atom for a protium atom has an effect on the rate, because that bond is broken in the rate determining step. The secondary isotope effect is where there is an effect on the rate, but, for example, the C–H/D bond does not break in the reaction. This is subdivided into α and β effects. In the latter, substitution of, for example, a deuterium for a protium β to the position of bond breaking slows the reaction due to the effect on hyperconjugation. In the former a replacement of, for example, a protium by a deuterium at the carbon containing the leaving group affects the rate. In this case, the greater the carbonium ion character the greater the effect.

I strain: Internal strain that results from changes in ring strain in going from a tetrahedral to a trigonal carbon or vice versa.

Jones reagent: A solution of chromic acid and sulphuric acid in water, used to oxidise secondary alcohols to ketones.

Ketal: A compound of the general type $R^1_2C(OR^2)_2$. *See also* acetal.

Ketene: A compound of the general type $R^1R^2C=C=O$. Where both R^1 and R^2 are alkyl groups it is more correctly called a ketoketene. *See also* aldoketene.

Ketene acetal: A compound with the general structure $R_2C=C(OR)_2$.

Ketene aminal: A compound with the general structure $R_2C=C(NR_2)_2$.

Ketenimine: A compound with the general structure $R_2C=C=N–R$.

Ketenimmonium salt: A compound with the general structure $R_2C=C=NR_2{}^+$.

Ketimine: *See* imine.

Keto: The prefix used to indicate the presence of a carbonyl group that is joined to two carbon atoms. Thus, for example, a keto acid is $RC(=O)CO_2H$.

Keto/enol tautomerism: The tautomerism between the keto and enol forms: $R_2CH–C(=O)R$ and $R_2C=C(–OH)R$. This type of isomerism requires at least one α-hydrogen atom.

Ketone: A molecule with a non-terminal carbonyl group, e.g. $R^1C(=O)R^2$ (where R^1 and R^2 are not hydrogen).

Ketoxime: The molecule $R_2C=N–OH$ (where R is not hydrogen).

Ketyl radical: The $R_2C\cdot(–OH)$ species.

Kiliani method: Used in carbohydrate chemistry to extend the length of the carbon chain by one carbon atom. Following the formation of a cyanohydrin of the corresponding ketone, the nitrile group is then hydrolysed to give the α-hydroxycarboxylic acid.

Kizhner reaction: The Russian name for the Wolff–Kishner reaction.

Knoevenagel reaction: A condensation reaction in which a methylene unit that has two adjacent $-M$ groups such as carbonyl, ester or nitro groups, reacts in the presence of base with an aldehyde or ketone, that usually does not have an α-hydrogen. Dehydration usually follows to yield the conjugated unsaturated compound.

Koble–Schmitt reaction: The carboxylation of sodium phenoxides by carbon dioxide in the *ortho* position. The potassium salt forms mainly the *para* substituted compound.

Koch reaction: The hydrocarboxylation of a double bond using CO and H_2O, in the presence of H^+ ions at elevated temperatures and pressures.

Kolbe reaction: The electrolytic decarboxylative dimerisation of the potassium salts of two carboxylate anions, used to prepare symmetrical R–R alicyclic compounds from $RCO_2{}^-K^+$.

Kornblum's rule: As the nature of a reaction changes from SN1 to SN2, an ambident nucleophile becomes more likely to attack with its less electronegative atom.

Lactam: A cyclic amide.

Lactide: A cyclic dimer formed from two molecules of α-hydroxyacids.

Lactone: A cyclic ester.

Lanthanide shift reagent: An organic complex of a lanthanide element that has the property of shifting the NMR signals of a compound with which it can form a co-ordination complex. Chiral lanthanide shift reagents shift the peaks of the two enantiomers to different extents.

Lasso mechanism: The groups that are going to react together are placed next to each other and then encircled, i.e. 'lassoed', to indicate that they have bonded, while the remaining truncated molecules fuse together. A small molecule, typically H_2O, formed by fusion of the two groups, is 'lassoed out'.

Le Châtelier's principle: A system at equilibrium will react to any perturbation so as to resist the change being forced on it.

Least motion: A theory proposed by Hine that those elementary reactions that will be favoured are those that involve the least change in atomic position and electronic configuration.

Levelling effect: If in any given solvent, two bases appear to be equally protonated within experimental error, then even though they may actually have very different pK_b values, they appear to have approximately the same pK_b in that solvent; i.e. the solvent is said to level the effect of the bases. Equally applicable to acids.

Lewis acid/base: A species that is capable of accepting/donating a pair of electrons; H^+ or $AlCl_3$ are examples of Lewis acids, while NH_3 is an example of a Lewis base.

Lewis structure: A representation of a molecule, ion, or radical, that shows the electrons as being contained in discrete localised orbits. If there are two or more such structures for any given compound, they may also be called canonical forms, which when combined give the resonance hybrid.

Lindlar catalyst: Pd on $CaCO_3$, partly poisoned with PbO. Used to hydrogenate partially, not partly, alkynes to alkenes.

London effect: The temporary creation of a dipole in a molecule caused by the random movement of electrons within a molecular bond.

Lone pair: A pair of electrons in the valence orbital, e.g. the non-bonding pair of electrons in NH_3.

Lossen rearrangement: The *O*-acyl derivative of a hydroxamic acid gives an isocyanate when treated with a base or, sometimes, just heat.

Lucas test: When an alcohol is mixed with concentrated HCl and $ZnCl_2$ at room temperature, tertiary alcohols form the alkyl chloride immediately, while secondary ones take about five minutes, and primary ones react far more slowly.

Maleic acid: *Cis*-butendioic acid. The *trans* isomer is fumaric acid.

Malic acid: 2-Hydroxybutandioic acid.

Malonic acid: The trivial name for propandioic acid, $CH_2(CO_2H)_2$, i.e. a *geminal* carboxylic acid.

Mannich reaction: The reaction of a ketone with formaldehyde in the presence of an amine or ammonia under acidic conditions to give the 3-aminocarbonyl adduct, which is referred to as a Mannich base. After methylation and heating with silver oxide, the α,β-unsaturated β-carbonyl compound may be formed.

Markovnikov's rule: The addition of the positive end of the addition molecule to the carbon that had the most hydrogens on it. This rule may be remembered by the saying that 'the rich become richer, while the poor remain poor' in relation to the number of hydrogen atoms attached to any particular carbon atom.

Meerwein–Ponndorf–Verley: The reduction of a ketone to a secondary alcohol using isopropanol and aluminium isopropoxide. The opposite of the Oppenauer oxidation.

Meisenheimer rearrangement: The rearrangement of a tertiary amine oxide on heating to give a substituted hydroxylamine. The migrating group is usually allylic or benzylic.

Meisenheimer salt: The salt that results from, for example, the reaction between a cyanide anion and 1,3-dinitrobenzene or 1,3,5-trinitrobenzene (at the 2 or 4 positions) in an aprotic solvent such as $CHCl_3$.

Mercaptal: The sulphur equivalent of an acetal, $RCH(SR)_2$, sometimes called thioacetal. The $RCH(OR)SR$ type compound tends to be called a hemimercaptol instead of the more correct hemimercaptal.

Mercaptan: A molecule containing the –SH functional group, i.e. the sulphur equivalent of an alcohol. Also called a thiol.

Mercaptide anion: The RS^- anion.

Mercaptol: A compound of the general type $R_2C(SR)_2$; also called mercaptole, dithioacetal or thioketal. An $R_2C(OR)SR$ compound is called a hemimercaptol.

Meso: A stereochemical term meaning that there is a plane of symmetry passing through the molecule, and hence it is not optically active.

Mesoionic: A molecule that cannot be satisfactorily represented by Lewis forms not involving charge separation. Nearly all known examples contain five membered rings of which the most common are the sydnones.

Mesomeric effect: The movement of π electrons in the π molecular orbital system.

Mesylate: A compound of the type $CH_3SO_2–O–R$.

Metaformaldehyde: The cyclic trimer of formaldehyde is usually called trioxane, but is sometimes called metaformaldehyde or trioxymethylene. The hydrated linear polymer form of formaldehyde, $(CH_2O)_n \cdot H_2O$, is called paraformaldehyde. The trimer of acetaldehyde is called paraldehyde, while its tetramer is called metaldehyde.

Metallocene: Also called a sandwich compound, it usually contains two cyclopenta-dienyl anions that form a sandwich around a metallic ion, but may contain a tropylium cation instead of one of the cyclopentadienyl anions.

Metathesis: The reaction when two alkenes react in the presence of a catalyst to interchange their alkylidiene, $R_2C=$, groups. Also called dismutation and disproportio-nation of olefins. The resulting mixture usually represents the statistical distribution of all possible products.

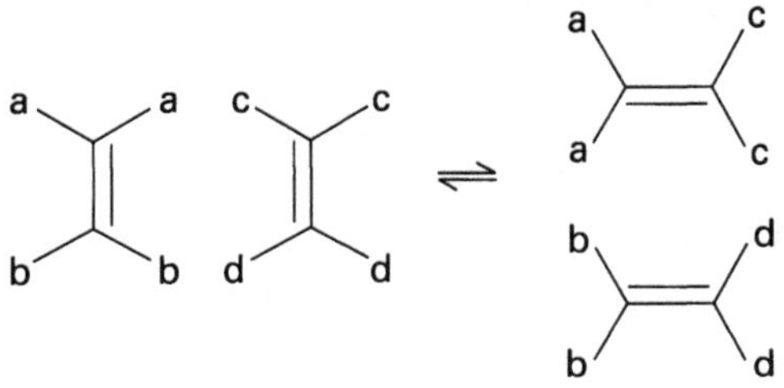

Methanonium ion: The CH_5^+ cation, very rare except in students' examination scripts, and chemical ionisation mass spectrometry.

Methoxide anion: The CH_3O^- anion.

Methylene unit: The $-CH_2-$ unit.

Michael addition: The addition to an α, β-unsaturated carbonyl compound by a carbanion, in particular if the anion is stabilised by a carbonyl group, or other $-M$ group such as nitrile or nitro.

Molecular formula: The formula which gives the actual number of atoms for each element in one molecule of the compound.

Molecular orbital theory: A theory used to describe the electronic structure of a molecule by the combination of atomic orbitals that give rise to molecular orbitals. The latter are orbitals that are centred on more than one nucleus.

Molozonide: The result of the first step of the 1,3-dipolar addition of ozone to an alkene, also called the initial or primary ozonide.

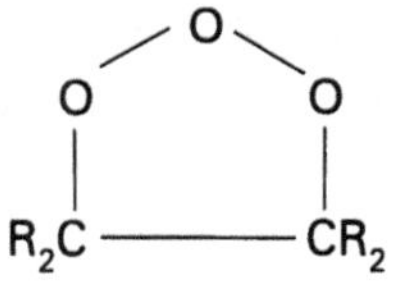

N-alkylation: *See* alkylation.

Naphthalene: The compound with the formula $C_{10}H_8$, a condensed aromatic species.

Nef rearrangement: Primary or secondary aliphatic nitro compounds may be hydrolysed to aldehydes or ketones, respectively, by treatment with a base followed by a mild aqueous acid which yields the aci form of the nitro compound. This can then be hydrolysed by concentrated sulphuric acid to give the carbonyl compound.

Neopentane: The molecule $C(CH_3)_4$; the neopentyl radical has the structure $(CH_3)_3CCH_2 \cdot$. More correctly written as *neo*-pentane.

Newman projection: A notation used to indicate the geometry around an sp^3/sp^2 bond.

Nitration: The replacement of a hydrogen atom with an NO_2 group, i.e. the formation of a nitro compound.

Nitrene: The univalent $-N$ species, i.e. the nitrogen equivalent of a carbene. The simplest nitrene, NH, is also called nitrene, as well as imidogen, azene or imene.

Nitrenium cation: A species of the form R_2N^+ or $R_2C=N^+$.

Nitrile: *See* cyanide.

Nitrile oxide: The $-C\equiv N^+-O^-$ functional group.

Nitrilium cation: The $R^1-C\equiv N^+-R^2$ cation.

Nitrite anion: The $NO_2{}^-$ anion.

Nitrite ester: A molecule with the general structure $R-O-N=O$.

Nitro: The $-NO_2$ group.

Nitro tautomer: *See* aci tautomer.

Nitrone: A molecule comprising the general structure $R_2C=N^+(O^-)R$.

Nitronic ester: The result of O-alkylation of the carbanion of the aci tautomer of a nitro compound and having the general formula $R_2C=N^+(O^-)OR$.

Nitronium cation: The $NO_2{}^+$ cation.

Nitrosation: The addition of the $-NO$ group.

Nitroso: The $-NO$ group.

Nitrosonium cation: The NO^+ cation.

Nitrous acid: The molecule HONO.

No-bond resonance: *See* hyperconjugation.

Norrish Type I reaction: The photolytic cleavage of an aldehyde or ketone to give $RCO\cdot$ and $R\cdot$.

Norrish Type II reaction: The photolytic cleavage of $R_2CH-CR_2-CR_2COR$ to give $R_2C=CR_2$ and R_2CHCOR.

Nosylate: A compound of the type $p\text{-}O_2NC_6H_4-SO_2-OR$.

Nucleofuge: A leaving group that departs with an electron pair.

Nucleophile: A species that attacks areas of positive charge or electron deficiency. Usually negatively charged or possessing a lone pair.

O-alkylation: *See* alkylation.

Octant rule: A rule that attempts to predict the conformation of a carbonyl carbon by cutting it into octants using three perpendicular planes each of which bisects the carbonyl carbon, and then looking at the orientation of the substituents within each octant. Used mainly on cyclohexanone derivatives.

Octet rule: Second row elements can only have a maximum of eight electrons, i.e. an 'octet of electrons', in their valence shell.

Olefin: A compound with a carbon/carbon double bond, more properly referred to as an alkene.

Oligomer: A polymer that contains only a few of the monomer, as opposed to several tens which would be a telomer, or hundreds of the monomer which would be a true polymer.

Onium salt: Formed when a Lewis base expands its valence, e.g. tetramethylammonium iodide, $(CH_3)_4N^+I^-$. Analogous to 'ate' complexes.

Oppenauer oxidation: The oxidation of a secondary alcohol to a ketone using $((CH_3)_3CO)_3Al$; the reverse of the Meerwein–Ponndorf–Verley reduction.

Optically active: A compound that rotates the plane of polarised light.

Orbital: The solution of a Schrödinger wave equation that is used to describe the behaviour of an electron is called an eigenfunction. When a graphical representation of such an eigenfunction is made, it is called an orbital. It represents the volume in space where an electron that occupies that orbital is most likely to be found.

Orbital symmetry: A property of orbitals that results from the wave nature of the equations used to define the properties of the electron.

Order of a reaction: The summation of those numbers that indicate the power to which each of the concentrations of the reagents are raised in the rate equation. It need not be an integer. It is an experimentally determined number.

Organometallic: A compound that contains an organic component that is covalently bonded to a metal atom.

Ortho carbonate: A molecule of the type $C(OR)_4$.

Ortho effect: In electrophilic substitution of aromatic rings that already have two substituents, one of which is a *meta* directing group and that positioned *meta* to an *ortho*/*para* directing group, then the new substituent tends to go to the *ortho* position. In mass spectrometry, the *ortho* effect is the loss of a neutral species, usually a stable molecule such as water, derived from the *ortho*-substituent in $M^{+\bullet}$. It defines the substitution pattern as *ortho*.

Ortho ester: A molecule of the general formula $R^1C(OR^2)_3$.

Ortho/para ratio: The ratio of electrophilic substitution at the *ortho*, as opposed to the *para*, position of a monosubstituted aromatic ring.

Osazone: A molecule of the general formula RC(=N–NHPh)–C(=N–NHPh)R; particularly important in carbohydrate chemistry.

Out–in isomerism: Found in those salts of tricyclic diamines that have the nitrogen atoms at the bridgeheads. If the bridges are longer than six atoms the N–H bond may be formed either inside the cage or outside.

Oxa-di-π-methane: The light-induced rearrangement of β,γ-unsaturated ketones to give an acylcyclopropane.

Oxalic acid: The molecule with the formula $(CO_2H)_2$.

Oxaphosphetane: The four-membered ring containing a phosphorous and oxygen atom, formed as an intermediate in the Wittig reaction.

Oxetane: A four-membered saturated ring comprising three atoms of carbon, one of oxygen.

Oxidation: The gain of an oxygen atom or atoms; the loss of a hydrogen atom or atoms; or the loss of an electron or electrons.

Oxidation number: The number associated with an atom, either alone or in combination with other atoms, that reflects the nominal loss or gain of electrons that has occurred. See the appendix on oxidation numbers for more detail.

Oxime: A compound of the general structure $R_2C=NOH$.

Oxime ether: A compound of the general formula $R_2C=NOR$.

Oxirane: Another name for an epoxide.

Oxonium cation: An R_3O^+ or $R_2C=O^+-R$ cation.

Oxo process: The commercial hydroformylation of an alkene by treatment with CO and H_2 over, usually, a cobalt carbonyl catalyst.

Oxy-Cope rearrangement: The rearrangement of a 3-hydroxy-1,5-diene to the 1-hydroxy-1,5-diene, which then tautomerises to the ketone or aldehyde.

Oxyamination: The formation of a carbon/oxygen and carbon/nitrogen bond from a carbon/carbon double bond.

Ozonide: The five-membered ring that results from the ozonolysis of carbon/carbon double bonds.

Ozonolysis: The formation of an ozonide, which may then fragment to give the corresponding aldehyde(s) or ketone(s).

378

π donor: A compound that donates a pair of electrons that were originally in a π orbital, and so forms a π-complex. For example, when benzene complexes with an electrophile, initially forming a π-complex, but which then rearranges to form a σ-complex or Wheland intermediate.

π orbital: A molecular orbital that has one nodal plane that includes the atomic centres.

π* orbital: An anti-bonding molecular orbital that has two nodal planes, one that includes the atomic centres, and one that is perpendicular and passes between the atomic centres.

Paenecarbanion: An E2 elimination reaction whose transition state resembles the E1cb mechanism.

Paenecarbonium: An E2 elimination reaction whose transition state resembles the E1 mechanism.

***Para*-Claisen rearrangement:** *See* Claisen rearrangement.

Paraffin: The trivial name for an alkane.

Paraformaldehyde: *See* metaformaldehyde.

Paramagnetic ring currents: *See* ring current.

Paratropic: Used to describe a compound that sustains paramagnetic ring currents.

Passerini reaction: The formation of an α-acyloxyamide when an isonitrile is treated with a carboxylic acid and an aldehyde or ketone. Related to the Ugi reaction.

Paterno–Büchi reaction: The photochemical addition of a ketone to an alkene to give an oxetane.

Pechmann reaction: *See* von Pechmann reaction.

Peptide: A polymer of amino acids.

Per: A prefix used to indicate that there is a maximum amount of the element under consideration, e.g. perchloroethane for C_2Cl_6; or, alternatively, more than the usual number of expected components, e.g. peroxide, in which there are two, rather than one, oxygen atoms involved.

Peracid: A molecule of the general formula $RC(=O)OOH$.

Perester: A molecule of the general formula $RC(=O)OOC(=O)R$.

Pericyclic: A reaction type in which the bond breakage and formation steps occur simultaneously without any charged or radical intermediates. There are three main sub-divisions, namely electrocyclic, sigmatropic and cycloadditions.

Perkin reaction: The reaction between an acid anhydride with an aromatic aldehyde in the presence of base (which is usually the corresponding carboxylate anion), followed by dehydration and hydrolysis results in the β-aryl-α,β-unsaturated carboxylate anion.

Glossary

Peroxide: A molecule with the general formula ROOR.

Peroxide effect: The use of peroxides to change the regiochemistry of addition from Markovnikov to anti-Markovnikov.

Peroxirane: An epoxide that has an oxygen atom attached to the oxygen in the ring, and so results in an ylid.

Peroxyacid: *See* peracid.

Peroxyester: *See* perester.

pH: Defined as $-\log_{10}[H^+]$. Strictly, may only be defined with water as the solvent, but is often used in a less rigid sense in different solvents.

Phantom atom: Used in the CIP notation to terminate the valency of all atoms except hydrogen, when the atom in question is involved in a multiple bond.

Phenol: The molecule C_6H_5OH.

Phenol–dienone rearrangement: The reverse of the dienone–phenol rearrangement.

Phenonium ion: The bridged carbonium cation that results from the departure of a leaving group that is β to the aromatic ring.

Phenoxide anion: The $C_6H_5O^-$ anion.

Phenoxy: The univalent group C_6H_5O-.

Phenyl: The univalent group C_6H_5-.

Phenyl cation: The $C_6H_5^+$ cation.

Phenylcarbene: The C_6H_5CH neutral molecule.

Phosgene: The molecule $COCl_2$.

Phosphine: The molecule PH_3.

Phosphonate: A compound of the general formula $(RO)_2P(=O)-R$.

Phosphonium cation: A cation of the type R_4P^+.

Phosphorane: An ylid of the type $Ar_3P^+-CH^--R$.

Photochemistry: The chemistry of reactions that proceed when exposed to light. Usually radical reactions or pericyclic processes, which often occur in either the gas phase or inert solvents, and which do not favour charged intermediates.

Photolysis: The cleavage of a molecule induced by light.

Pinacol–pinacolone rearrangement: The rearrangement of a 1,2-diol when treated with concentrated sulphuric acid. Named after the rearrangement of pinacol to pinacolone.

OH OH

$+H^+$
$\rightarrow$

O

Pinner reaction: The synthesis of the hydrochloride salt of an imidic ester when a mixture of a nitrile and an alcohol is treated with dry HCl. If hydrochloric acid is used instead of dry HCl, then the intermediate product is subsequently hydrolysed to give the ester.

Piperidine: A saturated six-membered cyclic compound that contains one nitrogen and five carbon atoms.

Pitzer strain: The strain that results from eclipsed conformations, as opposed to transannular strain.

pK_a value: A measure of the strength of an acid. The lower the value the stronger the acid. Negative values are possible for the strongest of acids. Defined as $-\log_{10}([A^-][H^+]/[HA])$.

pK_b value: A measure of the strength of a base. The higher the value, the stronger the base. Defined as $-\log_{10}([BH^+][OH^-]/[B])$.

Poisoning: A catalyst is said to be poisoned if it is working at less than maximum efficiency because of some chemical contamination, which is commonly a sulphur or lead species.

Polarisability: The ease with which the electron cloud surrounding a group may be distorted by the approach of an electrical charge. When the electron cloud is easily distorted, it is said to be very polarisable and so is also soft in nature.

Poly: A prefix meaning many.

Polymerisation: The addition of many monomers to each other to form a very long chain, which is called a polymer. Short polymers are called telomers, and even shorter ones are called oligomers.

p orbital: An atomic orbital that has one nodal plane, that passes through the atomic centre.

Prevost method: The overall *anti*-hydroxylation of an alkene achieved by treating the alkene with a 1:2 molar ratio of iodine and silver benzoate, followed by hydrolysis. Overall *syn*-hydroxylation may be achieved by the method of Woodward.

Prilezhaev reaction: The formation of an epoxide from the reaction of perbenzoic acid on an alkene.

Prochiral: A molecule of the general formula $CABX_2$: if one of the X groups were substituted for a group Y, then the molecule would become chiral.

Product spread: If a reaction gives rise to two products, then the greater the difference in yields of the two products the wider the product spread.

Propagation: A reaction in which the number of radicals formed equals the number of radicals consumed.

Propargyl: The group $HC \equiv C–CH_2–$.

Protic solvent: A solvent that is capable of forming hydrogen bonds with anionic solutes.

Prototropic: The isomerisation of a carbon/carbon bond, usually to give the thermodynamic equilibrium mixture of products. Often achieved by the addition of a strong base, and proceeds via a proton removal and an allylic rearrangement.

Puckered ring: When a ring system adopts a conformation that is not planar, it is said to be puckered, e.g. the boat or chair conformations of cyclohexane.

Pyramidal inversion: The inversion of the configuration of a molecule that has a trigonal pyramidal geometry, such as simple nitrogen compounds. Also called the umbrella effect.

Pyran: A doubly unsaturated six-membered ring compound that contains one oxygen and five carbons.

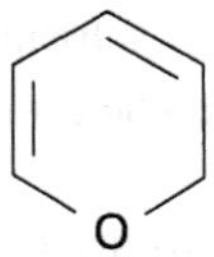

2H-pyran

Pyrazole: A doubly unsaturated five-membered cyclic compound that contains two nitrogen atoms.

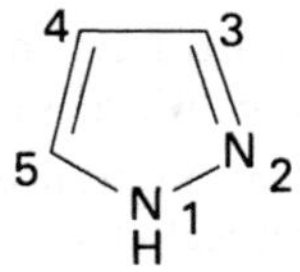

Pyrazoline: An unsaturated five-membered cyclic compound that contains two nitrogen atoms.

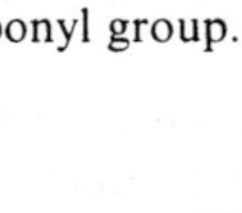

1-pyrazoline   2-pyrazoline

Pyrazolone: An unsaturated five-membered cyclic compound that contains two nitrogen atoms and a carbonyl group.

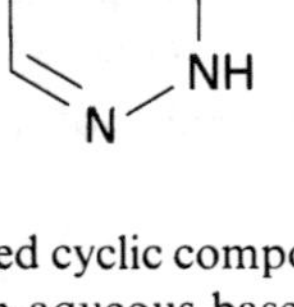

Pyridine: An aromatic six-membered cyclic compound, that contains only one nitrogen. Often used as a solvent, or as a non-aqueous base.

Pyridinium cation: The $C_5H_6N^+$ cation.

Pyridone: A doubly unsaturated six-membered ring compound with only one nitrogen and five carbons and one carbonyl group. Also known as pyrone. The hydroxyl tautomers are present only in small amounts.

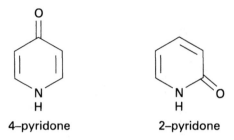

4–pyridone   2–pyridone

Pyrolytic: An elimination reaction that occurs on heating, requiring no other reagent.

Pyrrole: An aromatic five-membered cyclic ring containing a nitrogen atom and four carbons.

Pyrrolidine: A saturated five-membered cyclic ring containing four carbon atoms and a nitrogen.

Pyrroline: A five-membered cyclic ring containing four carbon atoms and a nitrogen atom with one carbon/carbon double bond.

3–pyrroline

Pyruvic acid: The compound CH_3COCO_2H.

Pyrylium cation: The six-membered aromatic cation which contains one oxygen and five carbons.

Quasi-racemate: A racemic compound formed between the R or S configuration of one compound and the S or R configuration of another, but closely related, compound.

Quinoline: Two condensed aromatic rings in which one CH unit is replaced by a nitrogen atom.

Quinone: The *ortho-* or *para*-dicarbonyl derivative of cyclohexadiene. Also known as benzoquinone.

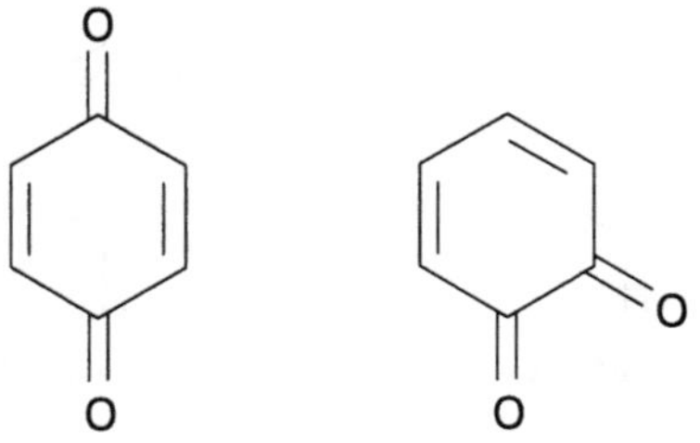

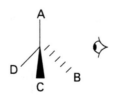

para–quinone   *ortho*–quinone

Racemic compound: A sample consisting of a one to one mixture of the *R* and *S* enantiomers of compound, in which there is an unusually strong van der Waals interaction between the species. They are often characterised by the melting point against molar ratio curve.

Racemic mixture: A mixture that contains equal concentrations of both enantiomers, and hence as a whole does not rotate the plane of polarised light.

Racemisation: The process whereby a pure enantiomer is converted to a mixture that contains an increasing proportion of the opposite configuration, until the point is reached where the concentration of the enantiomers is equal; the racemisation process is then complete.

Raney nickel: An activated nickel catalyst.

Rate determining step: The slowest step in a multi step reaction, and hence the one that determines the overall rate of the reaction. Also called the rate limiting step.

***R* configuration:** The arrangement of the substituents around an sp^3 hybridised centre, in which the group with the lowest priority according to the CIP rules, D, is furthest away from the viewer, and the remaining groups in order of decreasing priority, A > B > C, are arranged clockwise. If the arrangement is anticlockwise then it is the *S* configuration.

A
D   B
C

Rearrangement reaction: A reaction that involves a change in the skeleton of the molecule.

Reduction: A reaction in which the oxidation level of the starting material is reduced, usually by the addition of hydrogen or an electron, or the removal of oxygen.

Reed reaction: Chlorosulphonation using Cl_2 and SO_2 to give RSO_2Cl from RH.

***Re* face:** Used to describe the configuration around a trigonal atom where the groups are arranged clockwise in the order A $>$ B $>$ C according to the CIP priority rules. If the groups are arranged anticlockwise, then the configuration is said to be *si*.

Reformatsky reaction: The reaction of an α-bromoester in the presence of zinc with an aldehyde or ketone to give a 3-hydroxyester; or if dehydration subsequently occurs, then an α,β-unsaturated ester.

Regiospecificity: When a reaction can potentially give rise to two or more structural isomers, but actually only produces one, then the reaction is said to be regiospecific.

Reimer–Tiemann reaction: The formylation of aromatic rings using $CHCl_3$ and OH^-. Useful only for phenol, and a few heterocyclic compounds.

Relative atomic mass: The sum of the number of protons and neutrons in the nucleus of an element.

Relative configuration: A definition of the molecular co-ordinates relative to an arbitrary standard.

Relative molecular mass: The sum of the relative atomic masses in a molecule.

Resonance hybrid: The combination of two or more, hypothetical, structures called canonical forms, which represents the real structure of a molecule.

Retention: A situation in which the relative geometry of the chiral centre under investigation is the same at the end of the sequence of reaction steps as it was at the beginning. The opposite is inversion. If the geometry is randomised, then there is racemisation. Where racemisation occurs by proton exchange, and the racemisation takes place faster than the related isotopic exchange process, then the process is called iso-racemisation.

Ring/chain tautomerism: The tautomeric interconversion between the closed ring form and the open chain form of a molecule, e.g. in sugars, the interconversion between the closed ring hemiacetal and the open chain aldehyde.

Ring current: The electric current that may be induced to flow around a system. In an aromatic compound, i.e. one that is diatropic, the induced ring current is such that the induced magnetic field outside the ring increases the local field, and so shifts the NMR signal of protons outside the ring downfield. Conversely, anti-aromaticity is characterised by paramagnetic ring currents, which cause protons on the outside of the ring to be shifted upfield.

Ritter reaction: The addition of an alcohol to a nitrile to give an amide, in the presence of a strong acid.

Robinson annulation: The reaction of a cyclic ketone with an α,β-unsaturated ketone to form another cyclic ketone. It is the cyclic version of the Michael reaction.

Rosenmund reduction: The reduction of an acyl chloride to an aldehyde by hydrogenation using a palladium on barium sulphate catalyst.

Rotaxane: A molecule in which a linear portion is threaded through a ring and cannot become free because of bulky end groups.

Ruzicka cyclisation: The reaction of a long chained dicarboxylic acid with ThO_2 to form a cyclic ketone. The reaction requires high temperatures, and results in the overall pyrolytic elimination of CO_2 and H_2O.

σ bond: A molecular orbital that has no nodal plane.

σ* bond: An anti-bonding molecular orbital that has one nodal plane that passes between the atomic centres forming the bond.

σ complex: *See* arenium ion.

Sandmeyer reaction: The conversion of diazonium salts to aryl chlorides or bromides using cuprous chloride or bromide.

Sandwich compound: *See* metallocene.

Saponification: The hydrolysis of an ester with aqueous alkali to give an alcohol and the salt of the acid.

Saytzeff's rule: The German spelling of Zaitsev's rule. Also sometimes spelt Saytzev.

Schiff base: *See* imine.

Schlenk: *See* Grignard reagent.

Schmidt rearrangement: The rearrangement of a carboxylic acid to give an amine with one carbon less. It is achieved by treating the carboxylic acid with hydrogen azide, HN_3, followed by, for example, PCl_5. The reaction proceeds via an oxime, and is similar to the Beckmann rearrangement.

Schotten–Baumann reaction: The formation of an ester or amide from an acyl halide with an alcohol or an amine respectively when reacted with aqueous alkali.

***S* configuration:**   *See R* configuration.

Secoalkylation: The net addition of a $-CR_2CR_2CO_2R$ group to a carbonyl group.

Secondary isotope effect: *See* isotope effect.

Semicarbazide: The molecule $NH_2NHCONH_2$.

Semicarbazone: Compound of the general type $R_2C=NNHCONH_2$.

Semidine: An *o*- or *p*-arylaminoaniline.

***Si* face:**   *See re* face.

Side product: A product that is produced in minor amounts compared to the main product. However, this sometimes may be the desired product, or it may give very valuable information about the possible mechanistic pathways that have been followed.

Sigmatropic: *See* pericyclic.

Silyl enol ether: An enol ether in which the hydrogen of the hydroxyl group has been replaced by an alkylated silicon, e.g. the trimethylsilyl group (TMS).

Simmons–Smith reaction: The reaction between an alkene and CH_2I_2 with a Zn–Cu couple leads to the corresponding cyclopropane without the problem of having a free carbene in the reaction mixture.

Simonini reaction: A 2:1 mixture of the silver salt of a carboxylic acid, RCOOAg, reacting with iodine gives the ester RCOOR. A 1:1 mixture gives the iodo product, RI, which is similar to the Hunsdiecker reaction.

Singlet state: A species in which all the spins on the electrons are paired is said to be in the singlet state. In contrast, if two unpaired electrons have parallel spins, then the species is said to be in the triplet state.

Skew boat: *See* twist conformer.

Skraup reaction: The reaction between glycerol, sulphuric acid, a primary aromatic amine and an aromatic nitro compound to give a quinoline.

Small angle strain: The strain that results from the angles being much less than those resulting from normal orbital overlap. In three- and four-membered rings, the bonds that form the ring are called bent or banana bonds.

S number: The sum of the number of atoms in the bridges of a bicyclic system. *See* Bredt's rule.

Sodium amalgam: A solution of sodium in mercury, often used as a reducing agent.

Sodium borohydride: The compound $NaBH_4$.

Sodium copper cyanide: A mixed cyanide prepared from NaCN with CuCN, used to convert vinyl bromides to vinyl cyanides.

Sodium cyanoborohydride: The compound $NaBH_3CN$.

Soft electrophile/nucleophile: A species whose behaviour as an electrophile or nucleophile is mainly governed by the interaction of its frontier orbitals with those of the species being attacked, i.e. the highest occupied molecular orbital (HOMO) and the lowest unoccupied molecular orbital (LUMO).

Solvated electron: In the reaction between sodium and liquid ammonia, free electrons are formed that are solvated by the ammonia molecules. Used as a reducing agent.

Solventolysis: The splitting of a large molecule, usually into two fragments, by the overall addition of a molecule of the solvent. If the solvent is water, then the reaction is called hydrolysis.

Solvolysis: The substitution of a functional group with a molecule of the solvent.

Sommelet–Hauser rearrangement: The rearrangement of a benzyl quaternary ammonium salt, $C_6H_5CH_2N^+R_3X^-$, when treated with an alkali metal amide to give the *ortho*-methylbenzyl tertiary amine, $o\text{-}CH_3C_6H_4CH_2NR_2$. The tertiary amine may subsequently be alkylated, and the reaction repeated, with the substituent migrating by one position on each occasion until the *ortho* position is blocked.

Sonn–Müller method: The reduction of an aromatic amide to an aldehyde, in which the intermediate HCl salt of the imine is formed from the amide and PCl_5. This salt is then reduced to the aldehyde in a similar manner to the Stephen reduction.

s orbital: An atomic orbital that has no nodal plane.

Specific acid catalysis: In the two-step reaction, where the first step is the protonation of A by the conjugate acid of the solvent so as to form AH^+, and the second step is the reaction of AH^+ to form the products, if the first step is fast and the second step is rate determining, the reaction is described as specific acid catalysed. This reflects the fact that an equilibrium has been rapidly established between AH^+ and the strongest acid present in the solution, namely the conjugated acid of the solvent. Specific acid catalysis is in contrast to general acid catalysis.

Specific base catalysis: The same as specific acid catalysis, *mutatis mutandis*, but with the first step being the deprotonation of AH^+.

Specific rotation: The amount by which a chiral molecule rotates the plane of plane polarised light, $[\alpha]$, under standard conditions; it is defined for solutions as $[\alpha] = \alpha/lc$ and for pure compounds as being equal to α/ld, where α is the observed rotation, l is the cell length in decimetres, c is the concentration in grams per millilitre and d is the density in the same units.

Spirane: A compound that has two rings joined in such a manner that there is only a single carbon atom that is common to both rings, and so it has two bonds in one ring and two in the other, i.e. there is a quaternary carbon at the ring junction.

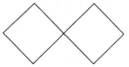

Spiroannulation: The overall conversion of a ketone, RCOR, to a di-α-substituted cyclobutanone. The reaction may be performed by the addition of diphenyl-sulphoniumcyclopropylide with the ketone RCOR to give the oxaspiropentane that, on treatment with a proton or a Lewis acid, rearranges to give the cyclobutanone.

Spiro dioxide: A compound that contains two epoxide groups that share a common carbon atom, which is bonded in a spiro manner.

sp^n orbital: An atomic orbital that results from the hybridisation of one s atomic orbital and n p atomic orbitals to give the hybrid orbital. sp is called diagonal hybridisation, while sp^2 and sp^3 are called trigonal and tetrahedral hybridisation, respectively.

Square planar: The name given to describe the geometry in which a general molecule AB$_4$ exists with the A atom centrally placed between four B groups that are directed towards the corners of a square and with all five atoms in a single plane. This species may exhibit geometric, but not optical, isomerism.

Squaric acid: The four-membered cyclic compound of the formula $C_4H_2O_4$. Two of the carbon atoms in the ring are part of carbonyl groups, while the other two are joined by a double bond and have a hydroxyl substituent. The dianion is aromatic.

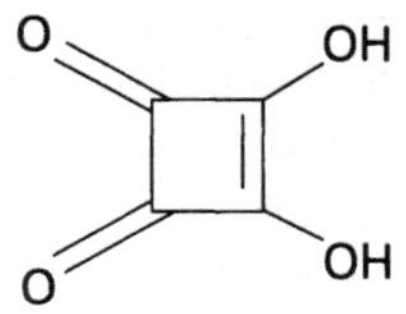

Staggered: *See* eclipsed.

Starting material: The principal carbon containing compound at the start of a synthetic route. If two or more carbon containing compounds are of approximately the same size, then the designation is usually based upon which molecule is the nucleophile.

Stephen reduction: The reduction of a nitrile to an aldehyde using HCl, followed by anhydrous $SnCl_2$, which produces the intermediate imine, which is then hydrolysed.

Stereoheterotopic: A term that includes both enantiotopic and diastereotopic atoms, groups and faces. Equivalent atoms, groups and faces would be homotopic.

Stereoisomers: Those isomers that are differentiated by the relative or absolute position of each atom within a molecule to every other atom. Each isomer has exactly the same number of each type of bond. The opposite of structural isomerism.

Stereoselective: A reaction in which a preferred stereochemistry is selectively produced.

Stereospecific: A reaction in which the product depends on the stereochemistry of the starting material.

Steric crowding: The unfavourable steric interaction caused by the close proximity of large groups, or the increase of such interactions, e.g. in going from sp^2 to sp^3 geometry.

Stobbe reaction: The addition of a 1,4-diester, i.e. a succinate derivative, under basic conditions, to a ketone, followed by an acidic work-up, resulting in an α,β-unsaturated ester and a β,γ-unsaturated carboxylic acid functionalities.

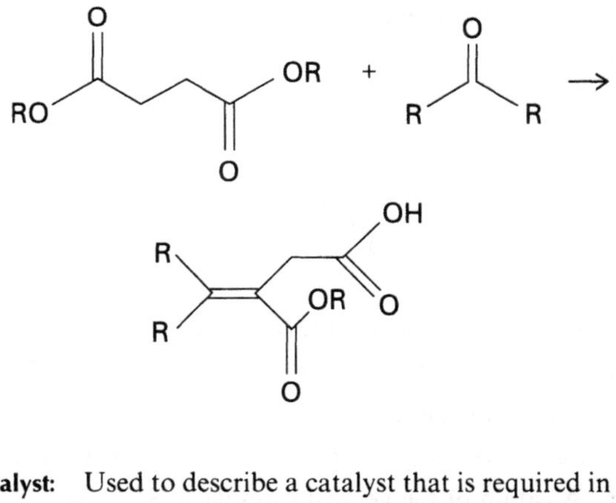

Stoichiometric catalyst: Used to describe a catalyst that is required in approximately the same concentration as the reagents it is catalysing, e.g. the $AlCl_3$ catalyst in the Friedel–Crafts reaction. In contrast to a trace catalyst, which is only required to be present in minute amounts, e.g. the catalytic hydrogenation of alkenes by Pt.

Stork enamine reaction: The reaction of an enamine with an alkyl halide that results in the alkylation of the β-carbon. Usually followed by hydrolysis of the resultant imine salt to give a carbonyl functionality at the α-carbon.

Strecker reaction: The treatment of a carbonyl compound with NaCN and NH$_4$Cl to give the α-aminonitrile derivative. It is a special case of the Mannich reaction. As the nitrile group may be easily hydrolysed, this is a convenient route to produce α-amino acids.

Structural isomerism: Isomers that are distinguished by having a different arrangement of bonds in the molecule.

Substitution reaction: A reaction where a heteroatom or group, i.e. not carbon or hydrogen, is replaced by another heteroatom or group.

Succinic anhydride: The cyclic anhydride of butane-1,4-dicarboxylic acid.

Sulphate ester: A compound of the general type (RO)$_2$SO$_2$.

Sulphene: A molecule of the general formula R$_2$C=SO$_2$.

Sulphenic acid: A molecule of the general formula RSOH, where R is an alkyl group.

Sulphenimide: A molecule of the general formula (RS)$_2$NR.

Sulphenyl chloride: A molecule of the general formula RSCl.

Sulphenylation: The addition of an RS– group.

Sulphinic acid: A molecule of the general formula RSO$_2$H, where R is an alkyl group.

Sulpholane: A dipolar aprotic solvent, which comprises a saturated five-membered ring containing one SO$_2$ group incorporated into a ring with four methylene groups.

Sulpholene: An unsaturated analogue of sulpholane, in which there is a double bond connecting C3 and C4.

Sulphonamide: A molecule of the general formula ArSO$_2$NHR.

Sulphonation: The replacement of a hydrogen atom with the SO$_2$OH group.

Sulphone: A compound of the general type R$_2$SO$_2$.

Sulphonic acid: A molecule of the general formula RSO$_2$OH, where R is an alkyl group.

Sulphonium cation: A cation of the general formula R$_3$S$^+$.

Sulphonylation: The replacement of a hydrogen atom with an SO$_2$R group.

Sulphonyl azide: A compound of the general type RSO$_2$N$_3$.

Sulphonyl chloride: A molecule of the general formula RSO_2Cl.

Sulphonyl hydrazide: A compound of the general type RSO_2NHNH_2.

Sulphoxide: A compound of the general type $R_2S=O$.

Sulphoximine: A compound of the general type $R_2SO(=NR)$.

Sulphur dichloride: The molecule SCl_2.

Sulphurisation: The addition of an RS– group.

Sulphuryl chloride: The molecule SO_2Cl_2.

Suprafacial: A cycloaddition reaction in which both of the new σ bonds are formed from the same face of the π system.

Sydnone: *See* mesoionic.

Syn: The addition or elimination of a molecule AB from the same side or face of another molecule.

Syn–anti dichotomy: An unusual result that occurs when alkenes are formed by an elimination reaction from large rings, in which the *cis* isomer is formed by *anti*-elimination, but the *trans* isomer is formed by *syn*-elimination.

Synclinal: *See anticlinal.*

Synfacial: *See* apofacial.

Syn isomer: In the case of trisubstituted C=N compounds, i.e. XYC=NZ, if Z is the same as one of the substituents on the carbon, i.e. X or Y, then the *syn* isomer is the one that has the two identical substituents on the same side; if they are on opposite sides, then it is the *anti* isomer. If all three substituents are different then the E/Z notation must be used.

Synperiplanar: *See anticlinal.*

Synthon: A hypothetical fragment generated by the disconnection of a molecule in a retrosynthetic analysis.

Taft equation: A structure/reactivity equation that correlates only field effects.

Target molecule: The desired compound at the end of a synthetic route.

Tautomerism: The relationship that exists between molecules with different structures, but yet are in rapid equilibrium with other. For example, the keto and enol forms of a carbonyl compound.

Tautomers: Compounds are structurally distinct, but are in rapid equilibrium. Usually, the difference is due to the position of a proton, but occasionally may be due to another element, e.g. a carbon atom such as occurs in a carbon valence tautomerism.

Telomer: Short polymeric molecule containing fewer monomeric units than a polymer, but more than an oligomer.

Termination: A reaction in which the product contains no radical species, but is formed from radical precursors.

Tetrahydrofuran: A saturated five-membered ring compound in which there is one oxygen atom and four carbon atoms.

Tetrahydropyran: A saturated six-membered ring compound in which there is one oxygen atom and five carbon atoms.

Tetrasulphide: A compound that contains four sulphur atoms in a chain, e.g. $R-S_4-R$.

Tetrazoline: A five-membered cyclic compound that contains four nitrogen atoms and one carbon.

Thiirane: *See* episulphide.

Thiiranium ion: An episulphide that is substituted on the sulphur to give a bridged sulphur cation.

Thioacetal: *See* mercaptal.

Thio acid: *See* thionic acid.

Thioamide: A molecule containing the group $R^1C(=S)NR^2R^3$.

Thiocarbamate: A molecule containing the group $RNHC(=O)SR$.

Thiocyanate: The univalent group $-S-C\equiv N$.

Thiocyanogen: The molecule $(SCN)_2$.

Thiocyanogen chloride: The molecule ClSCN.

Thioether: The sulphur analogue of an ether.

Thioisocyanate: The univalent group $-N=C=S$.

Thiol: *See* mercaptan.

Thiolic acid: A compound of the general type $RC(=O)SH$. *See also* thionic acid.

Thionic acid: A compound of the general type $RC(=S)OH$. Thiolic and thionic acids are sometimes not distinguished, and just called thio acids.

Thionyl chloride: The molecule $SOCl_2$.

Thiophen: The sulphur analogue of furan. Sometimes spelt thiophene.

Thiosulphate anion: The $S_2O_3{}^{2-}$ anion.

Thiourea: Compound of the general type $RNHC(=S)NHR$.

Thorpe reaction: The addition of the anion that results from the deprotonation of a nitrile compound, to another nitrile group. The resulting imine may be hydrolysed to yield a β-cyanoketone. If the reaction is performed internally it is known as the Thorpe–Ziegler reaction.

Thorpe–Ziegler reaction: *See* Thorpe reaction.

***Threo* isomer:**   A stereochemical term used to describe the configuration of a $C_{abx}C_{aby}$ system. *See also erythreo* isomer.

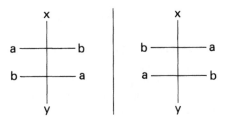

Tiffeneu–Demyanov ring expansion: The name given to the ring expansion when an aminomethylene group that is *vicinal* to a hydroxyl group, which in turn is attached to one of the ring carbon atoms, is treated with nitrous acid, resulting in production of the diazoium salt, which after the ring expansion gives the corresponding ketone. Thus, for example, 1-aminomethylcyclopentanol yields cyclohexone on treatment with nitrous acid.

Tishchenko reaction: When aldehydes are treated with aluminium ethoxide, one molecule is reduced while the other is oxidised, and the product is the resulting ester. This reaction involves a hydride anion transfer. With more basic alkoxides, aldehydes with an α-hydrogen undergo the aldol reaction.

Tollens reaction: The reaction of an aldehyde or ketone that has an α-hydrogen with formaldehyde, in the presence of $Ca(OH)_2$ to give initially a mixed aldol product; this is then usually reduced by a second molecule of formaldehyde to give a 1,3-diol, which is the result of a cross Cannizzaro reaction.

Topochemistry: The study of reactions that occur in the solid state.

Torsional effect: The interaction between non-bonded groups that are related by defining the dihedral angle between the groups.

Tosylate: A compound that contains the *p*-toluenesulphonate group, $-OSO_2C_6H_4CH_3$.

Trace catalyst: *See* stoichiometic catalyst.

Trans: A stereochemical term used to indicate that the two groups under consideration are attached one at each end of a double bond and are on opposite sides of the double bond, rather than on the same side as in the *cis* isomer. To characterise more fully a system in which there are more than two different groups, then the CIP rules need to be invoked, and the *E/Z* notation used instead.

trans   *cis*

Transacetalation: When acetals or ketals are treated with an alcohol of a higher molecular weight than that corresponding to the initial alkoxyl group(s), an exchange occurs and the heavier alkoxide replaces the lighter one.

Transannular strain: The strain that occurs due to the interaction of the substituents on C1 and C3 or C1 and C4, when a *gauche* conformation has been assumed by the molecule. It results from there being insufficient internal space to accommodate the groups in question.

Transesterification: The exchange of one alkoxyl group in an ester for another alkoxyl group, a reaction which may be catalysed by either acids or bases.

Transetherification: The exchange of one alkoxyl group for another in an ether.

Transition state: The transitory structure that exists at the mid-point in the reaction between the starting materials and the products. This corresponds to the structure that exists at the peak of the reaction co-ordinate diagram, which represents the configuration that has the highest energy along the reaction pathway. This hypothetical species cannot be isolated.

Transoid: A stereochemical term used to describe the conformation whereby four atoms form a plane with the first and fourth atoms being diametrically opposed, i.e. they adopt the same positions as would *trans* orientated groups around a double bond. The opposite of *cisoid*.

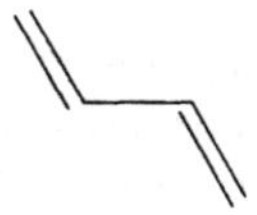

Transoximation: The conversion of a ketone to an oxime by treatment with another oxime.

Tresylate: A compound which contains the 2,2,2-trifluoroethanesulphonate group, $-OSO_2CH_2CF_3$.

Triazene: The $R_2N-N=NR$ configuration.

Triazine: A six-membered heterocyclic compound containing three nitrogen atoms. For example, 1,3,5-triazine is illustrated below.

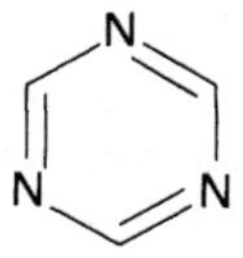

Triazole: An aromatic five-membered ring that contains three nitrogen atoms and two carbon atoms. Isomeric with osotriazole.

triazole   osotriazole

Triflate: A compound that contains the trifluoromethanesulphonate group, $-OSO_2CF_3$.

Trigonal hybridisation: The combination of one s atomic orbital with two p atomic orbitals to give three degenerate, i.e. equal in energy, hybrid orbitals which are labelled sp^2. They project from the centre towards the corners of an equilateral triangle, and so are all in the same plane.

Trioxane: A saturated six-membered ring in which there are three oxygen atoms. *See also* paraformaldehyde.

1,3,5-trioxane

Triplet state: *See* singlet state.

Tropone: The trivial name for cycloheptatrienone.

Tropylium cation: The $C_7H_7^+$ aromatic cation. Also called the tropenium cation.

Twist conformer: A conformation of a six-membered ring system that is halfway between the chair and boat conformations. Also called the skew boat form.

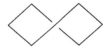

Udenfriend reagent: A reagent formed from a mixture of ferrous ions, oxygen, ascorbic acid, and ethylenetetraaminetetraacetic acid (EDTA) that is used to oxidise aromatic rings to phenol.

Ugi reaction: Also called the Ugi four-component condensation (4CC) in which an isonitrile is treated with a carboxylic acid and an aldehyde or a ketone in the presence of ammonia or an amine to give a bis-amide. Related to the Passerini reaction.

Ullmann ether reaction: The synthesis of diaryl ethers by the coupling of an aroxide anion with an aryl halide in the presence of copper.

Ullmann reaction: The coupling of aryl halides with copper to give the biaryl product.

Umbrella effect: *See* pyramidal inversion.

Unsaturated: Indicates the presence of multiple bonds in a molecule, which can undergo addition reactions, in particular with hydrogen in order to saturate all the available valences to single bonds.

Urea: The molecule of the formula $H_2NC(=O)NH_2$.

Ureide: The acyl derivative of urea, $RCONHCONH_2$.

Urethane: The molecule of the formula $H_2NC(=O)OC_2H_5$.

Valence isomer: A type of structural isomer that has the same basic skeleton, but differs in the exact bonding arrangement, e.g. the prismanes (or Ladenburg structure for benzene), and the Dewar and Kekulé structures for benzene (i.e. the bicyclo[2.2.0]hexadienes), are all valence isomers of each other.

| Ladenburg | Dewar | Kekulé |

Valence tautomerism: The rearrangement of the covalent bonds within a molecule, with the accompanying change in position of the nuclei that are involved in the bonds. Such molecules are said to be fluxional.

van der Waals forces: The electrostatic attraction between molecules that results from dipole/dipole interactions, whether permanent or induced.

Vicinal: Used to describe the configuration when the two functional groups in question are on adjacent carbon atoms in the chain. *See also geminal.*

Vilsmeier reaction: The use of disubstituted formamides with $POCl_3$ in order to formylate an aromatic ring. Also known as the Vilsmeier–Haack reaction.

Vinyl ether: *See* enol ether.

Vinylogue: Compounds related by the presence or absence of an extra $-CH=CH-$ unit.

von Braun reaction: There are two different reactions with this name: the reaction of *N*-alkyl substituted amides with PCl_5 to give the related nitrile and alkyl chloride; or the cleavage of tertiary amines by BrCN to give an alkyl bromide and a disubstituted cyanamide, R_2NCN.

von Pechmann reaction: The formation of a coumarin by the condensation of a phenol with a β-ketoester. The first step is a transesterification followed by a Friedel–Crafts ring closure. Sometimes simply called the Pechmann reaction.

von Richter reaction: The reaction of a *para* aromatic nitro compound with cyanide ions to give the *meta* carboxylic acid. This is a *cine* reaction.

Wagner–Meerwein: The rearrangement of a carbonium ion to give the most stable isomer.

Walden inversion: The inversion of the configuration at a chiral centre.

W formation coupling: A four bond coupling observed in the NMR spectrum, i.e. ^4J, e.g. between *meta* hydrogens in an aromatic ring.

Wheland intermediate: *See* arenium ion.

Whitmore 1,2-shift: The second step of a migration reaction, in which the migrating group moves from one centre to the adjacent one.

Wilkinson's catalyst: A homogeneous catalyst with the formula chlorotris(triphenyl-phosphine)rhodium.

Williamson reaction: The formation of an ether by the treatment of an alkyl halide with an alkoxide anion. *Gem*-dihalides react with alkoxides to give acetals, while 1,1,1-trihalides give ortho esters.

Wittig reaction: The reaction of an aldehyde or ketone with a phosphorous ylid (also called a phosphorane), $Ph_3P^+CR_2^-$, to give an unsaturated adduct.

Wohl–Ziegler bromination: The allylic bromination of an olefin using *N*-bromo-succinimide (NBS).

Wolff–Kishner reaction: The reduction of a carbonyl group to give a methylene group by means of hydrazine hydrate, $NH_2NH_2 \cdot H_2O$.

Wolff rearrangement: The rearrangement of a diazo ketone, $RCOCHN_2$, via the keto carbene, RCOCH, to the ketene, RCHCO, which may then be hydrated to a carboxylic acid, RCH_2CO_2H, as in the Arndt–Eistert synthesis.

Woodward–Hoffmann rules: These rules are based upon the principle of the conservation of orbital symmetry, and are used to predict which concerted reactions are allowed and which are forbidden.

Woodward method: The *syn*-hydroxylation of an alkene: the first step involves treating the alkene with iodine and silver acetate in a 1 : 1 molar ratio in wet acetic acid. The hydrolysis of the intermediate *anti*-β-halogenoester is via a normal SN2 reaction in which there is no neighbouring group effect because the ester function is solvated by the water. Overall *anti*-addition may be achieved by the method of Prevost.

Wurtz–Fittig reaction: The coupling of an alkyl and an aryl halide with sodium to give the alkylaromatic compound.

Wurtz reaction: The coupling of two alkyl halides, RX, with sodium metal to give the R–R compound. Usually of almost no synthetic value as there are many side reactions. However, the intramolecular reaction of 1-bromo-3-chlorocyclobutane with sodium gives bicyclobutane in about 95% yield.

Xanthate: A compound of the general type ROC(=S)–SR.

Xylene: A compound of the general type $C_6H_4(CH_3)_2$, which may exist in the *ortho*, *meta* or *para* isomeric forms.

Ylid: A molecule with a permanent positive and negative charge adjacent to each other in which the positive charge is on an atom from group V or VI of the Periodic Table, while the atom with the negative charge is carbon, e.g. a Wittig reagent. Also spelt ylide.

Ynamine: Compound of the general type $R^1-C \equiv C-NR^2_2$.

Ynediamine: Compound of the general type $R_2N-C \equiv C-NR_2$.

Zaitsev's elimination: The more highly substituted alkene derivative is formed. Sometimes spelt in the German manner, Saytzeff, or even Saytzev. Opposite of Hofmann elimination.

Ziegler alkylation: The alkylation of a (usually six-ring) heterocyclic nitrogen compound using an alkyllithium compound. It proceeds by the addition/elimination mechanism.

Ziegler catalyst: A mixture of nickel complexes and alkylaluminum compounds, which catalyses the addition of an alkene to another alkene.

Glossary

Zig-zag coupling: A five bond coupling observed in the NMR spectrum, i.e. ^5J, e.g. between H1 and H5 of naphthalene.

Zinin reduction: The reduction of a nitro compound to an amine using a sulphide or polysulphide.

Z isomer: See *E* isomer.

Zwitterion: A molecule that contains both positive and negative charges, but not on adjacent atoms as in an ylid, e.g. $RCH(NH_3^+)CO_2^-$.

Abbreviations

19.1 Atomic, group and molecular abbreviations

9-BBN	9-borabicyclo[3.3.1]nonane
Ac	acetyl group, $(CH_3CO)-$, group
acac	acetylacetone, $CH_3COCH_2COCH_3$
AcOH	ethanoic acid (acetic acid), CH_3COOH
AO	atomic orbital
Ar	a general aryl group
B:	a general base, usually in the B–L sense
B⁻	a general base, usually in the B–L sense
bis	two complex substituent groups joined to the main chain
B–L	Brönsted–Lowry acid/base
Bs	brosylate, $p-BrC_4H_6SO_2O-$, *para*-bromophenylsulphonyl
Bu	butyl group, C_4H_9-
Bz	benzyl group, $C_6H_5CH_2-$
CIP	Cahn–Ingold–Prelog rules
d	*dextro* rotatory
D	related to D-glyceraldehyde
di	two substituent groups joined to the main chain
DCC	dicyclohexylcarbodiimide
DMF	dimethylformamide, $HCON(CH_3)_2$
DMSO	dimethylsulphoxide, $(CH_3)_2SO$
E	*cis* geometry at rigid double bond, using CIP system

E^+	a general electrophile
Et	ethyl group, C_2H_5-
FG	functional group
HA	general acid
hex	hexane, C_6H_{14}
HOMO	highest occupied molecular orbital
$\pm I$	inductive effect
i-Bu, ^{i}Bu	*iso*-butyl group, $(CH_3)_2CHCH_2-$
In•	a general initiator for radical reactions
IPA	*iso*-propyl alcohol, $(CH_3)_2CHOH$
i-Pr, ^{i}Pr	*iso*-propyl group, $(CH_3)_2CH-$
ipso	disubstitution on the same carbon of a benzene ring
l	*laevo* rotatory
L	related to L-glyceraldehyde
LG	a general leaving group
LUMO	lowest unoccupied molecular orbital
m	*meta* disubstituted benzene, i.e. 1,3-
$\pm M$	mesomeric effect
$M^{+}_{\bullet}$	radical cation common in mass spectrometry
*m*CPBA	*meta*-chloroperbenzoic acid, $m\text{-}ClC_6H_4C(=O)OOH$
Me	methyl group, CH_3-
MO	molecular orbital
Ms	mesyl group, CH_3SO_2O-
n-Bu, ^{n}Bu	normal butyl group, $CH_3CH_2CH_2CH_2-$
NBS	*N*-bromosuccinimide
neo	used to describe a hydrocarbon with a quaternary carbon, e.g. *neo*-pentane is $(CH_3)_4C$
n-Pr, ^{n}Pr	normal propyl group, $CH_3CH_2CH_2-$
Ns	nosylate, *para*-nitrobenzenesulphonate, $p\text{-}O_2NC_6H_4SO_2O-$
Nu:	a general nucleophile
Nu^-	a general nucleophile
Nuc	a general nucleophile
o	*ortho* disubstituted benzene, i.e. 1,2-
p	*para* disubstituted benzene, i.e. 1,4-
Ph	phenyl group, C_6H_5 (also Φ symbol, the Greek capital 'phi')
PhH	benzene
Pr	propyl group, C_3H_7-
R	a general alkyl group, sometimes distinguished by raised suffix, e.g. R^1 or R^2
R	*recto* in the CIP rules
S	*sinister* in the CIP rules

sec-Bu	*sec*-butyl group, $CH_3CH_2(CH_3)CH-$
SM	starting material
t-Bu, ^tBu	tertiary butyl group, $(CH_3)_3C-$
THF	tetrahydrofuran
TM	target molecule
tri	three substituent groups joined to the main chain
tris	three complex substituent groups joined to the main chain
Ts	tosyl group, *para*-methylphenylsulphonyl, $p\text{-}MeC_6H_4SO_2O-$
TS	transition state
X	any halogen, or, more generally, any $-I$ group
Y	any general $-I$ group
Z	any $-M$ group
Z	*trans* geometry at a rigid double bond, using the CIP system

19.2 Mechanistic abbreviations

$\ddagger$	concerned with the transition state
β^-	electron
e^-	electron
:	lone pair of electrons
–	lone pair of electrons
hv	photon, i.e. light required
Δ	heat required
⇈	reflux
⌢	movement of one electron from the base of the fishhook to the head of the fishhook, e.g. a radical reaction
⌢→	movement of two electrons from the base of the arrow to the head of the arrow, e.g. a polar reaction
↜→	shorthand for a pair of electrons moving onto and then off an atom. Commonly found in carbonyl reactions
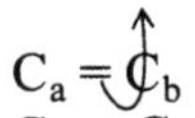 $C_a = \overset{\uparrow}{C_b}$	two electrons reacting through C_b
$C_a\text{---}C_b\text{---}C_c$	a partly broken $C_a\text{- - -}C_b$ bond in which those electrons are in the process of forming a new $C_b\text{- - -}C_c$ bond.
$\leftrightarrow$	resonance arrow linking two canonical forms
=	chemical equation, in which the reaction goes from left to right

$\longrightarrow$	reaction goes essentially to completion
$\rightleftharpoons$	equilibrium arrow, i.e. significant concentration of all compounds at equilibrium
 $\rightleftharpoons$	equilibrium is noticeably to one side, in this case to the right hand side
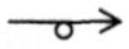 $\longrightarrow$	reaction proceeds with inversion of configuration at the chiral centre
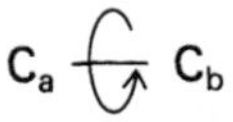 $C_a \text{ } C_b$	rotation about the C_a/C_b bond
$\Rightarrow$	retrosynthetic arrow, i.e. working backwards from the target molecule towards the starting material
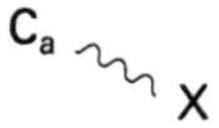 $C_a \sim X$	undetermined stereochemistry

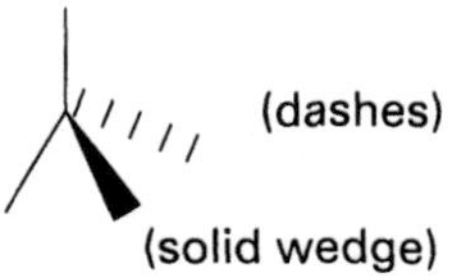 $////$ (dashes) (solid wedge)	stereochemistry determined in the manner shown with the solid line in the plane of the page, the dashed line behind, and the wedged line in front
C*	chiral centre or isotopic label
—	single covalent bond
=	double covalent bond
≡	triple covalent bond
.....	partial bond, i.e. bond being made or broken; alternatively, a bond which is partially delocalised
H $\longrightarrow$ Cl	bond polarised towards the arrowhead, i.e. the chlorine is the negative end
A → B	a dative bond, i.e. one in which both electrons originate from the A atom
+/−	full units of charge
(−)/(+)	partial charges, the modulus of which will add up to one or zero
δ+/δ−	small charges, which sum to zero, but whose modular sum is not one
D	dissociation energy
E	bond energy
ΔG	change in Gibbs free energy
ΔH	change in enthalpy
lp	lone pair of electrons
π	*pi* bonding molecular orbital
π*	*pi* anti-bonding molecular orbital
rds	rate determining step
σ	*sigma* bonding molecular orbital

σ*   *sigma* anti-bonding molecular orbital
ΔS change in entropy
upe   unpaired electron

19.3 Reaction type abbreviations

4CC	Ugi four-component condensation.
A1	*See* SN1cA.
A2	*See* SN2cA.
AAC1	Acid catalysed ester hydrolysis, acyl cleavage, unimolecular.
AAC2	Acid catalysed ester hydrolysis, acyl cleavage, bimolecular.
AAL1	Acid catalysed ester hydrolysis, alkyl cleavage, unimolecular.
AAL2	Acid catalysed ester hydrolysis, alkyl cleavage, bimolecular.
AdN-E	Tetrahedral mechanism of nucleophilic addition followed by elimination.
Ad3	Termolecular addition.
A-SE2	General acid catalysis of acetal in which the protonation of the substrate is the rds.
BAC1	Base catalysed ester hydrolysis, acyl cleavage, unimolecular.
BAC2	Base catalysed ester hydrolysis, acyl cleavage, bimolecular.
BAL1	Base catalysed ester hydrolysis, alkyl cleavage, unimolecular.
BAL2	Base catalysed ester hydrolysis, alkyl cleavage, bimolecular.
DA	Diels–Alder reaction.
E1	Unimolecular elimination.
E1cB	Unimolecular elimination conjugate base.
E2	Bimolecular elimination.
E2C	Bimolecular elimination with the base interacting primarily with the carbon.
E2cB	Bimolecular elimination conjugate base.
E2H	Bimolecular elimination with the base interacting primarily with the hydrogen.
ECO2	Bimolecular elimination forming a C=O bond.
Ei	Pyrolytic intramolecular elimination reaction.
FC	Friedel–Crafts.
FGI	Functional group interchange or interconversion.

HVZ	Hell–Volhard–Zelinskii.
MPV	Meerwein–Ponndorf–Verley reduction.
S$_E$1	Electrophilic substitution unimolecular.
S$_E$1(N)	S$_E$1 pathway in which a nucleophile (which may be the solvent) assists in the removal of the electrofuge.
S$_E$2	Electrophilic substitution bimolecular, front or back.
S$_E$2(co-ord)	*See* S$_E$C.
S$_E$2(cyclic)	*See* S$_E$i.
S$_E$C	Electrophilic substitution where there is initial bond formation between the leaving group (before it becomes detached) and the electrophile and then subsequent bond-breaking to yield the product. Also called S$_E$2(co-ord).
S$_E$i	Electrophilic substitution where a portion of the electrophile assists in the removal of the leaving group, forming a bond with it at the same time that the new C–Y bond is formed. Also called S$_F$2 or S$_E$2(cyclic).
S$_E$i'	Electrophilic substitution internal, but at the allylic position.
S$_F$2	*See* S$_E$i.
S$_H$1	Homolytic cleavage resulting in substitution.
S$_H$2	Bimolecular radical abstraction.
S$_N$1	Nucleophilic substitution unimolecular.
S$_N$1cA	Nucleophilic substitution unimolecular, conjugate acid, when the leaving group departs only after protonation. Also called A1.
S$_N$1cB	Nucleophilic substitution unimolecular conjugate base, when there is an initial deprotonation and then formation of a carbene.
S$_N$2	Nucleophilic substitution bimolecular.
S$_N$2'	Nucleophilic substitution bimolecular at the allylic position.
S$_N$2Ar	*See* S$_N$Ar.
S$_N$Ar	Nucleophilic aromatic substitution, also called S$_N$2Ar.
S$_N$i	Substitution nucleophilic internal.
S$_N$i'	Substitution nucleophilic internal at the allylic position.
S$_N$2cA	Nucleophilic substitution bimolecular, conjugate acid, when the leaving group only departs after protonation. Also called A2.

Molecular Notations

20.1 Non-structural notations

20.1.1 Empirical formula

This indicates the ratio of the constituent elements within a molecule using the smallest possible integers, i.e. CH_2O rather than $C_6H_{12}O_6$ for the glucose molecule. It conveys the least amount of information of any of the possible notations and hence it is of limited use. Normally, one only comes across an empirical formula in an examination question as an intermediary step in the calculation of the molecular formula from the elemental percentage composition.

20.1.2 Molecular formula

This formula indicates the actual number of atoms of each constituent element which is present in the molecule in question. It is derived from a knowledge of the empirical formula and the relative molecular mass (RMM). For example, ethyne and benzene each have the same empirical formula, namely CH. However, they have different RMMs, namely 26 and 78 respectively. Accordingly, the molecular formula for each compound is C_2H_2 and C_6H_6 respectively. Like the empirical formula, this notation contains only a limited amount of information. However, this residual ambiguity is used to advantage in two cases. First, in examination questions when setting problems, e.g. what are the possible isomers that correspond to a given molecular formula? Secondly, in indexes of organic compounds where it is useful to ignore the isomeric possibilities when trying to find a compound.

In indexes, the compounds are first listed in order of increasing carbon number and, for molecules with the same number of carbons, in increasing hydrogen number. The remaining elements are then listed in alphabetical order. The symbols D and T are used for the isotopes of hydrogen, i.e. deuterium and tritium respectively. Other non-standard isotopes are indicated by a preceding superscript, and listed in order of increasing atomic mass, e.g. ^{17}O and then ^{18}O. Thus, for example, the following compounds would be listed in the order indicated below:

CCl_3D	deuterotrichloromethane (or deuterochloroform)
$CHCl_3$	trichloromethane (or chloroform)
CH_2Cl_2	dichloromethane
CH_3Br	bromomethane
C_2H_4	ethene
C_2H_5Br	bromoethane
C_2H_6	ethane
C_2H_6O	ethanol or dimethylether

20.2 Two-dimensional structural notations

All structural notations attempt to give some indication of the real-life arrangement of the atoms within the molecule in question. The notations fall into two categories: two-dimensional and three-dimensional. The former type portrays the molecule as though it were flat, and leaves unresolved any explicit information about the three-dimensional structure. This type only indicates the connectivities of the component atoms to each other. In contrast, the latter type sets out to give as much information as is possible on a flat sheet of paper of the spatial configuration of the constituent parts of the molecule in question.

20.2.1 Dot and cross

This structural notation is very cumbersome. It is used only when it is necessary to illustrate the origin of each electron in a covalent bond. A dot is used to indicate an electron originating from one type of atom, while a cross is used for an electron originating from a different type of atom. It is used only for very simple molecules.

hydrogen, H_2 H ⁝ H

methane, CH_4 H ⁝ C ⁝ H

with H above and below the central C.

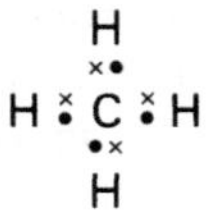

ethane, C_2H_6

```
       H  H
       ×•  •×
   H ×̇ C ː C ×̇ H
       •×  ×•
       H  H
```

× originally an electron from H
• originally an electron from C

ethene, C_2H_4

```
   H       H          H       H
    •× ×•                •  •
     C ×× C      or      C ː C
    ×•  •×             × •  • ×
   H       H          H       H
```

or

This notation can be used to represent the electronic configuration of a molecule that contains more than two different elements, but this can easily become confusing. Often the electrons that originate from the third element would be represented by a small square, or such like. If that element occurred only in an isolated part of the molecule, then it might be represented by either a dot or a cross, whichever was the more convenient.

The use of a third symbol may also be employed in order to distinguish between the origin of electrons from different atoms of the same element, e.g. the two carbon atoms in ethane.

ethane, $H_3C_a\!\!-\!\!C_bH_3$

```
       H  H
       ×•  •×
   H ×̇ Cₐː C_b ×̇ H
       •×  ×•
       H  H
```

× originally an electron from H
• originally an electron from C_a
• originally an electron from C_b

cyanoethane, CH_3CN

```
       H
       ×•
   H ×̇ C ː C ⫶⫶ N ː
       •×
       H
```

× originally an electron from H
• originally an electron from C
• originally an electron from N

```
       H
       ×•
or  H ×̇ C ː C ××× N ×
       •×
       H
```

× originally an electron from H or N
  depending on context
• originally an electron from C

Note that group abbreviations tend not to be used, because the principal object of this type of notation is to delineate as explicitly as possible, electron by electron, the bonding that is present. Accordingly, this notation is useful in ensuring that no electrons have been omitted by mistake.

20.2.2 Line notation

In this notation the chemical symbols are written on one line using the normal atomic symbols and relevant subscripts or superscripts. Each element is bonded directly to the one to the left, unless all the valences of that atom are satisfied, in which case it is bonded to the next one to the left, and so on as necessary.

Parentheses may be used to avoid ambiguities. Abbreviations may be used for common groups when certain points need to be highlighted.

CH_4	methane
R^1R^2CO	general ketone
$CH_3CH_2CH_3$	propane
CH_3COCH_3	propanone
$CH_3CHOHCH_3$ or $(CH_3)_2CHOH$	propan-2-ol
$CH_3CH_2CH_2OH$	propan-1-ol
$CH_3CHCHCH_3$	butadiene
CH_3COOH or CH_3CO_2H	ethanoic acid
$MeCOOH$ or $MeCO_2H$	ethanoic acid
$AcOH$	ethanoic acid
$(4\text{-}Me)C_6H_4CO_2H$	4-methylbenzoic acid
ΦX	monosubstituted benzene

This method is useful for portraying chemical formulae in typescripts and manuscripts, as a lot of information is conveyed in a condensed manner. However, it suffers from the problem that it does not give a pictorial representation of the stereochemistry of the molecule. In order to overcome this deficiency, it is possible to include the appropriate Cahn–Ingold–Prelog prefix, namely *R* or *S*, to indicate diastereo details, or one of the *cis*, *trans*, *ortho*, *meta* or *para* prefixes to indicate geometrical details.

$(2R,3S)\text{-}CH_3CHOHCHBrCH_3$	(2*R*,3*S*)-3-bromobutan-2-ol
$p\text{-}BrC_6H_4Me$	*para*-bromotoluene
cis-$CH_3CHCHCH_3$	*cis*-but-2-ene

20.2.3 *Expanded line notation*

The simple line notation is sometimes expanded by using two or three lines to indicate double or triple bonds as appropriate, e.g. $R^1R^2C=O$ for a general ketone. This notation highlights the position of the multiple bonds and so may be of use in distinguishing closely related compounds, e.g. $CH_3CH_2CH_2CH_3$ for butane, and $CH_3CH=CHCH_3$, rather than $CH_3CHCHCH_3$, for but-2-ene. The geometry of the double bond may be indicated by the use of the appropriate prefix, namely *cis* or *trans*.

Sometimes a single bond is indicated to clarify the structure, or to highlight certain aspects of it. Parentheses may also be used to ensure that there is no ambiguity.

$CH_3(=O)OH$	ethanoic acid
$CH_2=CH–CH=CH_2$	butadiene
$MeC\equiv N$	cyanomethane

This notation is useful when representing a structure in type, especially an aliphatic compound that has only a few substituents. However, multisubstituted alicyclic or aromatic compounds can rapidly become a little unclear, because of the number of parentheses that must be employed in order to resolve any ambiguity.

20.2.4 Stick notation

This notation uses atomic symbols or group abbreviations joined by single, double or triple lines to represent single, double or triple covalent bonds. All atoms are included in a stylised manner: at an sp^3 centre, right angles are used; at an sp^2 centre, a $120°$ angle is used; and at an sp centre, a $180°$ angle is used. This is a common notation, because it ensures that all the atoms are included, and that the molecule as depicted on paper corresponds to the molecular structure of the desired compound. Its weaknesses are that it is time consuming to draw, and, more importantly, complicated structures may become extremely cluttered, and so difficult to comprehend.

ethane, C_2H_6

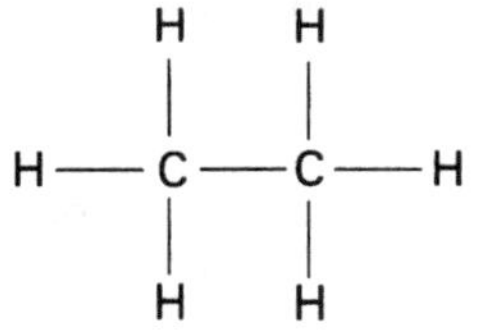

ethanonoic acid, CH_3CO_2H

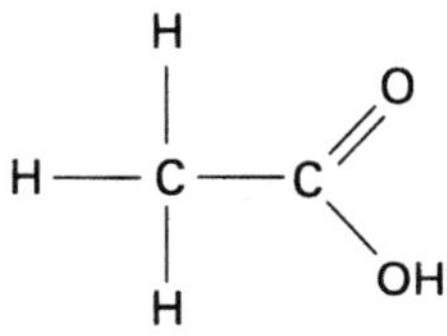

3-bromobutan-2-ol, $CH_3CHOHCHBrCH_3$

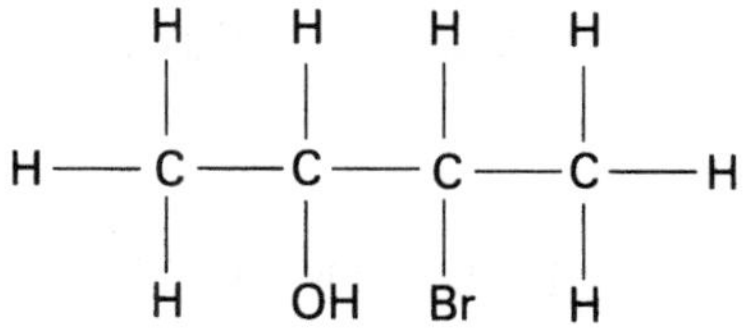

20.2.5 Two-dimensional skeletal notation

This is the normal method used by organic chemists, because it is fast, accurate, and the result is easy to comprehend rapidly. The method uses the following conventions. First, hydrogen atoms joined to carbons normally are not shown; they are assumed to be added as required to complete the normal valency of the carbon. Secondly, carbon atoms are not represented by the letter 'C', but instead are represented by either the end of a line, or the junction between two lines. Heteroatoms are represented by their normal atomic symbol. Thirdly, covalent bonds are represented by lines that connect the atomic symbols of the heteroatoms or the carbons joined to the heteroatoms, but with the above two provisos. Fourthly, the atomic symbol for hydrogen is included when it is bonded to a heteroatom, or when it is bonded to a carbon atom and is involved in that mechanistic step, e.g. the enolisation of a carbonyl group.

This method is of little use to portray very simple molecules: for example, methane is just a dot '•'; ethane is a line '–'; ethene is a double line '='; methanol, '–OH' and methanal, '=O'. However, the method becomes more valuable as the complexity of the molecule increases, e.g. hexane is represented by a zigzag of five parts, representing the six carbon atoms: '/\/\/'. Aromatic rings are sometimes represented by a regular hexagon with a circle inside to indicate the delocalisation of the π electrons. However, this representation can easily result in the miscounting of the electrons that are present in the aromatic ring. Hence, the alternative method, namely of representing each single and double bond within the aromatic ring separately, is to be preferred. The method really comes into its own for bigger molecules, e.g. 2-bromotoluene, hexatriene, cyclohexene, propylamine, ethylbutanoate, *cis*- and *trans*-butene. This notation is particularly good at representing the structure of aliphatic and aromatic compounds.

2-bromotoluene, 2-$BrC_6H_4CH_3$

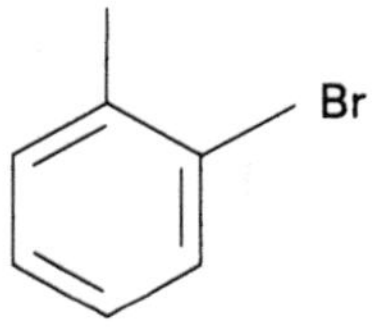

hexatriene, C_6H_8

cyclohexene, C_6H_{10}

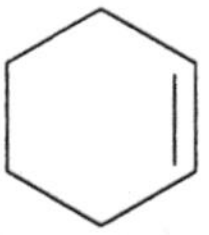

propylamine, $C_3H_7NH_2$

ethylbutanoate, $C_3H_7CO_2C_2H_5$

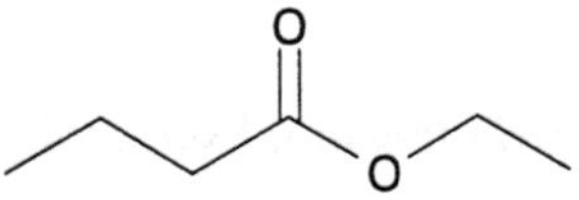

cis-butene, C_4H_8

trans-butene, C_4H_8

410

In addition to just indicating the connectivities of the constituent atoms, this notation is capable of readily imparting some stereochemical information, such as the geometric configuration around a multiple bond. No attempt, however, is made to indicate the configuration of the molecule, as the molecule is drawn as a flat figure. Hence, the enantiomers of butan-2-ol are not distinguished. Similarly, cyclohexane derivatives are represented by a flat, regular hexagon with side branches at appropriate junctions, with no attempt to indicate whether they are *cis* or *trans* to each other.

1,2-dibromocyclohexane, $C_6H_{10}Br_2$

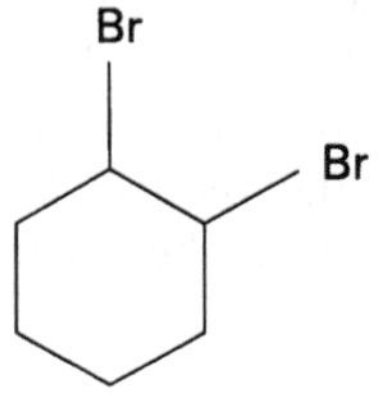

1-bromo-1-methylcyclohexane, $C_6H_{10}BrCH_3$

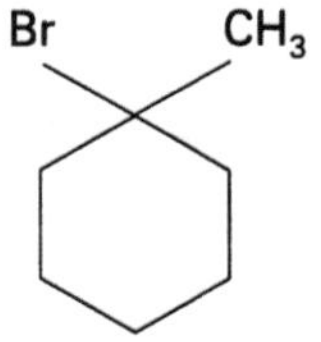

There is a common variation of this notation, which is used for fused ring systems such as decalin derivatives and steroids. In this variation, even though the rings are drawn as symmetrical polygons, the hydrogen that projects up from the page at the ring junction is represented by a small solid circle over the carbon at that junction, while the hydrogen that projects downwards is represented by an open circle.

trans-decalin, $C_{10}H_{18}$

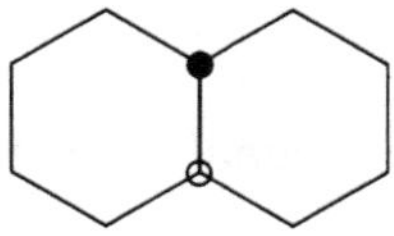

cis-decalin, $C_{10}H_{18}$

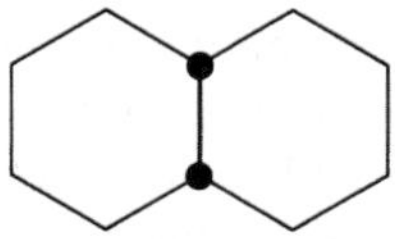

This variation is of very limited use, and, in order to convey more fully the three-dimensional structure of the molecule, more sophisticated notations must be employed.

20.3 Three-dimensional structural notations

20.3.1 Haworth notation

In this notation, a six-membered ring is represented by a regular hexagon, but as viewed from the side and slightly above. Those lines that represent the bonds that project towards the reader are thickened, while the substituents are positioned above and below the ring and are connected by lines of normal thickness. Alternatively, the bonds within the ring are represented by lines of equal thickness and the groups above and below the ring are represented by solid wedges or dashed lines, respectively. There is no attempt to indicate whether a bond is in front or behind another bond which it crosses from the viewpoint of the reader. Examples of this notation may be found in the literature; however, there are better notations that can be employed.

D-glucose, $C_6H_{12}O_6$

20.3.2 Three-dimensional skeletal notation

This is an extension of the skeletal notation. It is very good at indicating the structure of ring compounds and the cyclic transition states of aliphatic compounds, and so is particularly useful when writing organic reaction mechanisms that involve such species.

As an example, 1,4-dichlorocyclohexane is shown as a puckered ring with the positions of the chlorine atoms being indicated by placing them in conventional positions relative to each other. Vertical lines represent axial bonds, while lines at a slight angle to the horizontal represent equatorial bonds. Thus, this clearly shows whether the chlorine atoms are *cis* or *trans* to each other.

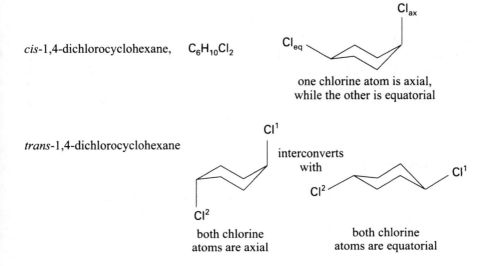

cis-1,4-dichlorocyclohexane, $C_6H_{10}Cl_2$

Cl$_{ax}$

Cl$_{eq}$

one chlorine atom is axial,
while the other is equatorial

trans-1,4-dichlorocyclohexane

Cl^1

interconverts
with

Cl^2

Cl^1

Cl^2

both chlorine
atoms are axial

both chlorine
atoms are equatorial

If a bond crosses another because of the viewpoint of the reader, then it is convenient to introduce a break in the bond that is further from the viewer so as to assist with the visual interpretation of the diagram. Both optical and geometrical isomers may be easily distinguished using this notation. For most purposes, this is the best notation to use for this type of molecule, because it is quick and easy to draw by hand; it is easily understood; and it indicates clearly the relative positions of the different parts of the molecule.

20.3.3 *Stereo projection*

This is yet another version of the two-dimensional skeletal notation, this time using the addition of dashed and solid wedge lines to indicate bonds that are either going away, or coming towards, the reader. Bonds shown with the usual thickness of line are in the plane of the paper. Hence, the stereochemistry at a chiral centre may be shown clearly and unambiguously. Consequently, this notation is particularly good at illustrating the change in geometry which a chiral centre undergoes in substitution reactions. For example, in a simple SN2 reaction, the inversion of the carbon atom that is substituted may be clearly seen.

Simple SN2 reaction

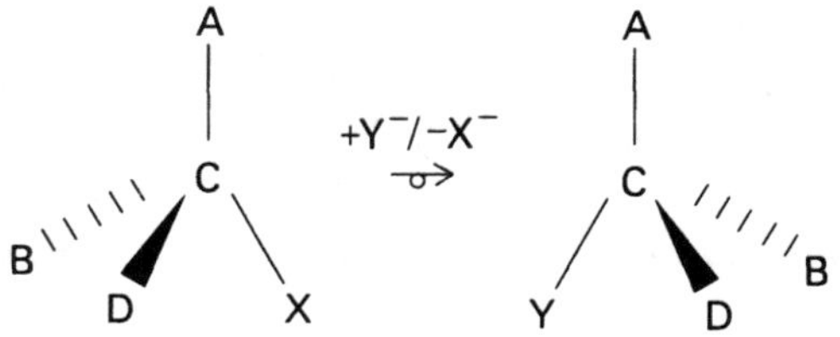

A

C

B

D   X

+Y$^-$/-X$^-$

A

C

Y   D

B

There are two common variations on this notation. Sometimes, at the chiral centre, only three bonds may be shown, two of which are of normal thickness, while the third is either dashed or a solid wedge; the fourth bond is to a hydrogen atom, and is not shown for it is assumed to be orientated as necessary to complete the usual geometry of the chiral centre. In the other common variation, three bonds to the chiral centre are represented with lines of normal thickness, and only the fourth uses a solid wedge or dashed line to indicate the stereochemistry. Neither of these variations are as clear as the normal version.

20.3.4 *Sawhorse projection*

This method is used to give a clear picture of the geometry along a carbon/carbon single or double bond, in particular one around which an addition or elimination reaction is proceeding.

E2 mechanism

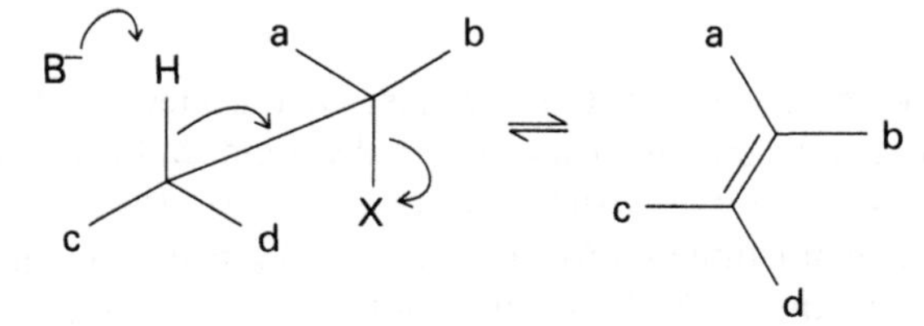

It is valuable when trying to illustrate the geometry of the formation of a three-membered ring, as in the Darzens condensation.

epoxide formation in the
Darzens condensation

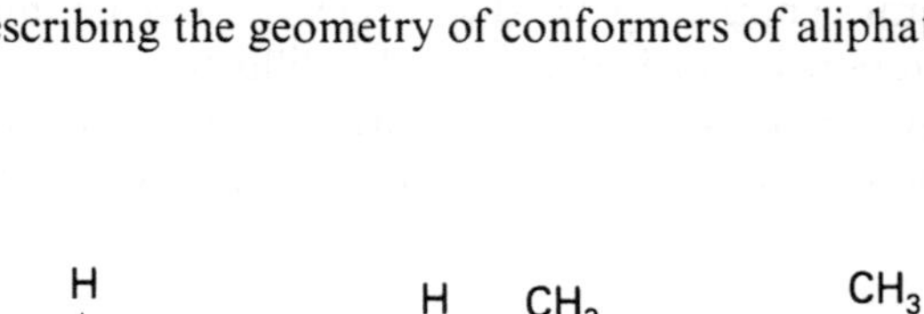

Lastly, it may be used when describing the geometry of conformers of aliphatic compounds.

conformers of butane

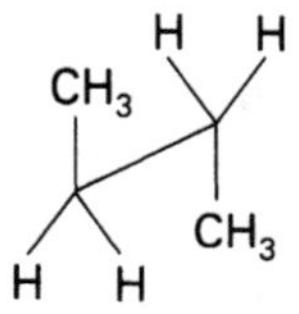

antiperiplanar 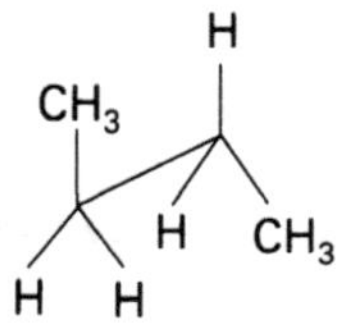   anticlinal 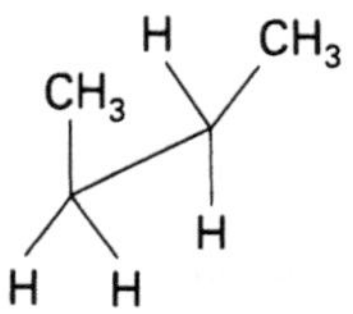   gauche or   synperiplanar
                                            synclinal

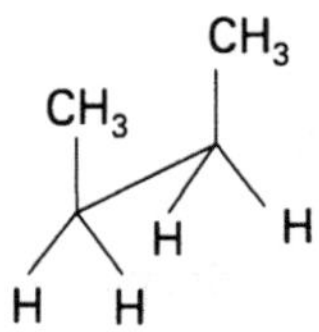

414

20.3.5 *Newman projection*

In this projection the reader looks along a single bond, so as to discern clearly the orientation of the substituents connected to the atoms that form that single bond. There are two situations in which such a notation is commonly used: first, when two sp^3 hybridised carbons are bonded together and one wishes to indicate the conformation of the structure, e.g. *antiperiplanar* or *eclipsed* or whatever; and secondly, when one of the carbons is also involved in a double bond to an oxygen atom, as occurs in carbonyl compounds. In the latter case, the α-substituents are placed on lines emanating from the rim of the circle representing the α-carbon that is furthermost from the reader. The carbon of interest is represented by a dot in the centre of the circle and is nearer the reader. Issuing from this central point are the double lines representing the double bond to the neighbouring atom, which is usually a carbonyl oxygen, and a single line that is the bond to the last remaining group.

It is used in particular to determine on which side an incoming nucleophile will attack a carbonyl group. Cram's rule states that the side of attack is the one that offers less steric hindrance. This allows the stereochemistry of the product to be predicted.

S, M, L : Small, medium and large groups attached to α-carbon

direction of attack of an incoming nucleophile

20.3.6 *Fischer projection*

This method was developed to illustrate the optical chemistry of small biological molecules, especially monosaccharides and amino acids. It rapidly becomes cumbersome for larger molecules. The vertical lines represent bonds that are going into the page, away from the reader, while the horizontal lines represent bonds that are coming out of the page, towards the reader. For molecules that contain more than one chiral centre, the arrangement must be vertical on the page, which means that it takes up a lot of space.

a general chiral molecule C_{abcd}

Stereochemical Terminology

21.1 Introduction

The elemental composition of a molecule can be unambiguously described by its molecular formula. In inorganic chemistry, the molecular formula is often sufficient in order to identify the species in question, because there is usually only one possible arrangement of the constituent atoms that are combined in that particular elemental ratio. Thus, for example, S_8 and S_n are two possible molecular formulae for elemental sulphur. There is only one structure in which the molecule that corresponds to the molecular formula S_8 can exist, namely the puckered crown, and so there is no need to give further information in order to characterise the molecule in question. One obvious exception to this general rule concerns elements, with the same molecular formula, that exist in various allotropic forms. For example, to identify precisely the type of phosphorus that is being used, it is necessary to indicate the 'colour', i.e. white, yellow, red or black. These terms refer to various allotypes of phosphorus, which differ in the relative spatial orientation of the component atoms within the lattice. A second major exception in inorganic chemistry would be the need to characterise the possible isomers of some co-ordination compounds of the transition metals, for these species may exhibit geometric or optical isomerism. However, apart from these limited exceptions, the general principle holds true, namely, the molecular formula of an inorganic molecule defines uniquely the compound in question.

In contrast, in organic chemistry, the situation is different for all but the very simplest of molecules. This is because, for any given elemental ratio of the constituent atoms, there might be many possible spatial arrangements that those atoms could assume relative to one another. So, even though the molecular formula conveys some useful information, it is often not sufficient in order to define uniquely the compound under consideration.

The problem may be easily illustrated by considering the molecules that have the molecular formula, $C_4H_{10}O$. There are eight isomers that have this molecular formula. Seven of these isomers may be distinguished by writing out the compounds using the line notation: $CH_3CH_2CH_2CH_2OH$, $(CH_3)_2CHCH_2OH$, $CH_3CH(OH)CH_2CH_3$, $(CH_3)_3COH$, $CH_3CH_2OCH_2CH_3$, $(CH_3)_2CHOCH_3$ and $CH_3CH_2CH_2OCH_3$. The first four compounds are alcohols, butan-1-ol, 2-methylpropan-1-ol, butan-2-ol, and 1,1-dimethylethanol; while the last three compounds are ethers, diethyl ether, *iso*-propylmethyl ether, and propylmethyl ether. However, there is a further ambiguity that has not been revealed by the line notation. For butan-2-ol there are actually two possible structures that differ only in the way the groups are arranged relative to one another in three dimensions, i.e. there are two stereoisomers.

the enantiomers of butan-2-ol, $CH_3CH(OH)CH_2CH_3$

In order to distinguish between these two possibilities, further information must be included in the name of the compound, i.e. stereochemical terminology is needed.

In the example given above, there were eight possible arrangements of the constituent atoms that corresponded to the same molecular formula. Each one of these possible arrangements is called an isomer. Isomers may be divided into two fundamental classes. The first type consists of those isomers which differ in the exact number of each type of bond that connects the atoms in the various molecules. These are called structural isomers, because the structure is different in each one. In the above example, the ethers and the alcohols are structural isomers of each other. In the second type, the isomers all have exactly the same connectivity, i.e. the manner and order in which the atoms are connected to each other are the same, but differ only in the precise position that each group occupies relative to all the others in the molecule. These are called stereoisomers. In the above example, the two isomers of butan-2-ol are stereoisomers.

21.2 Structural isomerism

Structural isomers may be divided into five sub-types: skeletal, positional, functional, tautomeric and meta.

There are also generic terms that are used to describe, and distinguish between, certain classes of alcohols, amines and amides, and we will consider these first.

Alcohols are classified according to the number of alkyl or aryl groups attached to the carbon to which the hydroxyl group is bonded. A primary alcohol has one such group attached to the carbon of the COH group, while a secondary alcohol has two, and a tertiary has three. Notice that methanol has no such carbon groups attached to the carbon of the COH group, and so strictly is not a primary alcohol, although it is often included in the study of such primary alcohols.

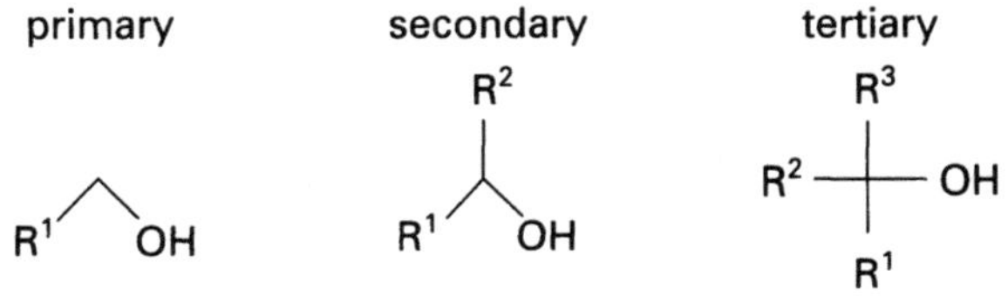

For amines, the classification depends on the number of alkyl or aryl groups that are directly attached to the nitrogen. The general formula of an amine is R_nNH_{3-n}, where n is the level of the amine. Accordingly, when $n = 1$, i.e. RNH_2, the compound is a primary amine. The compounds of the general formula R_4N^+ are called quaternary ammonium ions. Notice that, strictly, ammonia is not a primary amine (cf. methanol).

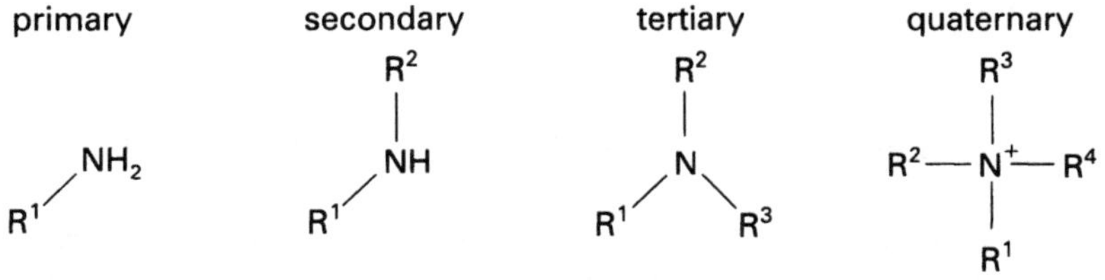

For amides, again the classification refers to the number of alkyl or aryl groups that are attached to the nitrogen, but now the general formula is $RCONH_{2-n}R_n$. Thus, if there are no such groups joined to the nitrogen, the molecule is a primary amide and so on.

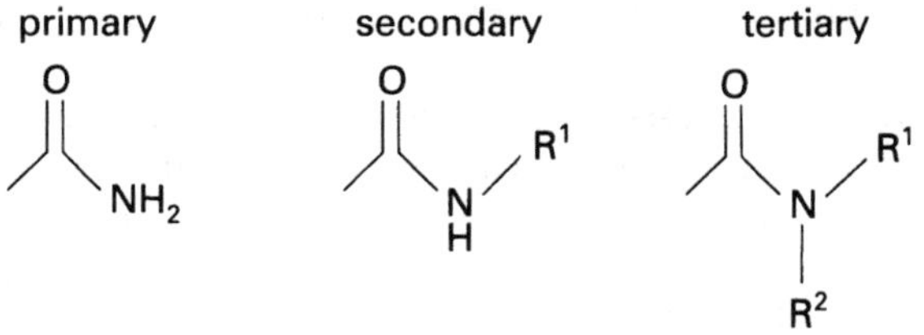

419

21.2.1 *Skeletal isomerism*

This type of isomerism is concerned with differences in the carbon skeleton of the molecule, namely whether it is straight or branched. It is also sometimes called chain isomerism. If one is only considering the relative location of the carbon atoms (each of which is joined to another carbon atom) then there must be at least four carbon atoms before this type of isomerism can occur.

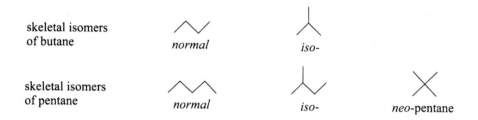

skeletal isomers
of butane   *normal*   *iso-*

skeletal isomers
of pentane   *normal*   *iso-*   *neo*-pentane

The best way to distinguish between these isomers is by naming them according to the IUPAC system, which bases the name upon the longest continuous carbon chain.

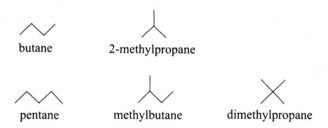

butane   2-methylpropane

pentane   methylbutane   dimethylpropane

However, it is still common to find references to an older system of nomenclature. In this system, the total number of carbon atoms in the chain is indicated by the stem of the name chosen; while the particular configuration that the carbon atoms adopted is indicated by the use of a distinguishing prefix. A straight chained molecule is given the prefix *normal*, which is abbreviated to *n-*, e.g. *n*-butane for $CH_3CH_2CH_2CH_3$. A compound that contains the $(CH_3)_2CH-$ group is called *iso*, which is abbreviated to *i-*, e.g. *i*-pentane for $(CH_3)_2CHCH_2CH_3$. If the compound contains a quaternary carbon then the prefix is *neo*, which is not abbreviated, e.g. *neo*-pentane for $(CH_3)_4C$.

This old system is also used to describe carbonium ions and radicals. Thus, primary, secondary, and tertiary carbonium ions are given the respective label, *normal, secondary* (abbreviated to *sec-* or just *s-*), and *tertiary* (abbreviated to *t-*). The term *neo* may be used to describe a primary carbonium ion that is immediately adjacent to a quaternary carbon, e.g. $(CH_3)_3CCH_2^+$.

n-pentyl    *sec*-pentyl    *t*-pentyl 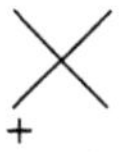   *neo*-pentyl

21.2.2 Positional isomerism

On a long straight-chained molecule that has a functional group, it is possible for this group to be placed at a number of different positions on the carbon chain. This is referred to as positional isomerism. At least three carbons are needed in the chain before this type of isomerism can occur. Again, as with all structural isomers, the best way to distinguish between these related compounds is to name them according to the IUPAC system.

However, it is still common to find references to an older system of nomenclature where the point of attachment of the functional group to the alkyl chain is indicated by a prefix, such as *iso* or *tertiary*. This bears a relationship to the prefix system that is used to distinguish between the different structural isomers of carbonium ions. These older terms have been included in parentheses under the examples given below.

possible isomers of C_3H_7OH

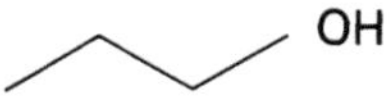

propan-1-ol
(*n*-propanol)

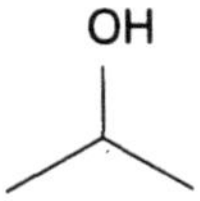

propan-2-ol
(*iso*-propanol)

possible isomers of C_4H_9OH

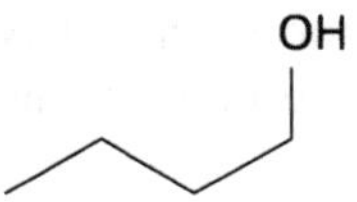

butan-1-ol
(*n*-butanol)

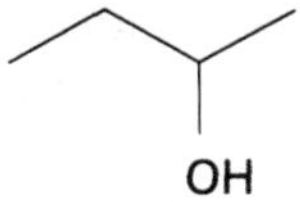

butan-2-ol
(*sec*-butanol)

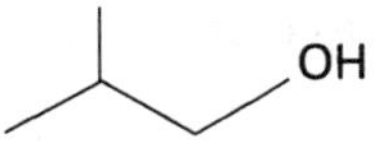

2-methylpropan-1-ol
(*iso*-butanol)

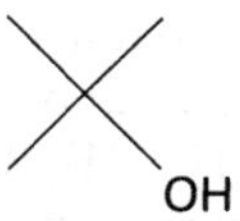

methylpropan-2-ol
(*t*-butanol)

possible isomers of $C_5H_{11}OH$

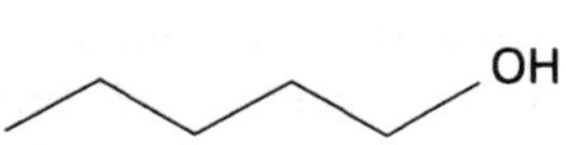

pentan-1-ol
(*n*-pentanol)

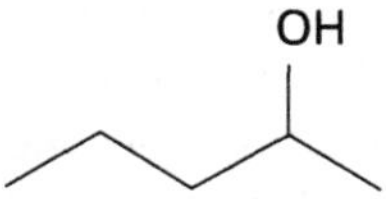

pentan-2-ol
(*sec*-pentanol)

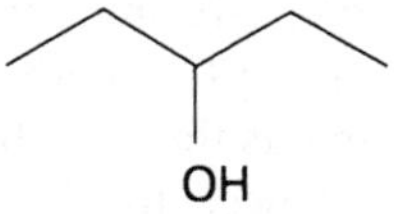

pentan-3-ol

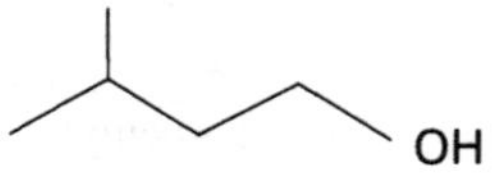

3-methylpentan-1-ol
(*iso*-pentanol)

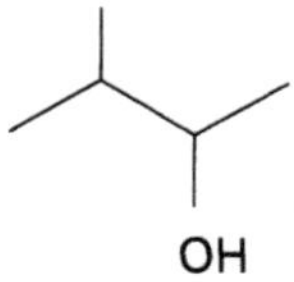

3-methylpentan-2-ol

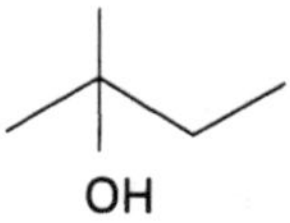

2-methylpentan-2-ol
(*t*-pentanol)

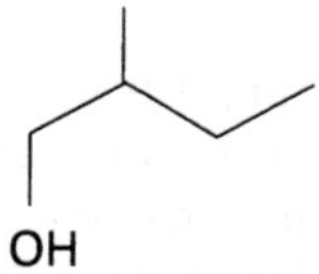

2-methylpentan-1-ol

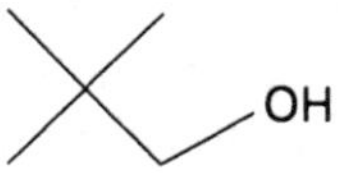

2,2-dimethylpentan-1-ol
(*neo*-pentanol)

It is also important to be able to refer to a particular carbon atom that is adjacent to the functional group under discussion. This is often done using small Greek letters. This system may be illustrated using 1-bromooctane as an example.

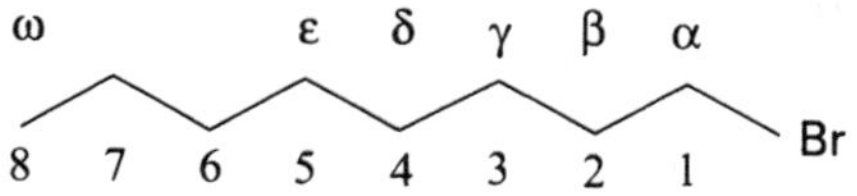

The ω position always refers to the terminal carbon, no matter how long the chain actually is.

If there are two identical function groups that are bonded to the same carbon atom, then they are said to be *geminal* to each other (abbreviated to *gem*) e.g. 1,1-dibromoethane, H_3CCHBr_2, contains *geminal* bromine atoms. If, however, the two functional groups in question are bonded to adjacent carbons, then they are said to be *vicinal* to each other (abbreviated to *vic*) e.g. 1,2-dibromoethane, $BrCH_2–CH_2Br$.

In an elimination reaction, if both groups are lost from the same atom, then it is called a 1,1- or α-elimination; if the groups are lost from adjacent atoms, then it is called a 1,2- or β-elimination; and if the groups are lost from atoms that are separated by another atom, then it is called a 1,3- or γ-elimination. The chain in which the elimination does not occur is indicated by a prime symbol, ''. For example, 4-bromopent-2-one gives pent-3-en-2-one on the elimination of HBr, and so the C5 methyl carbon has the β'-hydrogens, while the C3 methylene carbon has the β-hydrogens. Note that both the β and β'-hydrogens are defined with respect to the bromine substituent, which is eliminated in the HBr molecule.

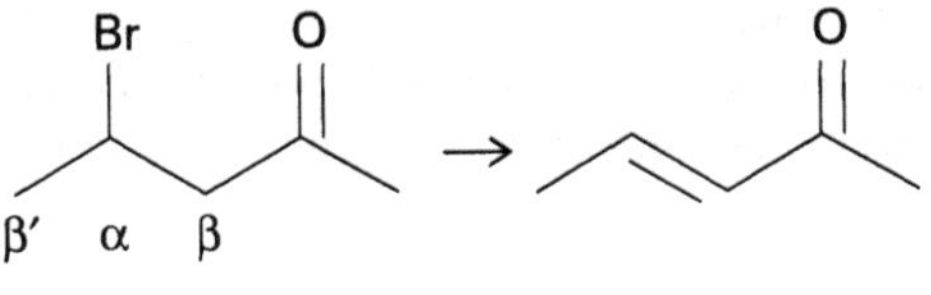

When the functional group is one that includes a carbon that is fully bonded to heteroatoms (or hydrogen atoms that are considered to be an intimate part of the functional group, as in, for example, an aldehyde grouping), then the lettering starts from the closest carbon atom in the chain to the fully bonded carbon.

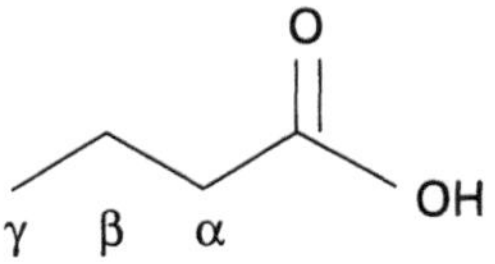

In disubstituted aromatic systems, the position of the second substituent with respect to the first functional group is indicated either by a number, with the first group having the 1 position, or by the prefix *ortho*, *meta* or *para* indicating the 2, 3 or 4 positions, respectively. These prefixes are often abbreviated to *o*, *m* or *p*, respectively. If a mono-substituted aromatic compound is attacked at a ring carbon that already bears a substituent, then this is called *ipso* attack.

If the nitrogen of an amine group is substituted, then the position of this substituent is indicated by the prefix *N*. If there are two relevant nitrogens which are substituted, then the second is indicated by the prefix *N'*. Similarly, oxygen substitution may be indicated by an *O*.

21.2.3 Functional isomerism

It is very common for a given molecular formula to be capable of representing molecules that have different functional groups. A simple example is C_2H_6O, which may exist either as an alcohol or an ether.

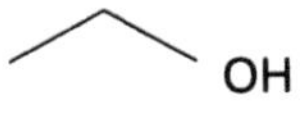

ethanol, C_2H_5OH
or EtOH

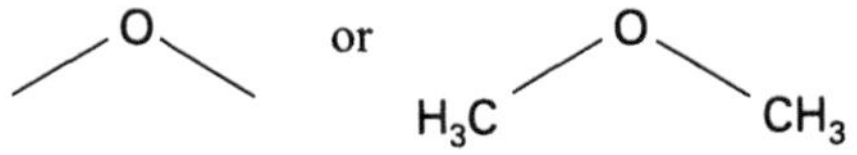

dimethylether, CH_3OCH_3

21.2.4 Tautomeric isomerism

This is a special type of functional isomerism. In this case, different functional groups are present, and these isomers, which are called tautomers, are often in equilibrium with each other. There are several types, but by far the most common involves a proton shift between an heteroatom and a carbon. The two most frequent examples of this type are the enol/keto and the aci/nitro tautomeric systems.

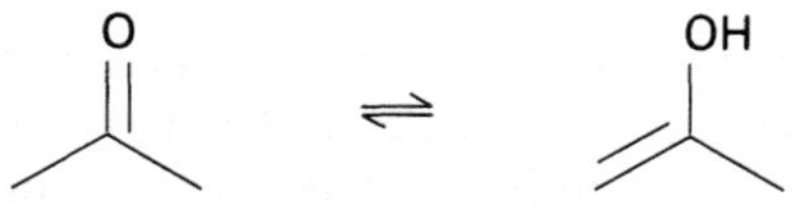

These are usually named according to the structure of the more stable tautomer, that is the keto and nitro versions, respectively. If reference needs to be made to the other tautomer, it is just called the enol or aci tautomer of the corresponding keto or nitro compound.

Sometimes, the enol tautomers may be isolated, but this is quite rare in practice. However, it is usually easy to make the enol tautomer *in situ*, and they can frequently be trapped.

There are other types of proton shift tautomers, such as phenol/dienone, nitroso/oxime and imine/enamine, but these are less often encountered. There are also valence tautomers that exhibit fluxional structures, which undergo rapid sigmatropic rearrangements. Molecules that exhibit this type of isomerism are very interesting, but are not often encountered in undergraduate courses. An example of a valence tautomer is illustrated by the Cope system.

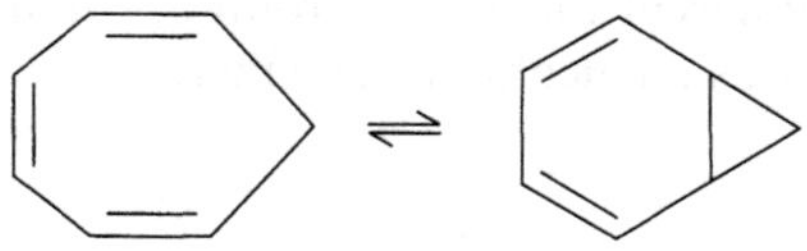

Sugars exhibit a different type of type of tautomerism called ring/chain, so named because it reflects the change from the cyclic hemiacetal derivative to the open-chain aldehyde derivative.

21.2.5 Meta isomerism

An unsymmetrical functional group such as an ester may have a different substituent on each end. If these groups are exchanged, then the two isomers are said to be meta isomers of each other. However, this classification scheme is rarely used.

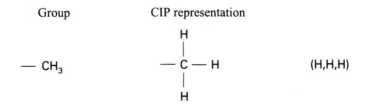

methylpropanoate   ethylethanoate

21.3 Stereoisomerism

While structural isomers differ in the number and type of bonds to be found in each isomer, in stereoisomers the only difference is the relative position of the groups with regard to each other. There are three main sub-types of stereoisomer: geometric, optical and conformational.

When discussing this type of isomerism, it is necessary to be able to identify and rank the various groups that are attached to the particular atomic centre of interest. In order to do this, the Cahn–Ingold–Prelog system is used, which works by ranking each group in accordance with the following rules:

1. Each group is ranked in order of decreasing atomic number of the atom that is directly joined to the carbon or group in question.

2. Where two or more of the atoms connected to the system in question are the same, then the atomic number of the second atom determines the order. The first difference is taken so (O,H,H) ranks higher than (C,C,C), even though it has an overall lower summation of atomic numbers. If this still does not resolve the issue, continue until there is a difference.

3. All atoms, except hydrogen, are formally given a valency of four. Where the actual valency is less than four, phantom atoms are added as necessary, with assigned atomic numbers of 0, which are represented by subscripted zeros.

4. Higher isotopes take precedence over lower ones.

5. All multiple bonds are split into single bonds, and phantom atoms are used to terminate the bonds.

Group	CIP representation	
— CH_3	$\begin{array}{c} H \\ | \\ -C-H \\ | \\ H \end{array}$	(H,H,H)

$$— CH_2 — CH_3 \qquad \begin{array}{c} H \quad\; H \\ | \quad\;\; | \\ —\; C — C — H \\ | \quad\;\; | \\ H \quad\; H \end{array} \qquad (C,H,H)$$

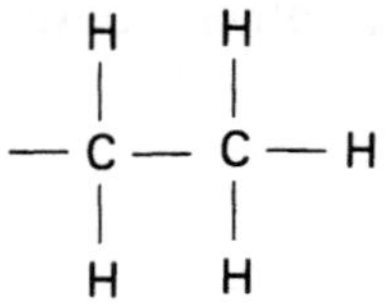

$$— CH(CH_3)_2 \qquad \begin{array}{c} H \\ | \\ H — C — H \\ | \\ — C — H \\ | \\ H — C — H \\ | \\ H \end{array} \qquad (C,C,H)$$

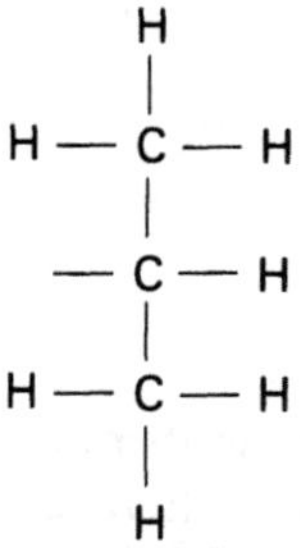

$$— CH_2OH \qquad \begin{array}{c} H \\ | \\ — C — O_{00} — H \\ | \\ H \end{array} \qquad (O,H,H)$$

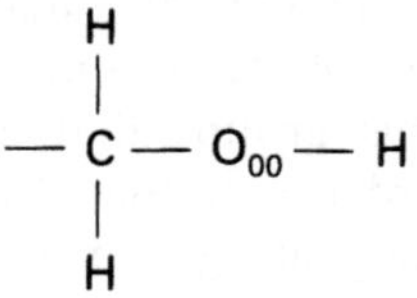

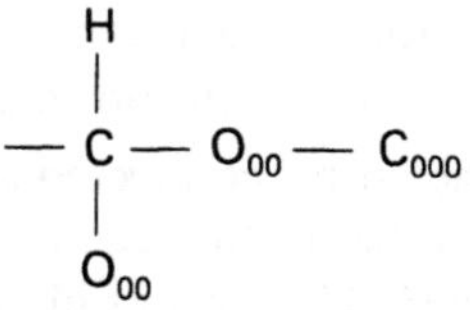

$$— CH = O \qquad \begin{array}{c} H \\ | \\ — C — O_{00} — C_{000} \\ | \\ O_{00} \end{array} \qquad (O,O,H)$$

$$— CH = CH_2 \qquad \begin{array}{c} H \quad\; H \\ | \quad\;\; | \\ — C — C — C_{000} \\ | \quad\;\; | \\ C_{000} \quad H \end{array} \qquad \begin{array}{l} (C,C,H) \\ \text{or more fully} \\ (C_{CHH}, C_{000}, H) \end{array}$$

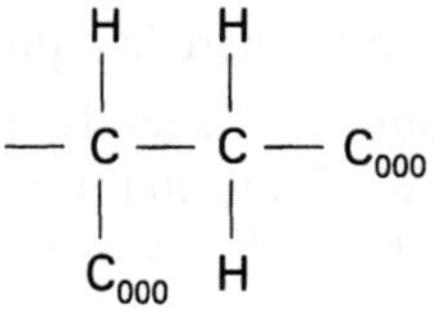

$$— CH \equiv CH \qquad \begin{array}{c} C_{000} \quad C_{000} \\ | \qquad\; | \\ — C — C — H \\ | \qquad\; | \\ C_{000} \quad C_{000} \end{array} \qquad \begin{array}{l} (C,C,C) \\ \text{or more fully} \\ (C_{CCH}, C_{000}, C_{000}) \end{array}$$

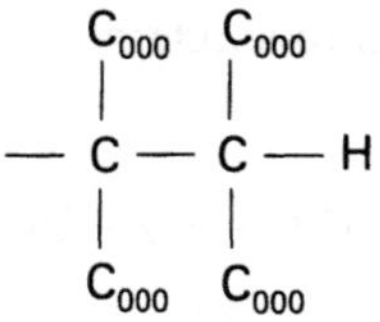

$$
\begin{array}{c}
\text{C}_{000} \\
| \\
\text{H} - \text{C} - \text{C} - \\
| \quad\quad | \\
\quad\quad\;\; \text{H} \\
\quad\quad\;\; | \\
- \text{C} - \text{C} - \text{C} - \\
| \quad | \quad | \\
\text{C}_{000} \;\; \text{C}_{000}
\end{array}
$$

(C,C,C)

or more fully

$(C_{CCH}, C_{CCH}, C_{000})$

21.3.1 *Optical isomerism*

If a compound rotates the plane of polarised light, it is said to be optically active. It has been found by experiment that any molecule that is optically active is not superimposable on its mirror image. This property is called chirality. There are three requirements for a molecule to be chiral. It must not have a centre of symmetry; it must not have a plane of symmetry; and it must not have an alternating axis of symmetry (i.e. rotation followed by reflection in the plane perpendicular to this axis of rotation). In all examples that will be encountered in common practice, only the first two types of symmetrical transformation need be considered.

For a compound that is chiral, there are two molecules that are related by the fact that one is superimposable on the mirror image of the other. These two molecules are called enantiomers, or sometimes enantiomorphs, of each other. Enantiomers have exactly the same physical and chemical properties except in two important respects. First, they rotate plane polarised light by opposite, but equal, amounts. The compound that rotates the plane anticlockwise is called the *laevo*, *l*, or (−) isomer, while the other is the *dextro*, *d*, or (+) isomer. The second difference is that they may react with other optically active compounds at different rates.

In organic chemistry, by far the commonest geometry that is chiral is the tetrahedral sp^3 hybridised carbon atom with four different groups attached to it, i.e. C_{abcd}.

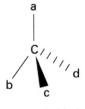

general chiral
molecule

The stereochemistry at every chiral centre may be identified by using the Cahn–Ingold–Prelog rules to rank the groups. Once the groups have been ranked in order, the lowest ranking group is positioned furthest away from the observer. If the remaining groups are arranged in a clockwise manner, when going from highest to lowest priority group, then that isomer is labelled the *R* enantiomer. In contrast, if the groups are arranged in the opposite order, i.e. they are arranged in an anticlockwise manner, then that is labelled the *S* enantiomer.

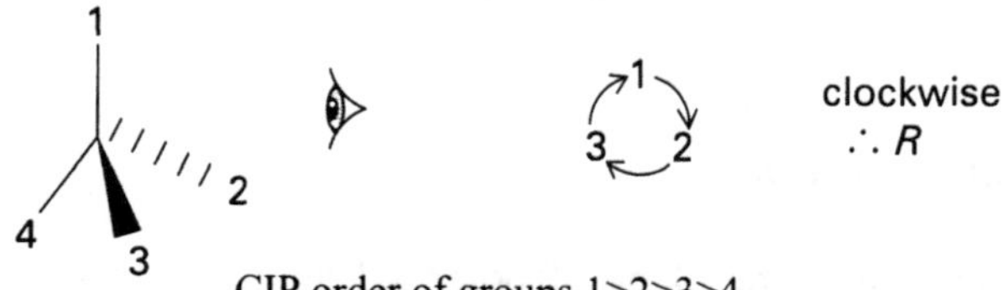

clockwise
∴ *R*

CIP order of groups 1>2>3>4

If instead of the general chiral molecule, C_{abcd}, two groups are the same, i.e. C_{abcc}, then the carbon under consideration will no longer be chiral. This is because it has a plane of symmetry, which contains the central carbon atom and the C–a and C–b bonds, and bisects the angle between the two C–c bonds.

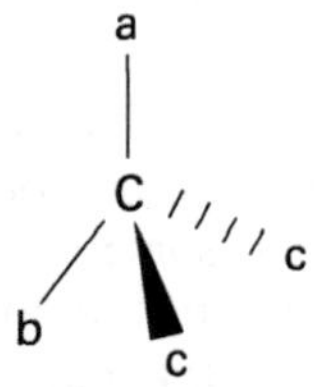

prochiral
molecule

However, if one of the c groups is substituted for a fourth group that is different from the others, then the molecule will become chiral again. The original molecule, C_{abc_2}, is said to be prochiral, and the two c groups are said to be enantiotopic, because the replacement of either of them would result in an enantiomer being formed. The two c groups will behave differently in an asymmetrical environment, and may be labelled *pro-R* and *pro-S* to distinguish them.

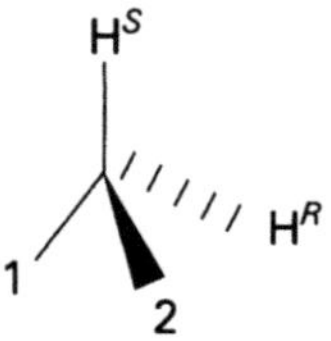

prochiral molecule, showing the
pro-R and *pro-S* hydrogens

The idea of enantiometric molecules may be extended from an atomic centre to a face of a trigonal molecule, by an extension of the Cahn–Ingold–Prelog system. Thus, when looking at one face of a trigonal system, if the groups are arranged in a clockwise manner, when moving from the highest to the lowest, then that face is called the *re* face. If the groups are arranged in an anti-clockwise manner, then that face is the *si* face.

If there is more than one chiral centre in a molecule, then the maximum number of possible *d,l* pairs of stereoisomers increases, and is given by the formula 2^{n-1}, where *n* is the number of chiral centres. So, if there are two chiral centres there are four possible stereoisomers, comprising two *d,l* pairs. Stereoisomers are called diastereomers if their relationship to one another is not enantiomeric. Where two diastereomers differ only in the stereochemistry at one chiral centre, then they are called epimers. If the two c groups in the prochiral example given above were in such positions that if they were replaced diastereomers would be formed, then the c groups would be called diastereotopic.

A molecule with two chiral centres may be represented by the totally general formula $C_{abc}C_{xyz}$. Such a molecule would have four isomers, namely two *d,l* enantiomeric pairs. If, however, the molecule has the symmetrical formula represented by $C_{abc}C_{abc}$, then there are only three isomers, namely one *d,l* enantiomeric pair, and an isomer that has an internal plane of symmetry, which accordingly is not optically active. This optically inactive compound is called the *meso* isomer.

stereoisomers of $C_{abc}C_{abc}$

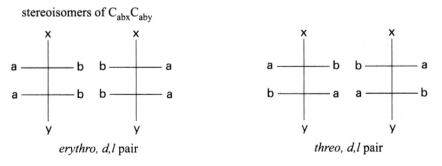

Another variation would occur when there are two groups that are common to each carbon, while the third is different, i.e. $C_{abx}C_{aby}$. In such a case, there are now two related *d,l* pairs, which are called the *threo* and *erythro* pairs.

stereoisomers of $C_{abx}C_{aby}$

This type of geometry is common in biological molecules, and that is why it is often illustrated using the Fischer projection, which was developed to illustrate the stereochemistry of small biological molecules such as sugars and amino acids.

If the two enantiomers of a compound are present in equal amounts, then the resultant mixture will not rotate the plane of polarised light because the effect of one isomer will cancel out the effect of the other. This is called a racemic mixture, and sometimes it has different properties from either of the enantiomers, e.g. different melting point and solubility.

The tetrahedral geometry is found most commonly at a normal sp^3 carbon centre. However, it may be found in other situations. For example, there is no need for the chiral centre to be a carbon atom, i.e. other tetravalent asymmetrical centres such as a quaternary ammonium salt would suffice. In passing, it ought to be noted that there needs only to be a very small difference in the four groups in order for optical activity to arise. This may be illustrated by the isotopically labelled sulphone, $R^1R^2S^{18}O^{16}O$, which is optically active.

Any trisubstituted, pyramidal, nitrogen atom bonded to three different groups is chiral. In such an example, the lone pair constitutes the fourth group. However, at room temperature there is a rapid inversion of the molecule, which means that effectively there is an equal concentration of both enantiomers, and so such a molecule does not exhibit any optical activity in normal circumstances. This inversion is called the umbrella effect.

The umbrella effect is also seen in other types of compounds, e.g. sulphoxides. However, the inversion process is slowed down sufficiently in substituted sulphoxides, R^1R^2SO, for the optical activity to be measured, e.g. $[\alpha]_{280} = +0.71°$ for $(+)$ $Ph^{12}CH_2SO^{13}CH_2Ph$.

In addition to the simple tetrahedral shape, other geometries can be chiral. For example, adamantane is a caged structure in which four bonds project to the corners of an expanded tetrahedron; consequently, suitably substituted adamantanes may be optically active. This example illustrates clearly that chirality is a molecular property, rather than an atomic one.

Other chiral geometries are also possible, e.g. those that possess perpendicular disymmetric planes. Optical activity will occur if there are two rings that contain an even number of atoms joined in a *spiro* manner, with each ring substituted at the carbon furthest from the *spiro* junction. In this case, an elongated tetrahedron is formed.

In order to be chiral, the groups at each end must be different from one another, but they do not need to be different from those at the other end. This elongated tetrahedral geometry may also be found where two double bonds are joined directly together as in an allene. Thus, $_{ab}C=C=C_{ab}$ will be chiral, but $_{ab}C=C=C_{cc}$ will not. This is because the latter has a plane of symmetry.

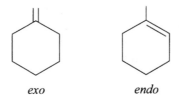

In general, an even number of double bonds joined directly together gives rise to a potentially chiral molecule, while an odd number gives rise to a molecule that displays geometric isomerism, which will be discussed in the next section.

In addition to the two ways of creating an elongated tetrahedral structure described above, it is possible to combine them, i.e. to have a double bond projecting from a ring containing an even number of atoms. This type of double bond is called an *exo*-cyclic double bond, in contrast to an *endo*-cyclic double bond. The isomer with the *exo*-cyclic double bond, has similar stereochemical properties to the two types of elongated tetrahedral structures discussed earlier.

exo   *endo*

It is possible to have a chiral centre about a single bond so long as there is restricted rotation. An example of this hindered rotation occurs in biphenyl derivatives, which have the same extended tetrahedral geometry as the *spiro* and allene compounds above. Isomers that may be resolved only because of hindered rotation about a single bond are called atropisomers.

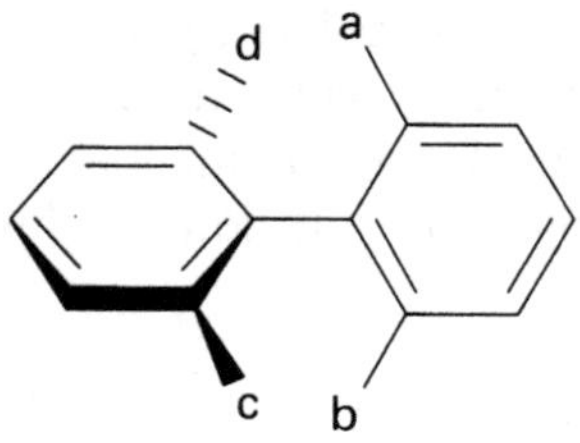

By using phase lagged X-ray crystallography it is possible to discover the absolute position of one group with respect to each of the others around a chiral centre. This was first done in 1951 by Bijvoet on a crystal of sodium rubidium tartrate. Up until that time, it had been impossible to assign the absolute configuration to any molecule, and so an arbitrary relative system had been suggested by Rosanoff. He proposed that the isomer of glyceraldehyde that was *dextro* rotatory be assigned the configuration shown below, and called it the D isomer. Glyceraldehyde was chosen because it is related to many sugars. It turned out that his suggested configuration was correct. This configuration would now be called the *R* isomer.

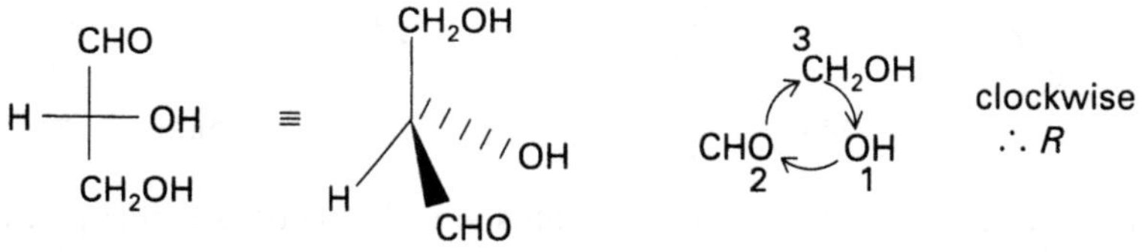

By relating one compound to another by synthetic routes that had a known stereochemical consequence, many optically active molecules could be designated either D or L. Now all compounds are named unambiguously according to the Cahn–Ingold–Prelog rules for each chiral centre.

A reaction in which only one set of stereoisomers is formed exclusively or predominantly is termed a stereoselective synthesis. In a stereospecific reaction, however, the particular isomer of the product molecule is determined by which isomer of the starting material was used initially. All stereospecific reactions must be stereoselective, but the converse is not true.

It is possible for other geometries to give rise to optical activity within a molecule, e.g. helical structures and knotted compounds. However, these geometries are not often encountered in practice.

21.3.2 Geometrical isomerism

Compounds in which rotation is restricted can exhibit geometric isomerism. These compounds do not rotate the plane of polarised light (unless they also happen to be chiral for other reasons). Furthermore, the physical and chemical properties of the various isomers are different.

The simplest example involves a molecule that contains a carbon/carbon double bond that has an identical group attached to each end of the double bond. In this case, there are two possible isomers, namely:

cis   *trans*

The isomer with the two groups on the same side is called the *cis* isomer, while the other one, where they are on opposite sides, is called the *trans* isomer. This type of isomerism is often referred to as *cis/trans* isomerism. However, this nomenclature system fails if there are four different groups on the double bond. Using the Cahn–Ingold–Prelog rules, the groups may be ranked, and if the groups with the higher priority at each end of the double bond are on opposite sides, then that is the *E* isomer; otherwise, if they are on the same side, then it is the *Z* isomer.

The idea may be extended to other rigid double bond systems such as an imine, $R_2C=NR$, or azo, $RN=NR$, in which case the terms *syn* and *anti* are used instead of *cis* or *trans*.

Hindered rotation may occur where there is no formal double bond. In such a case, it is sometimes possible to draw at least one canonical structure that has a double bond along the axis about which there is restricted rotation.

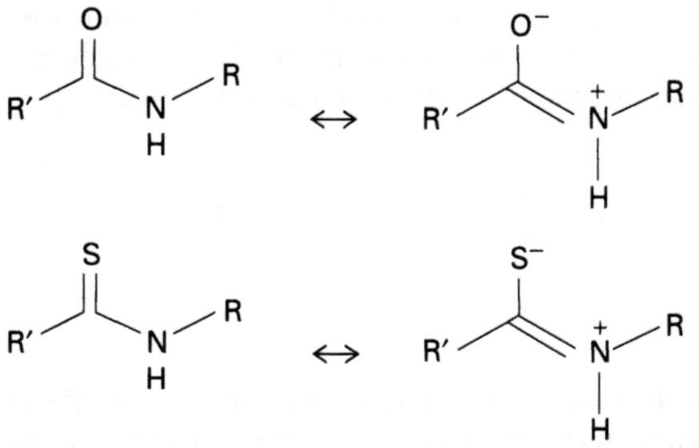

A further example of hindered rotation about a single bond occurs in conjugated dienes. The simple case of buta-1,3-diene exists in two forms, which are described as *transoid* and *cisoid*. The *transoid* form is the thermodynamically more stable form as there is less steric interaction between the terminal methyl groups.

cisoid   *transoid*

In other cases, the restriction to free rotation is imposed by the formation of a ring. For example, geometric isomerism may be exhibited by the three-membered ring of cyclopropane.

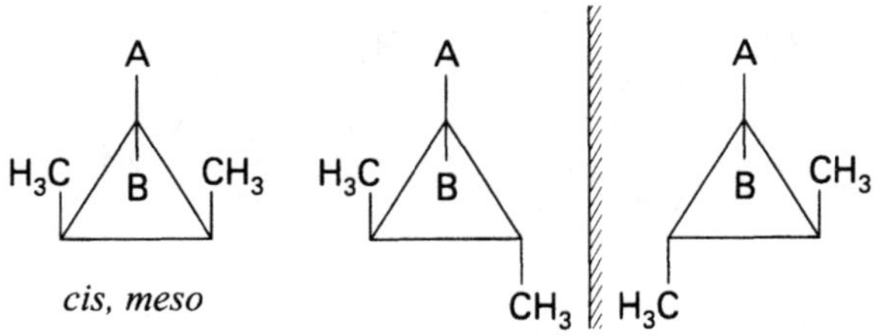

cis, meso   *trans, d,l* pair

Note that in this example the *trans* isomer also exhibits optical isomerism, while the *cis* one does not, because it has a plane of symmetry, i.e. it is *meso*.

In general, where there are two substituted carbons in the ring, e.g. C_{ab} and C_{cd}, there will be four possible isomers if a $\neq$ b $\neq$ c $\neq$ d, since neither the *cis*

434

nor the *trans* isomer is superimposable on its mirror image. This is true regardless of the ring size, or which two carbons in that ring are substituted, except that in rings containing an even number of atoms, where the two substituted carbons are opposite each other, then both the *cis* and the *trans* forms are *meso*.

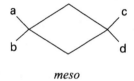

meso

When a = c and b = d, the *cis* isomer is always *meso*, while the *trans* isomer will have a *d,l* pair, except as above.

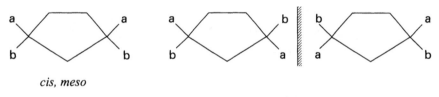

cis, meso

trans, d,l pair

Lastly, there are two types of bicyclic ring system that must be considered. The first is the fused ring system, e.g. decalin, which has two isomers, *cis* and *trans*.

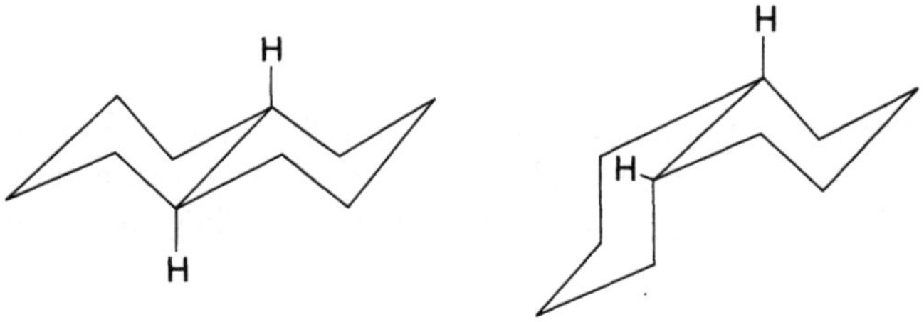

trans-decalin   *cis*-decalin

These isomers are rigid and cannot interconvert.

The second type of bicyclic ring system is the bridged ring system. If there is a substituent on one of the bridges, and it is closer to the longer unsubstituted bridge, then that is the *endo* isomer; while if it is closer to the shorter bridge, then that is the *exo* isomer. If the two unsubstituted bridges are of the same length, then this definition cannot be strictly applied. However, if there is another functional group on only one of the other bridges, then in practice it could still be used.

435

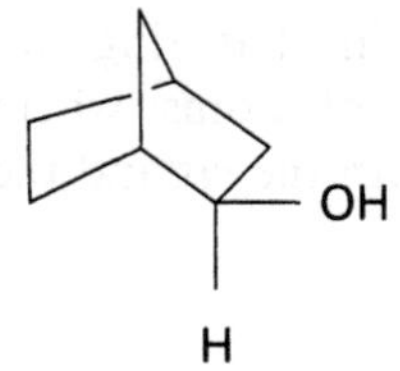

exo-2-norborneol 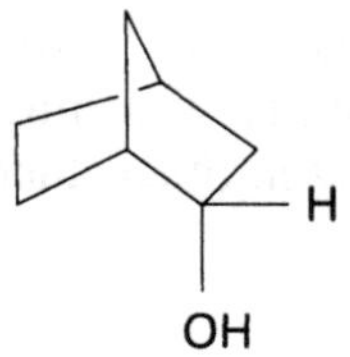   endo-2-norborneol 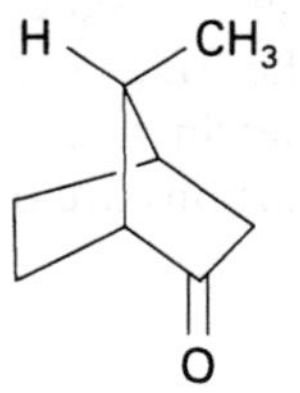   endo-7-methyl-2-norcamphor 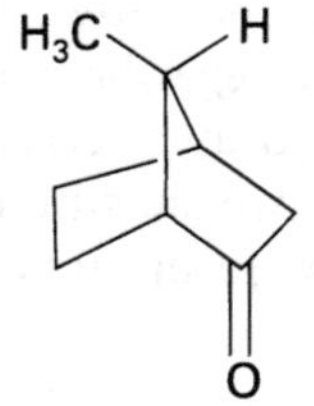   exo-7-methyl-2-norcamphor

The terms *endo* and *exo* are also used when describing the different orientations adopted by the reacting species in the Diels–Alder reaction just prior to the concerted bond formation.

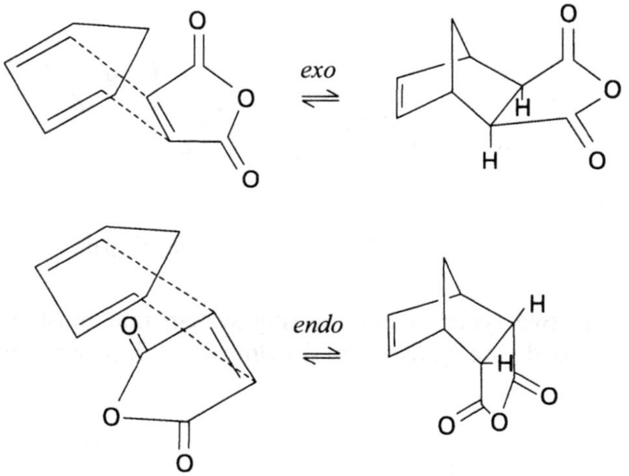

Lastly, the terms *syn* and *anti* are used when referring to the geometry of the neighbouring groups that are about to undergo α,β-elimination. In *syn* elimination the two groups are on the same side, while in *anti* they are on opposite sides.

21.3.3 *Conformers*

If two different three-dimensional arrangements of the constituent atoms can interconvert by free rotation about their component, single, bonds, then they are called conformers: if not, they are called configurations, i.e. to interconvert between the latter, bonds must be broken and reformed. Isomers that have different configurations may often be readily separated. In contrast, isomers that are conformations represent molecules that are rapidly interconverting at room temperature, and thus cannot be separated under normal circumstances, or distinguished by normal methods. We have already seen an example of conformers in the previous section, where the *cisoid* and *transoid* structures of dienes were mentioned.

Conformation analysis is particularly important in aliphatic systems. Taking butane as an example, there is obviously an infinite number of possible dihedral angles that the two methyl groups can adopt with respect to each other. However, two of the conformers are of particular interest in that they represent the states of lowest and highest interaction between the hydrogens on the adjacent methyl groups: these are the staggered and eclipsed conformers respectively.

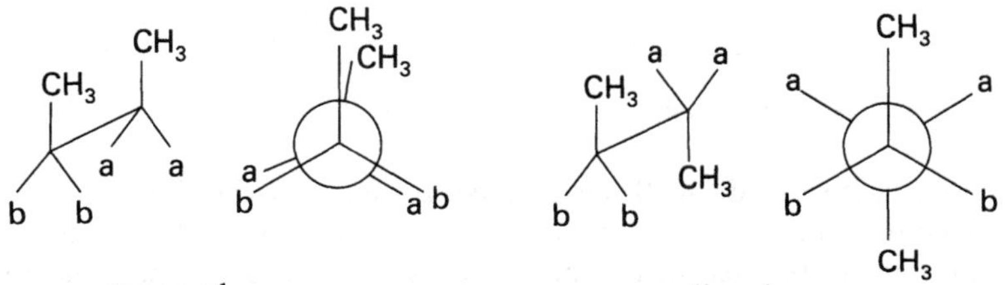

staggered   eclipsed

For the 1,2-disubstituted ethane, XCH_2-CH_2X, there are now four extreme conformers: namely, *antiperiplanar*, *anticlinal*, *synclinal* and *synperiplanar*.

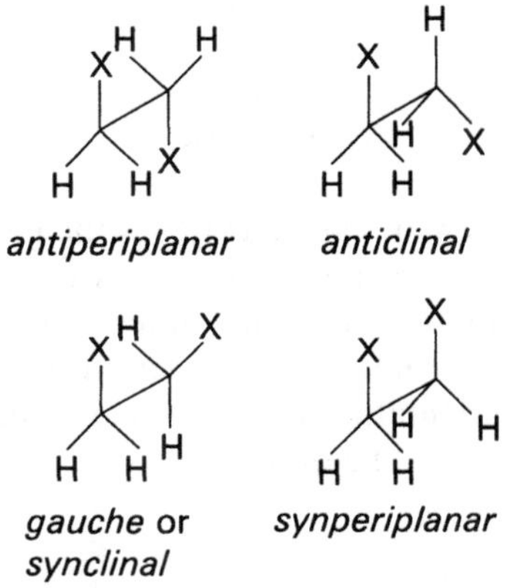

antiperiplanar   anticlinal

gauche or   synperiplanar
synclinal

About an sp^3-sp^2 bond there are also four extreme conformers, which are divided into eclipsing and bisecting.

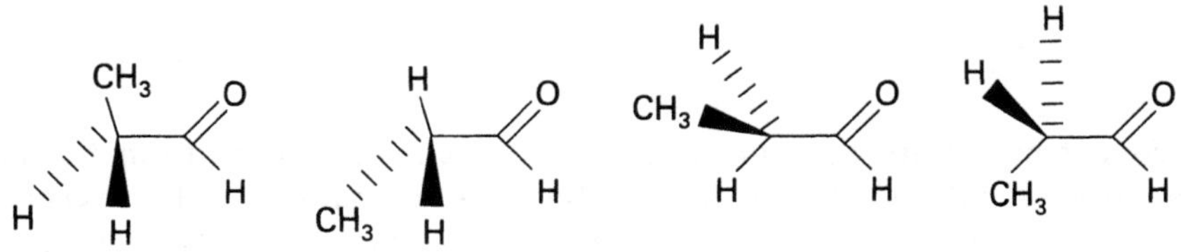

eclipsing   eclipsing   bisecting

437

In cycloalkane systems, completely free rotation is obviously no longer possible; however, there is still sufficient movement for a range of conformers to exist.

The cyclobutane ring exists in a non-planar, open-book conformation, which may flex further open or shut. The cyclopentane ring exists in two puckered conformations, the half chair and the envelope, both of which are non-planar. The position of the puckering may migrate around the ring, and this is called pseudorotation.

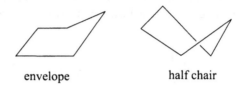

envelope   half chair

The cyclohexane ring adopts three main conformations: the chair, the boat and the skew boat or the twist. Each of these conformations is puckered, i.e. not flat. This is important, because of the implications this has for the possible angle of attack that an incoming reagent must follow.

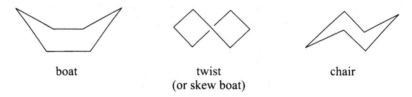

boat   twist   chair
(or skew boat)

The ring normally adopts a chair shape, as this is the most stable state. This is because the interactions between adjacent groups are minimised. There are, however, two chair conformations that may interconvert via the boat or twist conformation. In each chair conformation there are six axial bonds and six equatorial bonds, but the orientation of each bond changes on the conversion from one chair conformation to the other.

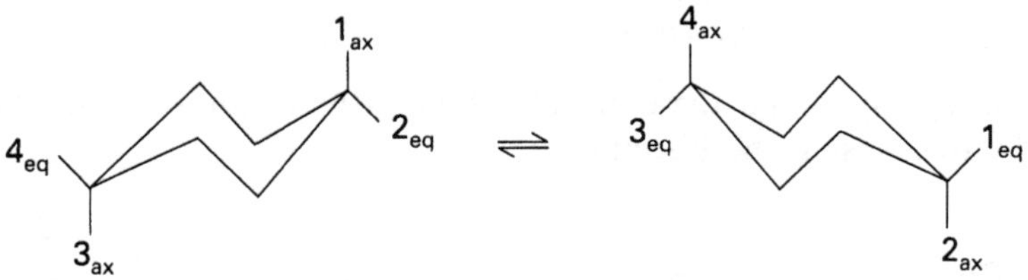

1_{ax} 4_{ax}

4_{eq} 2_{eq} $\rightleftharpoons$ 3_{eq} 1_{eq}

3_{ax} 2_{ax}

At room temperature, this transformation between one chair conformation and the other is rapid, so long as there is no bulky group that would effectively lock the molecule into one preferred conformation. This means that the cyclohexane system may be treated as if it were flat when trying to find how many stereoisomers exist for any given derivative.

For *cis*-1,2-disubstituted cyclohexanes, one group must be axial, while the other is equatorial. For *trans*-1,2-disubstituted cyclohexanes, the substituted groups must both be axial or both be equatorial. The same analysis is true for 1,4-disubstituted cyclohexanes. However, for 1,3-disubstituted cyclohexanes, the reverse is true: thus, in the *cis* isomer both the groups are axial or both are equatorial; while in the *trans* isomer one group must be axial and the other group equatorial.

For alkyl substituents, the most stable position is the equatorial one, which minimises the unfavourable axial/axial steric interactions. However, care should be exercised in deciding which conformation is the most stable, because the energy difference between the axial and equatorial states may be very small. Thus, for example, *trans*-1,2-dihalogenocyclohexanes exist predominantly in the axial/axial conformation. The ring may be locked into one conformation by using a very bulky group, such as *t*-butyl, which must adopt the equatorial position. Thus, such a group may be used to hold another group in the axial position. In such a derivative the two chair conformations do not rapidly interconvert.

When an oxygen is introduced into the ring, an alkyl substituent on the α-carbon remains stable in the equatorial position. However, if the substituent is polar, such as an alkoxide group, the reverse is true. This accounts for the greater stability of α-glucosides over β-glucosides. This is known as the anomeric effect.

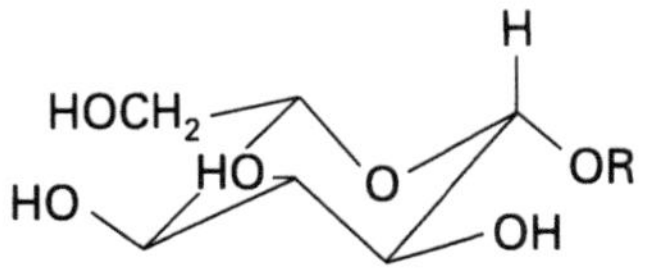

β-glucoside 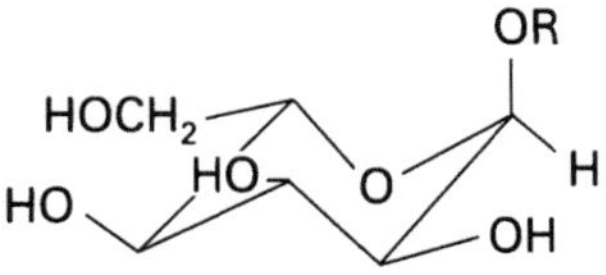   α-glucoside

Oxidation Numbers

22.1 Introduction

The oxidation number of an atom refers to the number of electrons that that atom has gained or lost. For any given atom, this number must be an integer, because it represents the hypothetical number of whole electrons that have been transferred to or from that atom. Thus, it corresponds to the whole number of electrical units of charge which that atom bears. The complete transfer of an electron only takes place in the most ionic of univalent compounds, e.g. KF, and as such is quite rare. As the degree of the ionic nature of the bond decreases, the artificial nature of the concept of oxidation numbers becomes more apparent. In covalent compounds, where the bonds between the nuclei are formed from electrons which are being shared much more evenly, the oxidation number bears no relationship to the charge that resides on any particular atom. However, so long as the limitations of the idea are remembered, the concept may still be used to advantage in covalent compounds, because the relative oxidation states for the same atom in different molecules may indicate the sort of reaction that might be used to interconvert the two molecules synthetically.

22.2 Oxidation numbers in ionic species

Inorganic chemistry may be divided into four types of reaction: acid/base, precipitation, complexation and redox. It is when considering redox reactions that oxidation numbers are used to greatest advantage. This is particularly true when ensuring that a reaction equation has been properly balanced, i.e. that the same amount of reduction as oxidation has taken place.

In simple ions the oxidation number represents the notional whole charge that resides on that ion. So in NaCl, the sodium has an oxidation number of $+1$, while the chlorine has an oxidation number of -1. In $MgCl_2$, the chlorine still has an oxidation number of -1, but the magnesium has an oxidation number of $+2$.

In Fe_2O_3, the oxygen has an oxidation number of -2, while the iron has an oxidation number of $+3$. Here, though, the artificial nature of the concept becomes apparent, because any ion with a $+3$ charge would have such a high charge density that it would polarise the electron rich cloud around the oxygen to some extent and hence reduce its own charge.

In complex ions such as nitrate, NO_3^-, and sulphate, SO_4^{2-}, this charge polarisation is acknowledged and there is no suggestion that the central atom would bear a $+5$ or $+6$ charge respectively. These numbers, however, are the values of the oxidation numbers assigned to the nitrogen and sulphur atoms respectively.

In inorganic chemistry the oxidation number is calculated by applying a number of simple rules:

1. All atoms in their elemental state have an oxidation number of zero.
2. Fluorine, when it is combined with other elements, always has an oxidation number of -1.
3. Hydrogen usually has an oxidation number of $+1$, except when present in metallic hydrides, when it has an oxidation number of -1.
4. Oxygen usually has an oxidation number of -2, except when present as a peroxide, when it has an oxidation number of -1, or when in combination with fluorine, when it has an oxidation number of $+2$.
5. Electropositive elements usually have positive oxidation numbers, i.e. they lose electrons readily; while electronegative elements usually have negative oxidation numbers, i.e. they gain electrons readily.
6. In complex ions the sum of the positive oxidation numbers and the negative oxidation numbers of all the constituent atoms gives the charge that resides on the complex ion.

In a complex ion, it is usually assumed that all the atoms of the same element have the same oxidation number. Where all the atoms of the same element occupy the same environment, then it is reasonable to assume that they have the same electron distribution and hence the same oxidation number. So, for example, in the peroxide anion, O_2^{2-}, it is reasonable to assume that each oxygen has an oxidation number of -1. Or, in the sulphate anion, it reasonable to assume that each oxygen has the same oxidation number, namely -2.

Usually, the oxidation number calculated by the application of the above rules leads to a whole number, which is in accord with the concept that it reflects the number of electrons transferred. However, in certain anions such as the superoxide, O_2^-, or the azide, N_3^-, one arrives at the fractional oxidation numbers of $-\frac{1}{2}$ and $-\frac{1}{3}$ respectively for each of the constituent atoms.

The problem is further highlighted by such complex anions as the thiosulphate anion, $S_2O_3{}^{2-}$, where if it is assumed that each sulphur bears the same oxidation number, one arrives at a value of $+2$ for each sulphur atom.

The problem becomes even more acute in the persulphate anion, $S_2O_8{}^{2-}$. In this example, if one starts by assuming that each of the oxygens has an oxidation number of -2, then the oxidation number of the sulphur is calculated to be $+7$. Alternatively, if the oxidation number of the sulphur is fixed first, at say $+6$, the value required for the oxygens is calculated to be $-1\frac{3}{4}$ each. Which answer is correct, or even useful?

In the four examples that have been given in the last three paragraphs all the problems may be sided-stepped by realising that the original rules that were employed initially, even though they worked well for simple ions, are now no longer sufficient to cope with more advanced examples. Hence, a new working definition is needed for molecular species that are bound together by covalent bonds.

22.3 Oxidation numbers in covalent species

The new working definition for calculating the oxidation number of an atom in species that contain covalent bonds may be stated simply. Every heteroatomic bond that is polarised towards the atom in question contributes a -1 unit to the oxidation number of that atom. Conversely, every heteroatomic bond that is polarised away from the atom in question contributes a $+1$ unit to the oxidation number of that atom. Further, for charged species, if an extra electron resides on the atom, then a -1 unit is added to the oxidation number; while if an electron is missing, then a $+1$ unit is added to the oxidation number. The sum of the oxidation numbers is still zero for a neutral molecule, or equal to the charge that the species bears if it is charged. Now, the emphasis is on finding the oxidation number for each atom individually, and not on calculating an average for all the atoms of the same element within a molecule.

Only heteroatomic bonds or any gain or loss of electrons are of concern when calculating the oxidation number, because it is only the gain or loss of electrons when compared to the electronic environment that is to be found in the elemental state that is of interest.

To illustrate how this new definition works in practice, consider the following examples. Diatomic oxygen, O_2, has an oxidation number of zero, as does elemental carbon, C_n. In methane, CH_4, each hydrogen is involved in a single heteroatomic bond that is polarised away from it, because the carbon atom has the greater electronegativity, and hence each hydrogen has an oxidation number of $+1$. The carbon has an oxidation number of -4, because it is involved in four heteroatomic bonds, each of which is polarised towards it.

In ethane, C_2H_6, each hydrogen has an oxidation number of $+1$, while each carbon has an oxidation number of -3, because each is involved in three heteroatomic bonds that are polarised towards itself. There is also one homoatomic bond, namely the C–C bond, but this does not contribute to the overall oxidation number, because there is no overall polarisation of this bond.

In ethene, C_2H_4, and ethyne, C_2H_2, the oxidation number of the carbon atoms is respectively -2 and -1, because they are now involved in two heteroatomic bonds or only one, respectively. The other bonds are homoatomic, and so do not contribute to the oxidation state.

In longer chain alkanes the oxidation number may be easily calculated for each atom. So, for example, in propane, $CH_3CH_2CH_3$, the terminal carbons have an oxidation number of -3, while the central one has an oxidation number of -2. Similarly for 2-methylpropane, $(CH_3)_2CHCH_3$, the methyl carbons have an oxidation number of -3, while the central carbon has an oxidation number of -1. In $(CH_3)_4C$, the central carbon has an oxidation number of zero, which is perfectly reasonable when it is remembered that it is in a similar environment to elemental carbon, C_n.

In organic compounds which contain oxygen, the oxidation number for each atom may be calculated in a similar manner. In methanol, CH_3OH, each hydrogen has an oxidation number of $+1$, for in each case the bond is polarised away from the hydrogen atom; in contrast, the oxygen has an oxidation number of -2, and the carbon has one of -2, giving an overall sum for the molecule of zero. The oxidation number of the carbon is the result of having three bonds that are polarised towards it, i.e. the three C–H bonds, and one bond that is polarised away from it, i.e. the C–O bond, giving a total of -2 units.

The oxidation numbers of the carbon atoms in the following oxygen containing compounds are: methanal, CH_2O (zero); methanoic acid, HCO_2H $(+2)$; carbon dioxide, CO_2 $(+4)$; ethanol, CH_3CH_2OH $(-3, -1$ respectively); ethanal, CH_3CHO $(-3, +1$ respectively); and, ethanoic acid, CH_3CO_2H $(-3, +3$ respectively).

Hence, it may easily be seen that as ethanol is converted to ethanal and then to ethanoic acid and finally to carbon dioxide, each step involves an oxidation of the carbon atom to which more oxygen atoms are being attached. Similarly, the addition of hydrogen to ethene to give ethane may readily be seen to be a reduction, as the oxidation number is reduced from -2 to -3.

If we now return to the compounds that caused the problems under the previous definition, we see that by analysing the oxidation state of each atom individually, and only considering heteroatomic bonds or any gain or loss of electrons, the problems are resolved.

Starting with diatomic oxygen, as there are no heteroatomic bonds and no electrons have been gained or lost, the oxidation number is zero. In this example, we will assume for simplicity's sake that molecular oxygen has a simple double bond between the two oxygen atoms under normal conditions. The first step is the addition of one electron to form the superoxide anion, O_2^-.

This results in one oxygen having two lone pairs, one homoatomic bond and one unpaired electron, which means that it has a share in the six electrons in the valence shell, exactly as it would have had in the elemental state, and so its oxidation number is zero. The other oxygen has three lone pairs and one homoatomic bond. Thus, it has a share in seven electrons, which is one more than the elemental state, and so it has an oxidation state of -1. This analysis resolves the problem of fractional oxidation numbers that was encountered above.

When a further electron is added to form the peroxide anion, O_2^{2-}, both oxygen atoms have three lone pairs and one homoatomic bond and so each atom now has an oxidation number of -1.

$$\left[\ddot{\underset{\cdot\cdot}{O}} - \ddot{\underset{\cdot}{O}} \right]^{-} \underset{}{\overset{+\beta^{-}}{\rightleftharpoons}} \left[\ddot{\underset{\cdot\cdot}{O}} - \ddot{\underset{\cdot\cdot}{O}} \right]^{2-}$$

The next example, the azide anion, N_3^{-}, is more complicated. It may be considered in three steps. First, a neutral nitrogen atom, which has five electrons accommodated in its valence shell, and hence an oxidation number of zero, gains a single electron to give it six electrons accommodated in three lone pairs. Thus, it now has an oxidation number of -1.

$$\ddot{\underset{\cdot}{N}} \overset{+\beta^{-}}{\rightleftharpoons} \left[\ddot{\underset{\cdot\cdot}{N}} \right]^{-}$$

Secondly, a triply bonded dinitrogen molecule, in which each nitrogen has an oxidation of zero, bonds to the atomic nitrogen anion to give rise to one of the canonical structures of the azide anion, namely, one terminal nitrogen has one lone pair and three homoatomic bonds, and so has an oxidation number of zero. The central atom has four homoatomic bonds, but one of which is polarised away from it, i.e. the dative bond, and so it has an oxidation number of $+1$. The other terminal nitrogen started with one extra electron and now also has a homoatomic bond, which is polarised towards it, and so now has an oxidation number of -2. This nitrogen atom has three lone pairs.

$$\left[\ddot{\underset{\cdot\cdot}{N}} \right]^{-} \quad :N \equiv N: \rightleftharpoons \left[\overset{2-}{\ddot{\underset{\cdot\cdot}{N}}} - \overset{+}{N} \equiv N: \right]^{-}$$

Thirdly, and on a slightly different note, one can hypothesise that there is a reorganisation of the electrons in the π molecular orbitals to give another canonical structure, where the central nitrogen still has an oxidation number of $+1$, but now each of the terminal nitrogen atoms has an oxidation number of -1. Again, this analysis resolves the problem of fractional oxidation numbers encountered above.

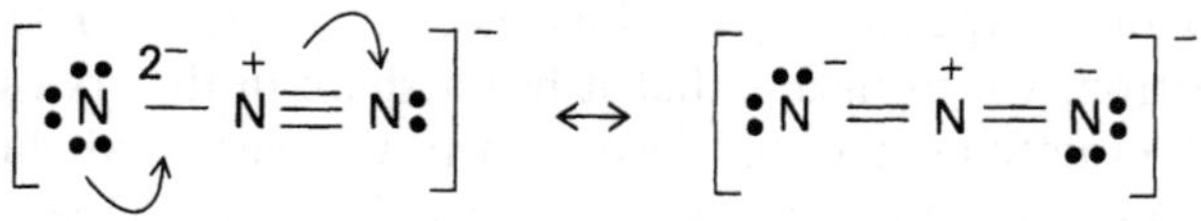

$$\left[\overset{2-}{:\!\ddot{N}} \!-\! \overset{+}{N} \!\equiv\! N: \right]^{-} \leftrightarrow \left[:\!\ddot{N} \!\overset{-}{=}\! \overset{+}{N} \!=\! \overset{-}{\ddot{N}}: \right]^{-}$$

In the thiosulphate anion, $S_2O_3^{2-}$, it is important to know how each sulphur is bonded.

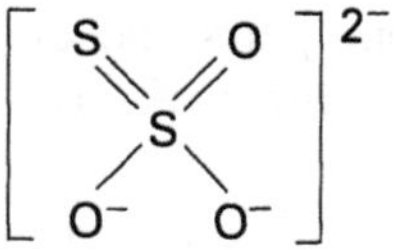

$$\left[\begin{array}{c} S \quad\quad O \\ \diagdown \quad \diagup \\ S \\ \diagup \quad \diagdown \\ O^{-} \quad O^{-} \end{array} \right]^{2-}$$

The sulphur that occupies a terminal position is in a similar configuration to one of the oxygens. However, it has two covalent homoatomic bonds to the central sulphur and so its oxidation number is zero. The central sulphur has four heteroatomic bonds to oxygen atoms and in each case they are polarised away from the sulphur, and so its oxidation number is +4. Thus, instead of each sulphur being given the same oxidation number of +2, as was the result above, a distinction is drawn between them which reflects the different chemical environments they occupy.

The persulphate anion, $S_2O_8^{2-}$, has the structure:

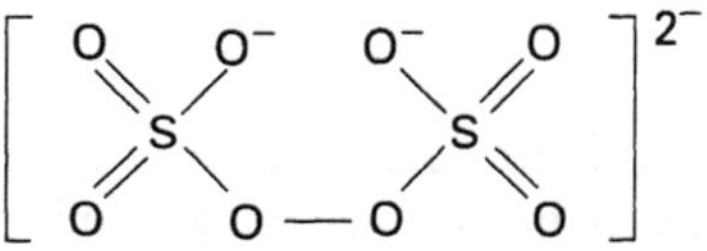

$$\left[\begin{array}{cccc} O \quad\quad O^{-} \quad\quad O^{-} \quad\quad O \\ \diagdown \quad \diagup \quad\quad\quad \diagdown \quad \diagup \\ S \quad\quad\quad\quad S \\ \diagup \quad \diagdown \quad\quad\quad \diagup \quad \diagdown \\ O \quad\quad O - O \quad\quad O \end{array} \right]^{2-}$$

Here each sulphur occupies a similar position and may be given an oxidation number of +6. The six terminal oxygens may be assigned an oxidation number of −2, while the two bridging oxygens, which form a peroxide type link between the two sulphurs, are only involved in one heteroatomic bond each and so each has an oxidation number of −1. It is this peroxide bond that is broken when the persulphate reacts as an oxidising agent, and concomitantly these oxygens are reduced to −2 as the two sulphate anions are produced.

So, in summary, it may be clearly seen that by examining the electronic environment of each atom in turn within a covalently bonded molecule, a sensible oxidation number may be calculated, which reflects the different environment of each atom.

23

Skeletal Index

This index is designed to help you find the name used to describe the common functional groups and other ubiquitous combinations of atoms in organic compounds. It is arranged in order of increasing complexity as perceived visually. Hydrocarbons are listed first; then, in the subsequent sections, compounds that contain oxygen, nitrogen, sulphur and phosphorus are introduced. Complexity is partly subjective; however, the following guidelines have been used when ordering the entries: increasing degree of unsaturation; more heteroatoms; larger rings; and more ring systems.

In this index, the old or trivial name for some compounds has been included, e.g. nitrous acid instead of nitric (III) acid for HNO_2. This is deliberate, because first, that is the name used in practice; and secondly, that is the name by which the compound is known in the older literature, which is where the most assistance is required.

23.1 Hydrocarbon compounds

C_nH_{2n+2}	saturated hydrocarbon, alkane: methane (1), ethane (2), propane (3), butane (4), pentane (5), hexane (6), heptane (7), octane (8), nonane (9), decane (10).
$R_2C=CR_2$	alkene (olefin)
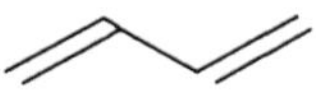	diene
	triene, and so on to a polyene
$R_2C=C=CR_2$	allene

$R_2C=C=C=CR_2$ cumulene
$R–C≡C–R$ alkyne
$R–C≡C^-$ acetylide
$R_2C:$ carbene
R_3C^+ carbonium ion
R_3C^- carbanion
$R_3C·$ alkyl radical
$R_3C–$ alkyl group
$R_2C=CR–$ alkenyl group (where $R=H$, then vinyl)
$RC≡C–$ alkynyl group
$–R_2C–(CR_2)–CR_2–$ alkylenyl group
$R_2C=CR–CR_2–$ allylic group
$R_2C=CR–CR_2–CR_2–$ homoallylic group

cyclopropane, generally cycloalkanes

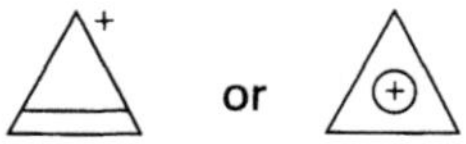

cyclopropenyl cation

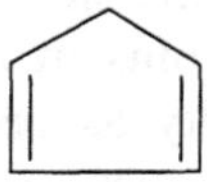

cyclopentadiene

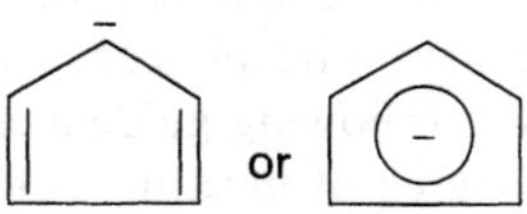

cyclopentadienyl anion

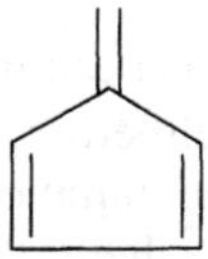

fulvenes

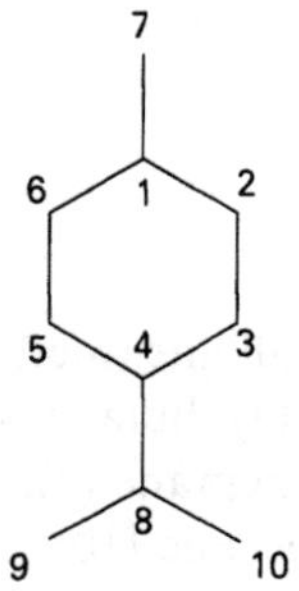

fully saturated monocyclic monoterpenoid, *p*-menthane

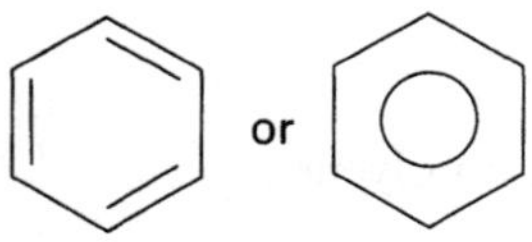

benzene

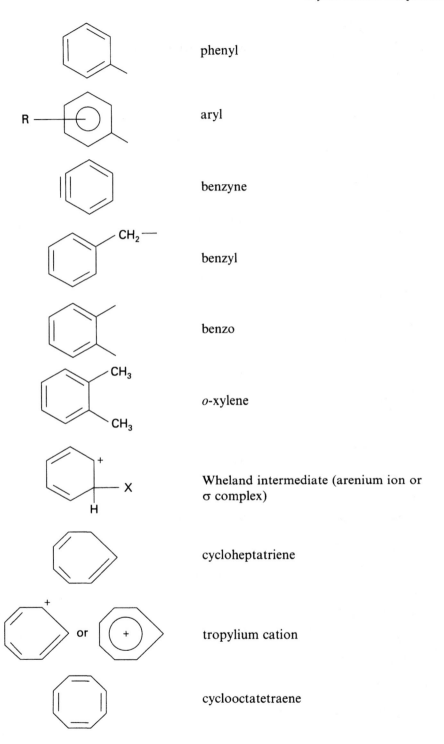

phenyl

aryl

benzyne

benzyl

benzo

o-xylene

Wheland intermediate (arenium ion or σ complex)

cycloheptatriene

tropylium cation

cyclooctatetraene

Skeletal Index

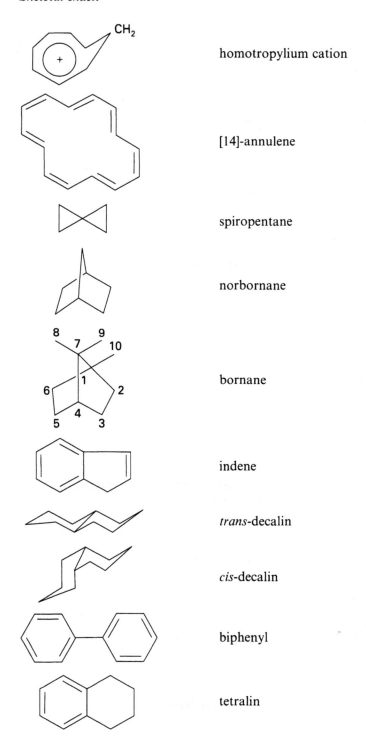

homotropylium cation

[14]-annulene

spiropentane

norbornane

bornane

indene

trans-decalin

cis-decalin

biphenyl

tetralin

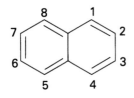

naphthalene

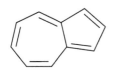

azulene

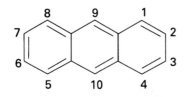

anthracene

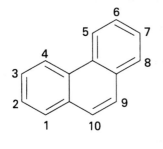

phenanthrene

phenalene

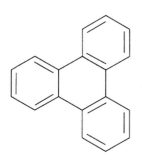

triphenylene

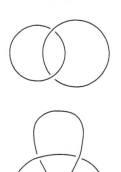

catenane

trefoil

23.2 Oxygen containing compounds

RCH_2OH	primary alcohol (R ≠ H)
R_2CHOH	secondary alcohol (R ≠ H)
R_3COH	tertiary alcohol (R ≠ H)
HO–	hydroxyl group
RO–	alkoxyl group
HO^-	hydroxide anion
RO^-	alkoxide anion
ROR	ether
H_3O^+	hydroxonium ion
R_3O^+	oxonium ion
H_2O_2	hydrogen peroxide
ROOH	hydroperoxide
ROOR	peroxide
RCH(OR)OH	hemiacetal
$R_2C(OR)OH$	hemiketal
$RCH(OR)_2$	acetal
$R_2C(OR)_2$	ketal
$RC(OR)_3$	orthoester
$C(OH)_4$	orthocarbonic acid
$C(OR)_4$	orthocarbonate

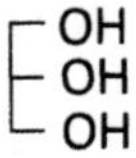

```
 ┌ OH
 ├ OH        glycerol
 └ OH
```

$C{=}O$ carbon monoxide

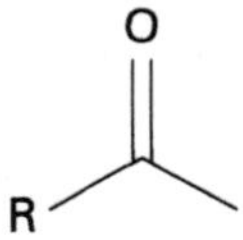

acyl group

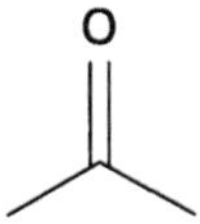

O‖ (structure)	carbonyl group
O‖ R—C—H or RCHO	aldehyde
O‖ R—C—R or RCOR	ketone
OH (structure)	enol
O⁻ (structure)	enolate
OR (structure)	enol ether
O‖ R—C—OH or RCO$_2$H	carboxylic acid
RCO$_2^-$	carboxylate anion
O‖ R—C—O—	alkanoate group (or acyloxy group)
O‖ —C—OH	carboxylic acid group

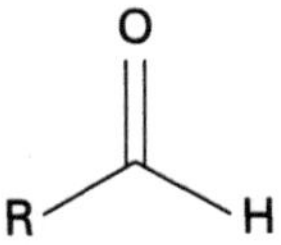

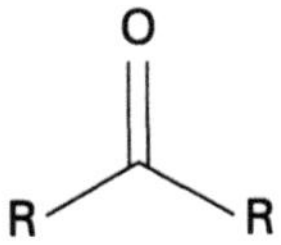

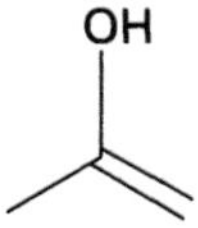

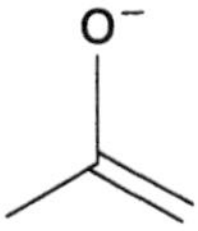

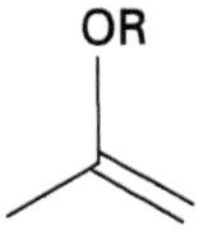

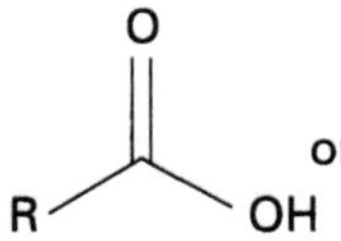

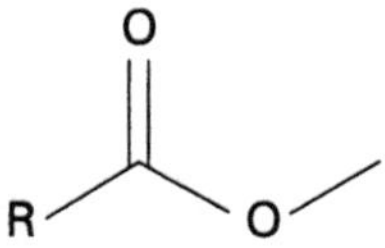

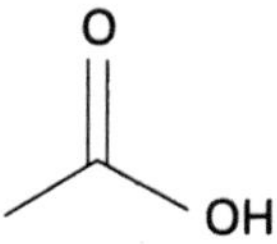

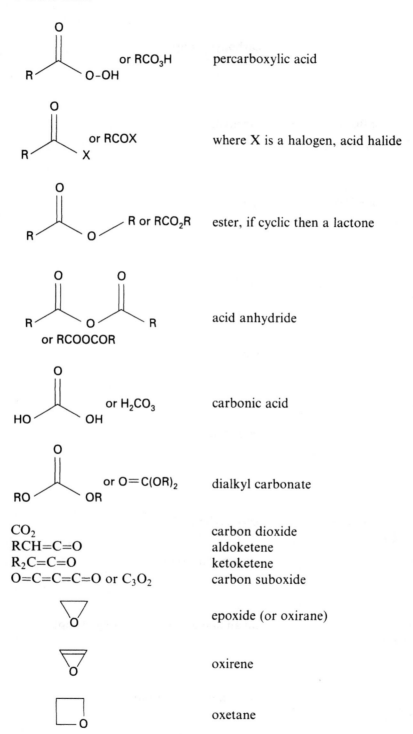

or RCO_3H — percarboxylic acid

or RCOX — where X is a halogen, acid halide

or RCO_2R — ester, if cyclic then a lactone

or RCOOCOR — acid anhydride

or H_2CO_3 — carbonic acid

or $O{=}C(OR)_2$ — dialkyl carbonate

CO_2	carbon dioxide
$RCH{=}C{=}O$	aldoketene
$R_2C{=}C{=}O$	ketoketene
$O{=}C{=}C{=}C{=}O$ or C_3O_2	carbon suboxide

epoxide (or oxirane)

oxirene

oxetane

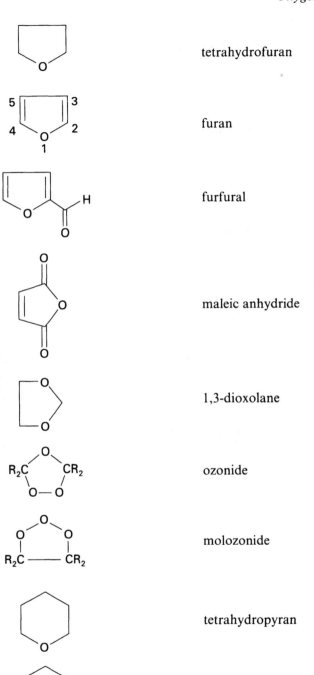

tetrahydrofuran

furan

furfural

maleic anhydride

1,3-dioxolane

ozonide

molozonide

tetrahydropyran

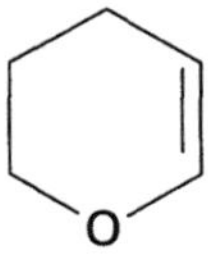

dihydropyran

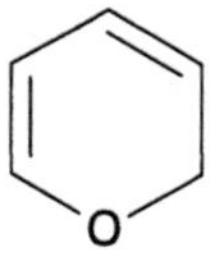

2-H (or α–) pyran

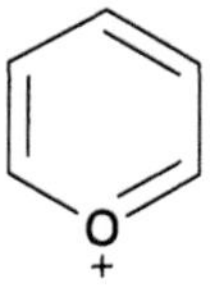

pyrylium cation

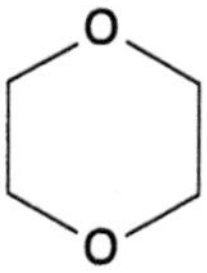

1,4-dioxan

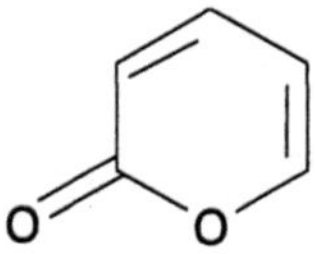

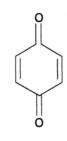

2-pyrone

p-benzoquinone

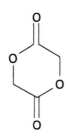

lactide

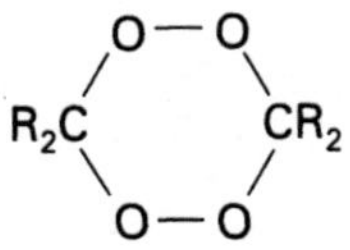

bisperoxide

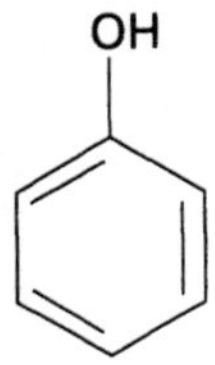

phenol

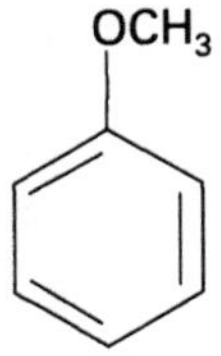

anisole

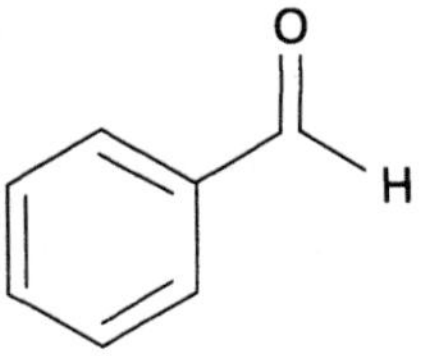

benzaldehyde

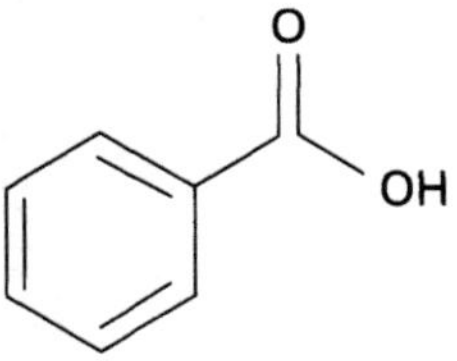

benzoic acid

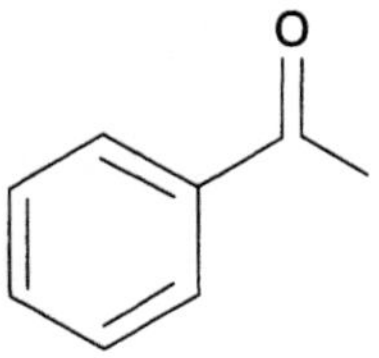

benzoyl group

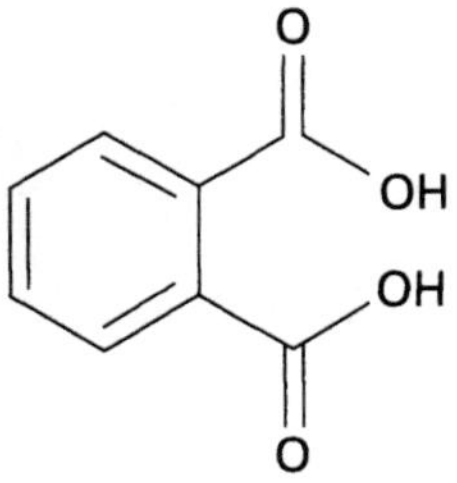

phthalic acid

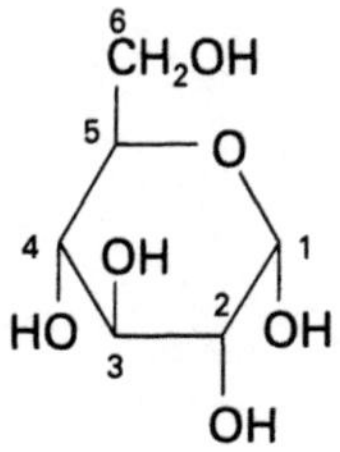

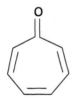

pyranose

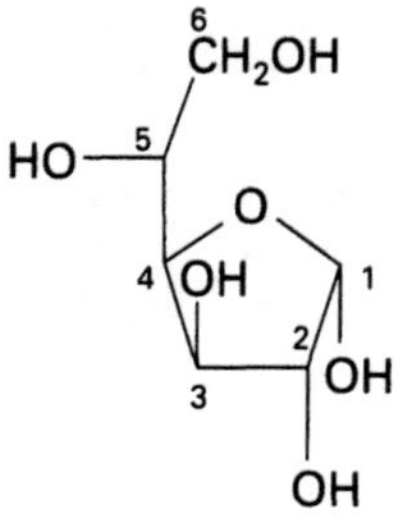

furanose

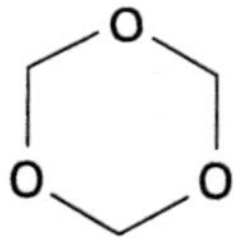

1,3,5-trioxane

tropone

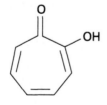

tropolone

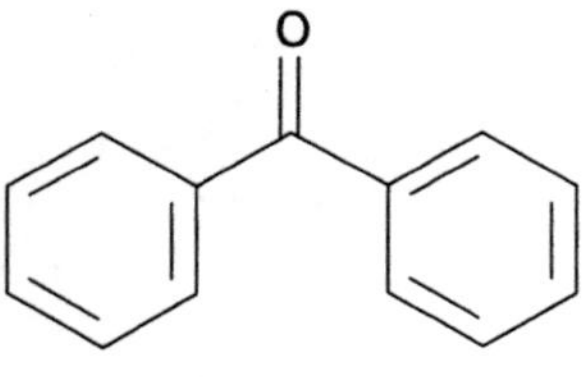

benzophenone

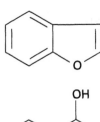

benzofuran

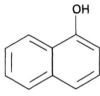

1-naphthol

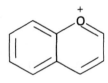

benzopyrylium cation

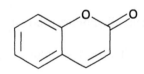

benzo-2-pyrone (or coumarin)

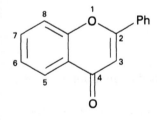

flavone

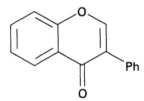

isoflavone

23.3 Nitrogen containing compounds

NH_3	ammonia
NH_4^+	ammonium cation
$R{-}NH_2$	primary amine
R_2NH	secondary amine
R_3N	tertiary amine

$R_4N^+X^-$ — quaternary ammonium salt

$RN:$ — nitrene

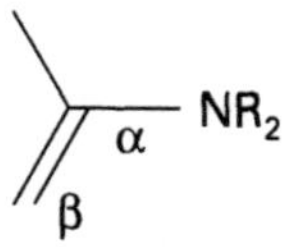

enamine

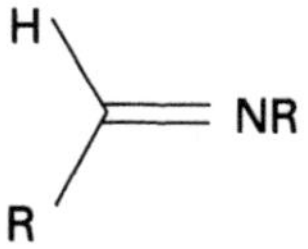

aldimine

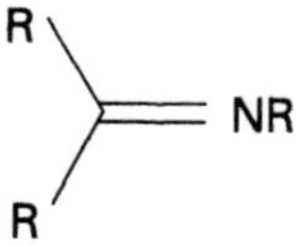

imine

$R_2C=NR_2^+$ — iminium cation (or immonium cation)

$R-C\equiv C-NR_2$ — ynamine

$R-N=C$ — isonitrile

$R-N=CX_2$ — where X is a halogen, alkyliminocarbonyl halide

HCN — hydrocyanic acid, or hydrogen cyanide

$R-CN$ — nitrile

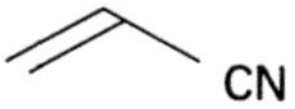

acrylonitrile

H_2N-NH_2 — hydrazine

$HN=NH$ — diimide

$Ar-NH-NH-Ar$ — hydrazo compound

$Ar-N=N-X$ — one nitrogen bonded to an atom other than carbon, diazo

$Ar-N=N-Ar$ — both nitrogens bonded to a carbon atom, azo

$R_2N-C\equiv C-NR_2$ — ynediamines

$R_2N-N:$ — aminonitrene

$R-N^+\equiv N$ or $R-N_2^+$ — diazonium cation

$R_2C=N-NH_2$ — hydrazone

R_2N-NH^- — hydrazide anion

$R_2C=N^+=N^-$ — diazoalkane

$ArCH=N-N=CArH$ — azine

$R-N=C=N-R$ — carbodiimide

H_2N-CN — cyanamide

$NC-CN$ — cyanogen

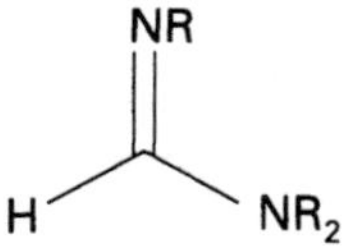

formamidine

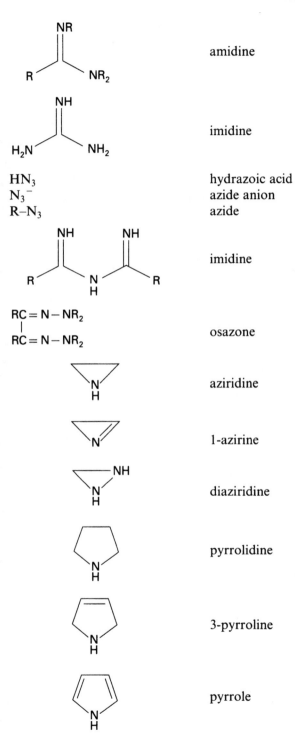

amidine

imidine

HN$_3$ hydrazoic acid
N$_3^-$ azide anion
R–N$_3$ azide

imidine

RC = N – NR$_2$
    |
RC = N – NR$_2$ osazone

aziridine

1-azirine

diaziridine

pyrrolidine

3-pyrroline

pyrrole

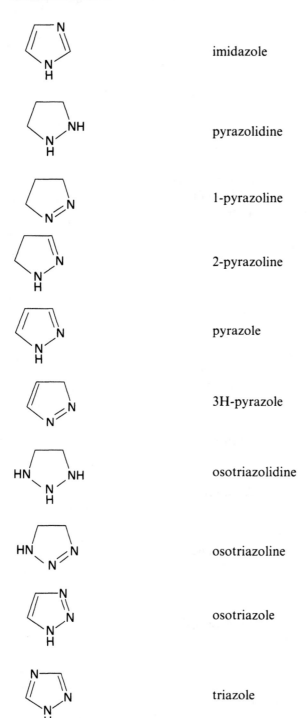

imidazole

pyrazolidine

1-pyrazoline

2-pyrazoline

pyrazole

3H-pyrazole

osotriazolidine

osotriazoline

osotriazole

triazole

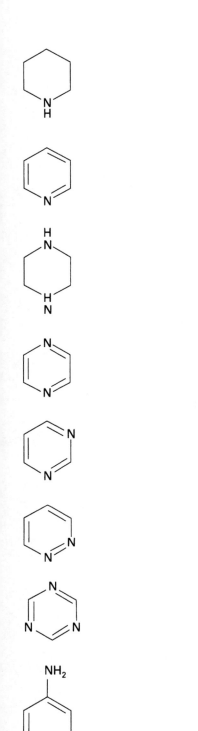

piperidine

pyridine

piperazine

pyrazine

pyrimidine

pyridazine

1,3,5-triazine

aniline

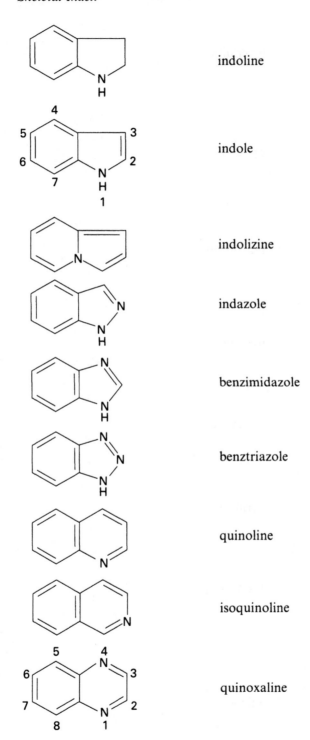

indoline

indole

indolizine

indazole

benzimidazole

benztriazole

quinoline

isoquinoline

quinoxaline

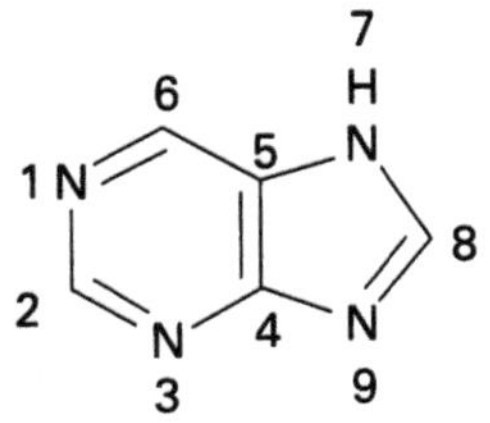

purine

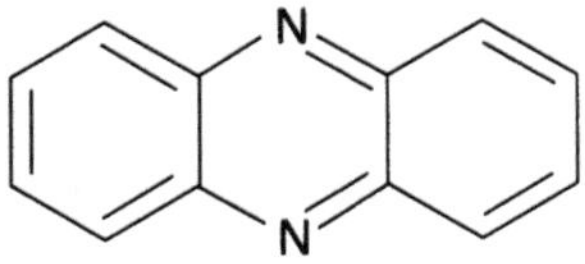

phenazine

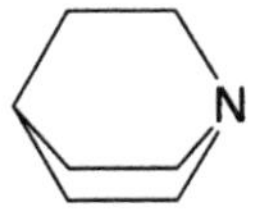

quinuclidine

23.4 Nitrogen and oxygen containing compounds

N_2O	nitrous oxide
NO	nitrogen oxide (or nitrous oxide)
N_2O_3	dinitrogen trioxide
NO_2	nitrogen dioxide
N_2O_4	dinitrogen tetroxide
N_2O_5	dinitrogen pentoxide
$H–O–N=O$ or HNO_2	nitrous acid
$H–O–NO_2$ or HNO_3	nitric acid
NH_2OH	hydroxylamine
$H–N=C=O$	isocyanic acid
$R–N=C=O$	isocyanate
$R–O–C≡N$	cyanate
$R–N=O$	nitroso

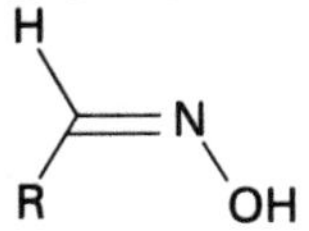

aldoxime

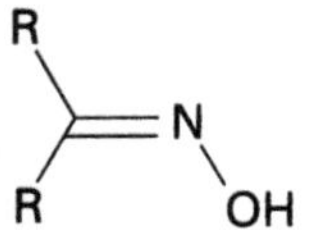

oxime

$R_3N^+–O^-$ amine oxide

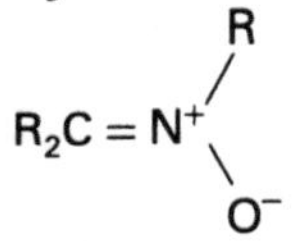

nitrone

Structure	Name
R–O–N=O	nitrite ester
$R-N^+$ (with O^- and $=O$) or RNO_2	nitro compound
$RCH=N^+$ (with OH and O^-)	nitronic acid
H–C(=O)–NR_2	formamide
H–C(=O)–NR_2	primary amide
H–C(=O)–NR_2	secondary amide, if cyclic then a lactam
H–C(=O)–NR_2	tertiary amide, if cyclic then a lactam
R–C(NH_2)–CO_2H	α-amino acid
R–C(=NH)–OR	imidic ester
R_2C(OH)(CN)	cyanohydrin
R_2N–NO	nitrosoamine
R–N^+(O^-)=N–R	azoxy compound
R_2–C(NO)(NO_2)	pseudonitrole

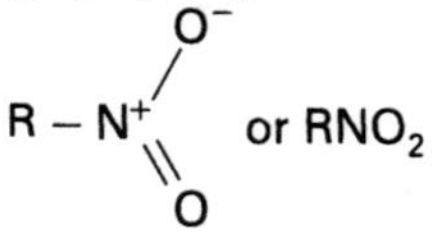

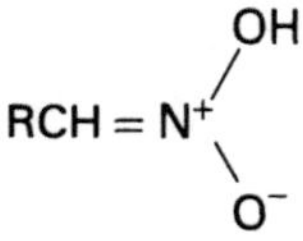

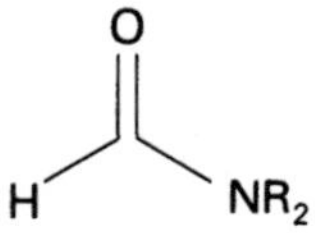

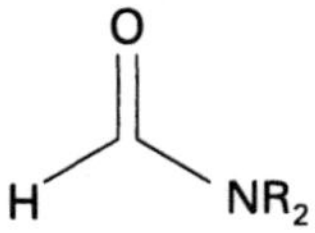

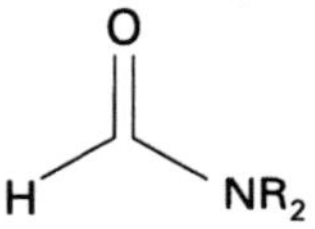

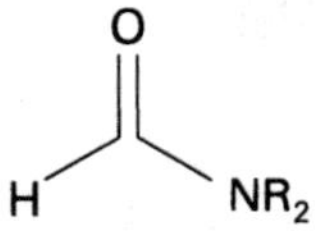

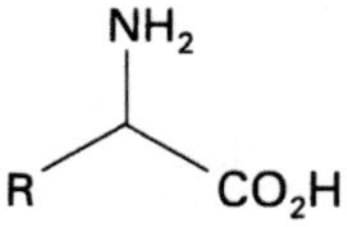

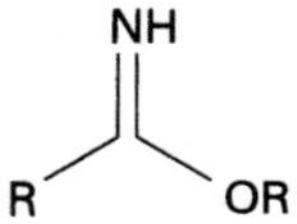

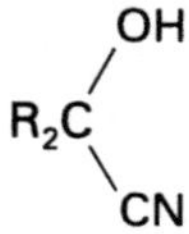

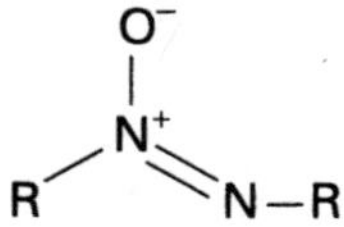

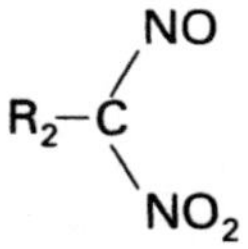

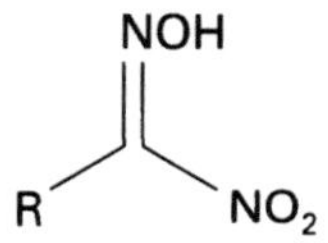

nitrolic acid

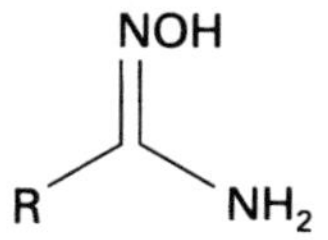

amidoxime

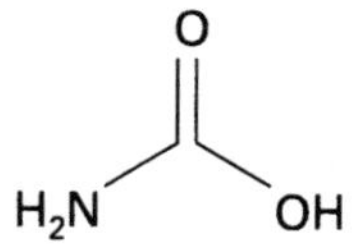

carbamic acid

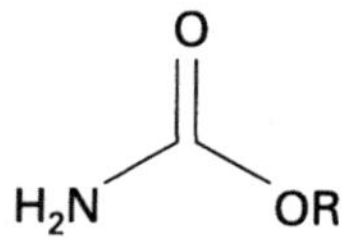

carbamate
(or where $R = C_2H_5$, then urethane)

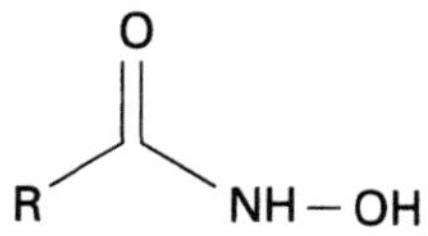

hydroxamic acid

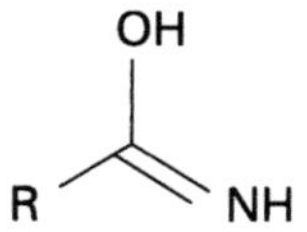

hydroximic acid

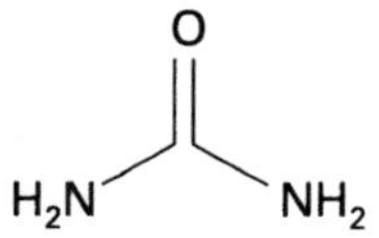

urea

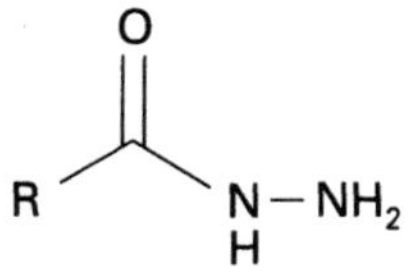

hydrazide

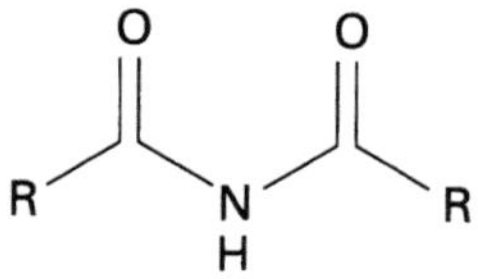

imide

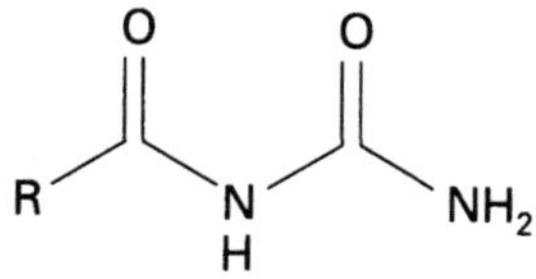

ureide

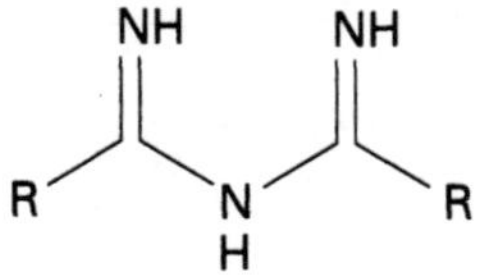

imidine

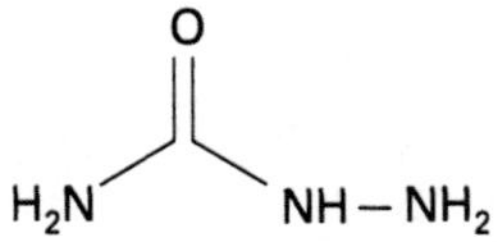

semicarbazide

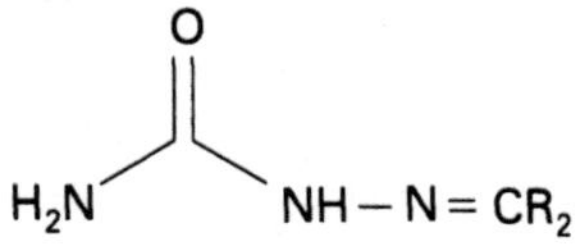

semicarbazone

oxazirane

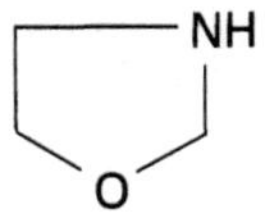

oxazolidine

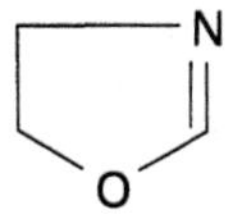

oxazoline

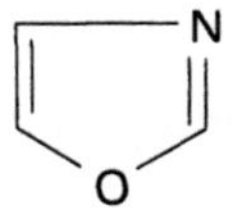

oxazole

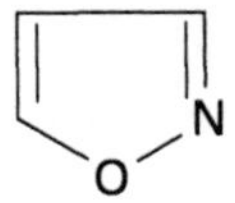

isoxazole

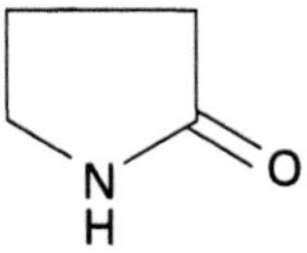

2-pyrrolidone

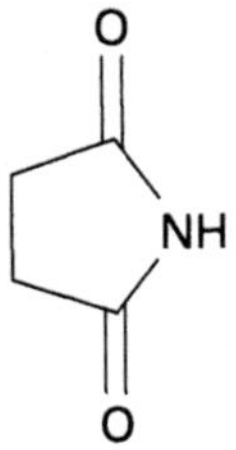

succinimide

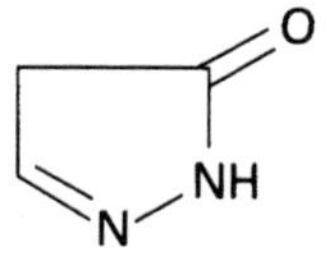

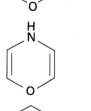 pyrazolone

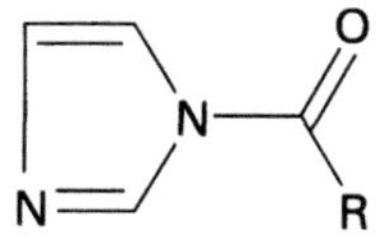

imidazolide

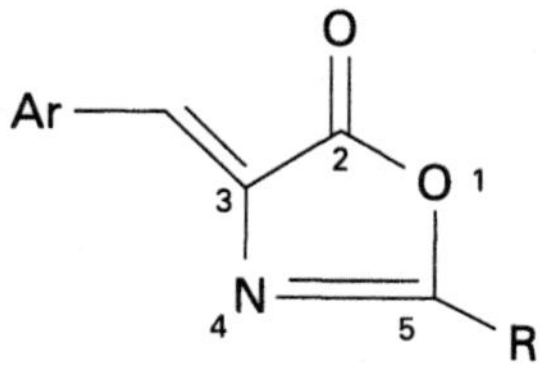

 azlactone

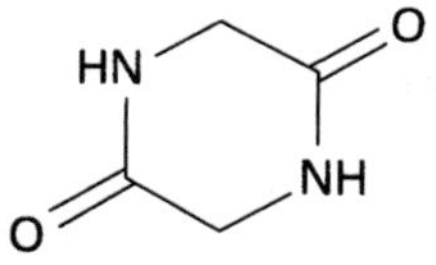

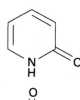 diketopiperazine

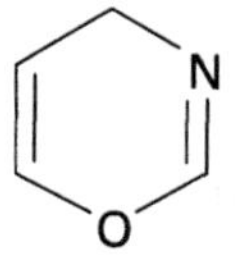

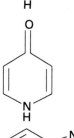 1,3-oxazine

1,4-oxazine

2-pyridone

4-pyridone

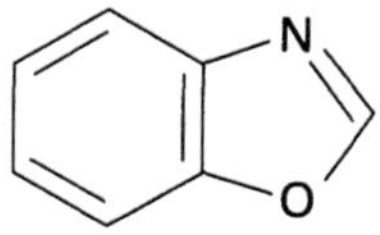

benzoxazole

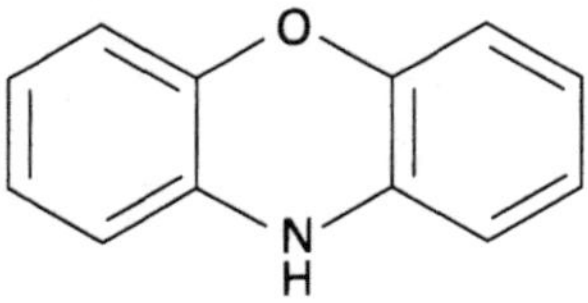

phenoxazine

23.5 Sulphur containing compounds

RSH	mercaptan (or thiol)
RS$^-$	sulphide anion
RSR	thioether
R$_3$S$^+$	sulphonium cation
RS–OH	sulphenic acid
R–S–S–R	disulphide
R–(S)$_n$–R	polysulphide
RCH(SR)$_2$	mercaptal (or thioacetals)
R$_2$C(SR)$_2$	mercaptol
R$_2$C(OR)SR	hemithioacetal

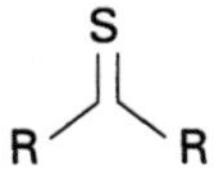

or RCSR — thiocarbonyl

R$_2$S $=$ CR$_2$ or R$_2\overset{(-)}{S}-\overset{(+)}{C}R_2$ — sulphur (or sulphonium) ylid

R$_2$S=O	sulphoxide
RSOCH$_2{}^-$	methylsulphinyl carbanion
R$_2$S(=O)$_2$	sulphone
R$_2$C=SO$_2$	sulphene
R–S–CN	thiocyanate
R–N=C=S	isothiocyanate
Ar–N=N–S–Ar	diazosulphide
R$_2$C=S=O	sulphine

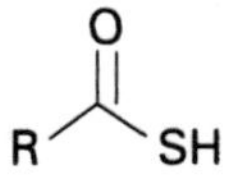

thiolic acid

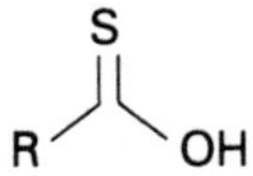

thionic acid

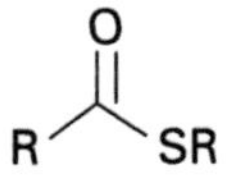

thioester

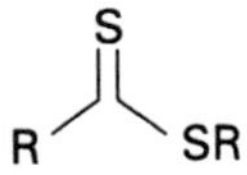

dithioester

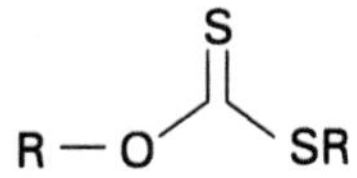

xanthate

$R–SO_2H$ sulphinic acid

$R–OSO_2H$ alkyl hydrogen sulphite

$R–SO_2OH$ sulphonic acid

$R–OSO_2OH$ alkyl hydrogen sulphate

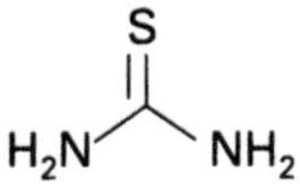

thiourea

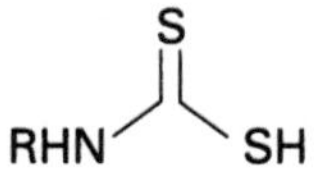

dithiocarbamic acid

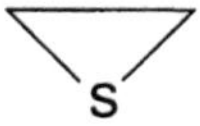

episulphide

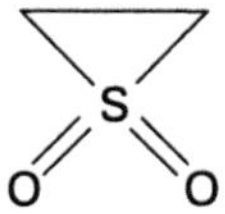

episulphone

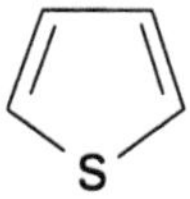

thiophen

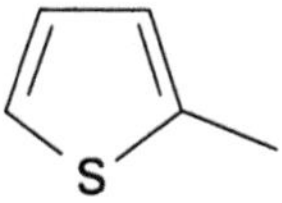

2-thienyl group

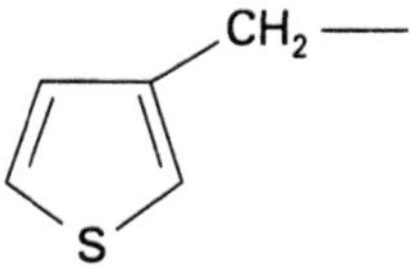

3-thenyl group

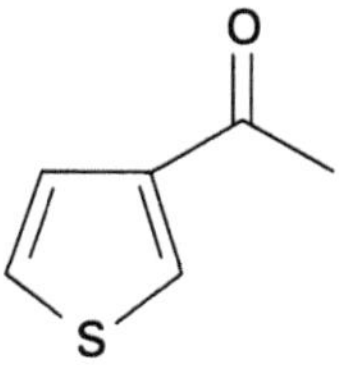

3-thenoyl group

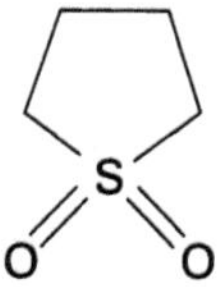

sulpholane

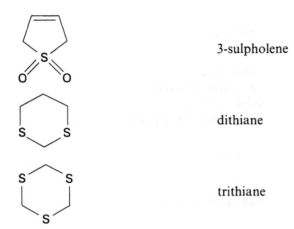

3-sulpholene

dithiane

trithiane

23.6 Phosphorus containing compounds

R_3P	phosphine
R_4P^+	phosphonium cation
H_3PO_3 or $P(OH)_3$	phosphorous acid
$P(OR)_3$	phosphite ester
$HP(OH)_2$	phosphonous acid
$HP(OR)_2$	phosphonite ester
$H_2P(OH)$	phosphinous acid
$H_2P(OR)$	phosphinite ester
H_3PO_4 or $O{=}P(OH)_3$	phosphoric acid
$O{=}P(OR)_3$	phosphate ester
$HP{=}O(OH)_2$	phosphonic acid
$H_2P{=}O(OH)$	phosphinic acid
$R_3P{=}CR_2$	phosphorane

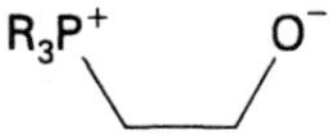

R_3P^+ ⌐ O^-

betaine

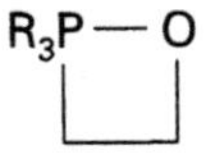

$R_3P - O$

oxaphosphetane

Index

Milton Keynes UK
Ingram Content Group UK Ltd.
UKHW021911071024
449327UK00022B/1651

9 780748 406418